ARTIFICIAL REEFS IN EUROPEAN SEAS

Artificial Reefs in European Seas

Edited by:

A. C. JENSEN
K. J. COLLINS
A. P. M. LOCKWOOD

*School of Ocean and Earth Science, University of Southampton, Southampton
Oceanography Centre, Southampton, SO14 3ZH, United Kingdom*

KLUWER ACADEMIC PUBLISHERS
DORDRECHT / BOSTON / LONDON

Library of Congress Cataloging-in-Publication data is available.

ISBN 0-7923-5845-7 (hardback)
ISBN 0-7923-6144-X (paperback)

Published by Kluwer Academic Publishers
PO Box 17, 3300 AA Dordrecht, The Netherlands

Sold and distributed in North, Central and South America
by Kluwer Academic Publishers, PO Box 358,
Accord Station, Hingham, MA 02018-0358, USA

In all other countries, sold and distributed
by Kluwer Academic Publishers,
PO Box 322, 3300 AH Dordrecht, The Netherlands

Printed on acid-free paper

Printed and bound in Great Britain by MPG Books Limited, Bodmin, Cornwall

Contents

,

Colour Plates 1–24 may be found between pages 242 and 243.

Acknowledgements

We would like to acknowledge colleagues involved in the initiation and production of the book. Bob Carling was the original commissioning editor for Chapman and Hall who provided the initial impetus in developing the project. The authors (and their close colleagues) of all the chapters worked hard to produce English language manuscripts and were patient and helpful with our editing procedures. Silvano Riggio provided the original artwork for the cover. Without the support of colleagues at Southampton such as Emma Mattey (who helped with the project in many ways from its inception), Jean Hart (who converted manuscripts to camera-ready copy and helped with editing), Kate Davies (responsible for many of the final figures and cover layout) and Phil Smith (proof reading and support) this book would not have been finalised. Henny Hoogervorst has guided the final production process for Kluwer.

Our sincere thanks to all. We could not have done it without you.

Antony Jensen
Ken Collins
Peter Lockwood

February 1999

Introduction and Background to 'Artificial Reefs in European Seas'

ANTONY JENSEN, KEN COLLINS and PETER LOCKWOOD
*University of Southampton, Southampton Oceanography Centre, Southampton SO14 3ZH, UK
School of Ocean and Earth Science*

It became apparent at the international artificial reef conference at Long Beach, California in 1991 that little was known of European artificial reef research outside Europe. The main cause for this lack of awareness seemed to be the relatively small number of English-language papers coming from Europe; workers, quite rightly, choose to publish mostly in their own languages. This effectively isolated European literature from most American and Japanese workers who comprise the numerical majority of active reef researchers. In addition European artificial reef scientists were, in the early 1990s, operating in relative isolation from each other: reef research was generally concentrated in the Mediterranean and there was only limited exchange between some research project teams in Italy, Monaco, France and Spain.

As a step towards ending this relative isolation, the European participants at the Long Beach conference agreed that a European network of artificial reef scientists should be formed. In 1994 this initiative was developed by one of the editors of this book (AJ) into a successful proposal to the European Commission (EC) 4th Framework AIR (Agriculture and agro-industry, including fishing) programme to financially support the European Artificial Reef Research Network (EARRN) for three years, commencing May 1995.

In preparing the bid to the EC for funding to support EARRN a questionnaire had been widely circulated in 1993 throughout geographical Europe to identify active reef scientists (Jensen, 1994). Those who replied from within the European Union (EU) formed the initial members of EARRN (EC funding for a network or 'concerted action' will only support activities of scientists within, the EU or those formally associated with the framework programmes). More generally, those who replied were considered to be potential contributors to a book aimed at providing an English-language synopsis of artificial reef research activity in European seas from its origins in Monaco during the 1960s (Bombace *et al.*, 1993) to the present date. This book represents the fulfilment of that aim.

Most of the contributors are current members of EARRN. In 1995 EARRN (EC contract AIR-CT94-2144) membership consisted of 51 scientists from 36 laboratories in Italy, France, Spain, Portugal, UK, Netherlands, Finland and Germany. The primary aims of EARRN were:

(1) To promote increased awareness of and collaboration between current artificial reef programmes throughout Europe, both marine and freshwater.
(2) To reach a consensus of opinion on given issues.

A.C. Jensen et al. (eds.), Artificial Reefs in European Seas, ix–xii
© *2000 Kluwer Academic Publishers. Printed in Great Britain.*

(3) To promote awareness of current issues (socio-economic and management) within the artificial reef scientific community, and encourage consideration of these aspects in developing future research proposals.
(4) To initiate a bibliography of European publications on artificial reef research.

As EARRN developed its activities (a major conference in 1996 (Jensen, 1997a) followed by four themed workshops (reef design and materials: Jensen, 1998a; research protocols: Jensen, 1997b; management of coastal resources and fisheries enhancement: Jensen 1997c; Socio-economic and legal aspects of artificial reefs: Whitmarsh *et al.*, 1997) associates were involved from Monaco, Poland, Romania, Israel, Denmark, Norway and Turkey. All were active in artificial reef research and contributed to the aims of the network, providing a summary of present research (Jensen, 1997) and recommendations for the direction of future research to the EC (Jensen, 1998b).

EARRN has (1999) entered a new phase and the network has expanded beyond its original 51 members. A website provides a focus for information (http://www. soc.soton.ac.uk/SUDO/RES/EARRN/contentsframe.htm) and an email discussion list facilitates communication. The next challenge for members is to develop Europe-wide research collaboration. At present there is little international European collaboration in research, projects tend to be based on national requirements. This is showing signs of change: workers in the Canary Islands (Spain) will be involved with artificial reef deployment research taking place off Madeira (Portugal) (R. Haroun, personal communication) over the next few years, and proposals for reefs in Greece are looking to incorporate Italian reef technology (C. Papaconstantinou, personal communication). Initiatives such as this show the way ahead.

Currently, most artificial reefs in European seas are deployed for research or contain a research component within the overall aims of the project. In Italy and Spain reef development has gone beyond this stage with reefs based on previous scientific experience being deployed for fisheries, mariculture and habitat protection purposes. Portugal is soon to deploy a 40 km² reef off the Algarve for fisheries enhancement based on research findings (M. Neves dos Santos, personal communication).

Members of EARRN, led by Professor Giulio Relini (University of Genoa, Italy), have played a key role in organizing the 7th International Conference on Artificial Reefs and related Aquatic Habitats (7th CARAH) in San Remo, Italy, 7–11 October 1999. This is the first time that this meeting has come to Europe, a recognition of the emergence of European influence within the global scene of artificial reef research.

Most European seas artificial reef programmes are represented in this book. Contributions for this book have been written by colleagues from Turkey, Israel, Italy, Monaco, France, Spain, Portugal, the UK, the Netherlands, Norway, Finland and Russia. Interests in artificial reefs are very varied, ranging from the 'expected' fishery enhancement through mariculture and ranching, nutrient removal and into environmental and habitat protection and nature conservation. The environmental factors vary as well. Artificial reefs have been placed in both oligotrophic and eutrophic waters of the Mediterranean Sea, off the Atlantic coast of southern Europe (including the Canary Islands and Madeira), the English Channel, southern North Sea and the Baltic Sea. This research activity reflects differing national and local priorities and the willingness of funding organizations to support research into the potential uses for artificial reefs as coastal zone and fisheries management tools.

New programmes, either in planning stages, awaiting licences or recently deployed, have yet to produce results and so do not feature in the following chapters. Examples of reef programmes include: the 1998/99 deployment of artificial reefs to protect seagrass habitat and promote coastal fisheries in Greek waters; an evaluation of artificial reef technology by the Danish Fisheries Department and the National Forest and Nature Agency (J. Stottrup, personal communication), which considers reefs to be an acceptable method of replacing habitat destroyed by 'stone fishing'; the deployment of cement stabilized quarry slurry reef blocks off the west coast of Scotland (M. Sayer, personal communication) in 2000, a project that aims to construct lobster habitat and provide an environmentally positive means of reusing fine rock cuttings from a remote quarry environment. Additional initiatives in the re-use of materials include recent deployment of tyre reef units to the Poole Bay reef (southern England) to evaluate the environmental acceptability of tyres in the marine environment and an application to deploy a steel lattice reef unit in the Moray Firth (eastern Scotland; G. Picken, personal communication). The latter is part of research proposed to evaluate the potential for artificial reefs to be constructed from the steel jackets of North Sea oil and gas production platforms (Aabel *et al.*, 1977).

Some reef deployments have only recently come to light. Such a programme is the trial of concrete reef units based on Japanese designs off the western Norwegian coast which are undergoing engineering evaluation (Per Jahren, personal communication).

Some research has been omitted because it was still underway whilst the book was being written. A good example of this is the evaluation of fish presence around a North Sea production platform by the Institute of Marine Research, Fish Capture Division, Bergen, which is being evaluated and awaits peer-review publication.

An omission from the book are reef studies undertaken in Romania. Artificial reefs have been developed in Romania but the severe political changes in recent years and subsequent economic problems have meant that little in the way of quantitative evaluation has been achieved. Romanian scientists are, however, convinced that artificial reefs are an important tool in the management of the serious eutrophication problem in the southern Black Sea.

The variety of research is a strength for the future, the recommendations for future research projects made by the EARRN to the EC (Jensen, 1998b) show that European scientists see reef technology developing to provide structures that are designed and targeted to specific aims in 'traditional' roles such as fisheries, environmental management and mariculture as well as in 'new' aspects such as tourism and impact mitigation.

References

Aabel, J.P., S.J. Cripps, A.C. Jensen and G. Picken. 1997. Creating artificial reefs from decommissioned platforms in the North Sea: a review of knowledge and proposed programme of research. Report to the Offshore Decommissioning Communications Project (ODCP) of the E and P Forum from Dames and Moore Group, RF-Rogaland Research, University of Southampton and Cordah, 115 pp.

Bombace, G., G. Fabi and L. Fiorentini. 1993. Census results on artificial reefs in the Mediterranean Sea. *Bollettino di Oceanologia Teorica e Applicata*. **11**: 257–263.

Jensen, A.C. 1994. Replies from the 1993 European Artificial Reef Questionnaire. Unpublished report. Department of Oceanography, University of Southampton.

Jensen, A.C. 1997a. *European Artificial Reef Research*. Proceedings of the first EARRN conference, March 1996 Ancona, Italy. Pub. Southampton Oceanography Centre, 449 pp.

Jensen, A.C. 1997b. Report of the results of EARRN workshop 1: Research Protocols. European Artificial Reef Research Network AIR3-CT94-2144. Report to DGXIV of the European Commission, SUDO/TEC/97/13, 26 pp.

Jensen, A.C. 1997c. Report of the results of EARRN workshop 2: Management of coastal resources and fisheries enhancement. European Artificial Reef Research Network, AIR3-CT94-2144. Report to DGXIV of the European Commission, SUDO/TEC/97/10, 33 pp.

Jensen, A.C. 1998a. Report of the results of EARRN workshop 4: European Artificial Reef Research Network AIR3-CT-94-2144. Report to DGXIV of the European Commission, SUDO/TEC/98/10.

Jensen, A.C. (1988b). Final report of the European Artificial Reef Research Network (EARRN). European Artificial Reef Research Network AIR3-CT94-2144. Report to DGXIV of the European Commission, SUDO/TEC/98/11.

Whitmarsh D, H. Pickering and A.C. Jensen. 1997. Report of the results of EARRN workshop 3: Socio-economic and legal aspects of artificial reefs. European Artificial Reef Research Network AIR3-CT94-2144. Report to DGXIV of the European Commission, SUDO/TEC/97/12, 21 pp.

1. Artificial Reefs off the Mediterranean Coast of Israel

EHUD SPANIER

The Leon Recanati Center for Maritime Studies and Department of Maritime Civilizations, University of Haifa, Mount Carmel, Haifa 31905, Israel.

Background

The south-eastern region has been shown to be the most oligotrophic part of the Mediterranean Sea (see Kimor and Wood, 1975; Lakkis and Lakkis, 1980; Kimor, 1983), which is generally considered to be the most impoverished large body of water known (Ryan, 1966). Berman *et al.* (1984) and Azov (1986) demonstrated the oligotrophic nature of the water, recording very low chlorophyll a concentrations (0.12–0.137 mg C m^{-3}) in the neritic zone 2 km off the northern coast of Israel. These low levels of primary production are reflected in the higher trophic levels of the food web. The marine fishery yield from the Israeli Mediterranean continental shelf is relatively low at less than 4000 tonnes per year (Grofit, 1993). Certain fish and marine invertebrates, however, are considered to be gourmet food and obtain high market prices in Israel. Many such fish are the groupers (family Serranidae) and sea breams (family Sparidae) that are associated with rocky substrata. Thus, most of the rocky seabed along the Mediterranean coast of Israel (mainly submerged sandstone or 'kurkar' ridges) provides habitat preferred by commercially important species. Israeli fishermen locate and fish these areas intensively. These fishing grounds, however, constitute less than 10% of the Israeli continental shelf (Adler, 1985). During the last two decades, fishing pressure on these sites has grown rapidly and some may have already been over-fished (S. Pisanty, personal communication). Increased spear fishing by SCUBA and skin divers and the activities of sport fishermen have additionally reduced the macro-fauna of these limited benthic habitats.

Preliminary observations of a few shipwrecks found along the Mediterranean coast of Israel revealed the presence of groupers and sea breams. Diamant *et al.* (1986) found that a small shipwreck off the Mediterranean coast of Israel contained 42% more fish species than a neighbouring, less complex, natural rocky patch reef of the same surface area.

In the early 1980s the Center for Maritime Studies at the University of Haifa and the Fishing Technology Unit, Israel Ministry of Agriculture initiated and funded the Israeli National Artificial Reefs project for the Mediterranean. Initially the programme was partially supported by a grant from the Israeli Ministry for Energy and Infrastructure. At the onset, the aim of the project was to determine the effectiveness of artificial reefs, made of different materials and in various configurations, in recruiting marine organisms, especially those of high commercial value. The project objectives centred on artificial reefs as possible solutions for the problems of limited habitat and food resources in the oligotrophic waters. The

A.C. Jensen et al. (eds.), Artificial Reefs in European Seas, 1–19
© 2000 Kluwer Academic Publishers. Printed in Great Britain.

solutions initially included artificial enrichment ('chumming') of the man-made reefs, identifying the type of artificial reef preferred by target commercial species, and investigating the effectiveness of artificial kelp in recruiting marine fauna and the role of fish cages in the framework of the artificial reef.

Deployment of the Reefs

The Israeli Mediterranean Artificial Reefs Research Group (IMARG) chose two sites for reef deployment. Site 1 was designated for smaller artificial reefs, with water depths between 18.5 and 26 m, about 1.8 km south-west of Tel Shiqmona (Israel Institute of Oceanography), Haifa (32°50′N, 34°56′E) (Fig. 1). Site 2 was approximately 3 km south-west of the first, off the coast at Kefar Zamir, Haifa (32°47′N, 34°55′E) (Fig. 1) in 32 m of water. A large (80 m × 30 m × 7–14 m) steel barge was sunk at this site. Permission and legal requirements for deployment were arranged through the local and national planning and building committees. Deployment was then approved by the Israeli national committee for coastal waters. The latter is now an inter-ministerial committee headed by the director of the Israel Ministry for the Environment and is the final authority for permitting dumping and building in the territorial waters of Israel. The Israeli Ministries of Agriculture and Law both prohibited unauthorized fishing within 1 km of the reef sites. Deployment of artificial reefs took place in four phases.

Phase One

In 1982 the first four small experimental reefs (two pairs) to be deployed by IMARG were placed 500 m apart in 26 m of water at Site 1. The reefs were formed from 3.3 × 3.3 × 3 m used car tyre reef units (Fig. 2). Tyres were connected with steel and/or reinforced fibreglass bars and weighted with concrete weights (Spanier *et al.*, 1985a). A control site of the same surface area as a reef unit, with a similar depth and type of seabed was fenced off with ropes. Each site was buoyed.

The first pair of reef units were constructed on-shore and deployed (Fig. 2) with the aid of a barge and crane. An alternative deployment method was developed which used empty oil drums attached to the tyre units for buoyancy which allowed the units (without the concrete ballast) to be towed to the site by a small fishing vessel. The latter method was used to place the second pairs of units. Advantages were seen in a lower cost and greater accuracy of placement. However, more diving operations were needed to move the concrete ballast and tie them to the tyres.

The following winter of 1982–3 was extremely severe. Storm-generated waves of more than 8 m maximum wave height (Rosen, 1983) damaged the reef units. A detailed examination of the biological and engineering findings led to the following conclusions:

(1) As most of the damage sustained was in the connections between the concrete weights and the tyres, these should be made either flexible or from one piece (of concrete, rubber, PVC, etc.).

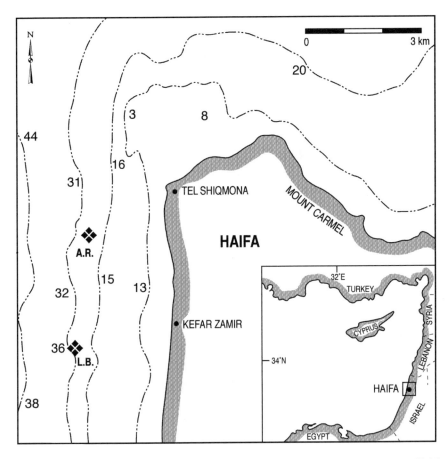

Figure 1. A map of the artificial reefs sites established by the Israeli Mediterranean Artificial Reefs Research Group. A.R. the site of the small reefs, L.B. the site of the large barge.

(2) The reef units should be reduced in height and made triangular in shape to reduce the drag effects of waves and currents.
(3) Lowering of reef unit height would be beneficial as most of the commercial species observed on the reef unit were in the lower part of the man-made structure.

Phase Two

These conclusions influenced the construction of the new IMARG used-tyre reefs placed in 1983 (Spanier *et al.*, 1985b, 1990) close to the site of the first reef. This time the units were placed at a depth of 18.5 m where the destructive force of the winter waves was considered to be minimal but SCUBA diving was possible for 50 minutes without divers having to undertake decompression stops. Nine structures were placed in this site. Four reef units were assembled from used car tyres (32 cm

Figure 2.   Schematic drawing of two artificial reef units. A plate of frozen fish flesh descends to the reef along the line from the surface buoy. (Reprinted from Spanier *et al.*, 1985a.)

inner diameter and 17 cm tyre width). The tyres were connected with 18 mm diameter steel bars. Four units were weighted with concrete poured into the bottom part of the lower layer of tyres, and one was weighted with heavy steel chains.

The tyres were arranged vertically ('S' type), horizontally in 2 ('5' type) or 3 ('9' type) layers or horizontal rows (forming tubes) perpendicular to each other ('C' type) (Figs 3,4). These units were floated with empty oil drums as buoyancy and towed to the site. Three other reef units were each made of two 'giant' tyres (1.3 m outer diameter) placed horizontally, one lying on the other, connected with steel bars and weighted with concrete poured into the lower tyre. An additional structure placed was 'artificial kelp', made of buoyant multifilament polypropylene ropes weighted with heavy steel chains (Fig. 5). The final element established in phase 2 was a small fish cage for the culturing of gilthead sea bream *Sparus aurata*. This cage was supported just above the seabed and supplied with fish-food from the surface by gliding a feeding device down the buoy line (for details, see Spanier *et al.*, 1985b, 1989). These last two components of the reef complex were transferred to the site by inflatable boats (Zodiac Mark III) and deployed by divers. The various components of the artificial reef complex were placed within 15–20 m of each other during this phase.

Phase Three

IMARG was involved in the placement of a 30 × 80 m steel barge at reef Site 2. Since the height of the barge was between 7 and 14 m, its upper deck, given that the depth of the seabed was 32 m, was similar to the depth of the smaller reefs. This barge was towed to Site 2 and sunk in 1983. The following year additional holes were made along its sides with the aid of explosives.

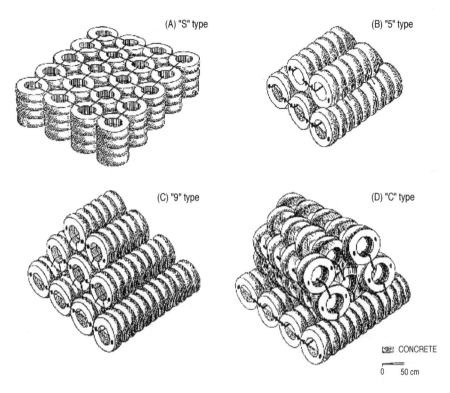

(A) "S" type

(B) "5" type

(C) "9" type

(D) "C" type

CONCRETE

0   50 cm

Figure 3. Schematic representation of the four types of the artificial reef units weighted with concrete in the second stage. A – 'S' type, B – '5' type, C – '9' type, D – 'C' type. (Reprinted from Spanier *et al.*, 1990).

Phase Four

IMARG's last activity in 1992 was the deployment of three small, low profile (<3 m high) barges (the first 15 m long with a 3 m beam, the second 4 m long with a 5 m beam and the third 3 m long with a 5 m beam) 50–70 m south of the tyre reef site. The barges were towed by tug boats to the site.

Research Methods

Fish species composition, abundance and size were determined from monthly trammel net sampling around Site 1, surveys started the year before placement of IMARG's phase one reefs (Spanier *et al.*, 1989a). Data on fish and macro-invertebrates associated with the reefs were obtained by SCUBA divers using a visual census technique (Russell *et al.*, 1978). Control sites and natural rocky outcrops were also censused using this method. Complementary information was obtained by underwater still and video photography.

Figure 4.   An underwater photograph of part of stage 2. (Photo by D. Syon.)

Figure 5.   An underwater photograph of the 'artificial kelp'. (Photo by S. Breitstein.)

Slipper lobsters were caught by hand, measured and tagged with numbered plastic 'spaghetti' tags as well as by puncturing holes in their telsons (for details see Spanier *et al.*, 1988; Spanier and Barshaw, 1993). Large fish, such as groupers, were caught underwater with hook and line and tagged in a similar manner.

The small reefs (Site 1) were censused weekly, weather permitting (but at least once a month). SCUBA divers equipped with lights recorded the presence, number and locations of lobsters and fish. Behavioural and ecological observations were noted. A census of the large barge was undertaken approximately every 3 months (Site 2).

Several small artificial reefs (Site 1) and a control site were enriched weekly by trash fish between 1982 and 1985 (Phases 1 and 2). The frozen fish, with weights attached, was released along a cable leading from the surface buoy to the site (Fig. 2; for details see Spanier *et al.*, 1985a, 1990). The effects of the enrichment on different sites and species were determined by extended observations and censuses with and without man-made enrichment. Behavioural and ecological experimentation was undertaken in the small reefs complex, focusing on the sheltering behaviour of the Mediterranean slipper lobster (*Scyllarides latus*) within the artificial reefs to determine if such behaviour was an adaptation to avoid predation. These investigations included tethering lobsters to stakes (with monofilament lines) in both the open area, and near the tyre reefs, and plugging sections of the reef openings with stones (for details, see Barshaw and Spanier, 1994; Spanier and Almog Shtayer, 1992). These experiments involved day and night censuses by SCUBA divers. Along with the biological observations of the reef itself, periodic underwater observations were made up to 50 m away from the reef units to assess their impact on the immediate surroundings. Special attention was also directed to the physical state of the different units. Any disappearance, movement or deformation of a reef component, or part of it, was noted and measured if necessary, especially after heavy storms.

Results

Biological Findings

The species composition of fish and macroinvertebrates recruited to the small artificial reefs was similar to the fauna sampled with trammel nets in the low profile biogenic substratum (20–50 cm high) which constitutes the shallow continental shelf off northern Israel (Spanier *et al.*, 1989a, 1990). Eighty-two percent of the 43 fish species sampled in the trammel net were also censused in the artificial reefs. The most common bony fish in the man-made reefs were *Epinephelus alexandrinus*, *E. marginatus (guaza)*, *Diplodus sargus*, *D. vulgaris* and *Pagrus coeruleostictus*, all species of Atlanto-Mediterranean origin, and *Siganus rivulatus*, *S. luridus* and *Sargocentron rubrum*, species of Red Sea origin (Lessepsian migrants; see Colour Plate 1). Although fish of Red Sea origin constituted only 13.5% of the fish species composition in the small artificial reefs in 1985, they contributed 64% of the large fish (total length (TL) ≥ 15 cm) abundance there. A similar trend was found in the trammel net samples (Spanier *et al.*, 1989a, 1989b, 1990). However, the organisms of high commercial value in the reefs were groupers, sea breams and lobsters, all of Atlanto-Mediterranean origin.

It is interesting to note that a similar study carried out in 1995 indicated that although only one additional species of fish was added to the list of Lessepsian migrants in the artificial reefs in the last 10 years their contribution to the biomass of large fish was already 94% (Spanier, 1995a; Spanier *et al.*, 1996). Other important invertebrate species noted in the artificial reefs were the bivalves *Venus verrucosa* and *Glycymeris pilosus* as well as the cephalopod *Octopus vulgaris*.

The first fish species recruited to IMARG's phase one reefs was the Mediterranean damselfish *Chromis chromis*, which was observed around the man-made structures less than 2 months after the placement of the units. The total number of fish and the number of species censused at the control site established in Phase 1 remained at the same low level during the months of observation but increased steadily at the two artificial reef sites. This process was even more pronounced on the enriched reefs, especially with large groupers (Spanier *et al.*, 1983; 1985a). Several groupers, especially of the species *E. alexandrinus*, which had been caught and tagged within the reef area, were repeatedly observed around the enriched reefs for up to 10 months after tagging. Forty-two species of fish and macroinvertebrates were censused in the (enriched and unenriched) artificial reefs during IMARG's second phase compared to 24 in natural rocky outcrops and only eight species in the enriched control site (Spanier *et al.*, 1985b, 1990). Fish and macroinvertebrates also appeared in much larger numbers in the man-made structure (Spanier *et al.*, 1985b).

Special attention was directed to the recruitment and location in and around the reefs of five common species of large (>20 cm TL) bony fish and of the Mediterranean slipper lobsters. None of these species was observed in the enriched control area but some were censused in natural rocky outcrops. The number of groupers associated with the reefs was related both to artificial enrichment and season (Fig. 6). Although their number fluctuated monthly, they were more abundant during the enrichment period. They showed a preference for shelters in the smaller horizontal holes (among the tyres) in the lower levels of the reefs (close to the seabed). The appearance of the white bream *D. sargus* was not clearly associated with enrichment (Fig. 7). They were observed in small shoals around the reef where they preferred the upper, more open structure, or appeared in large schools between or in the immediate vicinity of, the reefs. The presence of the rabbitfish *S. luridus* was apparently not affected by the artificial enrichment (Fig. 7). They appeared in small groups around the reef structures and rarely entered the reef holes during the day. The red squirrelfish, *S. rubrum*, on the other hand, was almost always found inside reef holes during the daytime censuses; they preferred larger holes (in the tyres; Colour Plate 1). Their abundance did not seem to be associated with enrichment (Fig. 7) but certain seasonal fluctuations were evident.

The Mediterranean slipper lobster *S. latus*, the reef organism with the highest commercial value (>$20 kg⁻¹), was first observed in the small tyre reef complex in the autumn of 1983. Since then this species has shown a very clear seasonal pattern of appearance in the man-made structures. They were abundant in the reefs from January to June and sparse from July to December (Spanier, 1991; Spanier and Almog Shtayer, 1992; Spanier and Lavalli, 1991; Spanier *et al.*, 1988; 1990; 1993). *S. latus* foraged during the day, mainly for bivalves (Spanier, 1987) and stayed in shelters in the artificial reefs during the night. The lobsters significantly preferred to dwell in smaller holes (between the tyres) than in the larger ones (middle of the tyres), and preferred horizontal shelters over vertical ones, and lower levels in the reefs over the upper parts. There were also differences in the preferences of lobsters for the various types of reefs (Fig. 8). When one shelter opening was blocked in 50% of the 'small opening' shelters in

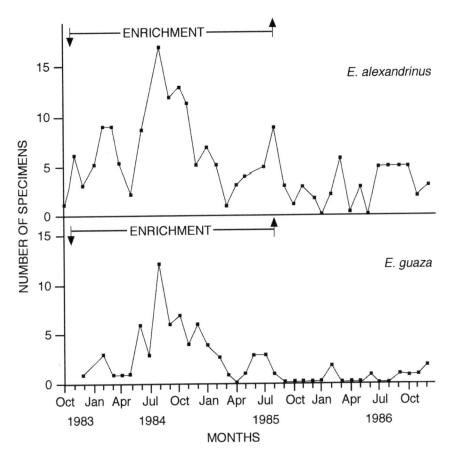

Figure 6. Appearance of the golden grouper, *Epinephelus alexandrinus*, and the dusky grouper, *E. marginatus (guaza)* in the man-made reef complex (Phase 2) with and without artificial enrichment. (Reprinted from Spanier *et al.*, 1990.)

the reef, 98% of the lobsters were censused in open-ended shelters. Lobsters also showed a tendency for gregarious sheltering (Spanier and Almog Shtayer, 1992). Live and open bivalves, mainly *G. pilosus* and *V. verrucosa*, food of the lobsters, were abundant close to shelters occupied by lobsters during the day in the lobster season. Thirty-two percent of the tagged lobsters were recaptured in the reef at least once. Time between recaptures was relatively short (1–17 weeks, mean 29 days) or long (10–37 months, mean 338 days).

Small fish such as schools of bougue, *Boops boops* and juveniles of other bony fish were recruited to the artificial kelp (Spanier *et al.*, 1990). Juvenile *S. aurata* grew well in the fish cage, for the period that it remained intact. Small fish in the reef complex were observed, under the cage, feeding on small fragments of fish food (and apparently also utilizing the excretion of the caged fish; Spanier, 1989; Spanier *et al.*, 1985b). The large barge was characterized by a large number of groupers, schools of *D. vulgaris* and schools of large amberjacks, *Seriola dumerili*,

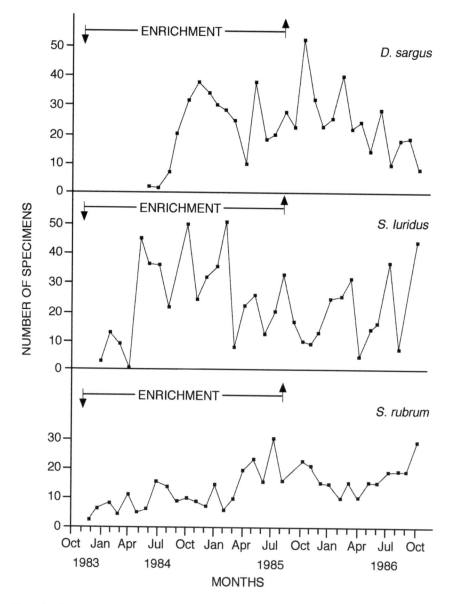

Figure 7. Appearance of the white bream *Diplodus sargus*, the rabbitfish *Siganus luridus* and the red squirrelfish *Sargocentron rubrum* in the man-made reef complex (stage 2) with and without artificial enrichment. (After Spanier *et al.*, 1990.)

swimming above the barge in season. Lobsters were recorded from holes in the lower parts of the barge and especially under the bow

The smaller barges recruited schools of *D. vulgaris*; *D. sargus* and some groupers were also observed there. Lobsters have not yet been noted in these structures.

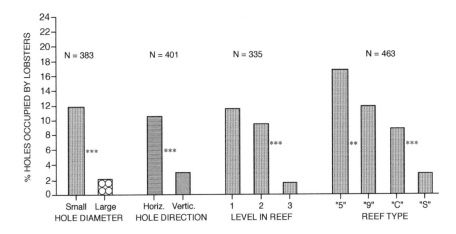

Figure 8. Shelter selection by the Mediterranean slipper lobsters. From left to right: preference for size of shelter openings; shelter direction; different levels in the reef; and type of reef. (After Spanier *et al.*, 1990; see Fig. 3.) Significance was determined using G-test for goodness of fit: ***$p < 0.005$, **$p < 0.01$.

Endurance of the Various Reefs

IMARG's four initial Phase 1 tyre reef units lasted only one year because of their square structure and the rigid connections between the tyres and the concrete ballast which could not withstand storm conditions. Of IMARG's nine Phase 2 reef units, those weighted with heavy steel chains were less stable than those ballasted with concrete. The triangular tyre reef with the chain ballast was initially moved more than 50 m during a heavy storm and was later swept away and lost. Similarly, the artificial kelp, weighted with the same type of chains, was initially compacted due to intense water movements associated with heavy storms (which made it biologically more effective; see Spanier *et al.*, 1985b, 1990). In later years it was swept away from the centre of the site. The fish cage net was torn from its steel frame during a storm the second winter after deployment. The 'S' shaped unit (Fig. 3) lost a considerable part of its concrete ballast and was finally swept away in a heavy storm two and a half years after placement. The remainder of the tyre units withstood even the heaviest winter storms (waves of >7 m significant wave height) with only minor movements of several metres. Yet, despite the relative immovability of these reef units for 15 years in the open sea, the 'C' type (Fig. 3) has started to collapse while only minor structural deformation, if any, has been noted in the other tyre units. The smaller barges have suffered only minor structural deterioration since placement; no movement has been detected. Approximately 2 m of sand has accumulated on the east (landward) side of the large barge while the currents have scoured under the western front of it, and created holes which arc populated by mobile macrofauna. These sediment movements, however, have not caused any movement or tilting of this large sunken vessel.

Conclusions and Discussion

The combination of man-made reefs and artificial enrichment was found to be effective in recruiting fish and macroinvertebrates The similarity of the species composition to that of natural rocky outcrops such as submerged sand stone ridges indicates that these man-made structures could substitute or mitigate for natural grounds 'lost' by over-fishing and other anthropogenic activities. Artificial reefs can be effective in recruiting commercial species in a low productive marine environment, such as the Levant basin of the Mediterranean, even without artificial enrichment. However, the rate of recruitment of predator fish, such as groupers, can be considerably enhanced by chumming.

The rabbitfish of the genus *Siganus* that feed on algae (Lundberg and Lipkin, 1979), the red squirrelfish *S. rubrum* and the white bream *D. sargus* which feeds on benthic invertebrates (Golani *et al.*, 1983; Hureau, 1987) are apparently not affected by the artificial enrichment but by the structural properties of the artificial reefs. The rabbitfish shelter in the reefs during the night and the nocturnal squirrelfish dwell there during the day (Colour Plate 1). The physical properties of the reefs also seem to be the major factor in the recruitment of large schools of fish such as *S. dumerili* and *D. vulgaris* to the barges and smaller fish to the artificial kelp. The latter may act as a 'bottom fish attracting device' (Spanier *et al.*, 1990). The seasonal recruitment of organisms to artificial reefs may be related to natural environmental fluctuations such as temperature. This seems to be the case with the Mediterranean slipper lobster (Spanier *et al.*, 1988). The best example of habitat selection in the reef was also demonstrated by this large crustacean. Spanier and Almog Shtayer (1992) suggested that the variety of shelter-related behaviours demonstrated by *S. latus* would effectively act to reduce predation of this lobster which lacks claws or spines for active defence. Foraging by night and sheltering by day, as well as transporting bivalve food items to their shelter may minimize exposure to a diurnal predator. Horizontal shelters supply shade and so reduce the chance of visual detection by diurnal predators. Small shelter openings can also increase shade but possibly their main function is to increase physical protection against large predators. Multiple shelter openings enable the lobster to escape through the 'back door', utilizing its fast tail flip swimming capability (Spanier *et al.*, 1991), should a predator be successful in penetrating the shelter. Finally, gregarious sheltering may be advantageous because of collective prey vigilance (the overall vigilance of a group is higher than in a single animal) and this may allow for an earlier alert to the approach of a predator (see Krebs and Davies, 1987). In addition, grouping together may have the benefit of concealment among cohorts, a single prey item is more difficult to pick out of a group than when isolated (the 'selfish herd' effect – the use of fellow group members as a physical shield against predation; see Hamilton, 1971).

Barshaw and Spanier (1994) have demonstrated some of the above-mentioned anti-predation advantages of sheltering in *S. latus*. In their field tethering experiments, more slipper lobsters were preyed upon during the day by the grey trigger fish *Balistes carolinensis* in the open, than in the artificial reefs where the lobsters mostly inhabited horizontal shelters with small openings. Natural rocky areas that include shelters with physical characteristics preferred by this species are limited in the Israeli Mediterranean continental shelf. Thus, as tagging experiments indicated, it seems advantageous for *S. latus* to return to a preferred site, such as an artificial reef, after short-term foraging and after long-term (years) movements (Spanier *et al.*, 1988). Similar preferences for physical characteristics of shelters have been demonstrated in juveniles of at least one species of spiny lobsters that have limited morphology for active defence (Spanier and Zimmer Faust, 1988; Zimmer Faust and Spanier, 1987).

Such behavioural ecological studies of key species in artificial reefs are crucial for designing effective man-made habitats not only for lobsters (Spanier, 1991, 1994, 1995b) but for many other commercial species as well. Since these studies are frequently associated with habitat selection and predator avoidance, artificial reefs designed accordingly can be used also to protect breeding populations of endangered marine species (e.g. Spanier, 1991).

A considerable portion of the biomass in IMARG's artificial reef was attributed to the Red Sea species that have migrated into the Mediterranean through the Suez Canal, a biogeographical phenomenon dominating the south-eastern Mediterranean (Spanier and Galil, 1991). Although their commercial value as food is presently less than that of the autochthonous Atlanto-Mediterranean species, many of these Lessepsian migrants are colourful and can be attractive to SCUBA divers. The relatively high water temperatures and the clarity of the south-eastern Mediterranean permit SCUBA diving during much of the winter. Thus, this portion of the fauna around the artificial reefs may be of high touristic value. Although IMARG's experimental reefs have been constructed from materials of convenience, most of the basic design could be used in commercial sized reefs. Such structures should be built of solid and durable material such as concrete and involve the various components that have been found to be effective in studies made thus far. The preliminary indication of leaching of chromium from the tyre reefs to the biofouling community (Spanier *et al.*, 1996) is another reason for exclusion of used car tyres as reef material.

The Future

In recent years there have been increasing demands in Israel, mainly by SCUBA clubs and sport fishermen, to establish new artificial reefs, particularly from ship wrecks. Without any clear national policy and criteria regarding placement and management of artificial reefs, there have been conflicting approaches by various users of the sea and national authorities. Consequently, at the end of 1993, the general director of the Israeli Ministry for the Environment nominated a national committee to formulate national policy and criteria for artificial reefs in the Mediterranean waters of Israel. The committee included members of IMARG, as well as representatives of the various authorities and organizations involved. The committee recommended the establishment of a limited number of sites along the Mediterranean coast of Israel for artificial reefs for fisheries and separate sites for man-made reefs for SCUBA diving. The recommendations included a list of environmental, safety and legal requirements for the placement of future reefs. Among the decisions that have already been adopted by the national committee for coastal waters, there is a recommendation to consider the placing of pre-fabricated artificial reefs on the shallow Mediterranean continental shelf for fisheries and other purposes. A similar committee dealt with artificial reefs in the Gulf of Eilat (Aquaba) in the Red Sea. Since the first committee submitted its report to the Israeli Ministry for the Environment (March 1994) several vessels have been sunk and established as artificial reefs for SCUBA divers along the Mediterranean coast of Israel. These include two ex-missile boats of the Israeli navy and two tug boats. All vessels were prepared and sunk in sites according to the recommendation of the national committee. The Israeli Diving Federation is responsible for these artificial reefs.

There is, however, a need to examine the extended utilization of artificial reefs to determine the ratio between the value of the yield and the investment (Milton, 1989). The effect of repetitive fishing on the rate of renewal of organisms needs to be studied, especially for species of high economic value. Such quantitative examination will yield a true estimation of the probable crop from these reefs, a knowledge essential for the calculation of the profitability of a large-scale artificial reef deployment. Another important factor for the calculation of the cost of an artificial reef field is the density of the reef units on the seabed. Each reef unit deployed has a given cost. Yet the number of fish able to live in a field of artificial reef units will not increase in simple proportion to the number of reef units per given area. The principle of decreasing marginal output will operate here. It is important to find the optimal reef unit density for the conditions of the eastern Mediterranean. In Japan it was found that a density of 3 m^3 of reef 1000 m^{-2} of seabed was the optimal density (White *et al.*, 1990) for reef unit deployments. As the productivity of the sea off Japan is higher than that off Israel, the optimal density for reef units in Japan should be selected as the highest density for the Israeli situation. The testing of three reef unit densities is suggested: those that were found best for Japan (3 m^3 of reef 1000 m^{-2} of seabed), an intermediate density of 1.5 m^3 1000 m^{-2} and a low density of 0.75 m^3 1000 m^{-2}.

Another aspect of artificial reef creation is to provide environmentally acceptable solutions to terrestrial problems. The recycling of fly ash, a product of coal burning which is an environmental nuisance could, in the guise of cement-stabilized fly ash blocks, convert a terrestrial annoyance to a beneficial material in the sea. The State of Israel uses increasing quantities of coal to produce electricity. At least 10% of the burned coal remains as fly ash. At present, this means approximately one million tonnes of fly ash is created per year. These enormous quantities of fly ash are directed in part to the cement industry (where it is used as a filler), and part is disposed of in the deep sea, a considerable distance offshore. This bulk disposal of fly ash in the deep sea is costly because of the high price of marine transportation and is ecologically undesirable. Recycling of the fly ash into blocks would be more economically and ecologically beneficial.

One of the possible uses of cement-stabilized fly ash is in the construction of artificial reefs. In recent years there have been several successful attempts to use the residues of coal burning for the construction of artificial reefs in the USA, the UK and Italy (e.g. Collins *et al.*, 1990, 1994, 1995; Collins and Jensen, 1995; Jensen *et al.*, 1994; Jensen *et al.*, chapter 16, this volume; Jensen and Collins, 1995; Parker, 1985; Relini, chapter 21, this volume; Woodhead and Jacobson, 1985). Since Israel's marine environment differs from that of the USA, Italy and the UK, and the coal used in Israel, and consequently the fly ash, differs in origin and so heavy metal content, the various facets of the subject in Israeli Mediterranean conditions should be examined. Not enough is known about how fly ash consolidates or, if included as a major component of marine concrete, how it will affect the marine biota or withstand the conditions of winter storms in the Mediterranean open sea off Israel.

Suzuki (1991) showed that marine concrete containing 54% fly ash and 15% cement was very stable and even strengthened in sea water. There is apprehension regarding the possible leaching of heavy metals and other chemicals into the sea, but laboratory experiments in Israeli Mediterranean marine conditions have indicated that the smaller the surface area of the fly ash (i.e., if it is consolidated), the smaller the probability of leaching becomes (Kress, 1991). These findings encourage the initiation of a scientific programme to investigate the feasibility of establishing 'fields' of artificial reefs utilizing fly ash as a major structural component, on the Mediterranean continental shelf of Israel.

The use of fly ash may well be beneficial to the national economy if the economic benefit resulting from the improvement of fishing grounds and the boost to diving and sport fishery tourism covers the cost of fly ash disposal and the establishment of fly ash reefs. This economic calculation is not possible today because of the lack of knowledge on the renewal rate of fish population with high commercial value in the reef. There are insufficient data regarding the cost of establishing a reef field as long as the desired density of the reef units in it is unknown. A 5-year project with three artificial reef fields at an estimated cost of $750 000 (construction, placement and research) has been proposed by IMARG to answer the above questions. The proposed artificial reefs are also planned to provide for the anchorage of cages for the intensive rearing of fish in the open sea (Fig. 9). The excreta of the fish together with the residue of the non-utilized food may enrich the reef complex. The reef units proposed for this future project will be constructed with holes and crevices based on previous findings, and will include 'artificial kelp' to recruit small fish and juvenile forms. The preliminary phase of the project, supported by the Israel Electric Company, started in October 1995 and will run for at least three years. The project is a collaboration between the University of Haifa and the Israel Oceanographic and Limnological Research Institute. Eighty 40 × 20 × 20 cm concrete blocks were made. Twenty are used as control (no fly ash) and in the other three groups coal fly ash formed 40%, 60% and 80% of the weight respectively. The blocks were arranged randomly on PVC tables 100 m off the tyre reefs. The durability of these blocks and possible secretion of toxic material to the marine biota settling on them are being investigated.

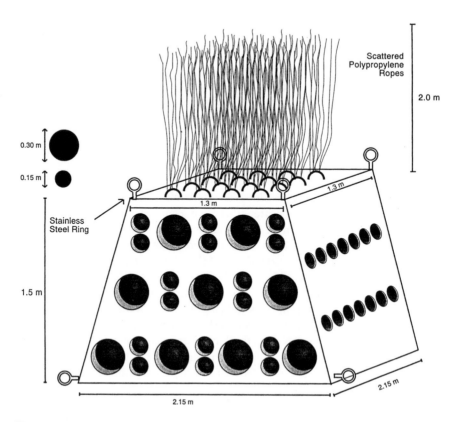

Figure 9.   Schematic drawing of a proposed reef unit to be made of concrete containing coal fly ash.

References

Adler, E. 1985. *The submerged kurkar ridges off the northern Carmel coast.* M.A. Thesis, University of Haifa, Israel, 106 pp. (In Hebrew; English abstract).

Azov, Y. 1986. Seasonal patterns of phytoplankton productivity and abundance in near-shore oligotrophic waters of the Levant Basin (Mediterranean). *Journal of Plankton Research.* **8**: 41–53.

Barshaw, D.E. and E. Spanier. 1994. Anti-predator behaviors of the Mediterranean slipper lobster *Scyllarides latus. Bulletin of Marine Science.* **55**: 375–382.

Berman, T., D.W. Townsend, S.Z. El Sayed, C.C. Trees and Y. Azov. 1984. Optical transparency, chlorophyll and primary productivity in the Eastern Mediterranean near the Israel coast. *Oceanologica Acta.* **7**: 367–372.

Collins, K.J. and A.C. Jensen. 1995. Stabilised coal ash studies. *Chemistry and Ecology.* **10**: 193–203.

Collins, K.J., A.C. Jensen and A.P.M. Lockwood. 1990. Fishery enhancement reef building exercise. *Chemistry and Ecology.* **4**: 179–180.

Collins, K.J., A.C. Jensen, A.P.M. Lockwood and W.H. Turnpenny. 1994. Evaluation of stabilised coal-fired power station waste for artificial reef construction. *Bulletin of Marine Science.* **55**: 1242–1252.

Collins, K.J., A.C. Jensen and S. Albert. 1995. A review of waste tyre utilisation in the marine environment. *Chemistry and Ecology*. **10**: 205–216.

Diamant, A., A. Ben Tuvia, A. Barnes and D. Golani. 1986. An analysis of rocky coastal eastern Mediterranean fish assemblages and a comparison with small adjacent artificial reef. *Journal of Experimental Marine Biology and Ecology*. **97**: 269–285.

Golani, D., A. Ben Tuvia and B. Galil. 1983. The feeding habits of the Suez Canal migrant squirrelfish, *Sargocentrum rubrum*, in the Mediterranean Sea. *Israel Journal of Zoology*. **32**: 194–204.

Grofit, E. 1993. The Fisheries and Aquaculture of Israel in Figures 1992. Israel Ministry of Agriculture, Department of Fisheries. 32 pp.

Hamilton, W.D. 1971. Geometry for the selfish herd. *Journal of Theoretical Biology*. **31**: 295–311.

Hureau, J.C. 1987. Sparidae. In Whitehead, P.J.P., M.L. Bauchot, J.C. Hureau, J.G. Nielsen and E. Tortonese. (eds.) *Fishes of the Northern Atlantic and the Mediterranean*. Vol. I, UNESCO, Paris; pp. 882–907.

Jensen, A.C. and K.J. Collins. 1995. The Poole Bay artificial reef project 1991–1994. *Biol. Mar. Medit.* **2**: 11–122.

Jensen, A.C., K.J. Collins, A.P.M. Lockwood, J.J. Mallinson and W.H. Turnpenny. 1994. Colonisation and fishery potential of a coal waste artificial reef in the United Kingdom. *Bulletin of Marine Science*. **55**: 1242–1252.

Kimor, B. 1983. Distinctive features of the plankton of the Eastern Mediterranean. *Ann. Inst. Oceanogr. Paris*. **59**: 97–106.

Kimor, B. and E.J.F. Wood. 1975. A plankton study in the Eastern Mediterranean. *Marine Biology*. **29**: 321–333.

Krebs, J.R. and N.B. Davies. 1987. *An Introduction to Behavioral Ecology*. Blackwell Scientific, London, 389 pp.

Kress, N. 1991. Chemical aspects of coal fly ash disposal at sea, predicting and monitoring environmental impact. In Adin, A., Y. Steinberger and J. Garty. (eds.) *Environmental Quality and Ecosystem Stability*. ISEEQS, Jerusalem, pp. 520–525.

Lakkis, S. and V.N. Lakkis. 1980. Composition, annual cycle and species diversity of the phytoplankton in Lebanese coastal water. *Journal of Plankton Research*. **3**: 123–126.

Lundberg, B. and Y. Lipkin. 1979. Natural food of the herbivorous rabbitfish (*Siganus* spp.) in the Northern Red Sea. *Bot. Mar.* **22**: 173–181.

Milton, J.W. 1989. Economic evaluation of artificial habitat for fisheries: progress and challenges. *Bulletin of Marine Science*. **44**: 831–843.

Parker, J.H. 1985. A five year study of building artificial reefs with waste blocks. *Bulletin of Marine Science*. **37**: 399.

Rosen, D.S. 1983. Wave recording analysis. P.N. 118/83, Coastal and Marine Engineering Research Institute, Technion, Haifa, 17 pp.

Russell, B.C., F.H. Talbot, G.R.V. Anderson and B. Goldman. 1978. Collection and sampling of reef fishes. In Stoddart, D.R. and R.F. Johannes. (eds.) *Coral Reef Research Methods*. UNESCO, Paris, pp. 329–345.

Ryan, W.B.F. 1966. Mediterranean Sea: physical oceanography. In Fairbridge, R.W. (ed.) *Encyclopedia of Oceanography*. Van Nostrand Reinhold, New York, pp. 492–493.

Spanier, E. 1987. Mollusca as food for the slipper lobster *Scyllarides latus* in the coastal waters of Israel. *Levantina*. **68**: 713–716.

Spanier, E. 1989. How to increase the fisheries yield in low productive marine environments. In *Ocean '89*, Vol. I. Seattle, pp. 297–301.

Spanier, E. 1991. Artificial reefs to insure protection of the adult Mediterranean slipper lobster, *Scyllarides latus* (Latreille, 1803). In Boudoresque, C.F., M. Avon and V. Gravez. (eds.) *Les Espèces Marines Protégées en Méditerranée*. GIS Posidonia, pp. 179–185.

Spanier, E. 1994. What are the characteristics of a good artificial reef for lobsters? *Crustaceana*. **67**: 173–186.

Spanier, E. 1995a. Artificial reefs in a biogeographically changing environment. Proceedings of ECOSET '95, International Conference on Ecological System Enhancement Technology for Aquatic Environments. Vol. II. Japan International Marine Science and Technology Federation, Tokyo, pp. 543–547.

Spanier, E. 1995b. Do we need special artificial reefs for lobsters? Proceedings of ECOSET '95, International Conference on Ecological System Enhancement Technology for Aquatic Environments. Vol. II. Japan International Marine Science and Technology Federation, Tokyo, pp. 548–553.

Spanier, E. and G. Almog Shtayer. 1992. Shelter preferences in the Mediterranean slipper lobster: effect of physical properties. *Journal of Experimental Marine Biology and Ecology.* **164**: 103–116.

Spanier, E. and D.E. Barshaw. 1993. Tag retention in the Mediterranean slipper lobster. *Israel Journal of Zoology.* **39**: 29–33.

Spanier, E. and B.S. Galil. 1991. Lessepsian migration: a continuous biogeographical process. *Endeavour.* **15**: 102–106.

Spanier, E. and K.L. Lavalli. 1998. Natural history of *Scyllarides latus* (Crustacea Decapoda): a review of the contemporary biological knowledge of the Mediterranean slipper lobster. *Journal of Natural History.* **32**: 1769–1786.

Spanier, E. and R.K. Zimmer Faust. 1988. Some physical properties of shelter that influence den preference in spiny lobsters. *Journal of Experimental Marine Biology and Ecology.* **121**: 137–149.

Spanier, E., S. Pisanty and M. Tom. 1983. Artificial reef in the Eastern Mediterranean: preliminary results. In Shuval, H.I. (ed.) *Development in Ecology and Environmental Quality.* Vol. II, Balaban, Rehovot/Philadelphia, pp. 317–325.

Spanier, E., M. Tom and S. Pisanty. 1985a. Enhancement of fish recruitment by artificial enrichment of man-made reefs in the Southeastern Mediterranean. *Bulletin of Marine Science.* **37**: 356–363.

Spanier, E., M. Tom, S. Pisanty, S. Breitstein, Y. Tur Caspa and G. Almog. 1985b. Development of artificial reefs for commercial species and open sea fish culturing in the Southeastern Mediterranean. In Mitchell, C.T. (ed.) *Diving for Science.* 85, La Jolla, Academy of Underwater Sciences, pp. 123–135.

Spanier, E., M. Tom, S. Pisanty and G. Almog. 1988. Seasonality and shelter selection by the slipper lobster *Scyllarides latus* in the southeastern Mediterranean. *Marine Ecology Progress Series.* **42**: 247–255.

Spanier, E., M. Tom, S. Pisanty and G. Almog Shtayer. 1989a. The fish assemblage on coralligenous shallow shelf off the Mediterranean coast of northern Israel. *Journal of Fish Biology.* **35**: 641–649.

Spanier, E., M. Tom, S. Pisanty and G. Almog Shtayer. 1989b. Comparison between Lessepsian migrants and local species in the fish community of the shallow shelf of northern Israel. In Spanier, E., M. Luria and Y. Steinberger. (eds.) *Ecosystem Stability and Environmental Quality.* Vol. IV/B. ISEEQS, Jerusalem, pp. 177–185.

Spanier, E., M. Tom, S. Pisanty and G. Almog Shtayer. 1990. Artificial reefs in the low productive marine environment of the Southeastern Mediterranean. *P.S.Z.N.I. Marine Ecology.* **11**: 61–75.

Spanier, E., D. Weihs and G. Almog Shtayer. 1991. Swimming of the Mediterranean slipper lobster. *Journal of Experimental Marine Biology and Ecology.* **145**: 15–32.

Spanier, E., G. Almog Shtayer and U. Fiedler. 1993. The Mediterranean slipper lobsters *Scyllarides latus* the known and the unknown. *Bios.* **1**: 49–58.

Spanier, E., K.J. Collins and J. Morris. 1996. Environmental changes in an artificial reef in the southeastern Mediterranean in the last decade. In Steinberger, Y. (ed.) *Preservation of our world in the wake of change.* Proceedings of the 6th International Conference of the Israeli Society for Ecology and Environmental Quality Sciences, Vol. VIB, Jerusalem, Israel, pp. 446–448

Suzuki, T. 1991. Application of high volume fly ash concrete. In Nakamura, M., R. Grove and C.J. Sonu. (eds.) *Recent Advances in Aquatic Habitat Technology*. Southern California, Edison Co, pp. 311–319.

White, A.T., L.M. Chou, M.W.R.N. De Silva and F.W. Gaurin. 1990. Artificial reefs for marine habitat enhancement in Southeast Asia. *ICLARM Education Series* **11**. ICLARM, Philippines, 45 pp.

Woodhead, P.M.J. and M.E. Jacobson. 1985. Epifaunal settlement, the process of community development and succession over two years on an artificial reef in New York Bight. *Bulletin of Marine Science*. **37**: 364–376.

Zimmer Faust, R.K. and E. Spanier. 1987. Gregariousness and sociality in spiny lobsters: implication for den habituation. *Journal of Experimental Marine Biology and Ecology*. **105**: 57–71.

2. Turkey: A New Region for Artificial Habitats

ALTAN LÖK and ADNAN TOKAÇ

Aegean University, Fisheries Faculty, 35100 Bornova, İzmir, Turkey

Introduction

The scientific study of artificial reefs is a comparatively new concept in Turkey, starting with the introduction of a reef at Hekim Island (Fig. 1) by the Aegean University Fisheries Faculty in 1991 and followed by the placing of a reef off Foça by Nine September University in 1993. During the 1980s small-scale reefs were deployed by diving clubs and universities, but results remain unpublished. There has also been an attempt by the İzmir Metropolis Municipality to prevent illegal trawling and to generate new fishing opportunities for sport anglers by depositing ten old trolley-bus bodies in İzmir Bay in 1989. Echo sounder studies have shown, however, that the effective height of these last structures has been reduced from 4 m to 1.5 m by sedimentation. A later project carried out by Çesme Municipality in 1995 to improve recreational fishing is too recent to have been fully evaluated. No artificial reef project in Turkey at present is being funded nationally or regionally; existing projects are all based on small-scale reefs.

This chapter presents results from the Hekim Island artificial reef. The aim of this study was to describe the fish population and the other effects of the artificial reefs which were deployed at two different depths on *Posidonia* beds.

Materials and Methods

The study was carried out off the north-east coast of Hekim Island in the Middle Bay of İzmir (38° 27' 00" N, 26° 47' 00" E) between January 1992 and March 1994. İzmir Bay has three different sections, referred to as the Inner, Middle and Outer Bay, which differ in topographic and hydrographic features. The Inner Bay is polluted, mainly with organics and heavy metals originating from factory (leather and textiles) and domestic discharges (sewage and other chemicals). These enter the Inner Bay through 28 outfalls and 10 streams. The seabed of the Middle Bay is sandy and sandy-gravelly, where shallower than 1 m, covered with *Posidonia oceanica* beds between 1 and 16 m depth and then mud down to 40 m. The pollution that originates from the Inner Bay periodically affects the artificial reef study area. Underwater visibility decreases and sedimentation rates increase when the polluted water is present.

Hollow 1 m^3 concrete blocks were deployed at two different depths: 15 blocks at 9 m and 15 blocks at 18 m depth. The blocks were arranged in a pyramidal shape (four as a base and one on top) by divers. The blocks were deployed at two

A.C. Jensen et al. (eds.), Artificial Reefs in European Seas, 21–30
© 2000 Kluwer Academic Publishers. Printed in Great Britain.

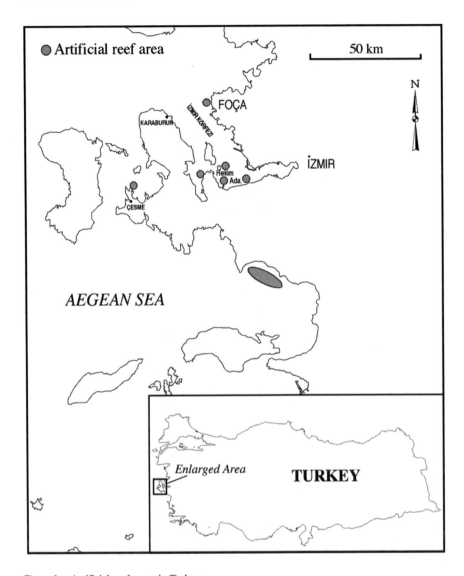

Figure 1.   Artificial reef areas in Turkey.

different depths to determine effects of depth differences on fish communities around the reefs.

Visual transect census techniques were used to determine species presence and density before and after reef deployment in the study area (Bortone *et al.*, 1992; Brock, 1982). Presence and behaviour of species around the reef were recorded with amphibious cameras (Nikonos V with 35 mm lens) and video (Sony CCD TR75).

Mann-Whitney U tests were used to determine whether there were any significant differences between mean species and number of individuals before and after reef deployment at the 0.05 probability level.

Results and Discussion

Mobile Fauna

The species that were recorded before and after reef deployment are given in Table 1 together with whole fish number counts carried out by the same divers at a similar time of day and within the same water volume. There is some similarity between Turkish fish species around the Hekim reef and those in the vicinity of northern Mediterranean reefs (Relini and Relini, 1989; Ramos and Bayle, 1990).

Eight species of fish were present before reef placement, increasing to 16 after deployment. The average number of species and individuals were 4.57 ± 0.61 and 134.28 ± 24.5 respectively before reef deployment changing to 7.28 ± 0.56 and 259.38 ± 55.9 respectively after reef deployment.

Differences between mean species number and number of individuals before and after reef deployment were significant at $p < 0.05$. Results show that the

Table 1. List of fish and other mobile species observed before and after reef deployment. Fish species were identified from UNESCO (1986) *Fishes of the North-eastern Atlantic and the Mediterranean.*

Species	Food	Present before reef placement	Present after reef placement
Atherinidae			
Atherina boyeri	Carnivorous	no	yes
Centracanthidae			
Spicara smaris	Herbivorous	yes	yes
Gobiidae			
Gobius niger	Carnivorous	no	yes
Labridae			
Coris julis	Carnivorous	yes	yes
Symphodus ocellatus	Carnivorous	yes	yes
Mullidae			
Mullus barbatus	Carnivorous	no	yes
Sparidae			
Diplodus annularis	Carnivorous	yes	yes
Diplodus vulgaris	Carnivorous	yes	yes
Boops boops	Herbivorous	yes	yes
Lithognathus mormyrus	Carnivorous	no	yes
Sparus aurata	Carnivorous	no	yes
Sarpa salpa	Herbivorous	no	yes
Spondyliosoma cantharcus	Carnivorous	no	yes
Pomacentridae			
Chromis chromis	Carnivorous	no	yes
Octopodidae			
Octopus vulgaris	Carnivorous	yes	yes

numbers of fish species and individuals doubled in the study area after reef deployment. It is thought that the presence of seagrass in the experimental reef area is an important factor contributing to this increase: Randall (1963) concluded that the presence of seagrass beds greatly increased the number and biomass of fish on an experimental artificial reef in the Virgin Islands. The number of species and individuals at the 9 m and 18 m reef varied seasonally (Figs 2, 3).

The number of species and individuals increased during the summer months and decreased in the winter months on the 9 m reef. At 18 m depth this was reversed, with an increase in the number of individuals during winter months. The difference between numbers of species was significant ($p < 0.05$) between the 9 m and the 18 m reef. *Spicara smaris, Boops boops, Chromis chromis* and *Sarpa salpa* were found around the 9 m reef only during summer and were totally absent from the deeper reef, while there was no significant difference between number of individuals ($p > 0.05$) present at the 9 m and 18 m reef. Bombace *et al.* (1989) suggested that the reduction in species number around artificial reefs in the Adriatic Sea during the winter months may be caused by fish migrating to deeper water. Many fish adapt to seasonal changes by moving to another habitat or area (Bohnsack *et al.*, 1991). The fact that the number of individuals did not differ statistically between reefs may have been the result of seasonal changes in population at the 9 m and 18 m reefs.

Coris julis, Serranus scriba, Diplodus annularis and *Diplodus vulgaris* can be described as resident species (54–100% presence; Table 1), *S. smaris, B. boops, S. salpa, C. chromis* and *Octopus vulgaris* as visitor species (42–56% presence) and *Atherina boyeri, Gobius niger, Symphodus ocellatus, Mullus barbatus, Lithognathus mormyrus, Sparus aurata* and *Spondyliosoma cantharcus* as transient species (42% presence) according to the classification of Bohnsack *et al.*, (1991). These results

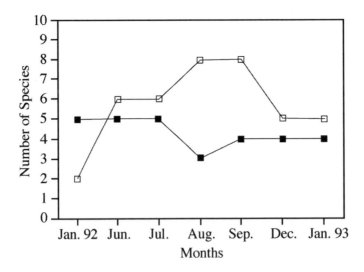

Figure 2.   The variation in number of fish species over 12 months on the 9 m (□) and 18 m (■) reefs.

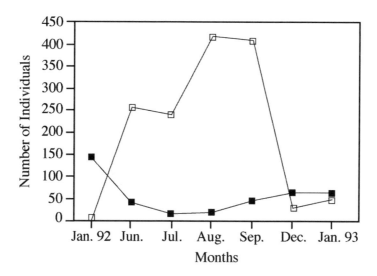

Figure 3.   The variation in the individual numbers of fish over 12 months on the 9 m (□) and 18 m (■) reefs.

suggest that depth and location are important in attracting visitor species but not as important for common resident species (Moffitt *et al.*, 1989).

Examination of the feeding regime of species found on the reefs shows that 80% of the species described were carnivores, the other 20% herbivores. Bohnsack *et al.* (1991) reported that carnivores tend to dominate the nektonic biomass on marine artificial habitats and that herbivores may be limited by the restricted food resource. The fact that *Posidonia* beds surrounding the reefs are good feeding areas for herbivorous species must not be overlooked.

Targeted commercial fish species such as *Oblada melanura* and *Diplodus sargus* were not observed on either of the artificial reefs but were present on a natural reef 200 m away. The reason for their absence from the reefs is not known but may be caused by a lack of a particular food resource and/or shelter characteristic.

Benthos

According to initial data, species richness of epibiota on the 9 m reef was greater than that on the 18 m reef. The lower number of species on the 18 m reef may be caused by high turbidity of the water, reducing light levels on the reef. The algal species identified on the 9 m and 18 m reefs are summarized in Table 2. Identification has been assisted by the work of Aysel *et al.* (1984) and Dural *et al.* (1992). The total number of taxa and percentage presence of algae at the 9 m reef were 20 and 40%, respectively. On the 18 m reef, the total number of taxa and the percentage presence were 10 and 19%.

It follows from these data that there are differences in algal cover in both the total number of taxa and the value of presence on the 9 m and 18 m reefs. Light levels were lower at 18 m than at 9 m and even the distribution of red and brown algae, tolerant of reduced light, had been negatively influenced by the high turbidity levels in İzmir Bay. The composition and number of taxa of blue-green algal species (which are not as influenced by turbidity) were increased. However, the number of taxa and the percentage presence of green algae were severely affected by low light levels and less red algae was present at 18 m than at 9 m.

Twenty invertebrate species belonging to 10 classes at the 9 m reef and 15 species belonging to eight classes at the 18 m reef were identified (Table 3).

Oyster spat settled on the 9 m reef blocks. This settlement was grazed by the sea-urchin *Sphaerechinus granularis*. A few months later, the starfish *Marthasterias glacialis* was seen preying on the remaining oysters (Fig. 4). Seventy to eighty percent of the oysters that settled were killed by this combined predation. Buckley and Hueckel (1985) reported this type of predation pressure on barnacles by *E. troschelii*.

The bivalve *Mytilus galloprovincialis* had settled and was observed on the 9 m reef in January 1994. This is a space-dominating species and it was expected that it would cover all reef surfaces, replacing the slowly decreasing *Ostrea edulis* population in the following few months. This turned out not to be the case. By August 1996 only a few adult mussels were present and there was no evidence of further settlement. Failure of the mussel settlement could be the result of algal

Table 2. Algal species identified on 9 m and 18 m reefs.

Taxon	9 m	18 m
Cyanophyta		
Lyngbya majuscula Harvey	+	+
Lyngbya martensiana Meneghini	+	+
Lyngbya sp.	+	+
Oscillatoria sp.	+	+
Phormidium sp.	+	+
Calothrix sp.	+	
Chlorophyta		
Ulothrix sp.	+	
Enteromorpha clathrata (Roth) Grev.	+	
Acetabularia acetabulum (L.) P.C. Silva	+	
Cladophora sp.	+	
Phaeophyta		
Feldmania sp.	+	
Asperococcus bullosus Lam	+	
Cutleria multifida (Sm.) Grev.	+	
Sphacelaria sp.	+	+
Dilophus sp.	+	+
Padina pavonina (l.) Lam	+	
Rhodophyta		
Ceramium sp.	+	+
Laurencia obtusa (Huds.) Lam.	+	
Polysiphonia sp.		+
Falkenbergia rufolanosa (Harv.) Schmitz	+	
Lophosiphonia sp.	+	+

Table 3. Invertebrate species identified on 9 m and 18 m reefs.

Species	9 m	18 m
Cnidaria		
Hydrozoa		
Eudendrium sp.	+	
Annelida		
Polychaeta		
Hydroides elegans (Haswell, 1883)	+	
Serpula vermicularis (Linne, 1767)	+	+
Pomatoceros triqueter (Linne, 1767)	+	+
Spirographis spallenzani (Viviani, 1805)	+	
Platynereis dumerlii (Audouin and Edwards, 1830)	+	
Nereis c. zonata (Malmeren, 1867)	+	
Arthropoda.		
Cirripedia		
Balanus perforatus (Bruguiére, 1789)	+	
Mollusca		
Bivalvia		
Ostrea edulis (Linne, 1758)	+	+
Mytilus galloprovincialis (Lamarck, 1819)	+	
Saxicava sp.		+
Anomia ephippium (Linne, 1758)	+	
Gastropoda		
Bittium reticulatus (Da Costa, 1778)	+	
Murex trunculus (Linne, 1758)		+
Bryozoa		
Cheilostomata		
Schizobrachiella sanguinea (Norman, 1868)	+	+
Schizoporella longirostris (Hincks, 1886)		+
Schizoporella errata (Waters, 1848)		+
Schizoporella sp.	+	+
Electra monostachys (Busk, 1854)	+	
Echinodermata		
Echinoidea		
Sphaerechinus granularis (Lamarck, 1816)	+	+
Paracentrotus lividus (Lamarck, 1816)	+	
Asteroidea		
Marthasterias glacialis (Linne, 1765)	+	+
Echinaster sepositus (Petz, 1783)	+	+
Holothuridae		
Holothuria tubulosa (Gmelin, 1788)	+	+
Ascidiacea		
Aplousobranchiata		
Didemnum sp.	+	
Phelbobranchiata		
Ascidia mentula (Müller 1776)		+

colonization of the reef surface (Fig. 5). The fact that not only the surface of concrete blocks but also the valves of dead and living mussels and oysters were covered by algae may have prevented further larval settlement of these species on the reef.

An alternative explanation could be that the mussels were killed by starfish, as happened to the *Ostrea*. Culture of mussels is considered to be more successful on 'mussel ropes' fastened between reefs rather than harvesting directly from reef surfaces, because the predation pressure is so high (Fabi and Fiorentini, 1989) on reef substrata. Ardizzone *et al.* (1989) suggested that reef blocks should be removed from the water periodically to kill epifauna and promote effective colonization by mussels for culture or that removable substrata are used for mussel culture. The benthic colonization of the reef will be, and is being, monitored for the foreseeable future.

The cephalopod *Octopus vulgaris* was observed on both artificial reefs. These animals moved inshore to reproduce as the water temperature fell in December and used the reefs as shelters. *O. vulgaris* migrates offshore when the water

Figure 4. The predation effect of *M. glacialis* on the oysters on the 9 m reef.

Figure 5. The oyster colonization and algal development on the reefs at 9 m.

Figure 6. A reef block at 18 m. *Serranus scriba* and benthic colonization on the reef (foreground), and high turbidity (background).

temperature begins to rise. Polovina and Sakai (1989) showed that artificial reefs in one region of Japan increased the production of octopus by providing suitable habitat.

Hundreds of egg bunches belonging to the squid *Loligo vulgaris* were recorded from the 18 m deep blocks in June 1993 and April 1994. This species attaches egg bunches to hard substrata in low current areas for incubation. Some Japanese artificial reefs are used as spawning reef for squids (Thierry, 1988).

Conclusions

Experimental artificial reefs doubled the number of species and individuals in the experimental area off Hekim Island, providing settlement surfaces, shelter and reproductive opportunities for some species. Long and detailed studies will reveal how artificial reef deployments affect the Turkish marine environment. *Posidonia* beds decrease in size day by day because of pollution and illegal trawl fisheries in Aegean and Mediterranean coastal waters (Ardizzone and Pelusi, 1984; Lök and Tokaç, 1993). Use of artificial reefs may be effective in preventing illegal trawl fishing and help in balancing the loss of this habitat in the marine environment.

The future of artificial reef applications is uncertain in Turkey, as in other Mediterranean countries (Bombace, 1989). There is no national artificial habitat project in Turkey at present. Only two universities and one municipality continue their artificial reef projects with the aim of experimenting with reefs and creating line fisheries. The results of these projects will directly affect any Turkish artificial reef deployments in the future.

References

Ardizzone, G.D. and P. Pelusi. 1984. Yield and damage evaluation of bottom trawling on *Posidonia* meadows. In Boudouresque, C.F., A.J. de Grissac and J. Oliver. (eds.) *International Workshop on Posidonia oceanica beds*. GIS Posidonia publ., France, pp. 63–72.

Ardizzone, G.D., M.F. Gravina and A. Belluscio. 1989. Temporal development of epibenthic communities on artificial reefs in the central Mediterranean Sea. *Bulletin of Marine Science*. **44**(2): 592–608.

Aysel, V., H. Güner, A. Sukatar and M. Öztürk. 1984. Check list of İzmir Bay marine algae: I. Rhodophyceae. *Aegean University Faculty of Science Journal Series B*. **7**: 47–56.

Bohnsack, J.A., D.L. Johnson and R.F. Ambrose. 1991. Ecology of artificial reef habitats and fishes. In Seaman, W. and L.M. Sprague. (eds.) *Artificial Habitats for Marine and Freshwater Fisheries*. Academic Press Inc., pp. 61–107.

Bombace, G. 1989. Artificial reefs in the Mediterranean Sea. *Bulletin of Marine Science*. **44**: 1023–1032.

Bombace, G., G. Fabi and L. Fiorentini. 1989. Preliminary analysis of catch data from artificial reefs in the central Adriatic. FAO Fisheries Report. **428**: 120–127.

Bortone, S.A., J. van Tassell, A. Brito and C.M. Bundrick. 1992. Visual census as a means to estimate standing biomass, length and growth in fishes. *Proceedings of the American Academy of Underwater Sciences, Diving for Science*. **12**: 13–21.

Brock, R.E. 1982. A critique of the visual census method for assessing coral reef fish populations. *Bulletin of Marine Science*. **32**: 269–276.

Buckley, R.M. and G.J. Hueckel. 1989. Analysis of visual transects for fish assessment on artificial reefs. *Bulletin of Marine Science*. **44**(2): 893–898.

Dural, B., H. Güner and V. Aysel. 1992. The comparison of marine flora of Çeşme and Eski Foça with Turkey and Mediterranean. *Aegean University, Faculty of Science Journal. Series B*. **14**: 65–67.

Fabi, G. and L. Fiorentini. 1989. Shellfish culture associated with artificial reefs. FAO Fisheries Report. **428**: 120–127.

Lök, A. and A. Tokaç. 1994. The role of artificial reefs used for the protection of sensitive marine zones and increasing fish stocks. *Aegean University, Faculty of Science Journal Series B Supplementary Issue*. **16**: 1061–1066.

Moffitt, R.B., F.A. Parrish and J.J. Polovina. 1989. Community structure, biomass and productivity of deepwater artificial reefs in Hawaii. *Bulletin of Marine Science*. **44**(2): 616–630.

Polovina, J.J. and I. Sakai. 1989. Impacts of artificial reefs on fishery production in Shimamaki, Japan. *Bulletin of Marine Science*. **44**(2): 997–1003.

Ramos, A.A. and J. Bayle. 1990. Management of living resources in the marine reserve of Tabarca Island (Alicante, Spain). *Bulletin de la Société Zoologique de France*. **114**: 41–48.

Randall, J.E. 1963. An analysis of the fish populations of artificial and natural reefs in the Virgin Islands. *Caribbean Journal of Science*. **3**: 31–46.

Relini, G. and L.O. Relini. 1989. Artificial reefs in the Ligurian Sea (Northwestern Mediterranean): aims and results. *Bulletin of Marine Science*. **44**(2): 743–751.

Thierry, J.M. 1988. Artificial reefs in Japan: a general outline. *Aquacultural Engineering*. **7**: 321–349.

3. Artificial Reefs in the Adriatic Sea

GIOVANNI BOMBACE, GIANNA FABI and LORIS FIORENTINI
Istituto di Ricerche sulla Pesca Marittima (IRPEM)–CNR, Largo Fiera della Pesca, Ancona
Italy

Introduction

Aims

Italian initiatives and experiments with artificial reefs, in particular in the Adriatic Sea, have been prompted by the requirement to effectively manage and enhance the coastal seas. The coastal zone is subject to inter- and intra-sector conflicts but has great potential for fishery and mariculture development.

The main aims for artificial reef construction in the Adriatic Sea were:

(1) To protect fish-spawning and nursery areas from illegal trawling in waters ≤ 50 m deep and/or within the Italian 3 mile limit.
(2) To create oases for repopulation by fish and invertebrates by deploying suitable structures that would provide shelter and protection for juveniles and adults.
(3) To redirect the flow of excess nutrients present in Adriatic coastal waters, currently causing eutrophication, into edible biomass through mariculture of filter-feeders.
(4) To protect small-scale artisanal set gear fisheries from damage by illegal trawling activity.

Choice of Artificial Reef Location

The artificial reef deployment sites were chosen by taking the following requirements into consideration:

(1) The presence of fishermen's associations which would manage, in a rational manner, the fishing and harvesting effort on and around the artificial reefs once the initial experimental phase was completed.
(2) Evidence of intra-sectorial conflicts between small-scale artisanal fisheries and trawling.
(3) Favourable environmental conditions for the settlement and development of sessile organisms. This still needs further investigation, as conditions which promote good growth and survivorship in filtering organisms may also be linked to high sedimentation rates. Such sedimentation is unfavourable for the subsequent settlement of filter feeders.
(4) The presence of both spawning and nursery areas of commercially exploitable species.

A.C. Jensen et al. (eds.), Artificial Reefs in European Seas, 31–63
© *2000 Kluwer Academic Publishers. Printed in Great Britain.*

These requirements indicated that the coastal zone within 5 km of the shore was a suitable area for reef deployment, and in the northern and central Adriatic Sea, artificial reefs were built in 11–14 m of water.

Physical and Ecological Features of the Adriatic Coast

The Adriatic coast (northern and central Adriatic Sea, from Trieste to the Gargano promontory) is characterized by:

(1)   The prevalence of a sandy-muddy or muddy seabed with scattered outcrops of limestone forming headlands (such as those at Conero and Gargano) or cliffs (Trieste). The seabed slopes gently: 5 km offshore of Ancona the water is only 15 m deep.

(2)   Heavy river runoff (an average of 7000 m^3 s^{-1} of freshwater), half from the Po river, which flows into the coastal area between the Isonzo river (near Trieste) and the Fortore river (near the Gargano promontory). This input carries a significant nutrient load (organic matter, phytoplankton, chlorophyll a) which causes eutrophication. The consequent high water column primary productivity provides ample food for planktivorous fish (e.g. sardines and anchovies) and filter feeders (mainly bivalves) and creates opportunities for mariculture.

(3)   A large seasonal variability in the physical parameters of the water column. Inshore, temperature can vary from 5°C in winter to 28°C in summer. During winter the coastal waters (as far as 10–12 km offshore) are well mixed, with temperatures as low as 5–6°C and salinity <37 ppt, whilst in the open sea water temperatures are approximately 10–12°C and salinity is >38 ppt. Consequently, a thermohaline front, vertically extended through the entire water mass, separates the coastal waters from the open sea. Prior to formation of this front, most mobile species migrate towards deeper, offshore waters where they can find warmer temperatures. In summer, stratification produces a horizontal thermohaline front which separates warmer surface waters of lower salinity from deeper and colder more saline waters.

These ecological factors influence Adriatic artificial reefs. Filter feeders, such as mussels (*Mytilus galloprovincialis*) and oysters (*Ostrea edulis* and *Crassostrea gigas*) settle abundantly on new artificial substrata. Food is plentiful which promotes survival, but subsequent biological settlement may be reduced by significant sedimentation on horizontal surfaces. During the winter months, because of the low water temperature, the growth rate of sessile organisms decreases and a part of the fish population attracted by the artificial reefs migrates offshore.

During strong winter storms, backwash and currents may erode the seabed, so undermining artificial reef units.

Funding

The first artificial reef initiatives were totally experimental and were carried out with the financial assistance of the National Council of Research (a government institution that supervises Istituto di Ricerche sulla Pesca Marittima (IRPEM) activity) and the Italian Merchant Marine Ministry, which regulated Italian fisheries up to 1993.

Subsequently all the reefs were constructed with the financial support of the Italian State and the European Economic Community (EEC). Most artificial reefs were established at the request of fishermen's co-operatives with scientific support from IRPEM.

Materials

At the beginning of the 1970s IRPEM researchers decided to deploy non-polluting, sea-water-resistant materials able to physically prevent trawling, offer shelter and protection for mobile fauna and provide suitable surfaces for settlement of sessile organisms. This led to the use of natural or 'naturalized' materials such as stones and concrete. Wood, e.g. old boats, and later steel, used as supports for submerged shellfish culture, were also integrated into reef designs. The original basic reef module was a 8 m^3 concrete block, weighing 13 tonnes which incorporated holes of different shapes and sizes (Bombace, 1981, 1982). It was called the 'IRPEM block' (Fig. 1).

This block efficiently prevented illegal trawling, thanks to its weight, whilst the holes provided habitats and shelters for marine biota. Its rough surfaces facilitated the settlement of sessile organisms and the cuboid shape allowed stacking of the blocks to create a 'pyramid', so increasing reef height. These cubic reef blocks were cast in steel moulds, making production efficient.

Initially, reef pyramids were made of 14 stacked blocks (nine at the base, four in the middle and one on the top). Later pyramids made of five blocks (four base and one on top) were constructed and the top block was fitted with an iron support for attachment of shellfish culture cables (Fig. 1; Bombace, 1981, 1982, 1987).

Description of the Adriatic Artificial Reefs

Porto Recanati

The first artificial reef in the Adriatic was deployed during 1974–75 at Porto Recanati, 5.5 km offshore, on a muddy seabed, 13–14 m deep (Fig. 2). The reef was constructed for IRPEM with the aims of preventing trawling, repopulating the area with marine life and developing new, sessile biomass, especially mussels and oysters. It has a central 'oasis' (3 ha) formed by 12 pyramids (Fig. 3), each one built from 14 IRPEM blocks. Each pyramid is 6 m high. The pyramids were placed on gravel 'mattresses' (to spread the weight and prevent subsidence) and were deployed in a rectangular arrangement about 50 m apart. Stone piles were placed between the pyramids to make the reef system continuous and two old vessels were sunk at the centre of the oasis. The oasis is surrounded by a protected area (2000 ha) where cubic blocks and other smaller anti-trawling blocks were dispersed.

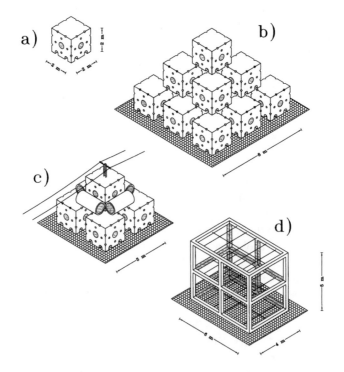

Figure 1. Structures designed by IRPEM as artificial reefs: (a) IRPEM block; (b) '14 block' pyramid; (c) '5 block' pyramid; (d) concrete cage for shellfish culture.

A total of 453 IRPEM blocks were used: 168 for the pyramid construction and 285 as isolated anti-trawling blocks, giving a total volume of 3624 m³. The stones placed under the pyramids provided a volume of 396 m³. In all, more than 4000 m³ of material was deployed.

Portonovo 1

The second experimental reef was built for IRPEM in 1983 in Portonovo Bay (Conero promontory) south-east of Ancona (Fig. 2). Portonovo Bay (about 2.2 km across with a maximum depth of 10 m) has a sandy seabed with scattered limestone outcrops and is bordered by rocky banks where mussels and oysters grow abundantly. The reef, situated approximately 0.8 km offshore at a depth of about 10 m, consists of four pyramids, each of five IRPEM blocks, placed at the corners of a square, about 20–25 m apart (Fig. 4) and represents a 'standard model' from which other larger reefs have been built. The top blocks of each pyramid are connected, one to the next, by steel wires from which cylindrical nylon nets, full of mussel seed, and baskets of oysters can be suspended (Fig. 4).

This small reef has been used by IRPEM researchers to carry out studies on shellfish culture.

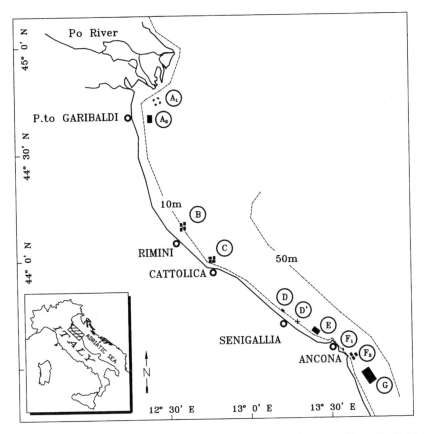

Figure 2. Location of the artificial reef deployed along the Adriatic coast: Porto Garibaldi 1 (A_1); Porto Garibaldi 2 (A_2); Rimini (B); Cattolica (C); Senigallia (D); Falconara (E); Portonovo 1 (F_1); Portonovo 2 (F_2); Porto Recanati (G). The unprotected site used as control is also reported (D').

Artificial Reefs Planned by IRPEM Between 1980 and 1990

Between 1987 and 1989 seven artificial reefs were deployed along the northern and central coast of the Adriatic Sea with assistance from the EEC regulations 2908/83 and 4028/86.

Five of these (Porto Garibaldi 1, Rimini, Cattolica, Senigallia and Portonovo 2 (Fig. 2)) were planned by IRPEM on the basis of previous experience at Porto Recanati and Portonovo 1. All five reefs were placed at the centre of a protected area (200 ha) defined by anchored corner buoys. Each reef consists of one or more oases formed by merging three or six standard reef groups (0.1 ha) each consisting of eight pyramids (Fig. 5; Table 1). The pyramids are formed by five IRPEM blocks and were deployed on gravel mattresses in a square, about 15 m apart. The top blocks of the pyramids are connected to each other by steel wires (allowing suspended shellfish culture) and surrounded by four concrete cylinders leaning on the base blocks, thus increasing the shelter available for marine organisms. Two

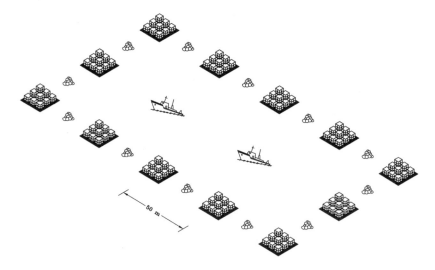

Figure 3.   Plan of the central oasis of Porto Recanati reef (not to scale).

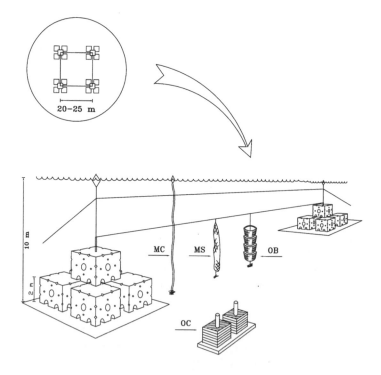

Figure 4.   Plan of Portonovo 1 artificial reef and diagram of the shellfish culture facilities: collectors for mussel seed (MC); nylon sock for suspended mussel culture (MS); oyster spat collectors (OC); plastic baskets for suspended oyster culture (OB).

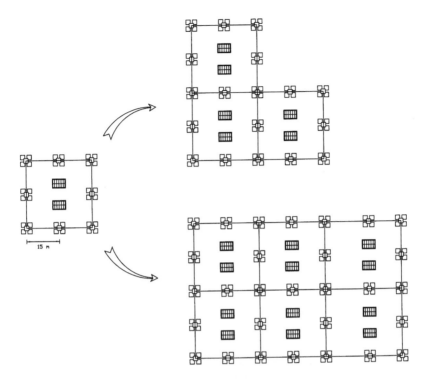

Figure 5. Diagram of the 'standard group' (eight pyramids and two concrete cages for shellfish culture) and the oases that constitute the five artificial reefs planned with the scientific support of IRPEM along the Adriatic coast at the end of the 1980s.

concrete cages (4 × 5 × 6 m; Fig. 1) for shellfish culture were placed at the centre of the pyramids and anti-trawling structures were scattered in the surrounding area.

Except for the Senigallia reef, which is an experimental reef constructed on behalf of IRPEM, all other reefs were established at the request of local fishermen's associations with the scientific support of IRPEM which carried out research before and after reef deployment.

Other Artificial Reefs

The following artificial reefs were established at the request of fishermen's co-operatives. IRPEM only gave their scientific support before and after deployment for the reef Porto Garibaldi 2.

Porto Garibaldi 2

The reef was completed in 1989 and is located about 5.5 km NE of Porto Garibaldi, 4.5 km offshore, at a depth of about 10 m, on a muddy seabed affected by Po river

Table 1. Characteristics of the five artificial reefs deployed along the Adriatic coast at the end of 1980s with the scientific support of IRPEM; weight increments of nekto-benthic fish in catches 3–4 years after reef deployment are reported as multiples of the catches obtained before reef deployment.

Location	Distance from coast (km)	Depth (m)	Number of oases	Distance between the oases (m)	Total number of immersed bodies	Total volume of immersed material (m³)	Covered area (ha)	Destination	Weight increment of nekto-benthic fish (times)
Porto Garibaldi 1	7.7	13	4	500	72 pyramids 24 cages	7500	1.1	Professional	10
Rimini	2.7	11	4	200	118 pyramids 48 cages	13000	2.4	Professional	42
Cattolica	2.7	11	4	200	118 pyramids 48 cages	13000	2.4	Professional	28
Senigallia	2.2	11	1	–	29 pyramids 12 cages	3250	0.6	Experimental	11
Portonovo 2	1.0	10	3	100	87 pyramids 36 cages	9750	1.8	Professional	–

sedimentation (Fig. 2). It covers a square area of about 1.8 ha and is placed at the centre of a protected area (169 ha) where 191 IRPEM blocks were scattered as anti-trawling structures. The reef is made of 100 especially designed structures (4.5 × 4.5 × 4.5 m), each formed by assembling concrete panels and cylinders (Fig. 6a). The structures were placed about 10 m apart.

Unfortunately these structures showed to be not very resistant to sea storms. One year after deployment some of them had already been destroyed and were found to be partially sunk into the seabed (IRPEM, 1993).

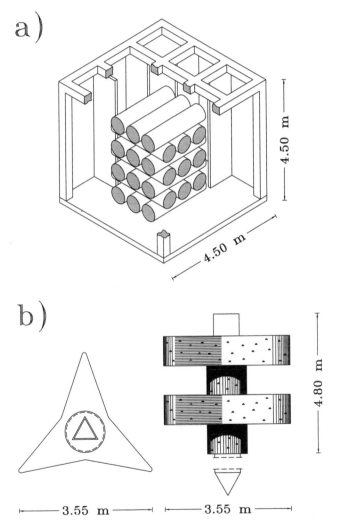

Figure 6. Concrete structures used to construct the reefs at Porto Garibaldi 2 (a) and Falconara (b).

Ancona

Unlike the other reefs described previously, which were designed as 'intensive multipurpose' artificial reefs, built to fulfil a range of aims (trawling prevention, re-population and mariculture), this reef is an 'extensive' artificial reef. It was established in 1988, at request of a co-operative of trawl fishermen, to repopulate an area of seabed and protect the coastal area from illegal trawling.

It is located north of Ancona and covers an area of 12 500 ha between 2.8 and 5.5 km offshore, on a sandy-muddy sea floor at depths ranging from 10 to 14 m (Fig. 2). One hundred and seventy-nine concrete structures called 'stars' (Fig. 6b) and 350 IRPEM blocks were scattered across the area.

Research Methods

Porto Recanati

Prior to reef deployment, records showed the seabed to be compact and able to tolerate the weight of '14 block' pyramids. Most of the structures are still undamaged 20 years after placement.

The aims of the research conducted after reef deployment were:

(1)   To evaluate the growth and biomass of mussels settled on the surfaces of the blocks by periodic scraping of standard 1600 cm^2 areas. Samples were collected on each level of the pyramids to evaluate the biomass at different depths (12–14, 10–12, 8–10 m).
(2)   To evaluate the fishery yields from the reef and to compare them with those obtained in surrounding unprotected areas. Data collection was carried out in the following ways:
   •   experimental catches using standard trammel nets at the reef and in areas placed at various distances from the reef;
   •   regular census of catch data obtained from professional fishermen using different set gear (pots, nets, etc.) in the protected and unprotected areas.

Portonovo 1

Research was focused on the practicality of this type of artificial reef for suspended mussel and oyster farming and the potential for the culture of bivalves in open sea where traditional farming methods were not suitable. A 5-year programme involved the following research:

(1)   tests on different materials (nylon, plastic and manilla) as mussel spat collectors.
(2)   Investigation into the time and depth of mussel settlement: nylon ropes were suspended each month from the sea surface to the seabed; a 10 cm portion of every metre was examined for mussel settlement 2 months after immersion.

(3) Assessment of different management techniques for suspended mussel culture: young mussels were transferred from collectors into nylon socks suspended from taut steel wires between the pyramids (Fig. 4). The growth rate of these mussels, subjected to a different number of thinning operations, revealed the influence, and the importance of these operations (Fabi *et al.*, 1986).

(4) Investigations into the timing of oyster spat settlement by monthly immersion of special plastic collectors (60 × 60 × 8 cm).

(5) Studies of oyster growth, under farming conditions, in plastic baskets suspended from the steel wires among the pyramids (Figs 4, 7).

(6) Planning of new structures to increase shellfish culture inside the artificial reef. A steel cage (4 × 8 × 5 m) was designed and deployed for suspension of nylon mussel socks and oyster baskets.

Porto Garibaldi 1 and 2, Rimini, Cattolica and Portonovo 2

A 4-year IRPEM research programme involving experimental fish catches, underwater observations and sampling of epibiota was carried out at each reef site, starting one year before deployment. The aims were to evaluate the influence of the artificial reefs on the fish assemblage and their effectiveness in terms of increasing fishing yield, the suitability of artificial reefs for settlement and on-growing of filter-feeding bivalves, and the influence of the surrounding habitat (e.g. closeness to natural hard substrata) on the fish populations and epibiota.

Each site was monitored once or twice a month using a bottom trammel net of the type used by professional fishermen (length 500 m; height 2 m; 70 mm inner

Figure 7. Suspended oyster culture at the Portonovo 1 artificial reef.

mesh size; 340 mm outer mesh size). The net was lowered into the water at dusk and pulled in at dawn, an average set of 12 hours (Bombace *et al.*, 1990a, 1994).

Biomass of mussels settled on the surfaces of both the base and top blocks of the pyramids was estimated by periodic scraping of standard 1600 cm² square areas. Two or more pyramids were randomly sampled each time.

Senigallia

Research carried out by IRPEM at this reef can be summarized as follows:

(1)   Mobile fauna
   (a)   Development of the fish assemblage living around and inside the artificial reef was studied.
   (b)   Techniques were developed to estimate the fish biomass.
   (c)   Studies aimed at understanding how much of the reef effect is production of new biomass caused by a greater availability of food and mortality reduction on the reef and how much is simple attraction from the surrounding areas.
(2)   Sessile organisms
   (a)   Evaluation of biomass and growth rates of mussels settled on the artificial substrata.
   (b)   Studies on the sessile community settled on the concrete blocks and in the surrounding soft substrata.
   (c)   Studies on coal ash substrata.
(3)   Mariculture
   (a)   Studies on mussel and oyster culture.
   (b)   Development of new mariculture techniques.

Mobile fauna

The fish population around the reef was surveyed using a bottom trammel net (as used at the artificial reefs of Porto Garibaldi, Rimini, Cattolica and Portonovo 2). The survey started 1 year before the deployment of the structures and continues to date (1998). Fishing has also been carried out in an unprotected control site about 5 km from the artificial reef on the same type of seabed and at the same depth (Fig. 2). In order to reduce bias caused by weather events the two sites were sampled either simultaneously or one immediately after the other (Bombace *et al.*, 1990b; Fabi and Fiorentini, 1994). Collections were also carried out during 1991–92 both by day (from dawn to dusk) and by night (from dusk to dawn). Because nocturnal sampling captured more species and larger numbers of individuals (Bombace *et al.*, 1997b) it became the standard method in the following years.

In addition to trammel netting, the day-time species composition of the fish community around and inside the artificial reef was surveyed monthly by visual census, starting immediately after deployment (Fig. 8). Fewer observations were conducted during the winter months because of rough sea conditions and frequent poor visibility (<3 m). Quantitative data derived from these visual censuses were

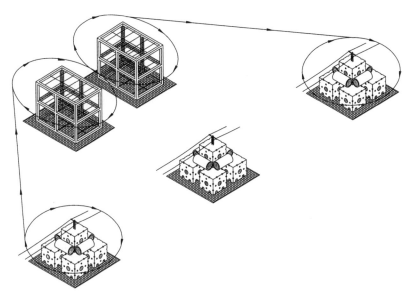

Figure 8. Graphic representation of visual transects swum inside the Senigalli artificial reef.

used to estimate the biomass of fish present at the reef together with information from catch (Bombace *et al.*, 1997b) and echo sounder surveys (Bombace *et al.*, 1998c).

Visual census and catches with trammel nets of different mesh size were also associated with ichthyoplankton sampling for 2 years. Data gathered described the presence of larvae, juveniles and reproductively active fish species around the reef. Some species reproduced inside the reef, an area associated with greater food and shelter availability, suggesting that the artificial reef really can contribute to an overall increase in total biomass (Bombace *et al.*, 1993c).

Sessile fauna

The biomass and growth rate of mussels, and of the overall benthic community settled on the artificial substrata, were estimated from samples taken periodically by scraping standard 1600 cm^2 square areas clear of all epibiota. Similar sampling was also carried out on coal ash blocks (1 m^3) to investigate the differences in the epibiotic colonization between coal ash and concrete.

A diver-operated suction sampler was also used to study the community settled on the horizontal surfaces of the concrete blocks and living in the soft sediment surrounding the pyramids.

Mariculture

Mussel and oyster culture

Besides the techniques previously applied at the Portonovo 1 reef, a new experimental technique for mussel farming was used to minimize the amount of

44   G. Bombace et al.

underwater work required. In a system similar to the Spanish 'bateas', nylon ropes, intersected at 50–70 cm by small plastic bars, were used for culture. The bars prevent the mussels from sliding down the ropes as the crop increases in weight. During the period of highest spat settlement, these ropes were suspended from the steel wires between the pyramids and from the concrete cages (Fig. 9). The mussel spat that settled was allowed to grow without further transfers or thinning out until commercial size (TL = 50 mm) was reached.

Fish farming
Trial farming of gilthead sea bream (*Sparus aurata*) started in 1993. Experimentation with different types of off-shore structures was driven by the necessity of finding suitable structures that can be used on the exposed coasts of the Adriatic Sea, where conventional marine fish farming is limited by absence of protected bays to provide shelter in adverse weather conditions.

An experimental steel cage (4 × 4 × 3 m) for fish on-growing equipped with an automatic food dispenser has been deployed on the seabed inside the reef complex (Fig. 10). A spherical, mid-water fish cage (7 m diameter) and a surface floating hexagonal cage (6 m sides) were also tested (Fig. 11) using the concrete blocks of the reef as anchorages (Bombace et al., 1996a, 1996b, 1997c, 1998b).

Lobster on-growing
From 1992 the on-growing of lobsters (*Homarus gammarus*) in captivity was investigated. Lobsters of various sizes (carapace length 49–98 mm; weight 100–700 g) were placed inside steel cages on the seabed (length 130 cm, width 110 cm, height 60 cm) each with a shelter formed by semi-cylindrical single-chamber units (Fig. 12; Bombace et al., 1996a, 1998a). The animals were not artificially fed. Lobster growth was followed over a 3-year period by periodically sampling and measuring the animals.

Figure 9.   Concrete cage for mussel culture at the Senigallia artificial reef.

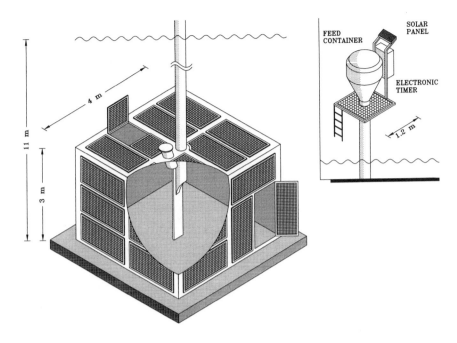

Figure 10.   Diagram of the seabed steel cage used for gilthead sea bream culture at the Senigallia artificial reef. The automatic food dispenser is shown in the box.

Pholas dactylus culture

Since 1992, settlement and growth of the boring edible bivalve *P. dactylus* in coal ash substrata has been followed by two-monthly collection of pieces of coal ash blocks (Bombace *et al.*, 1995) inhabited by *Pholas*.

Results

Porto Recanati

The biomass of mussels that settled on block surfaces was estimated at about 80–100 kg m^{-2} in the summer of 1977, two and a half years after reef deployment. Biomass stabilized at about 50 kg m^{-2} in the years 1977–80. The commercial production of the whole reef was about 200–250 tonne yr^{-1} in the first 3 years and 160 tonne yr^{-1} in the following 4 years. Mussels reached an average size of 6–7 cm total length (TL) at the end of the first year after their settlement, 8–9 cm at the end of the second year and 10–11 cm after 3 years.

Trammel net catches from the reef were dominated by commercially important species, mainly gilthead sea bream (*S. aurata*), sea bass (*Dicentrarchus labrax*), meagre (*Umbrina cirrosa* and *Sciaena umbra*) and scorpion fish (*Scorpaena porcus* and *Scorpaena scrofa*). A decrease in catch rates was observed with distance from

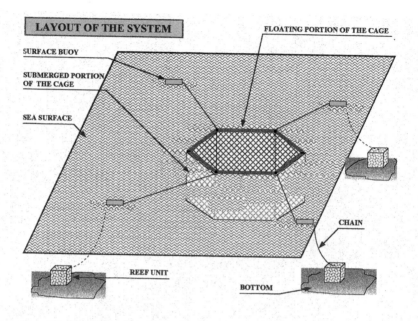

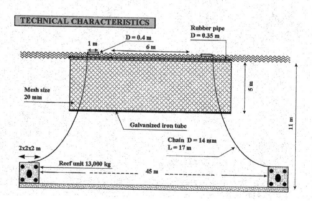

Figure 11. Diagram of the surface floating cage tested for gilthead sea bream culture at the Senigallia artificial reef.

the reef. Between 1977 and 1980 the average fishing yields obtained at the reef with different set gears (trammel nets and traps) were 20 tonne yr^{-1} of high-priced fish and cephalopods and 160 tonne yr^{-1} of *Sphaeronassa mutabilis*, a small edible gastropod presumably attracted to the artificial reef by the food availability and the presence of hard surfaces for egg case attachment (Bombace, 1982). The

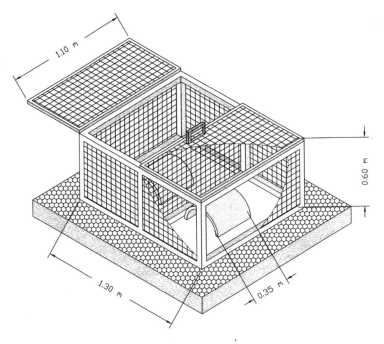

Figure 12.   Diagram of the cages used for the on-growing of *Homarus gammarus*.

income of professional fishermen operating in the area was about 2.5 times that of fishermen working in surrounding areas without artificial structures.

Ten years after deployment the total income obtained from the reef was about three times the building cost (Bombace, 1987).

The satisfactory results, mainly in terms of mussel and oyster settlement, obtained by the immersion of concrete blocks, led to further specialization of structures to create submerged shellfish farms inside the reef. The aims were to increase the reef productivity and to extend shellfish culture activities to open sea areas presently unsuitable for traditional farming techniques. To facilitate management of the shellfish farms the reefs were built as near as possible to the coast and the height of the structures was reduced. These factors helped to reduce the deployment costs.

All these ideas were applied in Portonovo 1 experimental reef construction.

Portonovo 1

Mussel culture

In the Adriatic Sea, mussel spawning generally occurs from early winter until late spring. Larval settlement peaks in late spring to early summer, which is the best period for immersion of collectors (Fabi *et al.*, 1989). Studies showed that nylon ropes were more suitable for this purpose than other materials, because they allowed high seed settlement and did not rot like the vegetable fibres (Fabi *et al.*,

1985). The highest settlement density was obtained between 1 and 5 m below the sea surface (Fabi et al., 1989). By autumn, the average total length of the seed mussels was 35–40 mm with biomass ranging between 10 and 15 kg m^{-1} of rope. This 'seed' was then harvested and transferred into nylon socks suspended from steel wires between the pyramids and/or from the steel cage, for growing-on. The growth rate of the farmed mussels decreased during the winter months because of low water temperatures (Fig. 13; Fabi et al., 1989; Fabi and Fiorentini, 1990). In spite of this by the following spring, about 50% of the mussels were marketable (TL ≥ 50 mm) with an average commercial yield of about 2.4 kg per kg of seed originally placed in the nylon socks.

If the mussels were thinned out once in late spring/early summer of the following year then by autumn all the specimens were over the commercial size (average TL 70 mm) and the yield increased to 7 kg per kg of original seed. Harvesting in autumn of the second year after settlement seemed to give the best yield; in fact no significant weight increase was obtained if mussels were kept for a second winter because of the reduction in growth rate and loss of condition associated with spawning. During the farming period an average mortality/loss of about 25% is considered usual, mainly because the smallest specimens are not effectively retained by the net mesh.

Oyster culture

Spawning in both oyster species (*Ostrea edulis* and *Crassostrea gigas*) occurs during the summer months. Spat recruited onto collectors from late summer to early autumn with only a short peak in settlement (Fabi et al., 1989). Large variations between years, both in the density and period of the maximum larval settlement,

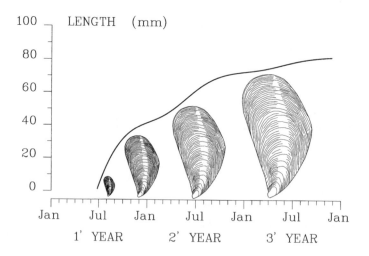

Figure 13. Seasonal growth curve of *Mytilus galloprovincialis* in suspended cultures at the Adriatic artificial reefs.

have been observed. When young oysters settled on plastic collectors reached about 20–30 mm in length they were removed from the collectors, placed into plastic baskets and then suspended from both wires between the reef pyramids and the cages. This transfer resulted in a 20% loss of, or damage to, specimens.

In the early summer of the following year, the oysters were transferred into baskets with larger holes which increased the water flow through the baskets and reduced the possibility of siltation and hole obstruction by mussels. *O. edulis* and *C. gigas* showed a similar reduction in growth during winter months (Fig. 14; Fabi *et al.*, 1989; Fabi and Fiorentini, 1990). At the end of the first year of suspended-basket growth, the *C. gigas* were a little larger than the *O. edulis*, though both species were over the minimum commercial size (TL 60 mm) and could be marketed at this time.

During suspended-basket cultivation, an average survival of about 75% may be considered normal. After one year of farming, the commercial yield obtained for *O. edulis* was about 3 kg per 100 spat (corresponding to a starting weight of 270 g) originally placed into baskets, and about 4 kg after a further year. In the case of *C. gigas* the yield was about 4 kg and 6 kg respectively, per 100 (380 g spat).

The biological information and the farming techniques derived from the work carried out at this reef have been applied to other experimental and commercial marine farms along the Adriatic coast and in other Italian seas (Fabi and Fiorentini, 1997). From tests on the steel cage as a structure for suspended and submerged shellfish culture, concrete cages were designed and introduced into the plans for new reefs.

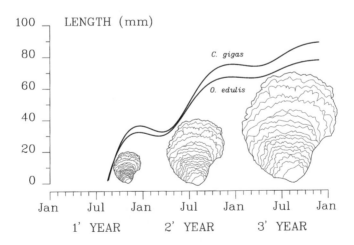

Figure 14. Seasonal growth curve of *Ostrea edulis* and *Crassostrea gigas* in suspended culture at the Adriatic artificial reefs.

Porto Garibaldi 1, Rimini, Cattolica and Portonovo 2

The scientific results obtained from the research carried out at these reef sites have been described by Bombace *et al.* (1990a, 1993a, 1993b, 1994) and Fabi and Fiorentini (1993) and can be summarized as follows:

(1)   Effects of artificial reefs on the original fish assemblage are more evident at sites distant from natural hard substrata (Fig. 15).

(2)   At sites far from natural rocky habitats species richness and diversity, as well as fish abundance, increase after reef deployment (Fig. 16). This increase is particularly notable in reef-dwelling nekto-benthic species such as *S. umbra*, *U. cirrosa* and a few sparids. The mean catch increments recorded for these species three years after artificial reef deployments were 10–42 times the initial values in weight. These increments seemed to be directly correlated to the volume of immersed material and inversely correlated to the distance between the oases (Table 1). Some crustaceans, such as the lesser spider crab (*Maja crispata*) and clawed lobster (*H. gammarus*), and molluscs (e.g. *Eledone moschata* and *Rapana venosa*), that were rare or completely absent in the original habitat, appeared in catches after the artificial structures had been placed.

(3)   Underwater observations and sampling showed rapid colonization of the concrete structures at all locations. The sessile community was dominated by filter-feeders, mainly mussels (*M. galloprovincialis*) and oysters (*O. edulis* and *C. gigas*). Except for Porto Garibaldi 1, where the lowest biomass values were recorded, the mean biomass of mussels settled on the structures ranged from 20 kg m^{-1} on the base blocks of the pyramids to 44 kg m^{-1} on the top blocks in the second year after reef deployment and from 32 kg m^{-1} to 55 kg m^{-1} respectively in the third year (Table 2).

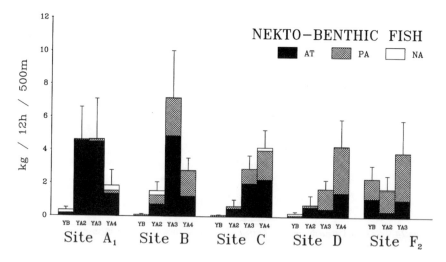

Figure 15.   Mean weight of nekto benthic fish per catch ± SE at five Adriatic artificial reefs in the year before the reef deployment (YB), and in the 2nd (YA2), 3rd (YA3) and 4th (YA4) year after deployment.

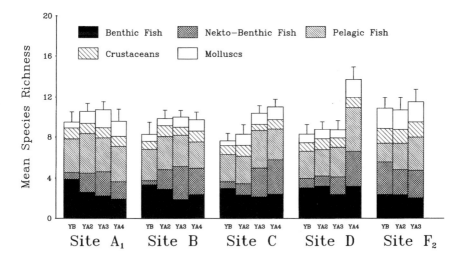

Figure 16. Mean species richness ± SE in the fish catches from five Adriatic artificial reefs in the year before the reef deployment (YB) and in the 2nd (YA2), 3rd (YA3) and 4th (YA4) year after deployment.

Table 2. Mean (± SE) biomass kg m^{-2} mussels (*Mytilus galloprovincialis*) settled on the base and the top blocks of the pyramids at the five artificial reefs built along the Adriatic coast. The commercial product (TL ≥ 5 cm) is indicated as percentage of the total.

	Porto Garibaldi 1	Rimini	Cattolica	Senigallia	Portonovo 2
2nd year					
Top block	11.4 ± 3.4 98%	44.0 ± 3.2 95%	34.4 ± 8.1 80%	60.0 ± 10.5 97%	43.2 ± 7.6 72%
Base block	0.5 ± 0.2 100%	20.8 ± 10.8 99%	20.6 ± 9.7 97%	46.7 ± 1.4 94%	20.6 ± 1.7 100%
3rd year					
Top block	16.2 ± 1.3 100%	50.8 ± 0.5 97%	54.8 ± 8.4 98%	46.9 ± 3.4 99%	42.1 ± 0.9 98%
Base block	0.0	51.4 ± 7.7 100%	32.6 ± 2.4 100%	39.1 ± 2.8 99%	32.9 ± 1.3 95%

About 200–300 tonnes of commercial mussels were collected from each of the reefs at Rimini, Cattolica and Portonovo 2 three years after deployment, a period when all fishing activities were forbidden inside the reefs by EEC regulation 4028/86.

Senigallia

The research programme conducted at this experimental site confirmed all the results obtained at the other modern reefs recently deployed in the Adriatic Sea (Porto Garibaldi 1, Rimini, Cattolica and Portonovo 2). The additional research carried out at this site has produced data to support the following points.

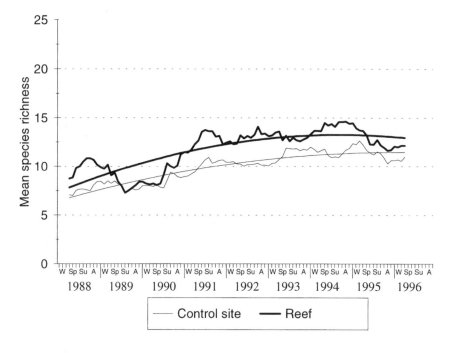

Figure 17.   Trend of mean species richness per catch and per season at the Senigallia artificial reef (D) and at the control site (D'). See Fig. 2 for location of D and D' respectively.

Figure 18.   Specimens of *Diplodus vulgaris* at the Senigallia artificial reef.

Mobile fauna

(1) Higher species richness and species diversity were always recorded in fish catches at the reef than at the control site (Fig. 17). The difference was mainly due to the occurrence at the reef of reef-dwelling species which were rare or absent from the original sandy-muddy seabed and the control site (Fig. 18; Bombace *et al.*, 1990b; Fabi and Fiorentini, 1994).

(2) Higher catch rates of these reef-dwelling species have generally been reported from the artificial reef than in the control site (Fig. 19). Catches of this group of fish have continued to increase during the 5 years since reef deployment (Panfili, 1997; Fabi *et al.*, in press).

(3) The fish assemblage around the artificial reef fluctuated seasonally on the reef in a similar fashion to populations around natural habitats. The lowest values in density and catches are generally recorded in winter, when most of the species migrate to deeper and warmer waters. However, the winter migration of many nekto-benthic fish from the reef seems to be less than that seen at the control site (Fig. 19).

(4) Eventual stock collapse of reef-dwelling species seems to be buffered for stocks inside the artificial reef in comparison with unprotected areas (Fabi and Fiorentini, 1993).

(5) Fewer species of fish, crustaceans and molluscs were recorded by visual observations than by trammel net (Fig. 20). Netting was found to be a more appropriate method to describe the population of the original sandy-muddy seabed. Conversely, the partially and obligatory reef-dwelling species are better recorded visually (Fabi and Fiorentini, 1994; Bombace *et al.*, 1997).

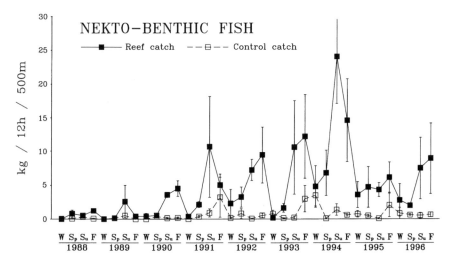

Figure 19. Mean catch per season ± SE of nekto-benthic fish at the Senigallia reef (D) and at the unprotected control site (D').

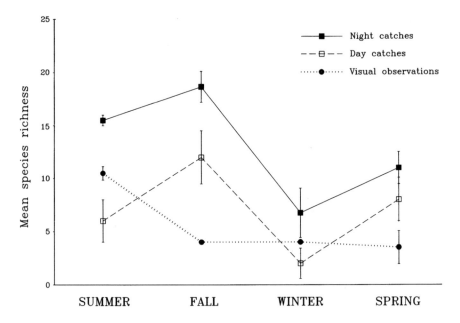

Figure 20.   Mean species richness per season ± SE recorded in day catches, night catches and visual observations carried out at the Senigallia artificial reef.

(6)   During the year some adult reef-dwelling fish were present on the reef at various stages of sexual maturity. Juveniles of various sizes were also present and their frequency of occurrence was greater than at the control site (Fig. 21). These observations and the information obtained from the ichthyo-plankton sampling provide strong indications that some species (e.g. *Diplodus annularis*) reproduce inside the reef (Bombace *et al.*, 1993c).

Sessile fauna

The biomass of mussels settled on the Senigallia reef increased in a similar fashion to that recorded at the other Adriatic artificial reefs. Three years after reef deployment mean mussel biomass was 47 kg m^{-2} on the top blocks of the pyramids and 40 kg m^{-2} on the base blocks with 99% of mussels being large enough to be termed a 'commercial product' (Table 2).

   Analysis of benthic colonization showed the lack of an algal component because of the high water turbidity and the prevalence of an animal community normally found in eutrophic waters. The epiphytic layer, settled just after the immersion of the substrata, was rapidly replaced by pioneer hard-substrata invertebrates mainly represented by serpulids (e.g. *Pomatoceros triqueter*), barnacles (e.g. *Balanus improvisus* and *Balanus a. amphitrite*), hydroids (e.g. *Clytia hemisphaerica* and *Bougainvillia ramosa*), bryozoans (e.g. *Bugula stolonifera* and *Schizoporella errata*) and oysters. The pioneer phase was followed by the settlement of *M. gallo provincialis* which became the dominant species in a short time.

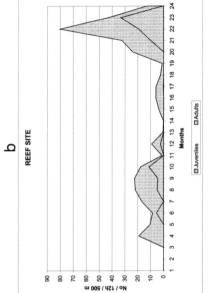

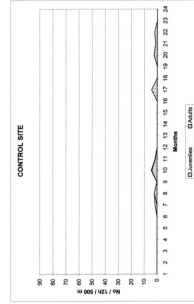

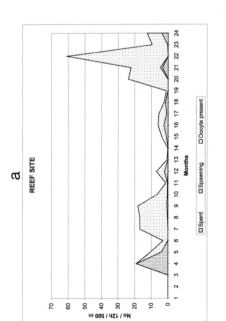

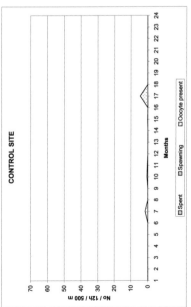

Figure 21. (a) Different stages of sexual maturity recorded in the specimens of *Diplodus annularis* caught at the Senigallia reef (D) and at the control site (D') (see Fig. 2) over a two-year period. (b) Mean number of juveniles and adults of *D. annularis* caught in the same period at the two sites.

Because of the sedimentation coming from riverine output and from the storms that move the surrounding soft seabed, the horizontal surfaces of the structures were gradually covered by a thin layer of sand and mud that favoured the settlement of soft-seabed species. As consequence, about 1 year after deployment, the benthic community settled on these surfaces was made up of a mix of hard and soft substrata species. Species richness and diversity reached higher values on horizontal surfaces in comparison with those recorded on the vertical surfaces, where the mud accumulation was less abundant and therefore allowed the hard-substrata organisms to dominate (Castriota et al., 1996).

Comparison between concrete and coal ash blocks did not show big differences in benthic colonization, although the latter were particularly suitable for oysters, which settled mainly on the vertical surfaces or the reef cubes, and for the boring bivalve Pholas dactylus, predominantly settling on the horizontal surfaces. No specimens of P. dactylus were ever observed on concrete blocks (Bombace et al., 1997a).

Mariculture initiatives

Mussel and oyster culture

During the first year mussels raised on nylon ropes at Senigallia grew a little faster (all specimens reached commercial size in the spring following settlement) than mussels farmed inside nylon socks, but this difference was negligible in the autumn of year 2 (TL 72 mm). Greater losses of mussels were observed as a result of mussels sliding down the ropes under the excessive weight of settlement. The production obtained after one year from immersion was about 8 kg m^{-1} of rope (Fabi and Fiorentini, 1990). Comparison with mussels farmed using the same technique inside Portonovo 1 reef showed a slower growth rate but higher production. The differences recorded between the two sites are thought to be related to the greater exposure of the Senigallia reef to waves and currents (Bombace et al., 1991). This increase in water movement enhances the supply of food and so favours mussel growth but at the same time, places greater forces on the mussel cluster and causes the mussels to slide down the ropes.

The settlement and growth of O. edulis and C. gigas were found to be the same in both areas.

Fish farming

The experimentation carried out confirmed that the on-growing of juvenile fish inside cages in the spring–autumn period can represent an additional activity in artificial reef systems along the Adriatic coast.

The steel seabed cage and the surface-floating, hexagonal cage satisfied the technological requirements: simple assemblage and maintenance, resistance to adverse sea conditions in shallow water and low costs.

The gilthead seabream farmed inside the seabed cage and fed with the automatic food dispenser showed good growth rates. From spring to autumn, their weight increased from 70–80 g to 300–350 g, with an average of 1.5–2.0 g/day and a conversion factor of 3.0–3.2. The condition factor was always high (1.7–1.9), confirming the good health of the animals. A reduction in the growth rate and an increase in the conversion factor were generally recorded starting from November, when the water temperature fell below 18°C. This decrease in growth rate and the possible losses that may be caused by storms, make it advisable not to continue fish farming during the winter months (Bombace *et al.*, 1996a, 1996b, 1997c).

Lower growth rates, lower condition factors and higher conversion factors were recorded in the trials carried out inside the surface-floating cage showing that, independently from the good technical characteristics of the structure, the animals did not find conditions for growth as favourable as in the seabed cage (Bombace *et al.*, 1997c, 1998b). The manual food supply certainly played an important role in this result. Moreover, in shallow water, the fish felt the effects of the waves and were subjected to a continuous stress. Therefore, this type of cage appears to be more suitable for fish culture in open sea areas with greater depth (>15 m).

Lobster on-growing
This study has produced data on the biology and growth of this species in the Adriatic Sea and has shown new possibilities for restocking inside artificial reefs (Bombace *et al.*, 1996a). Observations on lobsters farmed semi-intensively inside the steel cages have shown similar growth rates to those reported for tagged animals in Northern Europe (Hepper, 1967; Hewett, 1974; Thomas, 1958).

The majority of specimens under 80 mm carapace length (CL) (overall length about 240 mm) moulted at least twice per year, in spring (April–May) and late summer (September). Specimens over this size moulted once per year.

Males showed a greater carapace length increment at moult than females and this difference was constant with increasing size of the animals. Moreover, carapace length increment at moult was independent of the initial carapace length in both sexes even though, for females included in the size range $58 \leq CL \leq 78$ mm, the increment at moult tended to increase relative to the original carapace length. The average carapace length increment at moult was 9.7 mm (12.9%) for males and 7.2 mm (9.7%) for females (Fig. 22).

Differences between male and female weight increases were also recorded. In both sexes the growth increment at moulting was dependent on the pre-moulting weight but, given the same starting weight, males seemed to increase weight at moulting more than females. At 200 g initial weight the average weight increment per moult was about 93 g (46.5%) for males and 57.5 g (28.8%) for females.

Pholas dactylus culture
Coal ash from electricity generating power stations, mixed with other mineral components, creates hard substrata that, unlike concrete, are suitable for the settlement of boring bivalves which are found inside limestones or stiff clays when living in the wild. The common piddock (*P. dactylus*) is one of these bivalves. Harvesting in the wild, as in the case of date mussels (*Lithophaga lithophaga*), is forbidden by Italian law because of the destruction of natural hard substrata that is involved. An abundant settlement of common piddock spat (up to 200 indi-

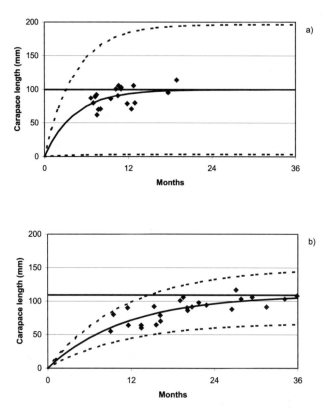

Figure 22. Growth curves computed for males (a) and females (b) of semi-intensively farmed lobsters (*Homarus gammarus*).

viduals dm^{-2}) was recorded on coal ash blocks from late summer to early autumn over a two year period (Colour Plate 2). Mortality was very high in the following 7–8 months, probably due to the competition for the space (Bombace *et al.*, 1995). Studies showed that this species has a fast growth rate and can reach 50 mm in length within the first year of life. Therefore, the use of coal ash substrata for mariculture could provide new opportunities for the fishery without any accompanying threat to natural rocky habitats. Experimentation is in progress and several biological, technological and economic aspects are still being investigated.

Porto Garibaldi 2

This artificial reef deployment produced an enrichment and diversification of the fish assemblage from that found on the original muddy seabed. Reef-dwelling species and some pelagic species were attracted by the submerged structures. During the first year after deployment the numbers of individuals of these two groups of fish caught increased by 5 and 1.6 times over that caught before deployment. The catch weight of the two groups rose by 23 and 1.6 times, respectively, during the same period. But, unlike data recorded at the other Adriatic reefs, over the following years these values gradually decreased (IRPEM, 1993). It is believed that this was caused by the reef structures falling apart and being buried.

Two years after reef deployment the mean mussel biomass settled on the hard substrata was about 26 kg m^{-2} and consisted mainly of commercial sized specimens. It was impossible to estimate the overall biomass because the condition of the reef structures prevented an accurate measurement of available surface area.

Conclusions

Results from the artificial reef initiatives along the Adriatic coast show that artificial reef deployment can produce positive effects on the environment, marine resources and fishery catches.

Environment

Reefs provide protection from trawling activities for benthic biocoenoses. This is very important in the case of particular biotopes of great ecological significance such as *Posidonia oceanica* meadows. Reefs and the associated biota also act as a sink for the great amount of nutrients, phytoplankton and particulate matter present in the inshore region of the Adriatic. The artificial reef system is thought to play a part in recycling organic material into food chains that, through filter and suspension feeders, lead to fish and eventually to man.

Resources

Reefs promote enrichment of the original habitat communities through the provision of hard substrata which leads to increased species diversity and richness. There is a reduction in the mortality of mature fish which come to the reef area to reproduce because the reef effectively prevents illegal trawling. Any increase in spawning stock may produce a long-term increase in overall fish biomass.

Reefs assist in the reduction of natural mortality by providing shelters for adults in vulnerable stages of their life (e.g. moulting crustaceans) and for post larvae and juveniles. This may also contribute to an overall increase in biomass.

It is thought that artificial reefs may provide an opportunity to rejuvenate fishery depleted stocks through mariculture and re-stocking enterprises. It is believed that such initiatives should be linked to hatchery production of target species. Surplus hatchery production could be released inside artificial reefs, where juveniles may find suitable conditions for survival. In this way artificial reefs may represent a link between hatcheries and a wild fishery.

Reefs provide a substratum which allows settlement of new sessile biomass that otherwise would be lost because of the lack of suitable settlement surfaces. In eutrophic waters the mass production of edible bivalves (mussels and oysters) is an important economic resource.

Fishery and Fishermen

The construction of artificial reefs has led to a reduction of conflict between trawling and small-scale artisanal fisheries in the coastal Adriatic. Artificial reefs have promoted the development of small-scale fisheries by providing areas where they can safely operate. Artisanal fishermen tend to join co-operatives to manage the protected areas and the farming systems. Reefs provide possibilities for replacing part of the small trawling fleet with new alternative activities, such as mariculture enterprises.

Reef Types

Three kinds of artificial reefs can be identified:

(1) Extensive artificial reefs: these cover very large coastal areas (thousands of hectares) and their main aim is to prevent illegal trawling. Simple anti-trawling structures are used and randomly deployed. Only administrative licences for the structure deployment are required and benefits are extended to all fisheries.
(2) Intensive artificial reefs: these cover limited areas and consist of particular structures. The aims are repopulation, attraction, shelter, protection and mariculture. The artificial structures are generally deployed close together. Administrative grant licences are necessary for the system managers.
(3) Semi-intensive artificial reefs: these represent a compromise between (1) and (2).

The Future

Twenty years of research experience shows that artificial structures represent an exceptional opportunity for interdisciplinary research. Many scientific and administrative aspects in this field remain in need of further investigation. Some of these are:

(1) The ratio between material surfaces and reef volume, and their relation to stock densities, type of structure, biological settlement and impact on fishing and mariculture must be examined.
(2) The relationship between sedimentation rates and colonization by sessile organisms has to be understood so that the optimum conditions for epifauna colonization may be recognized.
(3) Different techniques for biomass evaluation of mobile species have to be tested. Trophic webs inside artificial reefs and energy transfers have to be analysed in detail.
(4) Technological aspects, such as experimentation with new substrata and assessment of the impact of weather and hydrology on the submerged structures have to be investigated.

(5) Involvement of small-scale fishermen in the planning and development of any protected area must be developed.

(6) Simplification of the administrative procedures for the establishment of artificial reefs in Italy is required.

(7) Long-term plans for the development, management and evaluation of coastal areas must be organized and co-ordinated. These management plans must indicate areas of development, the time scale for any such activity and the human and financial resources to be involved.

Acknowledgments

The authors would like to thank all the IRPEM staff involved in the research: Dr L. Castriota, Mr G. Giuliani, Dr F. Grati, Mr V. Palumbo, Dr M. Panfili, Mr A. Piersimoni, Dr A. Sala, Dr A. Spagnolo and Dr S. Speranza for catch data collection, laboratory work and underwater operations; Mr W. Cipolloni and Mr L. Cingolani for their technical support in the construction of culture structures. Special thanks to Mr G. Gaetani and Mr A. Marziali who cooperated in the overall research.

Thanks also to the European Community – DGXIV, to the Italian Ministry for the Agricultural Politics – General Direction for Fishery and Aquaculture and to the Centro Ricerca e Valorizzazione Residui (ENEL) who funded a part of the research.

References

Bombace, G. 1981. Note on experiments in artificial reefs in Italy. *Stud. Rev. GFCM.* **58**: 309–324.

Bombace, G. 1982. Il punto sulle barriere artificiali: problemi e prospettive. *Naturalista Siciliano* **IV** (VI Suppl.): 573–591.

Bombace, G. 1987. Iniziative di protezione e valorizzazione della fascia costiera mediante barriere artificiali a fini multipli. Atti LIX Riunione SIPS, Genova, October 28–31, 1987, pp. 201–233.

Bombace, G., G. Fabi and L. Fiorentini. 1990a. Preliminary analysis of catch data from artificial reefs in Central Adriatic. FAO Fisheries Reports. **428**: 86–98.

Bombace, G., G. Fabi and L. Fiorentini. 1990b. Catch data from an artificial reef and a control site along the Central Adriatic coast. *Rapport de la Commission International de la Mer Méditerranée.* **32**(1): 247.

Bombace, G., A. Artegiani, G. Fabi and L. Fiorentini. 1991. Ricerche comparative sulle condizioni ambientali e sulle possibilità biologiche e technologiche di allevamento ottimale di mitili ed ostriche in mare aperto mediante strutture sommerse e sospese. Report to the Marine Merchant Ministry, General Fishery Division. 205 pp.

Bombace G., G. Fabi and L. Fiorentini. 1993a. Théorie et expériences sur les récifs artificiels. Actes du Colloque Scientifique 'Le Système littoral Méditerranéen', Montpellier, April 1993, pp. 68–72.

Bombace, G., G. Fabi and L. Fiorentini. 1993b. Aspects théoriques et résultats concernants les récifs artificiels realisés en Adriatique. *Bollettino di Oceanologia Teorica ed Applicata.* **11**(3–4): 145–154.

62 G. Bombace et al.

Bombace, G., G. Fabi and U. Giorgi. 1993c. Ricerche sull'ittioplancton e sulle forme giovanili di pesci in barriere artificiali. Report to the Marine Merchant Ministry, General Fishery Division. 83 pp.

Bombace, G., G. Fabi, L. Fiorentini and S. Speranza. 1994. Analysis of the efficacy of artificial reefs located in five different areas of the Adriatic Sea. *Bulletin of Marine Science*. **55**(2): 559–580.

Bombace, G., G. Fabi and L. Fiorentini. 1995. Osservazioni sull'insediamento e l'accrescimento di *Pholas dactylus* L. (Bivalvia, Pholadidae) su substrati artificiali. *Biologia Marina Mediterranea*. **2**(2): 143–150.

Bombace, G., G. Fabi, L. Fiorentini and A. Spagnolo. 1996a. Studi ed esperimenti su: a) Iniziative di allevamento ed ingrasso di astici e orate in mare aperto mediante gabbie; b) sistemi 'FADs' per l'attrazione e la concentrazione di pesci pelagici. Report to the Agriculture, Food and Fishery Ministry, General Direction for Fishery and Aquaculture. 91 pp.

Bombace, G., G. Fabi and G. Gaetani. 1996b. Sperimentazione di un prototipo di gabbia da fondo per l'ingrasso di pesce in medio Adriatico. *Biologia Marina Mediterranea*. **3**(1): 186–191.

Bombace, G., L. Castriota and A. Spagnola. 1997a. Benthic communities on concrete and coal-ash blocks submerged in an artificial reef in the central Adriatic Sea. In Hawkins, L. and S. Hutchinson with A.C. Jensen, J.A. Williams and M. Sheader. (eds.) *The Responses of Marine Organisms to their Environments*. Proceedings of the 30th European Marine Biological Symposium, Southampton, UK, September 1995. Southampton Oceanography Centre, pp. 281–290.

Bombace, G., G. Fabi, L. Fiorentini and A. Spagnolo. 1997b. Assessment of the ichthyofauna of an artificial reef through visual census and trammel net: comparison between the two sampling techniques. In Hawkins, L. and S. Hutchinson with A.C. Jensen, J.A. Williams and M. Sheader. (eds.) *The Responses of Marine Organisms to their Environments*. Proceedings of the 30th European Marine Biological Symposium, Southampton, UK, September 1995. Southampton Oceanography Centre, pp. 291–305.

Bombace, G., G. Fabi and A. Spagnolo. 1997c. Sperimentazione di strutture idonee all'allevamento di pesci in aree di mare aperto del medio Adriatico. Report to the Ministry for the Agricultural Politics, General Direction for Fishery and Aquaculture. *Biologia Marina Mediterranea*. 76 pp.

Bombace, G., G. Fabi, L. Fiorentini, F. Grati, M. Panfili and A. Spagnolo. 1998a. Maricoltura associata a barriere artificiali. *Biologia Marina Mediterranea*. **5**(3): 1773–1782.

Bombace, G., G. Fabi, F. Grati and A. Spagnolo. 1998b. Technologies for open-sea culture of *Sparus aurata* in the Central Adriatic. *Biologia Marina Mediterranea*. **5**(3): 439–449.

Bombace, G., G. Fabi, J. Leonori, A. Sala and A. Spagnolo. 1998c. Valutazione con tecnica elettroacustica della biomassa vagile presente in una barriera artificiale del medio Adriatico. *Biologia Marina Mediterranea*. **5**(3): 1844–1854.

Castriota, L., G. Fabi and A. Spagnolo. 1996. Evoluzione del poplamento bentonico insediato su substrati in calcestruzzo immersi in Medio Adriatico. *Biologia Marina Mediterranea*. **3**(1): 120–127.

Fabi, G. and L. Fiorentini. 1990. Shellfish culture associated with artificial reefs. FAO Fisheries Reports **428**: 99–107.

Fabi, G. and L. Fiorentini. 1993. Catch and growth of *Umbrina cirrosa* (L.) around artificial reefs in the Adriatic sea. *Bollettino di Oceanologia Teorica ed Applicata*. **11**(3–4): 235–242.

Fabi, G. and L. Fiorentini. 1994. Comparison of an artificial reef and a control site in the Adriatic Sea. *Bulletin of Marine Science*. **55**(2): 538–558.

Fabi, G. and L. Fiorentini. 1997. Molluscan aquaculture on reefs. In Jensen, A.C. (ed.) *European Artificial Reef Research*. Proceeding of the 1st EARRN conference, Ancona, Italy, March 1996, Southampton Oceanography Centre, pp. 123–140.

Fabi, G., L. Fiorentini and S. Giannini. 1985. Osservazioni sull'insediamento e sull' accrescimento di *Mytilus galloprovincialis* Lamk su di un modulo sperimentale per mitilicoltura immerso nella baia di Portonovo (Promontorio del Conero, Medio Adriatico). *Oebalia N.S.* **11**: 681–692.

Fabi, G., L. Fiorentini and S. Giannini. 1986. Growth of *Mytilus galloprovincialis* Lamk on a suspended and immersed culture in the Bay of Portonovo (Central Adriatic Sea). FAO Fisheries Reports. **357**: 144–154.

Fabi, G., L. Fiorentini and S. Giannini. 1989. Experimental shellfish culture on artificial reefs in the Adriatic Sea. *Bulletin of Marine Science*. **44**(2): 923–933.

Fabi, G., F. Grati, F. Luccarini and M. Panfili. (In press.) Indicazioni per la gestione razionale di una barriera artificiale: studio dell'evoluzione del popolamento necto-bentonico. *Biologia Marina Mediterranea*.

Hepper, B.T. 1967. On the growth at moulting of lobsters (*Homarus vulgaris*) in Cornwall and Yorkshire. *Journal of the Marine Biological Association of the UK*. **47**: 629–643.

Hewett, C.J. 1974. Growth and moulting in the common lobster (*Homarus vulgaris* Milne-Edwards). *Journal of the Marine Biological Association of the UK*. **54**: 379–391.

IRPEM. 1993. Rapporto definitivo sull'evoluzione della barriera artificiale a fini multipli realizzata dalla Coop. 'Tecnopesca' di Porto Garibaldi. 19 pp.

Panfili, M. 1996. Evoluzione del Popolamento Ittico in una Barriera Artificiale del Medio Adriatico. Degree thesis, Università degli Studi di Milano. 100 pp.

Thomas, H.J. 1958. Observations on the increase in size at moulting in the lobster (*Homarus vulgaris* M.-Edw.) *Journal of the Marine Biological Association of the UK*. **37**: 603–606.

4. Artificial Reefs in Sicily: An Overview

SILVANO RIGGIO[1], FABIO BADALAMENTI[2] and GIOVANNI D'ANNA[2]

[1]*Dipartimento di Biologia Animale, Università di Palermo, via Archirafi, 18, I 90123 Palermo, Italy;* [2]*Laboratorio di Biologia Marina, I.R.M.A.-CNR, Via G. da Verrazzano, 17, I 91014, Castellammare del Golfo (TP), Italy*

Perspective

Sicilian cave drawings from the Grotta del Genovese, Isle of Levanzo (west Sicily) ca. 12 000 B.C. show silhouettes of dolphins, tuna, groupers and bass which, together with remains of fish (tuna, groupers, bass and others), limpets and oysters from Grotta dell Uzzo, north-western Sicily, indicate the importance of fish and shellfish in the diet of coastal populations of that time (Villari, 1992a, 1992b). Remains of turtles (*Caretta caretta*), tuna and sharks are evidence of fishing activity during this period (Villari, 1995).

The setting up of structures to attract fish in the Mediterranean predates historical records of such objects in Japan (Simard, 1995), going back to the very beginning of the tuna fishery, garum and purple (Tyrian crimson dye obtained from shellfish) industries, nearly three millennia ago (Purpura, 1989, 1990).

The tuna fishery used huge nets, called *tonnare*, which were anchored in place by heavy boulders. At the end of each fishing season, when the *tonnare* were lifted, the ropes fastening these primitive anchors to the nets were cut and the stones left behind. Over the years deposition of these boulders made a new rocky habitat, which was populated by benthic fauna and fish. The fish were exploited by fishermen during the intervals between tuna fishing seasons. Improved yields from *tonnare* sites encouraged the sinking of wrecks as well as boulders so that the setting up of refuges and attraction sites for fish became common practice. In much the same way, new fishing grounds were created by dumping the ruins of ancient Greek temples in the sea, as occurred during the building of the *caricatori* quays, small harbours on the southern coast of Sicily that traded in wheat and sulphur produced on the Italian mainland. The harbours of Porto Empedocle and Gela are two examples.

In north-east Sicily some fish-attracting structures were made by the disposal of sandstone boulders anchoring the floating lath network (*cannizzi*) used to attract shade-loving fish, such as the dolphin fish *Coryphaena hippurus*, the pilot-fish *Naucrates ductor* and juvenile wreckfish *Polyprion americanum*. Other ancient fishing techniques, midway between fishing gear and fish attraction devices (FADs), still used in the shallow waters of Jerba and the Kerkennah Islands (Sahel, east Tunisia), are the *gargoulettes*, terracotta pots, by-products of local potteries, which are sunk at dusk to offer an attractive resting site to octopus. The pots are raised at dawn, most of them with an octopus inside (Daulon, 1978). Fishermen in Sicily, as in other parts of the south Mediterranean, are therefore aware of the historical

A.C. Jensen et al. (eds.), Artificial Reefs in European Seas, 65–73
© 2000 Kluwer Academic Publishers. Printed in Great Britain.

relationships between rocky seabed and good fishing, and have always been keen supporters of artificial reef-building schemes (Arculeo *et al.*, 1990a; Riggio, 1995a).

Planning and Implementing Artificial Reef Deployment in Sicily

The history of modern artificial reefs in Sicily started in 1974, when a law funding plans for the building and deployment of artificial reefs was approved by the regional government, but never implemented. It was 8 years later, in 1982, before an artificial reef project started. An experimental reef was laid off Terrasini, in the eastern sector of the Gulf of Castellammare (Fig. 1), made of four small pyramids of concrete cubes and pipes. This was followed 3 years later by a more complex reef, laid about 18 km west of Terrasini, made from concrete cube pyramids, surrounded by smaller reef units which were placed to provide a physical deterrent to trawling (illegal in water <50 m).

A number of amendments to the 1974 law enabled regional government, in 1978, to set up some *consorzi*, aimed at implementing existing plans and encouraging new ones. A *consorzio* is a public agency developed from the association of public and/or private organizations which pursues social and/or economic objectives. The Consorzio di Castellammare del Golfo (north-west Sicily) started in 1984 and 2 years later a series of artificial reef units made from concrete was laid in the mid-western part of the Gulf. Since the early 1990s reefs have been added in the Gulf of Palermo and Carini Bay. Contemporary with the Consorzio

Figure 1.   Location of the artificial reefs deployed and/or planned along the coasts of Sicily.

di Castellammare del Golfo is the analogous Consorzio di ripopolamento dei Comuni del Golfo di Patti, which has promoted the construction of artificial reefs in the Gulf of Patti, north-east Sicily. Other public agencies that have planned and placed artificial reefs are the Provincia Regionale (P.R.) di Palermo, the P.R. di Agrigento (southern Sicily) and the P.R. di Ragusa (south-east Sicily) (Fig. 1).

Major hindrances to artificial reef development in Sicily have been the excessively tedious and time-wasting procedures involved in obtaining the legal permission required for reef building and deployment. The absence of a regional plan for reef building and monitoring is regrettable, as has been the disregard of fishing restrictions in the Gulf of Palermo and the lack of policing of protected areas. This has led to unrestricted fishing; regulation of fishing has now been enforced in the three major Gulfs.

Most Sicilian artificial reefs are monitored to provide data on the biological colonization of their surfaces and fish assemblages in the surrounding waters. No information is available from reefs laid in the Gulf of Patti and along the southern coast of Sicily.

Past and Present Research

Artificial reef research in Sicily began before the first reef deployment, as a part of a national research programme, the National Research Council (CNR) 'Oceanography and Marine Bottoms – Biological Resources' programme, 1976 to 1981. This project allowed a long sequence of observations on biofouling to be carried out in Palermo Harbour (Mazzola and Riggio, 1976; Riggio, 1979; Riggio and Mazzola, 1976) and used available information to provide an insight into the processes of biotic colonization on artificial substrata. In addition the study focused on the biological colonization of concrete breakwaters sheltering some fishing harbours near Palermo, which were tested as readily available, extant mesocosms (Fig. 2), providing an easily accessible model of an artificial reef (Ardizzone *et al.*, 1977; Riggio and Ardizzone, 1979, 1981).

Comparison of biological colonization on natural and man-made substrata was also carried out to establish the role of larval recruitment and substratum selection (Costa *et al.*, 1983–84) on the establishment of benthic communities.

Survey of Artificial Reefs

The ecology of the first Sicilian artificial reef, deployed off Terrasini (Fig. 1), was studied and the sequence of colonization was described from immediately after deployment until a stable assemblage was attained (Badalamenti *et al.*, 1985; Badalamenti and Riggio, 1986; Riggio, 1982; Riggio *et al.*, 1985). An ecological survey was also carried out in the protected area off Trappeto (Riggio *et al.*, 1986).

The deployment of artificial reefs off Alcamo Marina and, more recently, Vergine Maria (north-western section of the Gulf of Palermo) and Carini Bay (Fig. 1), provided an opportunity for a thorough survey describing the patterns of

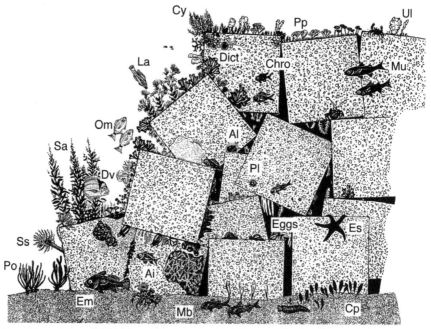

Figure 2.   An overview of the benthic and fish assemblages on the breakwater systems along the NW Sicilian coast. Ai = *Apogon imberbis*; Al = *Arbacia lixula*; Chro = *Chromis chromis*; Cp = *Caulerpa prolifera*; Cy = *Cystoseira* sp.; Dict = *Dictyota dichotama*; Dv = *Diplodus vulgaris*; Eggs = squid eggs; Em = *Epinephelus marginatus*; Es = *Echinaster sepositus*; La = labrids; Mb = *Mullus barbatus*; Mu = mugilids; Om = *Oblada melanura*; Pl = *Paracentrotus lividus;* Po = *Posidonia oceanica*; Pp = *Padina pavonica*; Sa = *Sargassum* sp.; Ss = *Sabella spallanzani*; Ul = *Ulva lactuca*.

biotic settlement and the efficiency of the reefs in providing food as well as acting as an attractant. Significant data on the ethology and feeding habits of fish living at or near the reefs, have been collected, together with information on artificial reef energy flows (see Badalamenti *et al.*, chapter 5, this volume). This research is still under way.

Conclusions from Feasibility Studies

Information, both quantitative and qualitative, from feasibility studies led to some preliminary conclusions that were used to make decisions on artificial reef design and deployment. These are summarized as follows:

(1)   Harbour breakwaters, wharves or coastal protection structures may be viewed as models of natural and man-made reefs. Information on their temporal biological character gives an indication of the feasibility for deployment and final success of artificial reefs, provided that such factors as surface texture,

wave impact, influence of depth and spacing of reef units are taken into account.

(2) A few species of fish colonized breakwaters and wharves. The most frequent colonizer was perhaps the herbivore *Sarpa sarpa*, which grazed *Ulotrichales*. Mugilidae became numerically dominant in eutrophic sites rich with organic sediments. Labrids were common in algae-rich reefs, which were also home to gobiid populations. Young groupers and lobsters had their refuges at the base of the reefs in clear waters. Shoals of goatfish fed in the area around breakwaters.

(3) Epilithic settlements on concrete did not differ generally from the colonization seen on nearby natural substrata. If siltation can be minimized, and the water is moderately eutrophic, a luxuriant biological community can develop within a few years. The rapidity and complexity of the colonization are likely to be directly related to the water trophism.

(4) The establishment of an endolithic community, as well as calcareous biokarst, was found to be a very slow process that appeared to take over a decade and needed water clarity and current to reach a state of full maturity. Speed of community development and complexity appeared to be directly related to the relative transparency of the water. The bioconcretions that develop in the southern Mediterranean Sea are typically coralligenous assemblages, characterized by a high diversity and aesthetic value, inhabited by a few species with high commercial value such as lobsters and groupers.

(5) The availability of exposed reef block edges was considered to be crucial for the successful settlement of planktonic larvae and disseminules. Seaweed and filter feeders colonized the exposed corners of the reef cube walls first, and then spread to the flat surfaces (Di Pisa and Riggio, 1982; Riggio and Di Pisa, 1982). Hollows, cavities and indentations acted as free edges, thereby enhancing colonization.

(6) Limestone was the material most suited to the settlement of a complex benthos, because of its erodability by the acid secretions of boring fauna. Steel-reinforced concrete was found to be unsuitable because the steel corroded in seawater, weakening the structure.

(7) Recruitment of benthos to artificial substrata appeared to be dependent on larvae and algal disseminules released from the local natural rocky reefs. Wherever these substrata were remote, the whole colonization process was slow and the ultimate settlement rather loosely structured.

(8) The final biological community was also less predictable. The presence of artificial reefs in Sicilian coastal waters promoted recovery of damaged *Posidonia oceanica* seagrass meadows, and played a part in further seabed colonization, provided that water transparency was high and siltation negligible (D'Anna *et al.*, chapter 6, this volume; Riggio, 1995b).

(9) The type and complexity of epibenthic assemblages influenced the species of mobile fauna living inside or near artificial structures. Labrids, serranids and lobsters favoured habitats in clear water; mugilidae and mullidae were more frequent in turbid waters.

Results from Later Artificial Reef Research

Many of these preliminary conclusions were supported to a large extent (Riggio, 1995b) by later observations from Sicilian reefs (see Badalamenti *et al.*, chapter 5, this volume; D'Anna *et al.*, chapter 6, this volume). The role of maritime structures as a model for artificial reefs was confirmed. Artificial reefs, and especially concrete substrata, acted as natural rocks in attracting selective meroplankton. This was especially true when limestone rocks were employed, either as discrete reef units, or when crushed and mixed with cement to yield concrete. Other substrata were definitely less suitable. A rough surface texture was found to be very important for larval settlement (Costa *et al.*, 1982–83).

Fish catches

The numbers of fish on and around reefs were higher than those in neighbouring seabed areas. Fish biomass fluctuated from being somewhat lower than that found in natural under-exploited rocky areas to much higher than that of exploited natural soft seabeds.

Artificial reefs unable to develop well-structured benthic communities because of high siltation rates deter trawling when they are placed to obstruct trawl nets (see Ramos Espla *et al.*, chapter 12, this volume). These have also been deployed to act as fish attraction devices (FADs) to ensure good fishing yields. Further studies on the feeding habits of reef fish could influence design parameters in the future, but it is prudent to diversify man-made substrata as much as possible until design criteria are satisfactorily tested.

Ecological significance of artificial reefs

All structures placed on the seabed cause an alteration in the prevailing ecological conditions that can either increase or lower the biodiversity and/or biotic productivity. Harbour breakwaters, wharves, oil platforms and other similar facilities, as well as artificial reefs, provide a very suitable habitat for benthic assemblages. This suggests that they play a role in the 'production' of fish, in the sense that they provide a physical resource that supports stages in the life-cycle of many nektonic species, a habitat that would not exist in such quantity without man-made submerged structures.

Some major features distinguish the colonization of harbour breakwaters from that of artificial habitats. The spontaneous colonization that follows the building of breakwaters is a beneficial effect that partly counterbalances the environmental disturbance brought about by harbour installations, namely reduced flushing and increased eutrophication and turbidity. The construction of yachting and fishing harbours in the Mediterranean often results in the smothering of a *Posidonia* meadow, replacing a diverse biocenosis with a reducing sulphide system populated by few microaerophilic organisms. The biotic advantage created by breakwaters can partly compensate for the loss of biodiversity caused by harbour construction.

Despite the massive use of concrete and the use of heavy machinery for deployment, artificial reef developers should aim neither to disturb the marine environment nor to cause adverse alterations of the ecosystem comparable to those following the construction of harbours or marinas.

Research shows that Sicilian artificial reefs have had a negligible impact on hydrology and water circulation, and after deployment become biologically similar to natural rocky substrata. Reef deployment appears to be an outstanding technique for improving degraded marine habitats and enhancing biological productivity.

Artificial reefs have been considered by many authors as 'islands' to be populated (Bohnsack and Sutherland, 1985; Matthews, 1985). From a systems point of view, they are interruptions of the spatial (and biocoenotic) continuum of soft seabeds. The effect of these rocky, fenestrated structures is to increase the area for water/seabed interaction. The newly added structures enhance energy transfer from plankton to benthos and *vice versa*; raise the environmental diversity (heterogeneity); create new spatial niches and ultimately increase the biotic diversity through recruitment of new taxa (fish and benthos on and around the hard substrata). The increase of biodiversity appears to be dependent on the total length of free reef block edges, niche diversity, local recruitment and water transparency.

Conclusions

Natural rocky substrata in Sicilian waters support diverse ecosystems and provide fishery yields for traditional, small-scale fisheries. Fishery yield and contribution to biodiversity are enhanced wherever rocky seabed is interposed with *Posidonia oceanica*. With this in mind, the development of artificial reefs in Sicily can be considered to be a multi-purpose technology for resource management within the coastal zone. Following initial research results Sicilian reef deployment has tried to promote the re-colonization of damaged seabed habitat, beyond the immediate scope of merely increasing the fishing yield. Its planned role encompasses such contrasting issues as environmental conservation and the improvement of fisheries and mariculture, thereby making artificial reefs a tool for integrated management of marine resources (Badalamenti *et al.*, chapter 5, this volume), the only possible solution to the many pressures affecting the coastal zone (Doumenge, 1995a, 1995b).

Later results have, to a large extent, confirmed previous expectations: artificial reefs have been successful in preventing near-shore trawling (despite attempts by fishermen to tow reef units away) and have greatly favoured the protection of the *Posidonia oceanica* beds and their re-colonization wherever sea grass had been cleared or damaged (D'Anna *et al.*, chapter 6, this volume; Riggio, 1995b). Artificial reefs have, subjectively, improved the quality of the benthic environment and increased the fishing yields (Arculeo *et al.*, 1990b), although the rise in catches is partly due to fish attraction and is by no means comparable to the fishery returns obtained in highly productive waters, as in the Adriatic (Bombace, 1982). Fishery yields from Sicilian reefs are quite 'normal' as they reflect the actual effectiveness of the Mediterranean artisanal fisheries. The lower exploitable biomass found in oligotrophic, southern Tyrrhenian waters is largely compensated for by the price obtained for the high-quality catch.

From a marine ecologist's point of view, artificial reefs in Sicily have indicated the biological potential of the island's coastal ecosystems; they have also been used as an experimental site. Results have revealed the complexity of a reef system, whose improved comprehension is posing more questions than it is answering, evidence of the vitality of this field of study and a powerful stimulus to penetrate more deeply into it.

In the following chapters (5 and 6) Badalamenti *et al.* and D'Anna *et al.* give more details of the results from reef deployment following the initial research and the conclusions made from the work around Sicily.

Acknowledgements

Research was funded by the Provincia Regionale di Palermo within the framework of the Progetto Mare. The authors wish to thank Antony Jensen (Southampton Oceanography Centre) for his contribution to the final writing of contributions to this volume and for correcting the English.

References

Arculeo, M., G. Bombace, G. D'Anna and S. Riggio. 1990b. Evaluation of fishing yields in a protected and an unprotected coastal area of N/W Sicily. FAO Fisheries Reports. **428**: 70–83.

Arculeo, M., G. D'Anna and S. Riggio. 1990a. An essay on the coastal fisheries of North and South Sicily. *Rapport de la Commission International de la Mer Méditerranée. CIESM.* **32**(1): 253.

Ardizzone, G.D., A. Mazzola and S. Riggio. 1977. Modificazioni nelle comunite incrostanti del Porto di Palermo in relazione a diverse condizioni ambientali. Atti del IX Congresso S.I.B.M., Lacco Ameno 19–22 Maggio 1977: 151–159.

Badalamenti, F., G. Giaccone, M. Gristina and S. Riggio. 1985. An eighteen months survey of the artificial reef off Terrasini (NW Sicily): the algal settlement. *Obelia, N.S.* **XI**: 417–425.

Badalamenti, F. and S. Riggio. 1986. An outline of the polychaete colonization of a small artificial Reef of the N/W Coast of Sicily. *Rapport de la Commission International de la Mer Méditerranée.* **30**(2): 417–425.

Bohnsack, J.A. and D.L. Sutherland. 1985. Artificial reef research: a review with recommendations for future priorities. *Bulletin of Marine Science.* **37**(1): 3–10.

Bombace, G. 1982. Il punto sulle barriere artificiali: problemi e prospettive. *Naturalista Siciliano* S. IV, VI (Suppl.). **3**: 573–591.

Costa, C., S. Riggio and G. Giaccone. 1983–84. Note bionomiche sulle comunità di substrati naturali ed artificiali lungo la costa di Vergine Maria (Golfo di Palermo). *Nova Thalassia.* **6** (Suppl.): 663–669.

Daulon, L. 1978. *Les Pêches Jerbiennes.* Association pour la sauvegarde de l'Ile de Jerba. Jerba, 125 pp.

Di Pisa, G. and S. Riggio. 1982. A mathematical model of the stability of harbur benthic communities. *Naturalista Siciliano* S. **IV**, VI (Suppl.): 661–666.

Doumenge, F. 1995a. L'interface pêche-aquaculture: coopération, coexistence ou conflit. *Norois*. **165**: 205–223.

Doumenge, F. 1995b. Les récifs artificiels: pourquoi et comment? *Biologia Marina Mediterranea*. **2**(1): 15–20.

Matthews, K.R. 1985. Species similarity and movement of fishes on natural and artificial reefs in Monterey Bay, California. *Bulletin of Marine Science*. **37**(1): 252–270.

Mazzola, A. and S. Riggio. 1976. Il fouling portuale di Palermo. II Contributo. *Memorie di Biologia Marina e di Oceanografia*. **VI**. (Suppl. 6): 41–43.

Purpura, G.F. 1989. Pesca e stabilimenti antichi per la lavorazione del pesce in Sicilia: III-torre Vindicari (Noto), Capo Ognina (Siracusa). *Sicilia Archeologica*. **69–70**. 25–37.

Purpura, G.F. 1990. Pesca e stabilimenti antichi per la lavorazione del pesce nella Sicilia occidentale: IV – Un bilancio. V Rassegna di archeologia subacquea, Giardini Naxos, 19–21 Ottobre 1990. Atti: 87–101.

Riggio, S. 1979. The fouling settlements on artificial substrata in the Harbour of Palermo (Sicily) in the years 1973–1975. *Quaderni del Laboratorio di Tecnologia della Pesca, Ancona*. **II**: 207–253.

Riggio S. 1982. The artificial Reefs of Terrasini (Northwestern Sicily) after one year of Submergence. *Journée Etude Recifs Artificielle et Mariculture suspend*. CIESM, Cannes. 67–71.

Riggio, S. 1995a. Géographie et pêche côtière en Sicile, pp. 107–130. In AA. VV. *La Pêche Côtière en Tunisie et en Méditerranée*. Cahier du CERES, série Géographique no. 10, Tunisie, 335 pp.

Riggio, S. 1995b. Le barriere artificiali e l'uso conservativo della fascia costiera: risultati dei 'reefs' nella Sicilia N/O. *Biologia Marina Mediterranea*. **2**: 129–164.

Riggio, S. and G.D. Ardizzone. 1979. Prospettive dell'impiego di substrati per il ripopolamento e la protezione dei fondali costieri della Sicilia nord occidentale. Atti del Convegno Scientifico Naz. P.F. Oceanografia e Fondi Marini, Roma 5–7 Marzo 1979: 157–184.

Riggio, S. and G.D. Ardizzone. 1981. Eutrofizzazione e comunità bentoniche su substrati artificiali. Indagine preliminare sulle coste della Sicilia nord occidentale. *Quadermi del Laboratorio di Tecnologia della Pesca, Ancona*. **3** (1 Suppl.): 587–603.

Riggio, S., F. Badalamenti, R. Chemello and M. Gristina. 1986. Zoobenthic colonization of a small artificial reef in Southern Tyrrhenian: Results of a three-year survey. General Fisheries Council for the Mediterranean. FAO Fisheries Reports. **357**: 109–119.

Riggio, S. and G. Di Pisa. 1982. Observations on the development of fouling on discontinuous surfaces in Palermo harbour. *Naturalista Siciliano* S. IV, **VI** (Suppl.)(1): 607–626.

Riggio, S., M. Gristina, G. Giaccone and F. Badalamenti. 1985. An eighteen months survey of the artificial Reef of Terrasini (N/W Sicily): the invertebrates. *Oebalia*, N.S. **XI**: 427–437.

Riggio, S. and A. Mazzola. 1976. Preliminary data on the fouling communities of the harbour of Palermo (Sicily). *Archives of Oceanography and Limnology*. **18** (Suppl. 3): 141–151.

Simard, F. 1995. Rèflexions sur les rècifs artificiels au Japon. *Biologia Marina Mediterranea*. **2**(1): 99–109.

Villari, P. 1992a. I molluschi marini nell'alimentazione preistorica e nei culti di età greca e romana in Sicilia orientale: i dati archeozoologici. *Animalia*. **19**(1/3): 67–77.

Villari, P. 1992b. I resti fossili di una fossa votiva del Tempio Ionico di Siracusa. *Animalia*. **19**(1/3): 79–89.

Villari, P. 1995. Le faune della tarda preistoria nella Sicilia orientale. Phoenix, collana di Ecologia diretta da Marcello La Greca, n. **5**. Ente Fauna Siciliana, Noto, Siracusa, 493 pp.

5. Artificial Reefs in the Gulf of Castellammare (North-West Sicily): A Case Study

FABIO BADALAMENTI[1], GIOVANNI D'ANNA[1] and SILVANO RIGGIO[2]

[1]*Laboratorio di Biologia Marina, I.R.M.A.-CNR, Via G. da Verrazzano, 17, I 91014, Castellammare del Golfo (TP), Italy.* [2]*Dipartimento de Biologia Animale, Università di Palermo, via Archirafi, 18, I 90123 Palermo, Italy*

Introduction

Most of the plans for fish stock replenishment recently undertaken in Sicily have focused on the Gulf of Castellammare. Reasons for choosing this biotope for a restocking plan include the size of the Gulf (300 km^2), the importance and traditional role of its fisheries and the existence of information describing the local marine environment. The Gulf of Castellammare is the widest bay in Sicily and fishing has always played a major role in the local economy. Today income from fishing complements that from tourism.

Small-scale artisanal and trawl-fisheries co-exist; the former are scattered along the coast, whereas the latter are based at ports in the Gulf. There was evident competition between artisanal fishermen and trawlers for fishery resources (Arculeo and Riggio, 1989) which was markedly reduced after artificial reef deployment (Badalamenti and D'Anna, 1985) and the trawling ban along the continental shelf (Pipitone *et al.*, 1996). Data on yields and biology of some commercial species have been available since the 1960s (Arculeo *et al.*, 1990a, 1990b; Arena and Bombace, 1970). The marine environment of the Gulf is a mosaic of Sicily's most representative coastal biotopes, where almost all of the regional ecological features are encountered. Good-quality environmental data provide clues that assist in better understanding the end result of artificial reef deployment in a variety of environmental conditions typical of the southern Mediterranean. A sound knowledge of the results from reef deployment within the Gulf of Castellammare makes it possible to develop a comprehensive model of artificial reef systems in the southern Mediterranean. With this purpose in mind the biological colonization of the reefs has been monitored since deployment in the early 1980s.

Study Area

The Gulf of Castellammare is a broad, crescent-shaped bay on the NW coast of Sicily; it lies in the South Tyrrhenian Sea at 38° 03' N, 12° 54' E, FAO fishing area 37.1.3 (Fig. 1). Its coastline is over 70 km long and its surface area is ca. 30 000 ha. Capo Rama, a tough dolomitic cliff, marks is easternmost boundary, and Capo San Vito peninsula forms the westernmost point. These headlands enclose an

A.C. Jensen et al. (eds.), Artificial Reefs in European Seas, 75–96

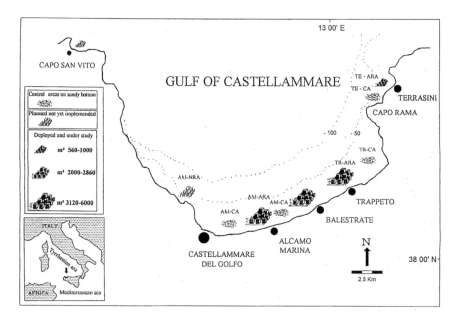

Figure 1.   Location of the study sites and of the artificial reefs deployed and/or planned in the Gulf of Castellammare. AM, Alcamo Marina; TE, Terrasini; TR, Trappeto; ARA, artificial reef area; CA, control site on sandy seabed; NRA, natural rocky area.

amphitheatre of terraced sandy-lime hills (ca. 100 000 ha) which slope down to the shore (Riggio *et al.*, 1992). The inland area is intensively cultivated and drained by a number of small streams which transport significant amounts of silt, suspended matter and nutrients into the sea (Calvo and Genchi, 1989).

The character of the coastal morphology varies between the east and west of the Gulf. The westernmost portion, facing north-east, is characterized by hard rocky substrata; the eastern sector is dominated by unstable soils in continuity with soft muddy marine sediments. Steep, calcareous cliffs are therefore the prominent landscape of the western coast, whereas high terraces of mixed sandy-clay soils and low sandy beaches, once sheltered by high littoral dunes, border the mid-eastern coastline. The seabed therefore changes from the west of the Gulf, where rocky cliffs plunge to considerable depths (ca. 80 m), to the mid-eastern portion, where the sedimentary seabed gently slopes to a depth of 60 m.

The prevailing winds blow from west to north-west during autumn and winter; southerly winds are more frequent in spring, whereas breezes from the east and north-east are a constant feature during high pressure weather in summer. The surface currents driven by the winds run eastward, tangential to Capo San Vito, and concentrate silt and organic matter in the area between Trappeto and Capo Rama (Fig. 1), where turbidity is intense for most of the year. The waters off Terrasini remain clear, protected by Capo Rama. The western portion of the Gulf is sheltered from this westerly wind; its waters are clean and oligotrophic despite some signs of pollution.

Man-made pollution from the densely populated urban belt, and especially from a number of factories, adds to the nutrient burden affecting the coastal waters in the eastern portion of the Gulf. The onshore disposal of dregs and sludge resulting from the distillation and processing of wine and grape-pips (Riggio *et al.*, 1992), together with other industrial and domestic output, has caused severe eutrophication of the coastal waters. Alcamo Marina and the Nocella stream outlet are the most endangered sites, with the water having a dramatically high BOD (Calvo and Genchi, 1989; D'Anna *et al.*, 1985; Riggio *et al.*, 1990, 1992).

Structure and Location of the Artificial Reefs

The artificial reefs were assembled on a sandy seabed at depths between 10 and 30 m. Most of the reefs were made from concrete blocks deployed to form pyramid-shaped reef units. Choice of the pyramid pattern was intended to maximize the surface area exposed to the planktonic disseminules and settling meroplanktonic larvae (Riggio *et al.*, 1992). The standard block weighs 13 tonnes, has a volume of 8 m³ (2 × 2 × 2 m) and has holes and cavities of different shapes and sizes, following the design of reef blocks used in the Adriatic Sea (Bombace, 1982; Bombace, *et al.*, chapter 3, this volume). Fourteen of these blocks were used to create the standard three-layer pyramid reef unit. A different block pattern (2.7 m³ concrete block, with four 20 cm² tunnels in each face) was chosen as the basic component of the 29-block, three-layer pyramid reef units at the Terrasini reef (Fig. 2).

Reef units to deter trawling, 'anti-trawling devices', were designed as prismatic concrete 'boulders' (nicknamed 'torpedoes') armed with iron spikes (Fig. 2) which were connected in groups of between 5 and 10 by a steel rope and randomly scattered around the pyramids (Bombace, 1989; Riggio and Provenzano, 1982). Each 'torpedo' measures 0.2 m³ (0.5 × 0.4 × 1 m) and weighs ca. 0.3 tonne.

The first experimental reef, composed of four pyramids, was laid in late 1981 on a sandy seabed in clear water off Terrasini Harbour, in the Terrasini artificial reef area (TE-ARA) (Badalamenti *et al.*, 1985; Riggio *et al.*, 1985). Special low-cost artisanal technologies were developed for the building and assembling of the cubes (Provenzano and Riggio, 1982).

A second artificial reef was deployed in 1984, 600 m off Trappeto town (in the Trappeto Artificial Reef Area, TR-ARA) on a muddy seabed affected by a very high siltation rate (Arculeo *et al.*, 1990a; Provenzano and Riggio, 1982; Riggio and Provenzano, 1982).

The third reef was placed off Alcamo Marina (in the Alcamo Marina Artificial Reef Area, AM-ARA) (D'Anna *et al.*, 1994): reef deployment started in 1986 and continued until 1995.

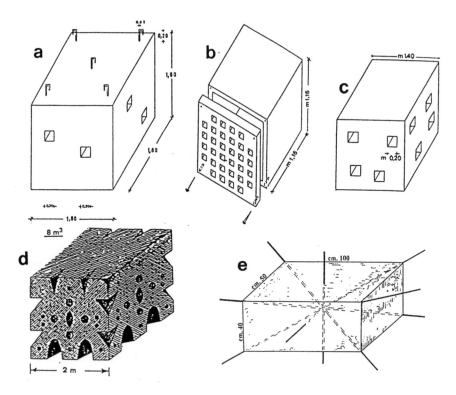

Figure 2. Shapes and dimensions of the main modules used for constructing the artificial reefs in the Gulf of Castellammare. (a, b, c, e) Terrasini and Trappeto reefs; (d,e) Alcamo Marina artificial reef.

Research Directions and Procedures

Two different lines of investigation were designed to provide an insight into the major elements which contribute to the success of a man-made reef (Fig. 3). They are, respectively, the ability to:

(1)   Mimic a natural rocky habitat by supporting successional benthic colonization;
(2)   Improve the fishing yields due to either
    a)   attraction of pelagic fish, or
    b)   supply of food, creation of a refuge and/or opportunities for spawning.

Methods

Analysis of Benthic Colonization

Seasonal sampling was carried out for 2 years in three artificial reef areas (Terrasini, Trappeto and Alcamo Marina; Fig. 1). Areas of 20×20 cm were

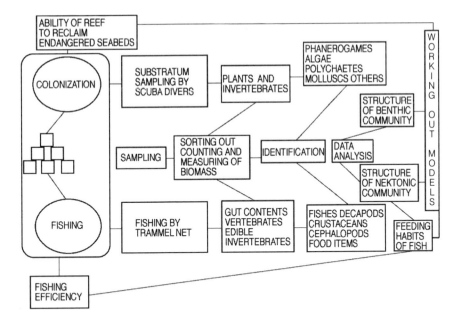

Figure 3. Main research directions for Castellammare artificial reefs.

sampled (epibenthos scraped off) by SCUBA divers from both horizontal and vertical surfaces of each pyramid layer. Fine debris and the small-sized mobile fauna were collected by means of a suction pump. Samples were preserved in 5% buffered formalin and sea water and taken to the laboratory. Sorting of fauna and identification to species level were restricted to polychaetes, molluscs, bryozoans and algae. Other groups were scarcely represented. A species–site matrix for each seasonal sampling was designed. Indices of community structure were calculated; multivariate statistical analysis, using the Jaccard similarity index, was applied to identify the sequential changes in the biotic assemblages. A dendrogram was obtained using the weighted average linking as the agglomerative algorithm (Pielou, 1984).

Evaluation of Fishing Yields and Description of the Nektonic Assemblages

Commercial catches from each reef were assessed the first year after deployment. Fishing yields were estimated by monthly samplings in each reef area (Terrasini, Trappeto and Alcamo Marina) and their respective control area (CA) (Terrasini CA, Trappeto CA and Alcamo Marina CA). The surveys lasted only one year with the exception of the Alcamo Marina sites where research continued until 1996. A 500 m long trammel net, 3 m in height, with a 54 mm inner mesh size, was used. The nets were fished (set) from sunset to sunrise with a 12-h mean fishing time (D'Anna *et al.*, 1992, 1994). The results of the fishing survey were analysed to evaluate the temporal variations of the mean yield during the year. Sampled species were identified and included in the pelagic (P), nektobenthic (demersal) (NB) and benthic (B) categories. The individual size and weight were recorded for each species. Species richness in terms of S (number of species), Margalef's d' (Margalef, 1958), Shannon-Weiner's H' (index of diversity) and evenness J' (Pielou, 1966) were calculated using total and benthic-nektobenthic taxa. Factorial correspondence analysis (FCA) was performed on a quantitative 6 × 60 station per species correlation matrix not inclusive of rare species (Benzecri, 1982), considering only the reef areas and their controls. Significance of axes was evaluated using the tables of Lebart (1975). Taxa were assigned to one of four groups according to their affinity for the substratum. The Kruskal-Wallis test (Siegel, 1956) was used to detect significant differences in fishing yields among sites.

Methods for Surveying Feeding Relationships

Fish samples were taken monthly, using a trammel net at night, from three sites: Alcamo Marina-ARA, Alcamo Marina-natural rocky area (NRA) and Alcamo Marina-CA. The gut contents of the following species were examined: red mullet *Mullus barbatus*, with a mean 16.5 ± 6.6 cm total length (TL), and *Mullus surmuletus* (TL 17.7 ± 1.4 cm); striped sea bream *Lithognathus mormyrus* (TL 20.1 ± 4.8 cm) and the sea bream *Diplodus annularis* (TL 13.9 ± 1.3 cm). Seasonal variations in feeding habits were assessed. Fish were measured, weighed and eviscerated. Stomachs were preserved in 5% buffered formalin in sea water. Remains of prey items were sorted out, weighed and examined under a binocular lens. Hydroids, polychaetes, molluscs, amphipods and polyzoans were, whenever possible, identified to species level. The following parameters were calculated: (a) percentage occurrence (F%); (b) wet weight (P%); (c) indices of stomach vacuity (SV), according to Hyslop (1980); (d) Levins' (1968) trophic niche-width (B); (e) Schoener's feeding overlap index (Schoener, 1970). The results were expressed graphically according to Costello (1990). After identification, the prey were classified according to their distribution on or within the substrata, the biotic community and trophic guild.

Results of the Benthic Colonization Survey

Species Richness and Distribution

To date, 303 benthic invertebrate species, including polychaetes (119), molluscs (136) and bryozoa (48) have been identified. The species distribution at the three stations was markedly different (Fig. 4). The number of taxa and the percentage cover of minor groups such as sponges and crustaceans have been negligible. As many as 174 algal species have been identified, belonging to the Rhodophyceae (103), Phaeophyceae (32), Chlorophyceae (33) and Cyanophyceae (8).

Development of Biological Assemblages

Despite the proximity of the artificial reefs, the patterns of colonization and the communities that settled during the first 2 years clearly differed between reefs. Settlement on newly deployed reefs was quite uniform, with a few pioneer species common to all pyramids (Fig. 5). As time progressed reef units in the Alcamo Marina ARA and Trappeto ARA persisted in their pioneer phases, whereas artificial reefs off Terrasini rapidly developed a mature population.

The differences in water clarity between sites are thought to have played a part in the subsequent community development; the water was at its most transparent in the Terrasini ARA, as determined by Secchi disk; to the west, at Trappeto ARA and Alcamo Marina ARA, underwater visibility declined and Secchi disk readings rarely exceeded 8 m (Fig. 6).

One year after deployment the Terrasini ARA pyramids were overgrown by a luxuriant canopy of brown algae which hid a diverse associated fauna which included gastropods, polychaetes and encrusting bryozoans. The highest values of species number, individual densities, algal percentage cover and biotic diversity were recorded here. An overall, although less marked than at Terrasini, increase in these factors was seen on the more eastern artificial reefs of Alcamo Marina and Trappeto (Figs 4, 7).

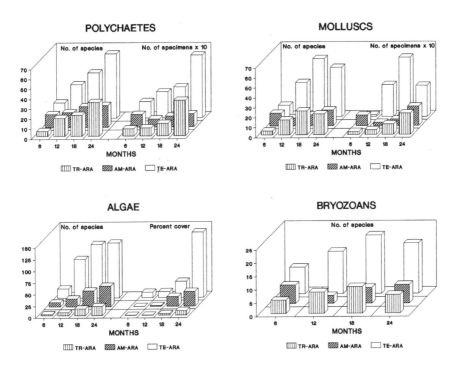

Figure 4. Temporal changes in the number of species, individuals and percent cover of the main faunal groups and algae. AM, Alcamo Marina; ARA, artificial reef area; TE, Terrasini; TR, Trappeto

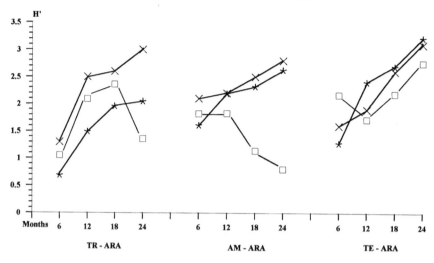

Figure 5. Monthly values of the Shannon-Weiner diversity index calculated for algae, polychaetes and molluscs. AM, Alcamo Marina; TE, Terrasini; TR, Trappeto; ARA, artificial reef area.

SECCHI DISK

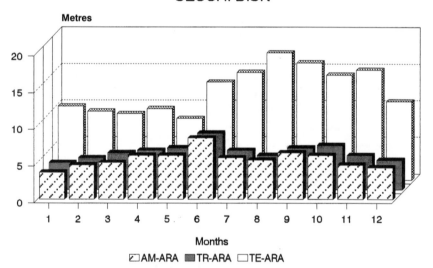

Figure 6. Monthly values for Secchi disk readings in the three artificial reefs under study. AM, Alcamo Marina; TE, Terrasini; TR, Trappeto; ARA, artificial reef area.

The reefs in the Alcamo Marina ARA supported a mixed assemblage of filter feeders with the occasional record of algae and encrusting faunal colonies. Massive colonies of the Madreporarian *Cladocora caespitosa* developed. A large

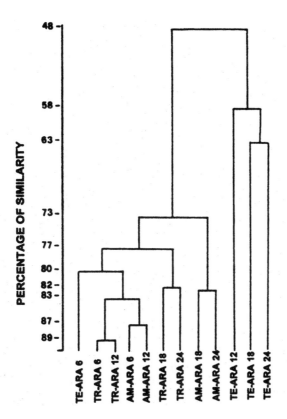

Figure 7. Dendrogram showing the similarities among the seasonal samples taken at the Alcamo Marina (AM), Terrasini (TE), and Trappeto (TR) artificial reefs at 6-monthly intervals.

amount of chlorophyll a was measured in the surface waters off Alcamo Marina but the high water turbidity and the intense siltation appear to have counterbalanced this high primary productivity and adversely affected the growth of communities on the Alcamo Marina ARA. Biological growth was stunted because of silt deposition and significant grazing pressure on the horizontal surfaces. A peculiar pattern of benthic settlement developed on this reef's vertical surfaces where the only noteworthy colonizers were encrusting groups like bryozoans, serpulids and bivalves. The benthic community settling at Alcamo Marina ARA, made up of a few macroalgae and a greater number of components than those recorded at Trappeto ARA, was intermediate between those of Terrasini and Trappeto.

The Trappeto ARA reef underwent a slow succession towards impoverished and unstable communities. The exclusive presence of taxa typical of sandy-muddy seabeds and the absence of macroalgae were the most significant features of the biotic settlements at Trappeto ARA.

An almost unique settlement of filter feeders (hydroids, sponges and bryozoa) was recorded.

Nektonic Assemblages in the Artificial Reefs and in the Control Areas

After 1 year of monthly trammel net sampling, 103 taxa were recorded (five crustaceans, six cephalopods, 92 fish) from the three artificial reef areas, their controls on sandy-muddy seabeds and on natural rocky areas. They were subdivided into 44 benthic, 42 nektobenthic (demersal) and 19 pelagic species.

Mobile Fauna in Artificial Reef Areas

The most frequent crustacean caught was the mantis shrimp (*Squilla mantis*). Squid (*Loligo* spp.) and cuttlefish (*Sepia officinalis*) were the most abundant cephalopods. Fish were especially represented by the Sparidae *Diplodus annularis, D. vulgaris, Lithognathus mormyrus* and *Pagellus erythrinus*. Some scorpaenids, serranids and one labrid species (most representative at Terrasini ARA) were also collected. The most frequent netted pelagic fish (P) were *Trachurus trachurus, Seriola dumerili, Spicara maena* and *Sardinella aurita*.

Mobile Fauna in Control Areas

The nektonic assemblages of control areas were well represented by species preferring a sandy seabed (e.g. the prawn *Penaeus kerathurus* and the pleuronectid *Bothus podas*). The Alcamo Marina natural rocky area site was especially populated by scorpaenids, labrids and red mullet. The highest values of S, d' and H' were recorded at this site in close proximity to the artificial reefs, whereas the values of the control sites were lower (Fig. 8). Evenness (J') calculated on the reefs was very close to the value calculated for the control sites.

Comparative Distribution of Fish

If only the reefs and their control sites are considered (Alcamo Marina, Trappeto and Terrasini) the first three axes of the FCA are significant ($P < 0.001$), accounting for as much as 90.6% of the total variance. A parabolic arrangement of the stations along a steep gradient from soft seabed (Fig. 9, left side) to natural hard substrata (Fig. 9, right side) appears. As is also evident from the graph, pelagic fish populations prevailed on the sandy-muddy seabeds of Alcamo Marina ARA and Alcamo Marina CA. Less mobile species related to rocky substrata were recorded most frequently at Terrasini ARA and Terrasini CA. The composition of the fish assemblages at Trappeto ARA and Trappeto CA was markedly heterogeneous.

Analysis of the Fishery Yields at Alcamo Marina over a 4-Year Period

Annual catches within the Alcamo Marina ARA were higher than at Alcamo Marina CA (except in 1991) while the highest fishing yields were recorded from Alcamo Marina NRA. Some fluctuations of fishery yield, mainly caused by the presence or absence of pelagic species, were recorded (Fig. 10). According to the

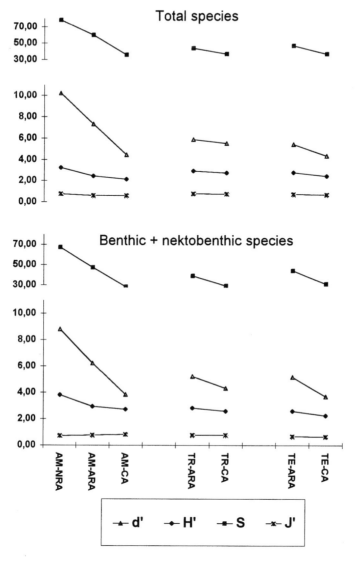

Figure 8. Community structure indices calculated separately for total and benthic nektobenthic species at each site. S = number of species; d' = species richness of Margalef; H' = Shannon-Weiner diversity index; J' = evenness index. For the other abbreviations see text.

Kruskall-Wallis test (performed on the annual mean yield of benthic, nektobenthic (demersal), pelagic and total species) significant differences among the three areas were only evident for benthic ($P < 0.05$) and nektobenthic (demersal) ($P < 0.05$) species whose yields from the NRA and ARAs were respectively higher than the CAs. No significant differences appeared as far as the total catch was concerned.

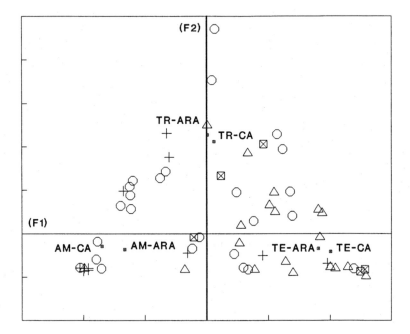

Figure 9.   Ordination of stations and species derived from the factorial correspondence analysis (FCA). Symbols: circles = sandy seabed species; triangles = pelagic species; + = rocky seabed species; barred squares = seagrass species. For the other abbreviations see text.

Fish Feeding

Results of Gut Content Analysis

Invertebrate deposit feeders, filter feeders, surface deposit feeders and carnivores were the most frequent prey items in the stomachs of red mullet (Mullidae) (Fig. 11). The proportion of empty stomachs was as high as 64% in *M. barbatus* from Alcamo Marina CA whereas it fell to 0% in *M. surmuletus* from the same site. As many as 180 different taxa, belonging to 20 main taxonomic categories, were identified from stomach content analysis. Niche-Breadth values are reported in Fig. 12. Shoener's index does not show any significant diet overlap (Badalamenti *et al.*, 1993).

Filter feeders were the most important prey of *L. mormyrus* whereas both filter feeders and carnivorous species made up the diet of *D. annularis*. Epibenthos on *Cymodocea nodosa* and *Posidonia oceanica* or other species associated with both sea grasses constituted up to 50% of *Diplodus annularis* prey items from all the surveyed sites (Badalamenti *et al.*, 1993).

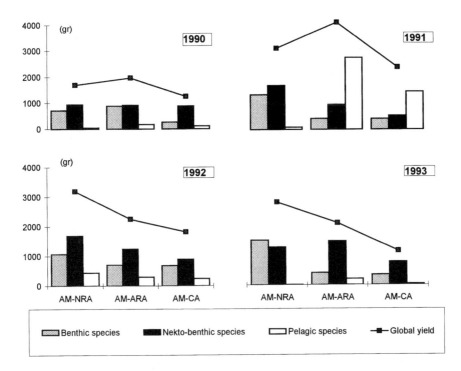

Figure 10. Annual mean yields obtained from the Alcamo Marina artificial reef area (AM-ARA) in comparison with those of natural rocky areas (NRA) and sandy seabed control areas (CA).

Feeding Habits of the Most Common Fish

Critical examination of the diets and feeding indices suggests that the four fish species most closely associated with the artificial reefs can be deemed as demersal micro-carnivores, feeding on small benthic invertebrates:

(1) The two Mullid species are sand-diggers. They search for food items in the infaunal assemblage;

(2) *L. mormyrus* is both a picker and a sand-digger, able to feed on both infaunal and epifaunal assemblages;

(3) *D. annularis* is a picker feeding on epibenthic organisms.

These observations on the feeding behaviour in relation to artificial reefs agree with data from surveys in natural habitats (Arculeo *et al.*, 1989a, 1989b). Modification of feeding habits can be induced by environmental changes following the place-ment of artificial substrata (Hückel *et al.*, 1989): diverse substrata very likely affect both the distribution and abundance of these species in the Gulf of Castellammare (D'Anna *et al.*, 1992, 1994).

Generalist fish like *D. annularis* can adjust their behaviour and adapt to changes in the environment and exploit the available food resources (Macpherson, 1981). *D. annularis* is, therefore, the only species from those examined strictly dependent

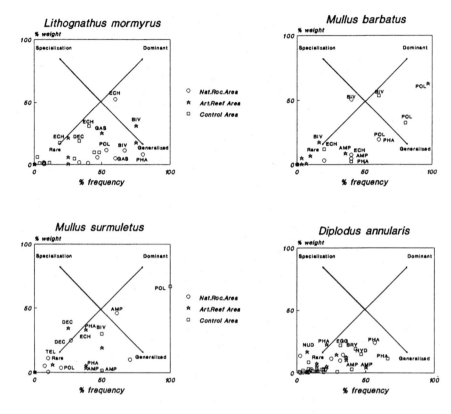

Figure 11. Graphic representation (Costello, 1990) of the percentage frequency of occurrence and of the weight of the main food items of the four target fish. AMP = amphipods; BIV = bivalves; BRY = bryozoans; DEC = decapods crustaceans; ECH = echinoderms; EGG = eggs of marine organisms; GAS = gastropods; HYD = hydroids; NUD = nudibranchs; PHA = phanerogames; POL = polychaetes; TEL = teleosts.

on the encrusting organisms (hydroids, polychaeta and bryozoa) settled on the artificial reef blocks (Badalamenti *et al.*, 1993). The other fish did not depend on the artificial reef as a primary source of food, although the increase of niche-width observed in *L. mormyrus* at Alcamo Marina ARA and the lower vacuity coefficient of *M. barbatus* at the same station could be linked to the presence of artificial reefs. In such a case, more indirect factors, like protection from predators (Bohnsack, 1989) or the organic enrichment of sediments around the reef (Ambrose and Anderson, 1990), presumably promoting infaunal productivity, would be responsible for the fish attraction and recruitment.

Research on the feeding habits of the two-banded sea bream, *Diplodus vulgaris*, was recently conducted at Alcamo Marina ARA (Pepe *et al.*, 1996). Most of the food items sorted out from 128 medium-sized individuals (ca. 10 cm total length) of the two-banded bream were tanaidacean crustacea (34%), serpulids (15%) and

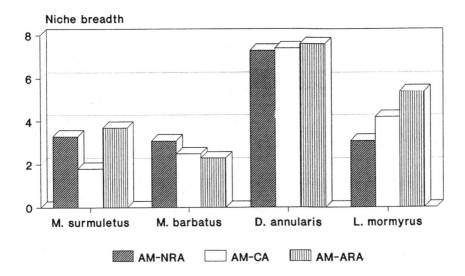

Figure 12. Levins' niche breadth values calculated for the four species in the artificial reef area (ARA), natural rocky area (NRA) and sandy seabed area (CA).

other polychaete worms (26%), which added up to as much as 75% of all prey (Fig. 13). These organisms had also been found in the artificial reefs and are strictly related to hard artificial substrata or photophilous seaweeds living on rocky substrata. A particularly significant find was 807 *Leptochelia savignyi*, a small tube-dwelling tanaidacean, which is frequently found on the concrete blocks. As a preliminary conclusion, the artificial reefs appear to provide foraging grounds for *Diplodus* spp., whereas Mullidae mainly feed from the sediment layer on horizontal cube surfaces, or on the soft-seabed communities adjacent to the reef sites. *Lithognathus mormyrus* apparently feeds on prey from sandy seabeds.

Nekton Assemblages and Fishery Yields

Biodiversity

The natural rocky area at Alcamo Marina showed the highest values of diversity and species richness, and the species were mainly those characteristic of the fish fauna associated with the sublittoral rocky seabeds of the Mediterranean (Harmelin, 1987). The higher species diversity of this area is very likely related to the seabed heterogeneity. Small stands of *Posidonia oceanica* attracted and concentrated labrid populations, an exclusive feature of this area, whereas flatfish populated the sandy patches scattered between the rocks.

The structure of the nekton assemblages in the control areas with sandy seabeds was significantly different. The low values of diversity and species richness were

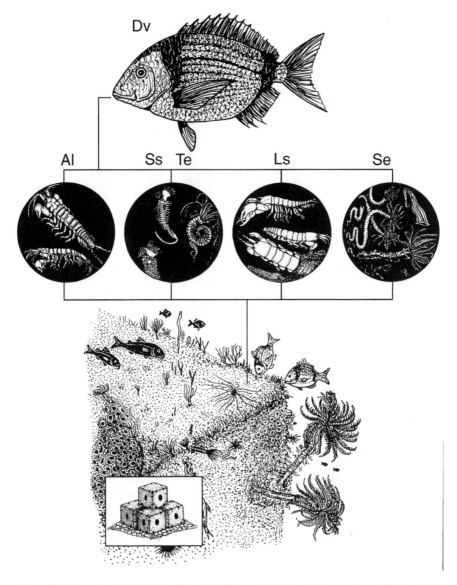

Figure 13. Synopsis of the feeding diet of *Diplodus vulgaris* in relation to the artificial reefs of Alcamo Marina. Al = *Apseudes latreilii*; Ls = *Leptochelia savignyi*; Se = serpulids; Ss = *Sabellaria spinulosa*; Te = terebellid worms.

typical of those from uniform soft seabeds, and the numerically dominant species were those normally found living on sandy sea floors.

The species composition and the indices of community structure in the artificial reef areas were intermediate between those calculated for natural rocky and sandy seabed areas. The higher values of species richness and diversity indices recorded from around the artificial reefs stress that there has been a marked increase in the variety and complexity of the fish assemblages from that of the control areas. On the other hand, the similar values for evenness calculated for the reefs and control areas highlight that there is only a slight difference in the homogeneity of their community structures.

Species Arrangement

The pattern displayed by factorial correspondence analysis is typical of those environmental situations heavily conditioned by a polarizing factor (in this case the type of substratum or seabed), as shown in the first axis (Fig. 9). The second axis is related to the first by a squared relationship and represents an intensity factor. Only the nektobenthic (demersal) component shows a significant increment of the fishing yields in the Alcamo Marina ARA as compared to Alcamo Marina CA, but, when the total catch is taken into account, no significant difference is seen.

Function and Fishing Efficacy

Fishing Yield

Fishing yields estimated in the Gulf of Castellammare are lower than those obtained from other Sicilian coastal areas (Arculeo *et al.*, 1990; Toccaceli and Levi, 1990), and this could be a consequence of the high siltation rates and input of fine particulate materials (clays) from the mainland. The artificial reefs in the Gulf of Castellammare have contributed positively in changing the fishing economy of the area. The reefs are a new topographic feature in the Gulf which have provided an additional type of habitat, increasing the area of hard substrata available for larval settlement. The large-scale reefs deployed off Alcamo Marina have halted illegal trawling, thereby conserving resources that are now exploited by small-scale fisheries. Moreover, Alcamo Marina ARA has proved to be an effective fish attraction device (FAD), especially for commercially important fish such as the amber jack, *Seriola dumerili*, schools frequently being found close to the reef units. The fishing yields from the artificial reef areas are definitely higher than from the control areas on sandy seabeds (Alcamo Marina CA), although they do not match the high yields reported from the Adriatic (Bombace, 1989; Bombace *et al.*, 1990; Fabi and Fiorentini, 1994) and the Ligurian sea (Relini *et al.*, 1989; Relini and Orsi Relini, 1989).

The statistically higher quantitative yields obtained in the nearby natural rocky areas suggest that artificial reefs are, in this case, less productive in a fisheries sense than expected, but these data need some critical examination to place them into context.

Fishery yield data in the Gulf of Castellammare are, to some extent, biased by a history of commercial over-exploitation by coastal and deep-sea trawling (Arculeo *et al.*, 1988; Arena and Bombace, 1970) and uncontrolled fishing with efficient set gear in artificial reef areas which dates back at least three decades.

Secondly, the damage caused by the intense organic pollution from factories and domestic sewers combined with the very high siltation rate has had a significant negative impact on naturally occurring communities (Riggio *et al.*, 1992). The yield and quality of fish taken in north-west Sicily appears to be related to the presence of complex seabed topographic features such as those found in stands of thick seaweed and *Posidonia* meadows (Arculeo and Riggio, 1989) or in shellfish beds (rare in Sicilian waters). Wherever the siltation rate is high, as is the case in the central area of the Gulf, fine sediments accumulate to form a continuously uniform, muddy layer. Frequent re-suspension of fine materials by currents and storms slows the process of colonization and may kill the benthos or prevent new recruitment. These phenomena have become more intense in recent years, enhanced by destructive management of the coastal strip, poor farming practices and dumping of waste material on the coast. As a consequence *Posidonia oceanica* has disappeared from a very large portion of the Gulf.

A reduction in pollution, coupled to a definitive curbing of trawling, could very reasonably enhance the biotic productivity (Pipitone *et al.*, 1996) of these areas.

A Critical Examination of Reef Fish Yield

The increment in fishery yields from the Alcamo Marina ARA is linked to the recruitment of nektobenthic (demersal) fauna, especially sparids, which joined the local fish fauna immediately after deployment of the artificial reef. The presence of sparids can account for the significant differences in the demersal assemblage near the artificial reefs when compared to the assemblage at sandy control areas (e.g. Alcamo Marina CA) and the communities of the natural rocky areas (D'Anna *et al.*, 1994). High-priced commercial fish, such as carangids and sparids, have been an important component of the fish community associated with the Alcamo Marina ARA. They provide significant income to fishermen who exploit the reef site, which manifests itself in the favourable attitude and enthusiasm of local fishermen towards the artificial reefs.

Biotic Settlements and Fish

Much in the same way as benthic communities are affected by hydrological conditions (Chang, 1985), fish assemblages are dependent on biotic structures. Tall macroalgae with erect finely branched thalluses, like the *Cystoseira* and *Sargassum*, can indeed compensate for the scarcity of refuges for both mobile invertebrate fauna and small fish (Spanier *et al.*, 1990), providing them with shelter from enemies and food (Badalamenti and Riggio, 1989; Harmelin, 1987). This conclusion is supported by the results of Gorham and Alevizon (1989), who simulated the stand of ribbon-like leaves in a seagrass forest using polypropylene rope strings attached to a frame laid on the seabed. The absence of labrids from Alcamo Marina (D'Anna *et al.*, 1994) and Trappeto reefs (Arculeo *et al.*, 1990) and their high frequency around the seaweed-overgrown installations off Terrasini (D'Anna *et al.*, 1992) support this argument. The huge numbers of peracarid crustaceans (namely amphipods, caprellids, Gammaridae and Tanaidacea), polychaete worms and encrusting serpulids living among the branching leaflets of *Cystoseira compressa* thalluses at Terrasini reef are a favourite prey for small, carnivorous fish.

As an initial conclusion, the experiments in the Gulf of Castellammare show that pyramid-shaped reefs provide effective habitats for algae, epibenthos and fish, but they should be installed in clear waters in order to attain their maximum efficacy. A solution to the problem of siltation on horizontal surfaces may well be to design and deploy reefs with surfaces at an angle of 45° to 135°, on the outer side of a reef to minimise sedimentation. Effective refuges for species intolerant of silting should also be built: some new reef units with the basic shape of a camping tent have been planned for this purpose (Fig. 14); their environmental efficacy is still to be tested.

Conclusions

Studying the artificial reefs in the Gulf of Castellammare has highlighted the positive role of pyramid reefs in the commercial fisheries of the Gulf. The study has

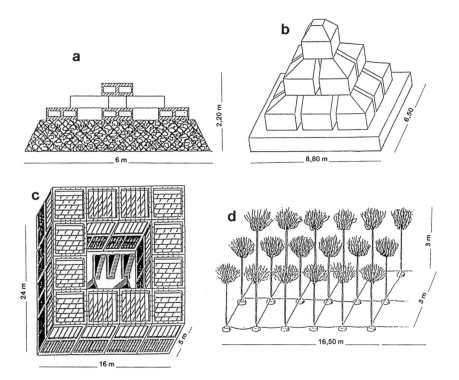

Figure 14. Artificial structures designed but not implemented: (a) module for protecting and enhancing populations of lobster and octopii; (b) pyramid planned with oblique surfaces to minimize mud deposition; (c) multipurpose module aimed at attracting pelagic species and hosting and feeding some sparids; (d) fish aggregation device (FAD) in polypropylene ropes to be deployed on the seabed.

also contributed to a better understanding of the biological colonization processes on reefs in Sicilian waters. Results have facilitated predictions of biological community development on artificial reefs due to be installed in clean water. Future scientific investigations will test these predictions as well as work towards a better understanding of the whole 'artificial reef system' and its 'carrying capacity'. Data already acquired will be integrated with new results to make the Gulf of Castellammare an area where the environment is sufficiently well understood to allow the testing of artificial reefs as a tool for the environmental management of the southern Mediterranean.

Acknowledgements

Part of this research was supported by the Consorzio "Ripopolamento ittico Golfo di Castellemmare". The authors wish to thank Drs R. Chemello, M. Gristine and M. Toccaceli for providing some of the data on benthos.

References

Ambrose, R. F. and T. W. Anderson. 1990. Influence of an artificial reef on the surrounding infaunal community. *Marine Biology.* **107**: 41–52.

Arculeo, M. and S. Riggio. 1989. Artisanal fishery in an area of Palermo Bay subjected to heavy environmental disturbance. Quaderni dell'Istituto per le Ricerche sulla Pesca Marittima del CNR. **5**(1): 51–65.

Arculeo, M., G. D'Anna and S. Riggio. 1988. Valutazione delle risorse demersali nell' area compresa fra Capo Gallo e Capo San Vito (Sicilia nord-occidentale): risultati delle campagne condotte nel 1985. *Atti Seminari Pesca e Acquacoltura. Publ. Min. Mar. Mercantile - C.N.R.* vol **IIIP**: 1413–1451.

Arculeo, M., G. D'Anna and S. Riggio. 1990a. An essay on the coastal fisheries of north and south Sicily. *Rapport de la Commission International de la Mer Méditerranée.* **32**(1): 253.

Arculeo, M., G. Bombace, G. D'Anna and S. Riggio. 1990b. Evaluation of fishing yields in a protected and an unprotected coastal area of N/W Sicily. FAO Fisheries Reports. **428**: 70–83.

Arena, P. and G. Bombace. 1970. Bionomie benthique et faune ichthyologique des fonds de l'étage circalittoral et bathyal des golfes de Castellammare (Sicile nord-occidentale) et de Patti (Sicile nord-orientale). *Journées Ichthyol.,* Rome, C.I.E.S.M.: 145–156.

Badalamenti, F. and G. D'Anna. 1995. Esperienze di barriere artificiali nel Golfo di Castellammare (Sicilia nord-occidentale). *Biologia Marina Mediterranea.* **2**(1): 165–173.

Badalamenti, F. and S. Riggio. 1989. I Policheti dei conenuti stomacali di *Mullus surmuletus* L. (Pisces Mullidae) nel Golfo di Palermo. *Oebalia, N.S.* **XI**: 79–87.

Badalamenti, F., G. Giaccone, M. Gristina and S. Riggio. 1985. An eighteen months survey of the artificial reef off Terrasini (NW Sicily): the algal settlement. *Oebalia N.S.* **XI**: 417–425.

Badalamenti, F., G. D'Anna, F. Fazio, M. Gristina and R. Lipari. 1993. Relazione trofiche fra quattro specie ittiche catturate su differenti substrati nel Golfo di Castellammare (Sicilia N/O). *Biologia Marina, Suppl al Notiziario S.I.B.M.* **1**: pp. 145–150.

Benzecil, J.P. 1982. *L'analyse des Données. II. L'Analyse des Correspondances.* Dunoud, Paris, 3rd edn. 632 pp.

Bohnsack, J.A. 1989. Are high densities of fishes at artificial reefs the result of habitat limitation or behavioural preference? *Bulletin of Marine Science*. **44**(2): 631–645.

Bombace, G. 1982. Il punto sulle barriere artificiali: problemi e prospettive. *Naturalista Siciliano*. **4**(6) (Suppl. 3): 573–591.

Bombace, G. 1989. Artificial reefs in the Mediterranean Sea. *Bulletin of Marine Science*. **44**(2): 1023–1032.

Bombace, G., G. Fabi and L. Fiorentini. 1990. Preliminary analysis of catch data from artificial reefs in central Adriatic. FAO Fisheries Reports. **428**: 86–98.

Calvo, S. and G. Genchi. 1989. Carico organico ed effetti eutrofici nel Golfo di Castellammare (Sicilia Nord occidentale). *Oebalia N.S.* **15**(1): 397–408.

Chang, K.H. 1985. Review of artificial reefs in Taiwan: emphasising site selection and effectiveness. *Bulletin of Marine Science*. **37**(1): 143–150.

Costello, M. J. 1990. Predator feeding strategy and prey importance: a new graphical analysis. *Journal of Fish Biology*. **36**: 261–263.

D'Anna, G., G. Giaccone and S. Riggio. 1985. Lineamenti bionomici dei banchi di mitili di Balestrate (Sicilia occidentale). *Oebalia N.S.* **XI**: 389–399.

D'Anna, G., F. Badalamenti and S. Riggio. 1992. Notes on the ecological significance of the fish fauna associated to artificial reefs in the southern Tyrrhenian. *Rapport de la Commission International de la Mer Méditerranée*. **33**: 378.

D'Anna, G., F. Badalamenti, M. Gristina and C. Pipitone. 1994. Influence of artificial reefs on coastal nekton assemblages of the Gulf of Castellammare (Northwest Sicily). *Bulletin of Marine Science*. **55**(2–3): 418–433.

Fabi, G. and L. Fiorentini. 1994. Comparison between an artificial reef and a control site in the Adriatic Sea: analysis of four years of monitoring. *Bulletin of Marine Science*. **55**(2–3): 662–665.

Gorham, J.C. and W. S. Alevizon. 1989. Habitat complexity and the abundance of juvenile fishes residing on small scale artificial reefs. *Bulletin of Marine Science*. **44**(2): 662–665.

Harmelin, J.G. 1987. Structure et variabilité de l'ichthyofaune d'une zone rocheuse protégée en Méditerranée (Parc national de Port-Cros, France). *Marine Ecology*. **8**(3): 263–284.

Hückel, G.J., R.M. Buckley and B.L. Benson. 1989. Mitigating rocky habitat loss using artificial reefs. *Bulletin of Marine Science*. **44**(2): 913–922.

Hyslop, E.J. 1980. Stomach content analysis: a review of the methods and their applications. *Journal of Fish Biology*. **17**: 29–37.

Lebart, L. 1975. Validité des résultats en analyse des données. Centre de Recherche et de Docum. sur le Consom., Paris. L.L/cd. **4465**: 157 pp.

Levins, R. 1968. *Evolution in Changing Environments*. University Press, Princeton, New Jersey.

Macpherson, E. 1981. Resource partitioning in a Mediterranean demersal fish community. *Marine Ecology Progress Series*. **4**: 183–193.

Margalef, R. 1958. Information theory in ecology. *Gen. Syst*. **3**: 36–71.

Pepe, P., F. Badalamenti and G. D'Anna. 1996. Abitudini alimentari di *Diplodus vulgaris* sulle strutture artificiali del Golfo di Casellammare (Sicilia nord-occidentale). *Biologia Marina Mediterranea*. **3**(1): 514–515.

Pielou, E. C. 1966. The measurement of diversity in different types of biological collections. *Journal of Theoretical Biology*. **13**: 131–144.

Pielou, E. C. 1984. *The Interpretation of Ecological Data*. John Wiley and Sons, New York. 263 pp.

Pipitone, C., F. Badalamenti, G. D'Anna and B. Patti. 1996. Divieto di pesca a strascico nel Golfo di Castellammare (Scilia nord-occidentale): alcune considerazioni. *Biologia Marina Mediterranea*. **3**(1): 200–204.

Provenzano, G. and S. Riggio. 1982. Technologie impiegate per la realizzazione di barriere artificiali lungo la costa palermitana. Atti UU. OO. sottoprogetti CNR 'Risorse Biologiche' e 'Inquinamento', Roma, 10–11 Novembre 1981: 119–154.

Relini, G. and L. Orsi Relini. 1989. Artificial reefs in the Ligurian Sea (Northwestern Mediterranean Sea): aims and results. *Bulletin of Marine Science*. **44**(2): 743–751.

Relini, G., M. Relini and G. Torchia. 1989. Fishes of the Loano artificial reef (Western Ligurian Sea). FAO Fisheries Reports. **428**: 120–127.

Riggio, S. 1982. The artificial reefs off Terrasini (Northwestern Sicily) after one year of submergence. *Journée d'études sur les aspects scientifiques concernant les récifs artificiels et la mariculture suspendue*. CIESM, Cannes. 67–71.

Riggio, S. and G. Provenzano. 1982. Le prime barriere artificiali in Sicilia: ricerche e progettazioni. *Naturalista Siciliano* S. IV (Suppl.) **3**: 627–659.

Riggio, S., M. Gristina, G. Giaccone and F. Badalamenti. 1985. An eighteen months survey of the artificial Reef of Terrasini (N/W Sicily): the invertebrates. *Oebalia N.S.* **XI**: 427–437.

Riggio, S., F. Badalamenti, R. Chemello and M. Gristina. 1990. Zoobenthic Colonization of a small artificial reef in Southern Tyrrhenian: Results of a three-year Survey. FAO Fisheries Reports. **428**: 109–119.

Riggio, S., G. D'Anna and M.P. Sparla. 1992. Coastal eutrophication and settlement of mussel bed in N/W Sicily: remarks on their significance. In Colombo, G., I. Ferrari, V.U. Ceccherell and R. Rossi. (eds.) *Marine Eutrophication and Population Dynamics*. Proceedings of the 25th European Marine Biology Symposium: pp. 117–120.

Schoener, T. W. 1970. Nonsynchronous spatial overlap of lizards in patchy habitats. *Ecology*. **51**(3): 408–418.

Siegel, S. 1956. *Nonparametric Statistics for the Behavioral Sciences*. McGraw-Hill, New York. 312 pp.

Spanier, E., M. Tom, S. Pisanty and G. Almog-Shtayer. 1990. Artificial reefs in the low productive marine environments of the Southeastern Mediterranean. *Marine Ecology*. **11**(1): 61–75.

Toccaceli, M. and D. Levi. 1990. Preliminary data on an experimental trammel net survey designed to estimate the potential of a planned artificial reef near Mazara del Vallo (Italy). FAO Fisheries Reports. **428**: 154–162.

6. Artificial Reefs in North-West Sicily: Comparisons and Conclusions

GIOVANNI D'ANNA[1], FABIO BADALAMENTI[1] and SILVANO RIGGIO[2]

[1]Laboratorio di Biologia Marina, I.R.M.A.-CNR, via A Da Verrazzano, 17, Castellammare del Golfo (TP), Sicilia, Italy. [2]Dipartimento di Biologia Animale, Università di Palermo, via Archirafi, 18, I 90123 Palermo, Sicilia, Italy

Introduction

The most notable features of the biotic colonization and fishing yield of artificial reefs are a straightforward response to local environmental conditions. When applying this assumption to the artificial reefs of north-west Sicily, the features that distinguish the reefs from each other are largely consistent with the major hydrographic characteristics of the Gulf of Castellammare, the Bay of Carini and the Gulf of Palermo. The differences in environmental conditions in the three biotopes have made it possible to compare the colonization of artificial reefs in unpolluted oligotrophic water (Bay of Carini), eutrophic water (Gulf of Palermo) and water with heavy siltation rates (Alcamo Marina artificial reef area in the Gulf of Castellammare).

Environmental Features of the Gulf of Palermo and Bay of Carini

The morphology of these areas is similar to that of the Gulf of Castellammare (Badalamenti *et al.*, chapter 5, this volume) (Fig. 1), but the terrestrial geology and anthropic impacts are very different. The Gulf of Palermo and the Bay of Carini are much smaller areas of water but the major difference from the Gulf of Castellammare is the sandstone platform which stretches along the coast forming a base for the massive dolomitic mountain ranges lying behind the coastal strip. Siltation rates are low, water transparency high; the effects of eutrophication are localized. As a result, algal settlements and biotic encrustations thrive on both natural and man-made substrata.

Bay of Carini

The seabed slopes very gently in continuity with the coastal plain. The prevailing soft sediments are colonized by *Posidonia oceanica* seagrass meadow which breaks into a 'récif barrière' offshore (Toccaceli and Alessi, 1989; Toccaceli, 1990). Hard rocky reefs occur along both sides of the Bay. Pollution is negligible and the offshore waters are oligotrophic. The reef site is underlain by a mixed sand-rock sea floor, and carpeted with stands of *Posidonia oceanica*.

Overall, 18 artificial reef pyramids, each made from five (2 × 2 × 2 m) concrete blocks, have been placed in 20 m of water and several hundred anti-trawling

A.C. Jensen et al. (eds.), Artificial Reefs in European Seas, 97–112
© 2000 *Kluwer Academic Publishers. Printed in Great Britain.*

'torpedoes' (Badalamenti *et al.*, chapter 5, this volume; Riggio *et al.*, chapter 6, this volume) have been scattered at depths between 40 and 50 m.

Gulf of Palermo

Disposal into the Gulf of untreated sewage from over one million people greatly contributes to coastal eutrophication. The sewer outfalls are located in the mid-western corner of the Gulf, but water currents disperse the pollution. Plumes tend to drift eastwards in winter, whereas they tend to converge towards the centre of the Gulf during weather associated with high barometric pressure in summer.

   An additional perturbation to the Gulf is the huge amount of particulate matter, organic and inorganic debris that erodes from the shoreline dumping grounds and is spread throughout the Gulf by currents. Such materials combine and precipitate on the seabed. The resultant layer of pelitic mud carpets the seabed of the whole Gulf, smothering the *Posidonia/Cymodocea* seagrasses and transforming the once rugged rocky seabed into a uniform muddy plain (Riggio *et al.*, 1995). In calm weather the precipitation of suspended fine material provides good water clarity.

Artificial reefs in the Gulf of Palermo

The six pyramid reefs of the Vergine Maria ARA (Artificial Reef Area) are laid on a terrace of coarse sand and silt. Each reef was made from 14 concrete blocks ($2 \times 2 \times 2$ m) in 20 m of water. One of the town sewers discharges nearby and so turbidity and eutrophication are often high, cloaking the reef pyramids in clouds of mud, silt and seston that block out the light. The particulate materials settle out onto the hard reef surfaces but they never reach the thickness attained at the Alcamo Marina ARA (Gulf of Castellammere). In calm weather the waters are clear, and light penetrates to a depth of 15 m, so promoting algal growth. In addition to the artificial reefs, several hundred anti-trawling 'torpedos' were scattered at depths between 20 and 50 m.

Research Methods

The same research techniques and methods outlined in Riggio *et al.* (chapter 4, this volume) and Badalamenti *et al.* (chapter 5, this volume) were employed at all reef sites.

The Ecological Approach to the Epibiotic Settlement

Distinctive patterns of colonization characterized the more eastern reefs (Vergine Maria ARA in the Gulf of Palermo, Carini Bay ARA (Toccaceli *et al.*, 1996)), Terrasini ARA in the easternmost portion of the Gulf of Castellammare (Badalamenti *et al.*, 1985; Riggio *et al.*, 1985) and those in the middle of the Gulf of Castellammare (Trappeto ARA and Alcamo Marina ARA (Badalamenti and

D'Anna, 1995)) (Fig. 1). The succession pattern previously described (Badalamenti *et al.*, chapter 5, this volume) for the eastern group of ARAs was representative, being repeated with minor variations at all sites. It was assumed to be a typical example of biotic succession in oligotrophic waters.

More complex, variable and often unpredictable stages of succession were observed in the group of reefs deployed in the Gulf of Castellammare (Trappeto ARA; Alcamo Marina ARA). These variations were interpreted as responses to a fluctuating environment. Colonization on these groups of artificial reefs will be dealt with separately.

Epibiotic Colonization of the Eastern Reefs (Terrasini, Carini Bay and Vergine Maria)

Three major stages could be recognized in the sequence of colonization:

(1) A rapid development of small erect species and encrusting colonies.
(2) A rapid growth of red algae followed by its sudden disappearance in mid-Summer.
(3) Establishment of a permanent stand of brown algae overlying two lower seaweed layers.

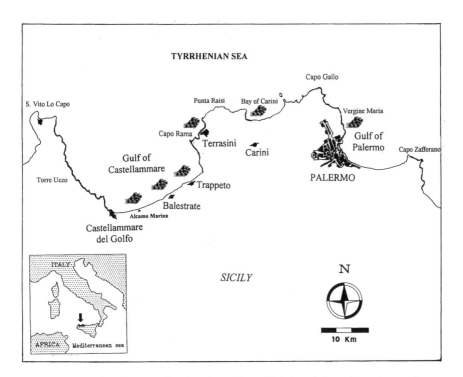

Figure 1. Map of study area showing the main artificial reef sites along the north-west Sicilian coast.

Stage 1

The very early biological settlement mainly comprised a thin covering of athecate hydroids, green algae and microbial colonies. These were followed by pioneer encrusting forms, most of which were the common genera encroaching on clear surfaces: the individual serpulid worms *Pomatoceros triqueter, Spirobranchus polytrema*, Spirorbinae (*Pileolaria militaris* and *P. pseudomilitaris*) and the colonial, round-shaped bryozoan *Cryptosula pallasiana*.

The duration of stage 1 was related to the season: it lasted longer on reefs deployed in autumn than those laid in summer.

Stage 2

This stage was characterized by the settlement and growth of the red alga *Lophocladia lallemandi*. This is a thermophilous seaweed which has emigrated from the eastern Mediterranean basin, originating in the Red Sea. It is a typical pioneer species, growing on shady sites, spreading quickly on rocky seabeds in the central and western Mediterranean. In a few days, heaps of 'powder-puff' shaped *Lophocladia lallemandi* bloomed, completely covering the walls of the reef cubes, filling the hollows and accumulating in the spaces between the reef blocks. As rapidly as it bloomed it then died, leaving space for further settlement. Thick mats of decaying algae then spread over the seabed forming a 'woolly' mucilage.

Stage 3

The initial phase of maturity was marked by the settlement and growth of brown algae of the genus *Cystoseira*. The first plants made their appearance on horizontal edges and soon spread onto all illuminated surfaces. Further growth resulted in a thick stand of diverse brown algae (*Padina pavonica, Dictyota dichotoma, Dictyopteris membranacea* and others) typically ascribed to the AP (photophilous algae) biocoenose (Bellan Santini, 1969; Pérès, 1982; Fig. 2). The most distinctive immigrants were species of the genus *Cystoseira*, with *C. compressa* dominant at all stations. Tough, bushy and photophilous elements made up the tallest layer, below which was an understorey of shade-tolerant red algae intermingled with animal colonies (*Bugula* spp.). These in turn were overgrowing a basal cover of creeping or encrusting photophobic taxa. This stage slowly evolved towards more advanced communities.

Stage 4

An expected final stage is the development of the *Cystoseira* association on the reefs. The exact character of this association will be influenced by the depth of a reef's horizontal surfaces (Badalamenti *et al.*, 1985; Giaccone and Bruni, 1972–3; Fig. 3). *Posidonia oceanica* colonization of the limestone gravel laid down prior to reef deployment will form meadows in the future (Riggio, 1995).

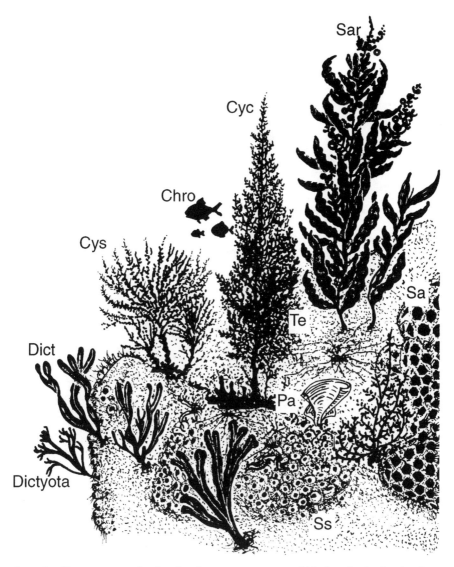

Figure 2. Upper corner of cube showing a mature stage of biotic colonization in clear waters (3 years). Chro = *Chromis chromis*; Cyc = *Cystoseira compressa*; Cys = *Cystoseira sauvageanana*; Dict = *Dictyopteris membranacea*; Dictyota = *Dictyota dichotoma*; Pa = *Padina pavonica*; SA = *Sabellaria alveolata*; Sar = *Sargassum* sp.; Ss = *Sabellaria spinulosa*; Te = *terebellids*.

Colonization of the Reef Interior

Succession inside the reef structure and in shaded niches was assessed by regular qualitative visual observation by scientific SCUBA divers. The reef interiors at all sites showed, within the first year after deployment, the development of significant

SUCCESSION IN CLEAR SICILIAN WATERS

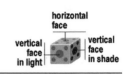

horizontal face
vertical face in light | vertical face in shade

1 month

- bacterial slime
- Hydroid colonies
- ciliate colonies

6 months

algal felt of photophilous brown algae with massive blooms of the red alga *Lophocladia lallemandii*

- sessile fauna dominated by *Cryptosula pallasiana*
- vagile fauna characterised by ubiquitous animal species

12 months

- brown algae
- *Lophocladia lallemandii*
- sessile fauna dominated by Serpulid worms
- vagile fauna dominated by Syllids and *Bittium* spp.

24 months

- settlements thriving on a sediment cover
- *Halopteris scoparia* and *Jania rubens* dominant

establishment of *Cystoseira compressa*

precorallicenous facies dominated by the *Udoteo-Peyssonnelietum* community

36 months

brown algae emerging from a sediment cover with dominance of *Cystoseira compressa* and *Sargassum* sp.

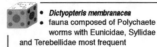

- *Dictyopteris membranacea*
- fauna composed of Polychaete worms with Eunicidae, Syllidae and Terebellidae most frequent

48 months

- photophilous algae dominated by *Cystoseira compressa* and *Sargassum hornschuchi*
- on an encrusted basal sub-stratum *Anomia* sp.

- *Cystoseira compressa*
- *Dictyopteris membranacea*
- *Peyssonnelia* sp. Ostreids - *Anomia* sp.

- Porifera and solitary Madreporaria

60 months

convergence towards a natural reef

Figure 3. Colonization stages of hard substrata in clear Sicilian waters.

colonies of encrusting filter feeders (bryozoans, hydroids and serpulids). Following this primary stage, Carini Bay reefs took 4 years and Terrasini ARA 5 years to develop a mature epibenthos community dominated by madreporarians and Demospongiae which replaced calcareous sponges 3–4 years after their first appearance.

Six years after placement the artificial reefs at Vergine Maria (Gulf of Palermo) and Alcamo Marina had not reached this stage of maturity. These two sites experienced the worst environmental conditions of all of the sites studied: the high rate of siltation led to the periodic smothering of horizontal surfaces by mud, slowing down the rate of successional colonization.

Full biological maturity of the reef's sedentary biota will be attained in the future, when an infauna is established in the blocks. This process, estimated to take about 20 years in clear waters, could, however, be enhanced by addition of calcareous pebbles to the concrete mixture, providing a suitable substratum for infauna (endolithic species).

Epibiotic Colonization of the Western Reefs

The biological colonization of the Trappeto ARA pyramids was slow, stunted and sparse because a thick layer of fine sediment was deposited on the horizontal surfaces of the reef cubes (Fig. 4). Colonization was restricted to isolated erect filter feeders protruding through the mud or growing inside the tunnels and hanging from the roofs of inter-reef cube hollows. *Sabella* (syn. *Spirographis spallanzanii*), terebellid worms and *Bugula* spp. were the most frequent colonizers, depicting a settlement typical of harbour fouling.

The colonization of the reefs off Alcamo Marina was more varied, differences being seen even between adjacent pyramids. The primary settlement was of either algae or filter feeders, but the former were not a constant feature and often died out suddenly, being rapidly replaced by a heterotrophic cover. Oyster beds were regularly established; erect or semi-erect colonies of filter feeders grew on most reef cubes. The patterns of colonization were different depending on the slope of the cube walls. There were marked differences between horizontal and vertical substrata. A noteworthy feature was the luxuriant stand of algae on partially collapsed pyramids, and particularly on lop-sided cubes (blocks that had partially subsided into the sediment).

The Fish and Other Mobile Fauna

The eastern and western reefs showed two distinct patterns of fish colonization. In the clear sites of Carini Bay and off Terrasini, as well as in the less clear waters off Vergine Marina, the commercially important (edible) species were qualitatively less abundant than in the turbid waters of the Gulf of Castellammare.

SUCCESSION IN TURBID SICILIAN WATERS

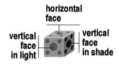

horizontal face

vertical face in light

vertical face in shade

1 month

- bacterial slime
- Vaucheriales
- *Zoothamnion colonies*
- Hydroid colonies

6 months

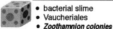

- Ubiquitous animal taxa
- serpulid colonisation
- *Bittium* sp.

12 months

- Sparse settlement of photophilous brown algae
- few filter feeders protruding from mud

- Filter feeder settling in crevices and holes sheltering from silting

24 months

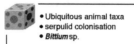

- Biotic settlement by taxa living in sandy-muddy sediment
- no more algae
- sea urchins very frequent

36 months

- Few taxa making up a typical fouling assemblage
- Large-sized filter feeders
- vagile fauna restricted to crevices

48 months

- Convergence towards impoverished biotic assemblages made up of taxa typical of non coherent substrata

Figure 4.   Colonization stages of hard substrata in turbid Sicilian waters.

The Eastern Reefs

The pomacentrid *Chromis chromis* was the first fish to appear, followed by labrids which aggregated between the reef cubes. Sparidae were especially frequent, represented by the white sea bream *Diplodus sargus*, the annular sea bream *Diplodus annularis* and the common two-banded sea bream, *D. vulgaris*. The grouper *Epinephelus marginatus* and the spiny lobster *Palinurus elephas* came at a later stage, the former as solitary young individuals, the latter as post-larvae. The last immigrant, observed more than two years after deployment, was the cardinal fish, *Apogon imberbis*. The tunnels and the 'ceilings' of spaces between reef cubes were a favoured place for squids and cuttlefish to lay their eggs.

The Western Reefs

Nektonic fauna was abundant from the first day of reef immersion, and included the scorpionfish (*Scorpaena porcus*), octopuses, the goatfish (*Mullus surmuletus*) and *Pagellus acarne*, some Serranidae, namely *Serranus cabrilla* and *S. scriba* and the Centracanthidae *Spicara maena*, *S. flexuosa* and *S. smaris*. Labrids were absent. Dense swarms of pelagic fish were recorded from around the pyramids off Alcamo Marina: among them were the greater amber jack (*Seriola dumerili*), the round sardinella (*Sardinella aurita*) and horse mackerel (*Trachurus trachurus*).

Fishing Yields at ARAs

As a rule, the fishing yields in proximity to artificial reefs have been at least as high, or higher, than those from unprotected fishing-grounds (D'Anna *et al.*, 1992). The highest catches have been recorded at Alcamo Marina ARA, the lowest at Terrasini ARA, as far as can be judged from these sparse data. Inside the reef zone at Trapetto ARA the fishing yields were 1.3 to 2 times those from nearby control areas (Arculeo *et al.*, 1990a, 1990b).

In only two cases have these positive results been less than conclusive: when the quality of the catch in terms of species caught (rather than its wet weight or biomass) has been critically considered, and when the survey has focused on oligotrophic areas.

Fishing Yields at Alcamo Marina ARA

The yields from the Alcamo Marina ARA were intermediate between those obtained from natural rocky substrata and the adjacent sandy seabed. Contrary to expectations, pelagic fish have accounted for most of the big catches recorded in the area. Sparidae are the second most frequent catch. Breams and some pelagic species have been observed feeding on and around the reefs. The trigger fish *Balistes carolinensis*, newly reported in the area after an absence of 20 years, appeared in shoals during the summer, staying for some days on the reefs, spending time feeding on the bryozoan colonies that carpeted the concrete reef walls. Only after the bryozoans had been completely grazed out did the schools of trigger fish leave the reef area.

Results obtained from a fishing survey carried out around the Alcamo Marina reef area showed a change in the fish community with distance from the reefs. Less motile species, such as the forkbeard (*Phycis phycis*), the two-banded sea bream (*Diplodus vulgaris*) and many others sampled around the reef cubes were missing from the neighbouring sandy substratum.

Peak yields from the fishing surveys came from the area within the artificial reef areas and from the natural rocky areas, but not from control sites on the sandy seabed. Such results from ARAs have been referred to as an example of an 'ecotone effect'; a confluence, or summation, of fish species occurring in the vicinity of a reef, each species benefiting in its own way from the food and/or shelter or just being attracted by the physical presence of the reef pyramids nearby.

Artificial Reefs and *Posidonia oceanica* Beds

An unexpected finding has been the frequent sprouting of *Posidonia oceanica* between the stones and pebbles making up the base of the reef site at Vergine Maria (Fig. 5). This has also been recorded at the Terrasini reef, where the seagrass has grown luxuriantly in previously uncolonized sandy seabeds, starting from inside the inner hollows of the sunken concrete frames (pot shaped structures) dispersed around both pyramids. Some shoots have been seen emerging from beneath the reef cubes. The *Posidonia* shoots are the end portions of rhizomes that have grown over exposed rock and have been detached from the parent plant by the spontaneous breaking of a weak segment in the middle of the rhizome. These propagula are free to drift in bottom currents, or may be dispersed by wave action during storm events. Once pushed into a cleft made by adjacent reef units or between rocks and/or pebbles, the rhizome twists into a fish-hook shape to anchor itself into the substratum. Small roots are produced, the plant becomes established *in situ* and develops new shoots (Fig. 6). This strategy of dispersion helps the phanerogame to colonize new rocky seabed areas and to spread from soft, unstable substrata. In this way artificial reefs may be a useful, although indirect, means of restoring endangered seagrass beds and promoting the spread of *Posidonia* into new areas (Fig. 7).

Comparison with Artificial Reefs in Other Mediterranean Sites

Sicilian artificial reefs show no affinities at all with the reefs laid in the western Adriatic (Bombace, 1989; Bombace *et al.*, 1994) and mid-Tyrrhenian (Ardizzone *et al.*, 1982, 1989) seas, probably because of the high siltation rates and nutrient loads in these localities.

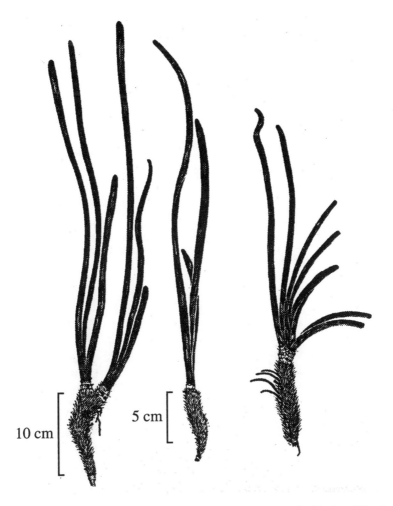

Figure 5. Woody *Posidonia oceanica* propagules implanted within the blocks of Vergine Maria artificial reef.

Artificial reefs in the Ligurian Sea (Balduzzi *et al.*, 1982; Relini, 1982; Relini *et al.*, 1995; Relini and Orsi Relini, 1989) and off the Provençal coasts (Charbonnel, 1990; Duval *et al.*, 1982) show similarities to Sicilian reefs. The resemblance seen in the benthic assemblages and fish fauna (also shared by biotic assemblages on natural sites) are consistent with the environmental characteristics of the coastal morphology in all three areas, and stresses the role played by the hydrography and substrata in larval recruitment and development of benthic communities.

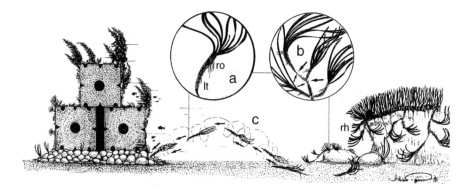

Figure 6. Spread of *Posidonia oceanica* and implantation in the artificial reefs. (a) free-living rhyzomatous propagulum (*lt* = lignified; *ro* = rootlets). (b) end portions of rhizomes ready for attachment (arrows point out tightening of rhizomes where breaks will occur). On the left, mechanism of colonization in reef base; right 'matte' growing on rocky base (*rh* = rhizome protruding from the 'matte').

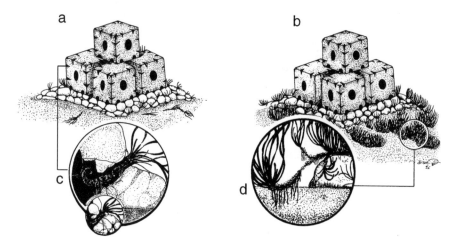

Figure 7. Model of colonization by *Posidonia oceanica* on artificial reefs. (a) early settlement; (b) advanced stage of colonization; (c) detail of the rhizome anchoring mechanism on the artificial reef's rocky base; (d) detail of sandy seabed colonization.

Significance of Artificial Reefs in Relation to Natural Reefs

Artificial reefs reproduce the physical features of a natural rocky system. This artificial system can be designed and placed to optimize biological colonization for man's benefit. Artificial reefs can therefore be considered as 'soft engineering or technology' to be used as a tool of marine environmental management, which has, according to design, a variety of outcomes.

Biomass Yields

From a Sicilian viewpoint the overall exploitable biomass of a reef is the sum of nekton production and fish attraction, and in turbid waters the latter effect has in many cases prevailed, masking the former.

The higher biomass values recorded from most turbid areas are therefore not surprising, as turbidity is often related to eutrophication (although in Sicily suspended fine sediment plays a significant role) and to a relatively higher primary productivity than in oligotrophic waters. This is of most benefit to animals at the end of a short food web, especially filter feeders. In the Adriatic and mid-Tyrrhenian seas, beds of mussels take advantage of the plentiful food and, in their turn, are a major component in the diets of fish and crustaceans (see Bombace *et al.*, chapter 3, this volume; Ardizzone *et al.*, chapter 7, this volume).

The least efficient artificial reef installations, in terms of exploitable biomass production, are seemingly those in clear waters, where the algal stands on the surfaces of the reef cubes form a dense biotic cover.

As a rule, low exploitable biomasses are counterbalanced by higher biodiversity values, a constant factor in Sicilian waters where more diverse fish communities are related to seabeds which provide low fishery yield. The small catch quantity of exploitable fish and crustaceans (groupers, brown meagres, breams and spiny lobsters) was compensated for by the high commercial quality (and price fetched) of the catch.

Significance and Role of the Biotic Colonization

Water clarity appears to be the prime factor influencing the settlement and succession of Sicilian benthic communities. Related factors are the amount of suspended sediment, the siltation rate and water movement. Two contrasting conditions can be envisaged (Fig. 8).

(1) An artificial reef placed in clear waters with minimum siltation: this will develop an autotrophic community, leading to thick stands of brown algae, together with the initial stages of *Posidonia* colonization. A diverse fauna will utilize the shelter and food provided by the seaweeds. Encrusting communities will develop in the sheltered galleries and crevices inside the reef structure. Biological succession will proceed and community maturity be reached 4–5 years after deployment. Maximum diversity values will be consistent with relatively low biomass levels. The fish will be mainly carnivorous species. The role of the artificial reef is to provide a solid substratum to support the settlement of epibiota.

(2) A reef placed in an eutrophic area with high siltation will have minimal biotic colonization on the outer reef surfaces; only some species of filter feeders will successively settle and survive. Encrusting fauna will be rather sparse. Fish biomass will increase, partly due to attraction of pelagic species. Most demersal fish will be omnivorous, the population mainly composed of goatfish and mullet. The role of the artificial reef is that of a refuge and a fish attraction device (FAD).

Figure 8.   Synopsis of benthic settlement in relation to an artificial reef in Sicilian waters. Turbid waters = left hand side, transparent waters = right hand side. Names of organisms populating the reef: 1 *Seriola dumerili;* 2 *Ostrea edulis;* 3 *Baslisties carolinensis;* 4 and 6 *Padina pavonica;* 5 and 8 *Sargassum* sp.; 7 and 10 *Oblada melanura;* 9 *Cystoseira* sp.; 11, 33 and 48 *Chromis chromis;* 12 and 13 *Loligo vulgaris;* 14 squid eggs; 15 *Paracentrotus lividus;* 16 *Thalassoma pavo;* 17 *Coris julis;* 18 *Sarpa salpa;* 19 gobids; 20 *Scorpaena* sp.; 21 and 41 madreporariens; 22 *Palinurus elephas;* 23 *Mullus barbatus;* 24 *Spondyliosoma cantharus;* 25 and 26 *Spicara maena;* 27 *Halocynthia papillosa;* 28 *Apogon imberbis;* 29 *Diplodus annularis;* 30 *Sphaerochinus granularis;* 31 *Scizobrachella sanguinea;* 32 *Caberea boryi;* 34 and 36 labrids; 35 mysids; 37 *Udotea* sp.; 38 *Sabella spallanzani;* 39 *Diplodus vulgaris;* 40 *Diplodus sargus;* 42 *Conger conger;* 43 *Sepiola* sp.; 44 *Sepia officinalis;* 45 *Epinephelus marginatus;* 46 *Sciaena umbra;* 47 and 51 *Posidonia oceanica;* 49 *Halopteris* sp.; 50 *Octopus vulgaris.*

All the intermediate conditions between these two extremes may be encountered. Predicting the precise biological outcome of a reef is not possible in many cases because slight deviations from the mean ecological conditions can trigger significant changes.

Acknowledgements

Thanks are due to Drs Michele Gristina and Marco Toccaceli for their friendly, long-lasting collaboration with our research plan, as well as for the valuable data and suggestions provided on the artificial reefs in the Gulf of Palermo.

References

Arculeo, M., G. D'Anna, C. Pipitone, S. Riggio and M.P. Sparla. 1990a. Note sull' efficienza della barriera artificiale di Alcamo Marina (Sicilia). *Oebalia* N.S. XVI 2: (Suppl.) 563–566.

Arculeo, M., G. Bombace, G. D'Anna and S. Riggio. 1990b. Evaluation of fishing yields in a protected and an unprotected coastal area of N/W Sicily. FAO Fisheries Reports. **428**: 70–83.

Ardizzone, G.D., C. Chimenz and A. Belluscio. 1982. Benthic Communities on the artificial reef of Fregene (Latium). *Journée d'études sur les aspects scientifiques concernant les récifs artificiels et la mariculture suspendue*, C.I.E.S.M. (Cannes): 55–57.

Ardizzone, G.D., C. Chimenz and A. Belluscio. 1989. The development of epibenthic communities on artificial reefs in the central Mediterranean Sea. *Bulletin of Marine Science*. **44**(2): 592–608.

Badalamenti, F. and G. D'Anna. 1995. Esperienze di barriere artificiali nel Golfo di Castellammare (Sicilia nord-occidentale). *Biologia Marina Mediterranea*. 2: 165–173.

Badalamenti, F., G. Giaccone, M. Gristina and S. Riggio. 1985. An eighteen month survey of the artificial reef off Terrasini (Sicily): the algal settlements. *Oebalia* N.S. **XI**: 417–425.

Balduzzi, A., S. Belloni, F. Boero, R. Cattaneo, M. Pansini and R. Pronzato. 1982. Prime osservazioni sulle barriere artificiali della riserva sottomarina di Monaco. *Naturalista Siciliano*. 3: 601–605.

Bellan Santini, D. 1969. Contribution à l'étude des peuplements infralittoraux sur substrat rocheux. *Rec. Trav. Stat. Mar. Endoume*. **63**(47): 295 pp.

Bombace, G. 1989. Artificial reefs in the Mediterranean Sea. *Bulletin of Marine Science*. **44**(2): 1023–1032.

Bombace G, G. Fabi, L. Fiorentini and S. Speranza. 1994. Analysis of the efficacy of artificial reefs located in five different areas of the Adriatic Sea. *Bulletin of Marine Science*. **55**: 559–580.

Charbonnel, E. 1990. Les peuplements ichthyologiques des récifs artificiels dans le département des Alpes-Maritimes (France). *Bulletin Société Zoologie Français*. **115**(1): 123–136.

D'Anna, G., F. Badalamenti, C. Pipitone and S. Riggio. 1992. Notes on the ecological significance of the fish fauna associated to artificial reefs in the southern Tyrrhenian. *Rapport de la Commission International de la Mer Méditerranée*. XXXIII CIESM. **33**: 378.

Duval, C., D. Bellan-Santini and J.G. Harmelen. 1982. Habitats artificiels immergés en Méditerranée nord-occidental: mise au point d'un module cavitaire expérimental. *Téthys*. **10**(3): 274–279.

Giaccone, G. and A. Bruni. 1972–73. Le Cistoseire e la vegetazione somersa del Mediterraneo. Atti dell'Istituto Veneto di Scienze, Lettere ed Arti. Tomo CXXXI. Classe di Scienze Matematiche e Naturali. pp. 59–103.

Pérès, S.J.M. 1982. Major benthic assemblages. In Kinne, A. (ed.) *Marine Ecology*. **5**(1): 373–521.

Relini, G. 1982. Le barriere artificiali nel Golfo Marconi (Mar Ligure). *Naturalista Siciliano*. 3: 593–599.

Relini, G. and L. Orsi Relini. 1989. Artificial reefs in the Ligurian Sea (N/W Mediterranean): aims and results. *Bulletin of Marine Science*. **44**(2): 743–751.

Relini, G., M. Relini and G. Torchia. 1995. La Barriera artificiale di Loano. *Biologia Marina Mediterranea*. **2**(1): 21–64.

Riggio, S. 1995. Le barriere artificiali e l'uso conservativo della fascia costiera: risultati dei 'reefs' nella Sicilia N/O. *Biologia Marina Mediterranea*. **2**(1): 129–164.

Riggio, S., M. Gristina, G. Giaccone and F. Badalamenti. 1985. An eighteen months survey of the artificial reef of Terrasini (N/W Sicily): the invertebrates. *Oebalia* N.S. **XI**: 427–437.

Riggio, S., S. Calvo, C. Fradà Orestano, R. Chemello and M. Arculeo. 1995. La dégradation du milieu dans le Golfe de Palerme (Sicile Nord-Ouest) et les perspectives d'assainissement. Villes des rivages et environnement littoral en Mediterranée, Montpellier. 82–89.

Toccaceli, M. 1990. Il recif-barriere di *Posidonia oceanica* (L.) Delile della baia di Carini (Sicilia occidentale). *Oebalia*. **XVI-2** (suppl.): 781–784.

Toccaceli, M. and M.C. Alessi. 1989. Cartografia biocenotica della praterie a fanerogame marine della baia di Carini (Sicilia occidentale). *Oebalia* N.S. **XV-1**: 341–344.

Toccaceli, M., M. Gristina and R. Parisi. 1996. Primi dati sugli insediamenti bentonici di due barriere artificiali sui fondali costieri della Provincia di Palermo, Sicilia Nord-Occidentale. *Biologia Marina Mediterranea*. **3**(1): 493–495.

7. Prediction of Benthic and Fish Colonization on the Fregene and other Mediterranean Artificial Reefs

GIANDOMENICO ARDIZZONE[1], ALESSANDRA SOMASCHINI[2] and
ANDREY BELLUSCIO[1]

[1]Dipartimento di Biologia Animale e dell'Uomo, Viale dell'Università, 32, 00185 Roma, Italy
[2]Museo Civico di Zoologia, Viale del Giardino Zoologico, 20, 00197 Roma, Italy

Introduction

In order to improve fishery yield, a general assumption of artificial reef research is that when reefs are placed on soft, sedimentary seabeds the new, available hard substrata tends to attract fish from neighbouring areas and/or to increase the area colonized by epifaunal species (Seaman and Sprague, 1991).

One of the main problems in planning artificial reef deployment is predicting equilibrium steady states in benthic and nektonic communities from different geographic areas. Both trophic conditions and biogeographic features are the main factors in controlling community structure and evolution (Bombace, 1989; Spanier, 1995). In general, the Mediterranean Sea is oligotrophic, except for some coastal areas that have been influenced by discharges and runoff from the terrestrial environment which causes local eutrophication (Ryan, 1966).

In Italian coastal seas artificial reefs have been deployed in both oligotrophic and eutrophic conditions. In the former the aims of reef deployment were to physically protect the sensitive *Posidonia* (seagrass) biotope from illegal trawling (Ardizzone and Migliuolo, 1982; Ardizzone and Pelusi, 1984; Guillèn *et al.*, 1994; Relini *et al.*, 1995a, 1995b), and increase habitat complexity with a hypothesized improvement in biomass and biodiversity. Artificial reefs in eutrophic waters were developed to utilize the excess nutrients in the water column to increase fishery yield (Bombace, 1989). The deployment of artificial reefs led to the development of a new, hard-substratum biological community on the reef surface. The biomass provided a food source for commercially important species which resulted in an increased fishery yield. This yield was dependent on the presence of the reef and its food source (Bombace, 1986, 1989; Relini and Orsi Relini, 1989; Riggio, 1989).

About 25 artificial reefs have been deployed and described in the scientific literature along the European coastline of the Mediterranean Sea, whilst in the Levant basin (SE Mediterranean Sea) only one reef is documented off the coast of Israel (Spanier, 1995; Spanier *et al.*, 1989, 1990). Along the Northern Mediterranean coastline, artificial reefs have been deployed in oligotrophic waters within or below the range of *Posidonia oceanica* beds (Monaco, in the Larvotto Marine Reserve; Spain, in the Tabarca Marine Reserve and Balearic Islands; France, in Marseille Gulf, Juan Gulf and Beaulieu sur Mer) (AMPN, 1982; Balduzzi *et al.*, 1987; Bayle-Sempere *et al.*, 1994; Bellan, 1982; Bellan and Bellan

A.C. Jensen et al. (eds.), Artificial Reefs in European Seas, 113–128
© *2000 Kluwer Academic Publishers. Printed in Great Britain.*

Santini, 1991; De Bernardi, 1989; Duval, 1982; Duval and Cantera, 1982; Gòmez-Buckley and Haroun, 1994; Lefevre *et al.*, 1986; Moreno *et al.*, 1994; Sànchez-Jerez and Ramos-Esplà, 1995). Around the Italian coast, reefs have been deployed in a variety of situations providing a gradient from oligotrophic waters in the Southern Tyrrhenian Sea to the eutrophic conditions of the central Adriatic Sea (Ardizzone and Bombace, 1982; Bombace, 1989; Relini and Orsi Relini, 1989; Riggio, 1989). Ten years of results from the Fregene artificial reef in the central Tyrrhenian Sea allows conclusions and comparisons to be drawn with other Italian artificial reefs deployed in differing water conditions.

Study Area

The Fregene artificial reef is located in the central Tyrrhenian Sea (Latium) 9 km north of the mouths of the Tiber River and 2.7 km offshore (Fig. 1). The water has a high organic content, carried into the sea by the River Tiber that brings waste from Rome (Blundo *et al.*, 1978; Pagnotta *et al.*, 1992). These conditions result in eutrophication around the reef that is quite different from the typically oligotrophic waters of the Tyrrhenian sea. Values of chlorophyll a range between 5 and 30 mg m^{-3} and mean transparency is 4.04 ± 2.26 m, whereas oligotrophic water is characterized by a value of chlorophyll a < 5 mg m^{-3} (Ardizzone and Giardini, 1982).

The artificial reef was deployed in 10–14 m of water on a silty-sand seabed in March 1981. It is composed of 280 concrete cube-shaped blocks (each having 2 m sides) arranged in groups of five (four at the bottom and one on the top) to form a pyramid, and covering an area of 6 ha. The blocks contain hollows and cavities moulded into the shape in order to increase the surface/volume ratio (Bombace, 1977).

Materials and Methods

Samples of epibiota were collected by SCUBA divers between May 1981 and October 1985 and from January 1991 to September 1992. Samples of epibenthos from two standard areas of 400 cm^2 each were removed from a vertical reef surface at various time intervals (Ardizzone *et al.*, 1989, 1994). The collected material was fixed and preserved in 10% buffered formalin. After sorting, macrobenthic species were identified and specimens of *Mytilus galloprovincialis*, which proved to be the most abundant species at least initially, were measured (shell length) and their biomass was estimated as dripped wet weight (Ardizzone *et al.*, 1996). The population dynamics of this species were studied by means of length–frequency distribution analysis. The growth curve was calculated using the Von Bertalanffy equation whereas the temporal trend of the linearized catch curve was used to define the population stability rate (Ardizzone *et al.*, 1996).

Polychaetes were consistently the most diverse macrobenthic taxon and so they were studied in detail. Description of polychaetes by 'trophic status' allowed an

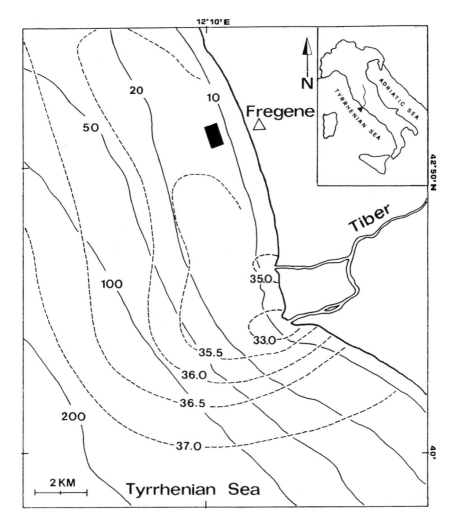

Figure 1. Map showing the study area and location of the Fregene artificial reef (black rectangle). Solid lines are isobaths in meters and dashed lines indicate mean surface salinity in parts per thousand.

understanding of the temporal trend of the community as a whole (Gravina *et al.*, 1989; Somaschini *et al.* 1997).

Qualitative observations of the fish assemblage were undertaken from 1981 up to 1985. After a kind of benthic community equilibrium was observed in 1991–1992, seasonal sampling of fish together with visual censuses were carried out to quantify the presence of fish species during the benthic equilibrium. Stomach contents were examined and prey species identified in order to reveal the preferred feeding sites of the sample fish species (Ardizzone *et al.*, 1997).

Results

General colonization pattern over 11 years

Over the 11-year period macroalgae were not observed on the reef, although both macrobenthic and fish communities showed profound changes (Ardizzone *et al.*, 1989, 1994, 1996, 1997; Gravina *et al.*, 1989; Somaschini *et al.*, 1997; Fig. 2).

Pioneer settlement period (1981)

The reef was first colonized by diatoms, hydroids (*Obelia dichotoma, Bouganvillia ramosa*), serpulid polychaetes (*Pomatoceros triqueter, Pomatoceros lamarckii, Hydroides pseudouncinata*), barnacles (*Balanus eburneus, Balanus perforatus*), and molluscs (*Anomia ephippium, Mytilus galloprovincialis*). Numbers of planktonic mussel larvae peaked in spring, the first plantigrades settling on the reef in June 1981, 2 months after the hydroids. The mussels attached to the filamentous hydroids. Thereafter, plantigrades moved to the hard surface by means of their active foot (secondary settlement). They first colonized spaces around holes, cavities and corners more exposed to water movement; only later did they colonize remaining surfaces (Ardizzone and Chimenz, 1982).

The first fish species to concentrate around the reef, only a few days after deployment, were juveniles of *Pagellus bogaraveo* and some specimens belonging to genera *Spicara* and *Trachurus*.

Mussel dominance phase (1982–1983)

During this period mussels dominated the benthic community, reaching a density of 13 525 individuals m^{-2} (approx. 120 kg wet weight m^{-2}) and influencing the settlement of other epibenthic species. Mussel valves together with barnacles and oysters altered the topography of the substrata, increasing its heterogeneity. Epibenthic species settled on this new, available biogenic substrata, and in turn new plantigrades settled on epibiotic hydroids so increasing community complexity. Sediment particles and 'mussel mud' progressively accumulated among valves and byssus filaments, which allowed the settlement of new typically 'soft seabed' species such as *Papillicardium papillosum, Pagurus prideauxi, Macropodia longirostris* and *Sabella penicillus*.

The fish assemblage also increased with new species being observed, such as *Parablennius rouxi, Diplodus annularis, Dicentrarchus labrax* and, particularly, the mussel-eating species *Balistes carolinensis*, and *Sparus aurata. Octopus vulgaris* was also recorded.

Regressive period (1983–1985)

Both over-sedimentation and 'mussel mud' production caused a decrease in the availability of hard substrata for epibenthic species to colonize. Mussel larvae did

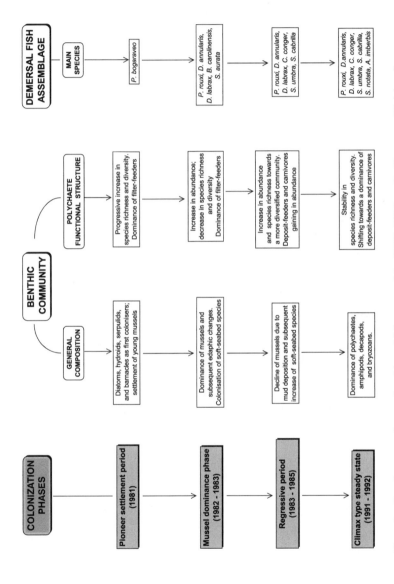

Figure 2. Development of macro-benthic and fish communities (1981–1992) on the Fregene artificial reef.

not find suitable surfaces for settlement and a gradual reduction of mussel density was observed, leading finally to their disappearance from the reef. The abundance of mussels changed from 13 525 individuals m^{-2} recorded in previous periods to about 3000 individuals m^{-2} in October 1985. In spite of this, the number of species increased due to an expansion in soft seabed species.

With the disappearance of mussels the mussel eating fish became rare, while *Conger conger, Sciaena umbra* and *Scyllarus arctus* increased in abundance. *Gaidropsarus mediterraneus, Parablennius gattorugine, Scorpaena notata,* and *Apogon imberbis* also became part of the fish assemblage.

Climax type steady state (1991–1992)

A new biological equilibrium, characterized by the numerical dominance of deposit feeders and carnivores, established on the hard and very muddy artificial reef substratum 11 years after deployment. Mussels, barnacles and hydrozoans previously recorded as the most abundant taxa had largely disappeared, dramatically changing the benthic community. Polychaetes (47 species and 910 individuals m^{-2}), crustaceans (15 species and 28 750 individuals m^{-2}), were dominant together with bryozoans. This last taxon was represented by *Schizoporella errata* and *Scrupocellaria reptans,* the only sessile organisms extensively distributed on the reef. Only a few previously recorded typically epifaunal species were still present, such as *Ostrea edulis, Hydroides pseudouncinata, Pomatoceros triqueter* and *Serpula concharum.* Some other organisms such as the molluscs *Striarca lactea* and *Nassarius incrassatus,* the polychaete *Polydora ciliata,* and the amphipods *Corophium sextonae, Corophium acutum* and *Stenothoe eduardi,* were previously rare but had become fairly common.

During this last steady-state phase, species such as *Apogon imberbis, Scorpaena notata, Serranus cabrilla* and *Parablennius rouxi* were found in all censuses, so they were considered as resident species on the reef.

The feeding habits of 26 fish species recorded around the reef were examined at this time, but none of them had benthic reef species in their stomach contents, suggesting that the preferred feeding sites were the soft seabed around the reef and not the reef itself (Ardizzone *et al.*, 1997).

Mytilus galloprovincialis *population dynamics*

Only after the settlement of some pioneer species did *Mytilus galloprovincialis* start to colonize the reef and become numerically important. Peaks in biomass (120 kg m^{-2}) and abundance (13 525 individuals m^{-2}) were attained during the second and third year after deployment but began to decrease afterwards. The length frequency distributions of mussels recorded in some of the sampling months are shown in Fig. 3. Only one cohort was present during the year 1981, after recruitment in spring. The recruitment of the second cohort occurred in spring 1982 while the third cohort settled in 1983 giving a more complex population structure than previously seen. These three cohorts partially overlapped each other until 1985, but no mussels have been recorded since June 1986.

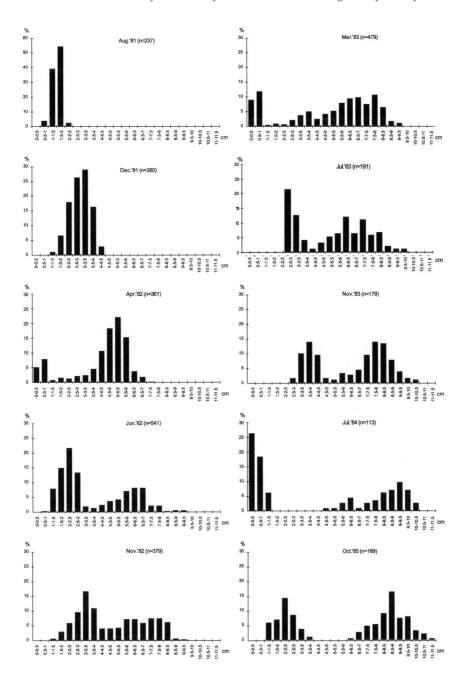

Figure 3. Length–frequency distribution of mussel *M. galloprovincialis* on the Fregene artificial reef.

The abundant food supply allowed rapid mussel growth, individuals attaining a shell length of 30 mm 5 months after settlement and 70 mm in the second year. Thereafter, the growth rate decreased and mussels attained a mean length of 80 mm during the third year. Parameters of the Von Bertalanffy curve were: $L\infty = 111.65$; $K = 0.68$; $T_0 = 0.75$, that is the highest $L\infty$ reported for the Mediterranean Sea (Ardizzone et al., 1996).

The analysis of the temporal trend of the total mortality Z which in this context is equal to natural mortality M (due to the absence of fishing mortality F) revealed that the main cause responsible for the rarefaction and extinction of the mussel population was a progressive reduction in the abundance of younger cohort compared to the older ones. The reduction of recruitment led to the ageing and, finally, the disappearance of the population. Recruitment was affected by an excessive silt and 'mussel mud' deposition which was responsible for both a reduction of hydroid cover (needed for first settlement), and a progressive decrease in the amount of hard substratum necessary for secondary settlement.

Polychaete community history

Polychaetes were one of the most important macrobenthic groups throughout the study period both in abundance and in species richness (81 species and 5970 individuals). Consequently, their temporal development was studied in detail (Gravina et al., 1989; Somaschini et al., 1997; Figs. 4, 5).

During 11 years of study, four periods of development in the benthic community can be identified, each divided from the next by changes in ecological parameters, such as number of species and individuals, frequency of rare species and diversity index (H) (Somaschini et al., 1997). During the first period (from June 1981 to January 1982), the number of species increased more than the number of individuals; many species settled on the new substratum and were present with a very low number of individuals. At that time filter feeders such as serpulids were the most important trophic group and burrowers were represented by the opportunist *Capitella capitata* (Gravina et al., 1989). During the second period (January 1982 to July 1983) the number of species was steady and the number of individuals increased; competitive interactions became more important so that fewer competitive species were gradually displaced by a few species which became very abundant (*Ceratonereis costae, Harmothoe extenuata, Adyte pellucida, Sabellaria spinulosa*). Although filter feeders were still important, carnivores and surface deposit feeders became more abundant. During the third period (July 1983 to October 1985) the number of polychaete species and individuals initially increased but during the last sampling month numbers decreased. The initial increase was caused by edaphic changes consisting of an increase in substratum complexity which lead to a progressive diversification of trophic pattern. Deposit feeders (*Polydora ciliata, Cirratulus chrysoderma, Aphelochaeta marioni, Terebella lapidaria*) and carnivores (*Lysidice ninetta, Syllis truncata cryptica*) became more abundant to the detriment of filter feeders. A disturbance event identified as the disappearance of the mussel bed is thought to be responsible for the impoverishment of the poly-chaete community observed in June 1985. Nevertheless, a new equilibrium seems

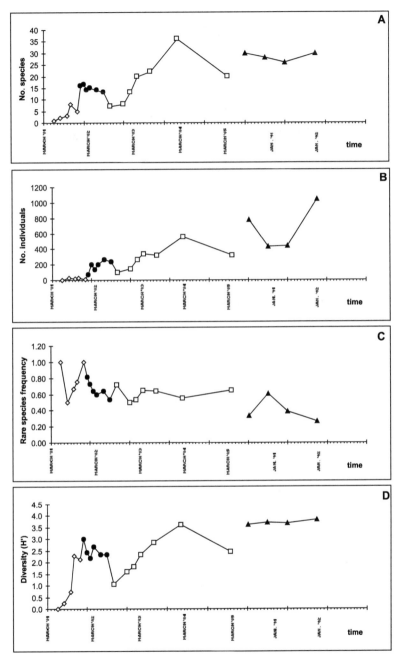

Figure 4. Temporal trend in the number of polychaete species (A), individuals (B), frequency of rare species (C), and diversity (D) on the Fregene artificial reef. The four main temporal changes are identified by different symbols. ◊: First period (from June 1981 to January 1982); ●: Second period (from February 1982 to September 1982); □: Third period (from November 1982 to October 1985); ▲: Fourth period (from January 1991 to September 1992).

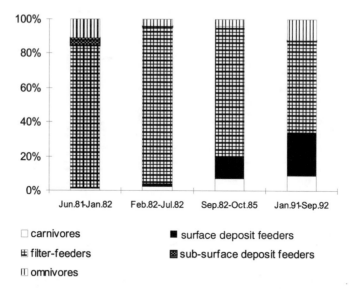

Figure 5. Trophic composition of the polychaete community in four main periods identified over 11 years.

to have been reached in the years 1991–1992, as shown by the stability of species number and diversity. The number of deposit feeders had increased from October 1985, but carnivores and omnivores were present in a greater abundance. A similar shift from a dominance of filter feeders to one of carnivores and deposit feeders due to sediment accumulation was also observed on horizontal artificial substrata in Northern Sicily (Riggio *et al.*, 1989).

Discussion and Conclusions

Forecasting benthic community development on artificial substrata has been discussed for a long time and opinions differ (Sarà, 1987). Some authors describe the successional sequence towards a climax as a predictable event (Clements, 1916), whereas for some, other stochastic larval recruitment is a critical factor which influences the process of colonization towards one of many possible multiple stable points of equilibrium (Sutherland, 1981).

Although many Mediterranean studies differ in the aim and context of the experiences, some processes of regular succession on artificial substrata have been observed in oligotrophic and eutrophic waters respectively.

Despite the difficulty in finding data describing temporal colonization of benthic communities and in comparing results using various taxa as descriptors of a complex situation, some generalization can be achieved. Oligotrophic conditions characterize artificial reefs studied in the Ligurian Sea (Loano (see Relini, chapter 8, this volume) and Marconi artificial reefs), and in Northern Sicily (some experiments carried out in the Gulf of Castellammare) (Badalamenti *et al.*, 1985, chapter 5, this volume; Relini *et al.*, 1986, 1994; Riggio *et al.*, 1985, 1989). Outside of Italy analogous trends have been observed along the French and Monaco coast (Balduzzi *et al.*, 1987; Duval and Cantera, 1982). In oligotrophic waters the gradient of light penetration and consequently water depth is responsible for changes in community structure and evolution. It is believed that, even if hydrozoans, serpulids, and bryozoans (*Schizoporella errata*) were important pioneer species, algae would be the dominant group thereafter. In shallow, well-lit waters ecological succession tends to an equilibrium dominated by the alliance of the order *Cystoseiretalia* which includes photophilic algal associations of upper and lower infralittoral fringes exposed to water movement (Badalamenti *et al.*, 1985; Petrocelli and Saracino, 1992; Relini *et al.*, 1994; Riggio *et al.*, 1989). Time required to reach an equilibrium, as in the case of the *Cystoseiretum spinosae* or *Cystoseiretum sauvageauanae* climatic association, depends on various edaphic factors such as substratum orientation, border effect, abrasion by drifting sand, and sediment deposition (Badalamenti *et al.*, 1985; Falace and Bressan, 1990; 1995). On vertical surfaces exposed to water movement three main successional steps have been described: first is the development of an encrusting blanket consisting of algae and micro-organisms, second is the growth of small-sized algae and branching colonies of filter feeders, and, finally, the settlement of large-sized algae (*Cystoseira* and *Sargassum*) with the establishment of a persistent long-iving algal cover (Riggio and Ardizzone, 1979). This equilibrium can be prevented if a biological disturbance such as the grazing activity of sea urchins occurs. On less exposed shaded parts of the reef and at greater depth, algae are less important as light penetration is reduced. In this situation zoobenthos is characterized by many colonial encrusting taxa such as bryozoans and sponges as well as by serpulids, spirorbids and barnacles (Relini *et al.*, 1994). Mussels are always absent but oysters can be an important component of the fauna. The development of sciaphilic 'precoralligenous' and 'coralligenous' assemblages can be expected (Relini *et al.*, 1994; Riggio *et al.*, 1989).

As far as fish colonization in oligotrophic waters is concerned, all Mediterranean studies show the dominant species to be *Chromis chromis* (D'Anna *et al.*, 1994, 1995; Relini *et al.*, 1995). Furthermore, a group of species, such as *Apogon imberbis, Diplodus sargus, D. vulgaris, D. annularis* and *Serranus cabrilla*, is very common on Italian artificial reefs in oligotrophic waters. *D. annularis* is always linked to artificial reefs in both oligotrophic and eutrophic conditions. Even if this group of species is considered to be 'typical', a set of more than 50 nektonic species has been recorded in oligotrophic waters.

Eutrophic conditions, reported for the Fregene artificial reef and the Adriatic Sea, lead to a rapid and abrupt invasion of filter feeders, such as mussels (*Mytilus galloprovincialis*) and oysters (*Ostrea edulis* and *Crassostrea gigas*) (Ardizzone and Chimenz, 1982; Bombace *et al.*, 1995). Over about 11 years, the Fregene artificial reef showed a series of biotic changes through time, reaching a kind of biological equilibrium. During this time mussels settled, dominated the community but finally disappeared after 5 years. The final equilibrium state is characterized by the absence of mussels, due to the progressive deposition of mud. Alteration of the substrata prevented the first and second settlement of young mussels so that the biological community progressively evolved towards an equilibrium mainly characterized by deposit feeders and carnivores, and with a greater biological diversity than that seen when mussels dominated.

Pioneer species on artificial substrata in eutrophic waters were hydroids, bryozoans and serpulid polychaetes. After a few months of submersion, mussels became numerically dominant; oysters were generally abundant on surfaces near the bottom in lower densities (Bombace *et al.*, 1994). Mean biomass of about 50 kg m^{-2} was reported for mussels after three years on four artificial reefs located in the central Adriatic at 10–14 m depth (Bombace *et al.*, 1994). Mussels settled on artificial substrata in the Adriatic are generally harvested as soon as they reach commercial size. Because of this perturbation, it is impossible to judge whether these mussels represent a climax community or only a successional stage that is continually renewed by colonization after harvesting. Permanent mussel beds are found on rocks in much shallower depths (0–5 m) close to the artificial reef sites, but the presence of mussels on natural rocks has been observed up to 10 m depth (Fabi, personal communication). It is interesting that the artificial reefs support populations of *M. galloprovincialis* in water deeper than the accepted preferred depth range for that species (Ardizzone *et al.*, 1996). Oil production platforms in the Adriatic experience mussel fouling on jacket legs between 5 and 10 m. It is thought that these populations do not experience the problems of siltation found in shallower water as they are further from soft seabed sediments (Ardizzone *et al.*, 1980; Relini *et al.*, 1976; Relini, 1977; Taramelli *et al.*, 1980). It is postulated that the harvesting operations on the Adriatic artificial reefs reduces the sediment accumulation sufficiently to allow mussels to settle and achieve harvestable size.

Characteristic fish species (not considering pelagic species) around artificial reefs in eutrophic waters, as shown by reefs in the Adriatic Sea, are mainly *Sciaena umbra, Umbrina cirrosa, Dicentrarchus labrax, Lithognatus mormyrus* and *Diplodus annularis*. All these species are significantly more abundant in the vicinity of artificial reefs than on neighbouring soft seabeds (Fabi and Fiorentini, 1994). Moreover, an important set of species such as *Conger conger, Oblada melanura* and the mollusc *Octopus vulgaris* has been reported in oligotrophic conditions. Fish colonization observed at the Fregene reef can be considered as intermediate between the two extreme trophic conditions. *D. annularis* was the main fish species, even if other important species have been observed both typical of eutrophic (*Sciaena umbra* and *Dicentrarchus labrax*) and oligotrophic waters respectively (*Serranus cabrilla* and *Apogon imberbis*) (Ardizzone *et al.*, 1997). Furthermore, the absence of *C. chromis* and the initial abundance of *M. galloprovincialis* are the most important factors in interpreting the reef fish assemblage as one of eutrophic water. The comparison among various Italian reef studies points out that the two species *C. chromis* and *M. galloprovincialis* can be considered as indicators of very low and rich trophic levels respectively. The alternative dominance, presence or absence of one of them at the beginning of the experimental deployment can be considered as important information about the future colonization pattern.

The temporal evolution of a reef benthic community is also affected by water quality. Progress towards an equilibrium state proved to be a very slow and gradual process in oligotrophic waters, where a large number of species and a high value of diversity characterize the benthic community in spite of the low values of biomass recorded (Relini *et al.*, 1994; Tumbiolo *et al.*, 1995). A dissimilarity in species composition between the artificial reef and the nearest hard seabed was pointed out at Monaco 10 years after reef deployment (Balduzzi *et al.*, 1987). In eutrophic conditions a rapid progress to a low diversity, high biomass community is expected (Pearson and Rosenberg, 1978). This is not necessarily the case as can be seen at Fregene artificial reef where the mussel community developed quickly but was finally replaced by a polychaete dominated community. This may not be typical as the mussels were prevented from settling by a build-up of sediment, so preventing the mussel population reaching an equilibrium. This sedimentation is a critical factor in the development of artificial reef biological communities (Ardizzone *et al.*, 1989; Badalamenti *et al.*, 1985; Falace and Bressan, 1995). In eutrophic conditions the slow process from initial colonization to stability is caused by the speed of changes to the atypical hard substratum of an artificial reef placed in a soft seabed area, which is naturally subjected to high sedimentation rates. The sedimentation rate effectively controls the character of biological community that can survive.

References

AMPN. 1982. Les récifs artificiels de la reserve sous-marine de Monaco. *Journée d'études sur les aspects scientifiques concernant les récifs artificiels et la mariculture suspendue.* Cannes, C.I.E.S.M. pp. 79–81.

Ardizzone, G.D. and G. Bombace. 1982. Artificial reefs experiments along a Tyrrhenian Sea Coast. *Journée d'études sur les aspects scientifiques concernant les récifs artificiels et la mariculture suspendue.* Cannes, C.I.E.S.M. pp. 49–51.

Ardizzone, G.D. and C. Chimenz. 1982. Primi insediamenti bentonici della barriera artificiale di Fregene. Atti del Convegno Progetto Finalizzato C.N.R. *Oceanografia e Fondi Marini*, Rome November 1981: pp. 165–181.

Ardizzone, G.D. and M. Giardini. 1982. Annual cycle of nutrients and chlorophyll-*a* of an artificial reef area (Middle Tyrrhenian Sea). *Journée d'études sur les aspects scientifiques concernant les récifs artificiels et la mariculture suspendue.* Cannes C.I.E.S.M. pp. 53–54.

Ardizzone, G.D. and A. Migliuolo. 1982. Modificazioni di una prateria di *Posidonia oceanica* (L.) Delile nel Medio Tirreno sottoposta ad attività di pesca a strascico. *Naturalista Siciliano.* **3**: 509–515.

Ardizzone, G.D. and P. Pelusi. 1984. Yield and damage evaluation of bottom trawling on *Posidonia* meadow. In Boudouresque C.F., A.J. de Grissac and J. Oliver. (eds.) *International Workshop on* Posidonia oceanica *beds*. GIS Posidonie, Marseille, France. **1**: 63–72.

Ardizzone, G.D., C. Chimenz and A. Carrara. 1980. Popolamenti macrobentonici di substrati artificiali al largo di Fiumicino (Roma). *Memorie di Biologia Marina e Oceanografia.* suppl. **10**: 115–120.

Ardizzone, G.D., M.F. Gravina and A. Belluscio. 1989. Temporal development of epibenthic communities on artificial reefs in the central Mediterranean. *Bulletin of Marine Science.* **44**(2): 592–608.

Ardizzone, G.D., M.F. Gravina and C. Gusso Chimenz. 1994. A ten years research on marine artificial habitat for fishery purposes. In Argano R., C. Cirotto, E. Grassi

Milano and L. Mastrolia. (eds.) Contributions to Animal Biology. Halocynthia Association, Palermo, pp. 47–53.

Ardizzone, G.D., M.F. Gravina, A. Belluscio and A. Somaschini. 1996. Colonization and disappearance of *Mytilus galloprovincialis* Lam. on an artificial habitat in the Mediterranean Sea. *Estuarine Coastal and Shelf Science.* **43**(6): 665–676.

Ardizzone, G.D., A. Belluscio and A. Somaschini. 1997. Fish colonisation and feeding habits on a Mediterranean artificial habitat. In Hawkins, L.E. and S. Hutchinson with A.C. Jensen, J.A. Williams and M. Sheader. (eds.) *The Responses of Marine Organisms to Their Environments.* Proceedings of the 30th European Marine Biological Symposium, Southampton, September 1995, pp. 265–275.

Badalamenti, F., G. Giaccone, M. Gristina and S. Riggio. 1985. An eighteen months survey of the artificial reef of Terrasini (N/W Sicily): the algal settlements. *Oebalia. N.S.* **11**: 417–425.

Balduzzi, A., F. Boero, R. Cattaneo Vietti, M. Pansini and R. Pronzato. 1987. La colonisation des structures artificielles immergées dans la Réserve sous marine de Monaco. In A.M.P.N. *Conception et construction des récifs artificiels.* Note de presentation à la IV Conference Internationale sur les habitats artificiels pour les pêches. University of Florida Miami, USA, pp. 1–3.

Bayle-Sempere, J.T., A. Ramos-Espla and J.A. Garcìa Charton. 1994. Intra-annual variability of an artificial reef assemblage in the marine reserve of Tabarca (Alicante, Spain SW, Mediterranean). *Bulletin of Marine Science.* **55**(2–3): 824–835.

Bellan, G. 1982. Annelides Polychetes (Serpulides exclues) de deux types de récif artificiels immergés dans la région marseillaise. *Journée d'études sur les aspects scientifiques concernant les récifs artificiels et la mariculture suspendue.* Cannes, C.I.E.S.M. pp. 93–98.

Bellan, G. and D. Bellan Santini. 1991. Polychetous Annelids (Excluding Serpulidae) from artificial reef in the Marseille area (French Mediterranean coast). *Ophelia.* Suppl. 5: 433–442.

Blundo, C., T. La Noce, R. Pagnotta, M. Pettine and A. Puddu. 1978. Distribution of nutrients off the mouth the Tiber river and its relationship with biomass. *Thalassia Jugoslavica.* **14**: 339–355.

Bombace, G. 1977. Aspetti teorici e sperimentali concernenti le barriere artificiali. In Cinelli, F., E. Fresi and L. Mazzella. (eds.) Atti del IX Congresso della Società Italiana di Biologia Marina. Ischia (Naples), pp. 29–41.

Bombace, G. 1986. Introduction générale sur le thème 'Récifs artificiels'. FAO Fishery Reports. **357**: 51–64.

Bombace, G. 1989. Artificial reefs in the Mediterranean Sea. *Bulletin of Marine Science.* **44**(2): 1023–1032.

Bombace, G., G. Fabi, L. Fiorentini and S. Speranza. 1994. Analysis of the efficacy of artificial reefs located in five different areas of the Adriatic Sea. *Bulletin of Marine Science.* **55**(2–3): 559–580.

Bombace, G., G. Fabi and L. Fiorentini. 1995. Artificial reefs and mariculture: the Italian experiences. *Proceedings ECOSET 1995.* Published by Japan International Marine Science and Technology Federation. **2**: 830–835.

Clements, F.E. 1916. *Plant Succession.* Carnegie Institute, Washington: 242 pp.

D'Anna, G., F. Badalamenti, M. Gristina and C. Pipitone. 1994. Influence of artificial reefs on coastal nekton assemblages of the Gulf of Castellammare (Northwestern Sicily). *Bulletin of Marine Science.* **55**(2–3): 418–433.

D'Anna, G., F. Badalamenti, R. Lipari, A. Cuttitto and C. Pipitone. 1995. Fish assemblage analysis by means of visual census survey on an artificial reef and on natural areas in the Gulf of Castellammare (NW Sicily). *Proceedings ECOSET 1995.* Published by Japan International Marine Science and Technology Federation. **1**: 221–226.

De Bernardi, E. 1989. The Monaco Underwater Reserve. Design and construction of artificial reefs. *Bulletin of Marine Science.* **44**(2): 1066.

Duval, C. 1982. Bilan de la faune mobile de petits modules artificiels immergés dans la zone de Marseille. *Journée d'études sur les aspects scientifiques concernant les récifs artificiels et la mariculture suspendue*. Cannes, C.I.E.S.M. 105–108.

Duval, C. and J. Cantera. 1982. Données préliminaires sur la faune de Mollusques de modules artificiels immergés dans la région de Marseille. *Journée d'études sur les aspects scientifiques concernant les récifs artificiels et la mariculture suspendue*. Cannes, C.I.E.S.M. 89–92.

Fabi, G. and L. Fiorentini. 1994. Comparison between an artificial reef and a control site in the Adriatic Sea. Analysis of four years of monitoring. *Bulletin of Marine Science*. **55**(2–3): 538–558.

Falace, A. and G. Bressan. 1990. Dinamica della colonizzazione algale di una barriera artificiale sommersa nel Golfo di Trieste: macrofouling. *Hydrores*. **7**(8): 5–27.

Falace, A. and G. Bressan. 1995. Adapting an artificial reef to biological requirements. *Proceedings ECOSET 1995*. Published by Japan International Marine Science and Technology Federation. **2**: 634–639.

Gòmez-Buckley, M.C. and R.J. Haroun. 1994. Artificial reefs in the Spanish coastal zone. *Bulletin of Marine Science*. **55**(2–3): 1021–1028.

Gravina, M.F., G.D. Ardizzone and A. Belluscio. 1989. Polychaetes of an artificial reef in the Central Mediterranean Sea. *Estuarine Coastal and Shelf Science*. **28**: 161–172.

Guillén, J.E., A.A. Ramos, L. Martìnez and J.L. Sànchez Lizaso. 1994. Antitrawling reefs and the protection of *Posidonia oceanica* (L.) Delile meadows in the Western Mediterranean Sea: demand and aims. *Bulletin of Marine Science*. **55**(2–3): 645–650.

Lefevre, J.R., J. Duclerc, A. Meinesz and M. Ragazzi. 1986. Les récifs artificiels des établissements de pêche de Golfe Juan et de Beaulieu-sur-mer, Alpes Maritimes – France. *Journée d'études sur les aspects scientifiques concernant les récifs artificiels et la mariculture suspendue*. Cannes, C.I.E.SM. pp. 109–111.

Moreno, L., I. Roca, O. Renones, J. Coll and M. Salamanca. 1994. Artificial reef program in Balearic waters (Western Mediterranean). *Bulletin of Marine Science*. **55**(2–3): 667–671.

Pagnotta, R., M. Zoppini and A. Guzzini. 1992. Chemical and hydrobiological characteristics of a riverine-coastal water system: the case of the Tiber River mouth. *Memorie Istituto Italiano Idrobiologia*. **51**: 185–200.

Pearson, T.H. and R. Rosenberg. 1978. Macrobenthic succession in relation to organic enrichment and pollution of the marine environment. *Oceanography and Marine Biology Annual Review*. **16**: 229–311.

Petrocelli, A. and O. Saracino. 1992. Alghe di substrati artificiali immersi nel Mar Ligure. *Biologia Marina*, suppl. al Notiziario S.I.B.M. **1**: 303–304.

Relini, G. 1977. Possibilità di sfruttamento del fouling di strutture off-shore nei mari italiani: i mitili di Ravenna. Atti del VII Simposio Nazionale sulla Conservazione della Natura, Bari, 20–23 Aprile 1977: 179–185.

Relini, G. and L. Orsi Relini. 1989. Artificial reefs in the Ligurian Sea (Northwestern Mediterranean): aims and results. *Bulletin of Marine Science*. **44**(2): 743–751.

Relini, G., S. Geraci, M. Montanari and V. Romairone. 1976. Variazioni stagionali del fouling sulle piattaforme off-shore di Ravenna e Crotone. *Bollettino di Pesca, Piscicoltura e Idrobiologia*. **31**: 1–30.

Relini, G., A. Peirano, L. Tunesi and L. Orsi Relini. 1986. The artificial reef in the Marconi Gulf (Eastern Ligurian Riviera). *FAO Fishery Reports*. **357**: 95–103.

Relini, G., N. Zamboni, F. Tixi and G. Torchia. 1994. Patterns of sessile macrobenthos community development on an artificial reef in the Gulf of Genoa (Northwestern Mediterranean). *Bulletin of Marine Science*. **55**(2–3): 745–771.

Relini, G., Trentalance, M. Relini, G. Torchia and F. Tixi. 1995a. Stabilized harbour muds for artificial reef blocks. *Proceedings ECOSET 1995*. Published by Japan International Marine Science and Technology Federation. **1**: 114–118.

Relini, G., G. Torchia and M. Relini. 1995b. The role of F.A.D. in the variation of fish assemblage on the Loano Artificial Reef (Ligurian Sea – NW Mediterranean). *Proceedings ECOSET 1995*. Published by Japan International Marine Science and Technology Federation. **1**: 1–5.

Riggio, S. 1989. A short review of artificial reefs in Sicily. FAO Fishery Reports. **428**: 128–137.

Riggio, S. and G.D. Ardizzone. 1979. Prospettive dell'impiego di substrati artificiali per il ripopolamento e la protezione dei fondali costieri della Sicilia nord-occidentale. Atti del Convegno Scientifico Nazionale P.F. Oceanografia e Fondi Marini. Roma: pp. 157–184.

Riggio, S., M. Gristina, G. Giaccone and F. Badalamenti. 1985. An eighteen months survey of the artificial reef of Terrasini (N/W Sicily): the invertebrates. *Oebalia N.S.* **11**: 427–437.

Riggio, S., F. Badalamenti, R. Chemello and M. Gristina. 1989. Zoobenthic colonization of a small artificial reef in Southern Tyrrhenian: results of a three year survey. FAO Fishery Reports. **428**: 138–153.

Ryan, W.B.F. 1966. Mediterranean Sea: physical oceanography. In Fairbridge, R.W. (ed.) *The Encyclopedia of Oceanography.* Van Nostrand Reinhold, New York. p. 312.

Sànchez-Jerez, P. and A. Ramos-Esplà. 1995. Influence of spatial arrangement of artificial reefs on *Posidonia oceanica* fish assemblage in the Western Mediterranean Sea: importance of distance among blocks. *Proceedings ECOSET 1995*. Published by Japan International Marine Science and Technology Federation. **2**: 646–651.

Sarà, M. 1987. Persistence and changes in marine benthic communities. *Nova Thalassia.* **3** suppl.: 7–30.

Seaman, W. Jr and L.M. Sprague. (eds.) 1991. *Artificial Habitats for Marine and Freshwater Fisheries*. Academic Press Inc., San Diego, California. 285pp.

Somaschini, A., G.D. Ardizzone and M.F. Gravina. 1997. Long-term changes in the structure of a polychaete community on artificial habitats. *Bulletin of Marine Science.* **60**(2): 460–466.

Spanier, E. 1995. Artificial reefs in a biogeographically changing environment. *Proceedings of ECOSET 1995*. Published by Japan International Marine Science and Technology Federation. **2**: 543–547.

Spanier, E., S. Pisanty and G. Almong. 1989. Artificial habitat for fisheries in the Southeastern Mediterranean: A model for low productive marine environments. *Bulletin of Marine Science.* **44**(2): 1070.

Spanier, E., M. Tom, S. Pisanty and G. Almong-Shtayer. 1990. Artificial reefs in the low productive marine environment of the Southeastern Mediterranean. *P.S.Z.N.I. Marine Ecology.* **11**: 61–75.

Sutherland, J.P. 1981. The fouling community at Beaufort, North Carolina: a study in stability. *American Naturalist.* **118**: 499–519.

Taramelli, E., C. Chimenz and C. Berna. 1980. Popolamento a molluschi di una piattaforma al largo di Fiumicino (Roma). *Memorie di Biologia Marina e Oceanografia* Suppl. **10**: 319–325.

Tumbiolo, M.L., F. Badalamenti, G. D'Anna and B. Patti. 1995. Invertebrate biomass on an artificial reef in the Southern Tyrrhenian Sea. *Proceedings ECOSET 1995.* Published by Japan International Marine Science and Technology Federation. **1**: 324–329.

8. The Loano Artificial Reef

GIULIO RELINI

*Laboratorio di Biologia Marina ed Ecologia Animale, Università degli Studi di Genova,
DIP.TE.RIS, Via Balbi, 5, 16126 Genova, Italy*

Introduction

In the past 40 years the coastal marine environment in the Western Ligurian
Riviera has been affected by three major human activities:

(1) Illegal fishing, especially otter-trawling, in the shallow (< 50 m) coastal zone
    where this type of gear is banned. The problem is often compounded by the
    use of undersize net mesh in the cod end of the trawls.
(2) Pollution, mainly sewage discharges into the sea. Such discharges increased
    considerably in recent decades following an increase in the number of inhab-
    itants in the Italian Riviera and the success of the tourist industry.
(3) Siltation caused by badly planned beach replenishment schemes designed to
    repair sandy shores. These schemes have used fine silty material rather than
    sand or gravel with the result that the seabed has become muddy and marine
    flora has been destroyed. Where the result is the decline or disappearance of
    the seabed-stabilizing *Posidonia* (seagrass) meadows a chain reaction may be
    involved in which storms become even more destructive to beaches and so
    more material has to be discharged to maintain the sandy shore for tourism.
    The result is that the natural rocky seabed is disappearing while the sandy
    seabed is becoming abiotic at a macroscopic level. The effect of this situation
    on small coastal fisheries has already resulted in a significant decrease in
    quality and quantity of fish caught.

It was felt that artificial reefs could play an important role in the protection and
management of shallow-water marine environments in the Ligurian Sea, because
they provide settlement substrata for epifauna which are able to utilize at least
some of the waste organic matter discharged into the sea. The physical presence
of a reef will protect marine biocoenosis, such as *Posidonia* beds, from trawling
and, at the same time, protect artisanal fisheries from damage. Such reefs will also
offer shelter and protection to adult marine animals.

Previous Artificial Reef Building Experience in Italy

Prior to reef building in Loano the first artificial reef in Italy was deployed in the
Ligurian Sea in December 1970 near the town of Varazze (Fig. 1) by a sport-
fishing club which obtained permission, without any scientific support, to sink
1300 car bodies in order to prevent trawling and to improve sport fishing in an

A.C. Jensen et al. (eds.), Artificial Reefs in European Seas, 129–149
© 2000 Kluwer Academic Publishers. Printed in Great Britain.

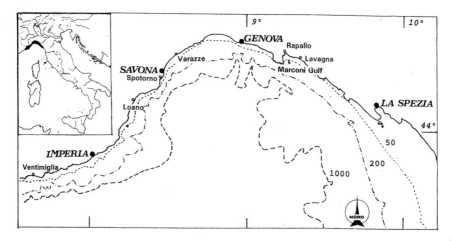

Figure 1.   Location of the main artificial reefs in the Gulf of Genoa (Ligurian Sea, Mediterranean).

area between 35 m and 50 m depth. The reef was not successful, the cars rusted badly, sank into the muddy seabed and were covered in silt (Relini and Relini Orsi, 1971). Understandably the settlement of organisms and sport fishing remained poor (Relini and Wurtz, 1977; Relini, 1979, 1982b, 1983a, 1983b).

On the basis of this experience, which also highlighted pollution problems, the use of car bodies and similar waste materials for the purpose of creating artificial reefs was banned by the Ministero Marina Mercantile. In 1979, 16 wooden barges were sunk close to the car bodies. These proved to be more effective than car bodies in preventing trawling, were more attractive to fish and provided a more suitable substratum for epibiotic settlement.

Artificial reefs were constructed in the Marconi or Tigullio Gulf (Fig. 1) between 1980 and 1985 (Relini, 1982a, 1984). Wooden barges, concrete blocks, dock-gates made of timber and iron, and loose materials (sand, gravel, coarse pebbles) were used. By 1984 the artificial reefs had reached a volume of 16 185 m³ with a surface available for organism settlement of 9234 m² covering 21 km² of seabed and protecting at least another 6 km² of the Gulf (Relini *et al.*, 1986).

The annual summer sport fishing competition showed an increase in catches in terms of weight of fish per fishermen and per trip from 1982 to 1986, despite the increased number of participants in each competition (Relini and Orsi Relini, 1989). In 1987 a marked decrease in catches occurred although some individual results were still good. The positive trend reappeared in 1988.

SCUBA diver observations confirmed a high diversity of communities settled on different substrata, but they also reported damage to the wooden barges by wood borers (teredinids, etc.).

The Loano Artificial Reef

The Town Council of Loano, with the needs of the local professional artisanal fishery and sport fishermen in mind, requested funds for the construction of an artificial reef in an area near the port of Loano. The main aims of the project were to protect natural habitats from further damage and if possible to restore some of them and, in particular, to attract and produce marine resources suitable for fishermen. A major goal of the Loano project was to protect the *Posidonia* meadow and if possible to extend it, because of its great importance in the maintenance of biological diversity, the fishery economy and protection against coastal erosion. The restocking of fishery resources was also an important aim as these had been exploited in a deteriorating environment with increasing competition between professional and sport fishermen as well as between trawling and the artisanal fishery.

The project, planned by ECG (Engineering Contractor's Group, Rome) which also drew on previous Italian experience (Bombace, 1977, 1981; Relini 1982b, 1984), was approved in 1985 and received financial support from the EEC and from the Ministero Marina Mercantile (Fig. 2).

The Marine Environment Around Loano

The shallow seabed along the Ligurian coast comprises mainly thin sand and mud partially covered by *Posidonia oceanica* and *Cymodocea nodosa* meadows. To the east of the Port of Loano, rocks and pebbles occur at a depth of 2.5 m. These, together with rocks forming the Port wharf, pebbles of zone R and wrecks W (Fig. 2) comprised the total hard substrata in the area before the artificial reef was built. In water between 3 m and 12 m deep most of the seabed is covered by *Cymodocea nodosa* and related organisms. At depths > 25 m occurs the Coastal Detritus biocenosis seabed (CD) which is gradually changing into the biocenosis of coastal terrigenous muds (CTM). The *Posidonia* bed extends between 9 m and 20 m water depth, but is retreating eastward and currently ends off the small pile south of Loano Harbour (see Fig. 2). In the past the *Posidonia* meadow extended along all the coast in the direction of Pietra Ligure, as can be seen on old fishing maps. At present the density of plants is not uniform; in the central part of the existing meadow there are 120–150 shoots m^{-2}, while near the border no more than 50 occur. Some environmental parameters are given in Table 1.

The Structure of the Artificial Reef

The artificial reef at Loano (LAR) was deployed in the summer of 1986, covering an area of about 350 ha in water depths ranging from 5 m to 45 m. The artificial reef (Relini and Moretti, 1986) was initially composed of two parts. The first was a central main group of 30 pyramids, 25 m apart, positioned along the sides of and inside a 100 × 200 m rectangle (Fig. 3a). Each pyramid comprised five blocks, four blocks (cubic block of 2 m side, Fig. 3b,d,e) forming the base and one block placed

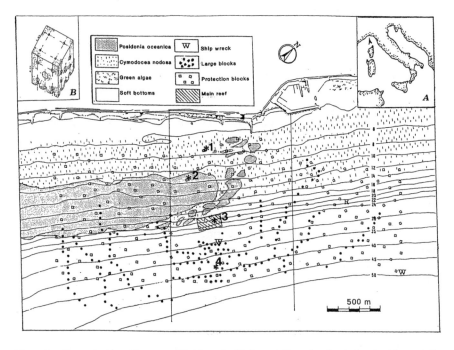

Figure 2. General view of Loano artificial reef with four stations for the study of settlement (1, 2, 3, 4). The main reef composed of pyramids of large blocks (B) and the distribution of blocks protecting *Posidonia* beds, immersed in 1986 are indicated. Additional large blocks (B), represented by black dots, were immersed in 1989.

Table 1. Maximum, average and minimum values of some environmental parameters measured at different depths off Loano during 1989, 1990 and 1991.

	Temperature (°C)			Conductivity (mmhos cm⁻¹)			Salinity (PSU)		
	Max	Av.	Min	Max	Av.	Min	Max	Av.	Min
Surface	26.1	18.4	12.2	56.8	49.2	43.1	38.8	37.1	34.2
5 m	25.0	18.2	12.2	56.8	49.0	43.1	38.0	37.3	36.1
10 m	25.0	17.9	12.2	57.0	48.8	43.0	38.0	37.2	36.2
18 m	24.8	17.6	12.0	56.7	48.5	43.0	37.9	37.5	36.6
36 m	24.2	17.4	12.0	54.8	48.2	42.9	37.9	37.2	36.0

	$N\text{-}NO_2(\mu g\ l^{-1})$			$N\text{-}NO_3(\mu g\ l^{-1})$			$P\text{-}PO_4(\mu g\ l^{-1})$		
	Max	Av.	Min	Max	Av.	Min	Max	Av.	Min
Surface	38.5	12.5	1.5	10.6	5.1	0.8	4.2	1.5	0.6
5 m	36.0	12.9	2.8	10.6	4.9	2.3	3.1	1.4	0.3
10 m	17.9	8.3	3.8	13.6	5.4	1.5	3.4	0.9	0.6
18 m	17.4	8.3	3.4	8.3	4.5	0.8	3.4	1.4	0.3
36 m	15.0	9.2	3.8	16.6	6.8	1.5	2.8	1.4	0.3

on top of the base to form the apex of the structure. Each pyramid was placed on a 50 cm thick layer of stone to prevent the structure from sinking into the muddy seabed (Figs 3d,e; Fig. 4).

The second part was composed of a grid of 200 single blocks (cubic blocks of 1.2 m side, Fig. 3c) distributed in rectangle of about 3 × 1.5 km in such a way as to prevent any passage wider than 250 × 50 m. The aim was to prevent trawling in the area while allowing the use of artisanal gear. Unfortunately, however, although each of these blocks weighs 3 tonnes, they have been moved by trawlers.

In the Summer of 1989, 150 2 m³ blocks were placed in such a way as to reinforce the protection of the central restocking zone and the most external part of the area where the small blocks were positioned.

On the orders of the Savona Port Authority the whole area is covered by a ban on any form of sampling with the exception of that carried out for scientific research.

Benthic Studies

Observations of the colonization patterns of benthos and fish have been made since 1986 by SCUBA divers, who photographed settled organisms and took samples from standard surfaces (20 × 30 cm) of macrobenthos present on the concrete blocks of the pyramids.

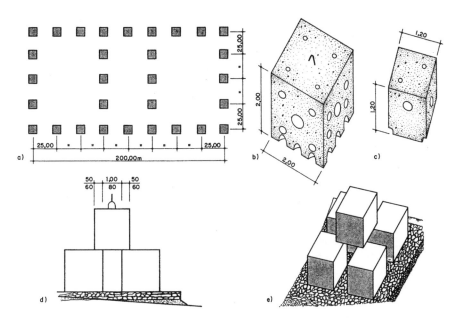

Figure 3. The structure (a) of the main reef (200 × 100 m) composed of 30 pyramids. Each pyramid (d, e) consists of five large blocks (b) with 2 m sides. The small blocks (c) have 1.2 m sides.

Figure 4.   Diver 'inside' a Loano reef pyramid. (Photo A. Jensen).

Studies of the settlement and development of communities in the waters off Loano have been carried out since May 1987 using asbestos panels (20 × 30 × 0.4 cm). This is a well-developed technique, originating from fouling studies, that makes it possible to take the settled organisms into the laboratory whilst knowing the exact period of field exposure. In the central part of the Loano artificial reef four monitoring stations at different depths were chosen (Fig. 2) for the deployment of racks of settling panels: station 1 at 5 m depth; station 2 at 10 m depth; station 3 at 18–20 m depth and station 4 at 30–36 m depth. SCUBA divers took photographs of block colonization of and took samples from the standard surfaces.

Species Recorded

Seventy-six algae (41 *Rhodophyta*, 18 *Phaeophyta* and 15 *Chlorophyta*) (Cecere *et al.*, 1993) and 120 species of sessile animals were identified (Relini *et al.*, 1995b) on panels immersed at four depths. The most diverse animal group was bryozoans (58 species) followed by serpulids (20 species) and hydroids (19 species). Forty more species were found only on concrete blocks. When the species numbers of the main animal taxa found during the 3-year investigation at the four depths were compared, a generally consistent pattern of species richness between years (Relini *et al.*, 1994a) was revealed. Algae decrease and the number of animals increased with depth.

Biomass

The biomass of settled organisms was evaluated as non-decalcified wet weight. The monthly values per panel at the four stations are available only for the first year of measurement and they range from 2 to 51 g 12 dm^{-2} (Relini and Cormagi, 1990). In general the biomass was low if compared with settlement in eutrophic environments. Differences in biomass between seasons were not as large as found in Ligurian harbours (Montanari and Relini, 1973). During a 1-year exposure, biomass, percentage cover and number of sessile species increased as the period of panel immersion lengthened. Decreasing trends of biomass and cover were observed in relation to depth and in particular between station 3 (18 m) and station 4 (30 m). Clear seasonal settlement occurred, although the biomasses were similar. On the Loano artificial reef, mussels were never dominant at any depth, as has been described for some other artificial reefs in Italian seas (Ardizzone *et al.,* 1989; Bombace, 1977, 1981, 1989).

Development of the Biological Community on Settlement Panels and Concrete Blocks

Detailed data on the development of community on panels exposed for 3, 6, 9, and 12 months at four stations from May 1988 to May 1989 are given in Relini *et al.* (1990b).

After one year the settlement on the concrete blocks described from SCUBA diver observations, photos and samples, was similar to that described on panels exposed for 12 months.

Two years after deployment all the structures were well colonized both from a quantitative and from a qualitative point of view. In particular at reef station 1 (5 m) the settlement consisted of a well-developed algal coating. Brown algae dominated, in particular *Dictyota.* Also abundant were bryozoans (especially *Schizoporella*) and serpulids (*Salmacina* sp., *Serpula* sp., *Pomatoceros* sp.). At reef station 2 (10 m) the algal component diminished while the animal component increased. There were numerous ascidians, sponges and hydroids. The whole surface appeared to be colonized. At station 3 algae were reduced even further. Serpulids, hydroids, bryozoans and good-sized oysters were dominant. Large groups of serpulids were noted and egg-capsules of cephalopods recorded, usually in the vicinity of the holes in the reef blocks. At station 4 there was a dominance of bryozoans, serpulids and barnacles, but the association was rather impoverished.

Two phenomena observed during the second year after deployment are worthy of particular attention: the presence of large oysters (8–10 cm in size) and the grazing activity of sea urchins (*Paracentrotus lividus* and *Arbacia lixula*). Large surfaces were completely cleaned and freed for new colonizers; only large barnacles (*Balanus perforatus*) and oysters (*Ostrea edulis*) could withstand the activity of these echinoids.

Re-colonization of these base surfaces occurred during the third year. In particular, holes and the shaded surfaces were thickly covered by a layer of organisms, mainly hydroids (*Eudendrium racemosum, Bougainvillia ramosa, Obelia* spp.), bryozoans (the largest colonies were *Pentapora ottomulleriana, Schizoporella errata* and *Sertella* sp.) and sponges (*Clathrina coriacea, Leuconia aspera, Crambe crambe, Oscarella lobularis*). Some gorgonians (*Alcyonium palmatum, Callogorgia verticillata*) and hard corals (*Cladocora, Caryophyllia*), rare in coastal waters in the Ligurian Sea, also settled. In some places heavy settlement of oysters (*Ostrea edulis*) occurred on the vertical surface of the blocks. This is a possible resource that could be cultivated and exploited in the future, despite considerable predation by the sea star *Marthasterias glacialis*.

After four years (in 1990) large brown algae (*Sargassum vulgare, Dictyota dichotoma, Cystoseira* spp., *Dictyopteris membranacea, Padina pavonica*) developed on the upper horizontal surface of the blocks as the result of ecological succession. In particular, 60–80 cm long *Sargassum vulgare* occurred. These algae do not settle on newly immersed surfaces but need substrata 'prepared' by other organisms.

During 1991 the large algae were drastically reduced to small plants or disappeared. This was probably because of silt deposits and the high turbidity of the water caused by beach replenishment works along the Loano coast (Relini *et al.,* 1994a). The influence of sedimentation on colonization was studied by Relini *et al.,* 1998a.

The differences found in the colonization between stations (and also between depths) are greater than the seasonal differences at the same station. Data obtained on the Loano artificial reef using panels immersed for three 1-year cycles, and observations carried out on blocks over a period of 5 years provide the most detailed and wide-ranging information available in the Mediterranean and in Europe with regard to studies on the colonization processes of sessile macrobenthos on artificial reefs. The communities which developed on the Loano artificial reef showed considerable (species) richness and a slow process of community development. They are similar to those described for the modules at Monaco (Balduzzi *et al.,* 1982, 1985, 1986) and in the north of Sicily (Riggio *et al.,* 1986, 1990) but are very different both qualitatively and quantitatively from the communities on artificial reefs constructed at Fregene, in the central Tyrrhenian Sea (Ardizzone *et al.,* 1989) and at Ancona (Bombace, 1981) in the middle Adriatic Sea. On Fregene and Ancona reefs there was a clear dominance of sestonophagous filters-feeders (mussels and oysters), a dominance reached in a very short time.

On the concrete blocks of the artificial reef pyramids the different exposure of each face to light, currents and sedimentation strongly influenced settlement and the evolution patterns of the community.

On the upper side, the one most exposed to light, a true ecological succession was observed, which tended towards a mature stage dominated by *Cystoseiretalia*, as described by Badalamenti *et al.* (1985) for Sicily and partially by Falace and Bressan (1990) for Trieste. On the Loano artificial reef, large plants of *Sargassum* sp. and *Cystoseira* sp. occurred during the third year and then disappeared, probably due to heavy silting. On the inside faces of the blocks of the pyramids typical skiophilous organisms settled, giving rise to a complex community in which sponges, coelenterates, large bryozoans such as *Pentapora ottomulleriana* were dominant (see Fig. 4). This community is still developing towards a 'cave community' or coralligenous assemblages. After 10 years the process of succession in all communities settled on the different faces of the blocks still continued. The process of reaching a steady-state community at the Loano artificial reef has been complicated and slowed by unpredictable factors, such as hydro-dynamism, sedimentation and predation.

The grazing by sea urchins (*Paracentrotus lividus* and *Arbacia lixula*) was a strong interference factor in the development of the community on the blocks. It is well known (Relini *et al.*, 1994a; Riggio *et al.*, 1986) that the efficiency of an artificial reef in providing food and refuge for fish and other organisms of commercial interest is related to colonization phases and in particular to achieving a mature stage in terms of biomass, species richness, ecological niches and food webs.

Fish Studies

The study of the mobile fauna associated with an artificial reef includes a great number of interests which range from ethology of the organisms to the problems of production and exploitation and restocking of fish populations.

Studies on and around the Loano artificial reef were carried out using a direct census method at monthly intervals from February 1989 up to January 1994 and by limited fishing activity using long-lines and trammel nets (Relini *et al.*, 1990, 1993, 1994; Relini and Torchia, 1993). From 1994 only visual censuses were undertaken (Relini *et al.*, 1997a,b) and the role of a FAD attached to a reef pyramid was studied (Relini *et al.*, 1995c).

Although direct censuses do not record the entire fish population, underwater observation can be an effective method in the process of understanding the dynamics of population establishment and seasonal variations on artificial reefs.

Species Observed

A total of 76 species were recorded (Table 2); 67 fish (see Colour Plates 3 and 4), four crustaceans (see Colour Plate 5) and five cephalopods from SCUBA diver observations and trammel net catches. Among the species listed there are some of high commercial value, such as sea bream (*Sparus aurata*), sea bass (*Dicentrarchus labrax*) and brown meagre (*Sciaena umbra*). Some species recorded, such as the brown meagre, the dusky grouper (*Epinephelus marginatus*) and the comb grouper (*Mycteroperca rubra*, only the second record of the species in the Ligurian sea) have been quite rare along the Ligurian coast for some decades.

Underwater Censuses

A total of 45 species of fish were observed in over 70 dives from February 1989 to January 1994. Eight of these were only seen during nocturnal dives or during surveys not carried out for the census. Some species were found in all surveys, others only periodically or without any precise regularity. The number of fish counted each month is shown in Fig. 5. Additional censuses carried out in nearby

Table 2. Species recorded by SCUBA divers (+) and trammel net catches (*). V = high commercial value, R = rare.

Comment		SCUBA	Trammel
	Fish		
	Anthias anthias (Linnaeus, 1758)	+	
	Apogon imberbis (Linnaeus, 1758)	+	*
	Arnoglossus laterna (Walbaum, 1792)		*
	Arnoglossus thori (Kyle, 1913)		*
	Boops boops (Linnaeus, 1758)	+	*
	Bothus podas (Delaroche, 1809)		*
	Buglossidium luteum (Risso, 1810)		*
	Chromis chromis (Linnaeus, 1758)	+	*
	Conger conger (Linnaeus, 1758)	+	
	Coris julis (Linnaeus, 1758)	+	*
V	*Dentex dentex* (Linnaeus, 1758)	+	*
V	*Dicentrarchus labrax* (Linnaeus, 1758)	+	
	Diplodus annularis (Linnaeus, 1758)	+	*
	Diplodus puntazzo (Cettim, 1777)	+	
V	*Diplodus sargus* (Linnaeus, 1758)	+	*
	Diplodus vulgaris (E. Geoffroy St. Hilaire, 1817)	+	*
	Engraulis encrasicolus (Linnaeus, 1758)	+	*
V/R	*Epinephelus marginatus* (Lowe, 1834)	+	
	Gobius cruentatus (Gmelin, 1789)	+	
	Gobius niger (Linnaeus, 1758)	+	
	Labrus bimaculatus (Linnaeus, 1758)	+	
	Labrus merula (Linnaeus, 1758)	+	*
	Labrus viridis (Linnaeus, 1758)	+	
	Lepidotrigla cavillone (Lacépède, 1801)		*
V	*Lophius piscatorius* (Linnaeus, 1758)	+	
	Merluccius merluccius (Linnaeus, 1758)		*
	Microchirus variegatus (Donovan, 1802)		*
	Monochirus hispidus (Rafinesque, 1814)		*
	Mugil sp.	+	
	Mullus barbatus (Linnaeus, 1758)		*
V	*Mullus surmuletus* (Linnaeus, 1758)	+	*
R	*Mycteroperca rubra* (Bloch, 1793)	+	
	Oblada melamura (Linnaeus, 1758)	+	*
	Ophidion barbatum (Linnaeus, 1758)		*
	Ophidion rochei (Muller, 1845)		*
	Ophisurus serpens (Linnaeus, 1758)	+	
V	*Pagellus erythrinus* (Linnaeus, 1758)	+	*
	Pagellus acarne (Risso, 1826)		*
V	*Pagrus pagrus* (Linnaeus, 1758)		*
	Parablennius incognitus (Bath, 1968)	+	
	Parablennius rouxi (Cocco, 1833)	+	
V	*Phycis phycis* (Linnaeus, 1766)	+	
	Sardinella aurita (Valenciennes, 1847)		*
	Sarpa salpa (Linnaeus, 1758)		*
V/R	*Sciaena umbra* (Linnaeus, 1758)	+	
	Scomber japonicus (Houttuyn, 1780)		*
	Scorpaena notata (Rafinesque, 1810)	+	*
	Scorpaena porcus (Linnaeus, 1758)	+	*
	Scorpaena scrofa (Linnaeus, 1758)	+	*
V	*Seriola dumerili* (Risso, 1810)	+	*
	Serranus cabrilla (Linnaeus, 1758)	+	*

Continued on next page

Table 2.   *Continued*

Comment		SCUBA	Trammel
	Serranus hepatus (Linnaeus, 1766)	+	
	Serranus scriba (Linnaeus, 1758)		*
	Solea vulgaris (Quensel, 1806)		*
V	*Sparus aurata* (Linnaeus, 1758)	+	
	Sphyraena sphyraena (Linnaeus, 1758)		*
	Spicara fluxuosa (Rafinesque, 1810)	+	*
	Spicara maena (Linnaeus, 1758)	+	*
	Spicara smaris (Linnaeus, 1758)	+	*
	Spondyliosoma cantharus (Linnaeus, 1758)	+	*
	Symphodus cinereus (Bonnaterre, 1788)	+	*
	Symphodus mediterranus (Linnaeus, 1758)	+	
	Symphodus tinca (Linnaeus, 1758)	+	*
	Trachurus meditarraneus (Steindachner, 1863)		*
	Trachurus trachurus (Linnaeus, 1758)		*
R	*Umbrina cirrosa* (Linnaeus, 1758)	+	
	Uranoscopus scaber (Linnaeus, 1758)		*
	Cephalopods		
V	*Illex coindetii* (Venary, 1839)		*
V	*Loligo vulgaris* (Lamarck, 1798)	+	*
V	*Octopus vulgaris* (Cuvier, 1797)	+	*
V	*Sepia officinalis* (Linnaeus, 1758)	+	*
	Todarodes sagittatus (Linnaeus, 1758)		*
	Crustaceans		
V/R	*Homarus gammarus* (Linnaeus, 1758)	+	
V	*Palinurus elephas* (Fabricius, 1787)	+	
V	*Scyllarus arctus* (Linnaeus, 1758)	+	
	Squilla mantis (Linnaeus, 1758)		*

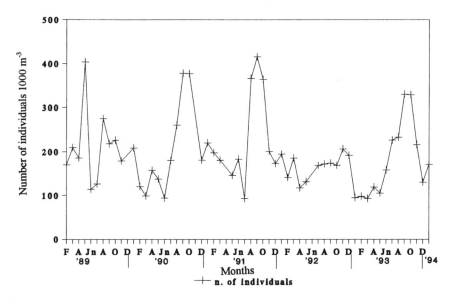

Figure 5.   Number of fish counted in each month from February 1989 to January 1994.

zones where there were no rock formations or *Posidonia* meadows produced low values of both in terms of numbers of species (a maximum of seven) and of individuals observed.

The overall fish population of the Loano artificial reef was lowest in the period between May and June and highest between August and November due to the recruitment of juveniles.

Percentages of occurrence (out of a total of 62 censuses) on the four-point scale previously employed by Harmelin (1987) were used:

- 1st level of frequency > 75%,
- 2nd level = 50–74.9%,
- 3rd level = 25–49.9%,
- 4th level < 25%.

Fourteen taxa belonging to the first and second levels of occurrence (Table 3), and *Conger conger,* can be considered as forming part of the base population of the Loano artificial reef. *C. conger* has been added although it has a low frequency of occurrence due to its cryptic behaviour.

Seasonal Changes

It is possible to identify a clear seasonal change in the number of individuals, and the species composition of the censused population (Fig. 6). Two distinct populations, one present from summer to autumn, the other from winter to spring, seem to occur. The differences between the fish populations present in summer and winter and those between summer and spring are often very marked (Fig. 7).

Temporal Changes and Macrobenthos Development

There were different components to the fish population in each of the 4 years 1989–90, 1990–91, 1991–92 and 1992–93. The Shannon-Wiener diversity index, calculated for each annual population composition, ranged from 3.33 in 1989–90 and 3.17 in 1990–91, dropping to 2.18 in 1991–92 and returning to a higher level (3.19) in 1992–93.

Temporal developments in the fish community appeared closely linked to changes in the macrobenthic settlement on the blocks of the Loano artificial reef. In particular, in 1990 there was a development in the algal (*Dictyota* sp., *Sargassum vulgare*) cover of the pyramids' horizontal surfaces, making the environment more complex and richer thus favouring an increased species richness. There was a considerable increase in the number of Labridae living on the reef, especially in the number of juveniles. In 1991 the disappearance of the large *Sargassum* and the decrease in other algae led to a notable reduction in the number of species recorded (Fig. 8). This disappearance can probably be linked to severe muddying and the subsequent turbidity of the water. The number of species increased only very slowly over the following years with a progressive and more complex macrobenthic colonization of the concrete blocks, particularly in shaded zones.

Table 3. Percentage of presence and relative levels of frequency for the species surveyed over five years (February 1989 to January 1994). Frequency level 1, > 75% occurrence; 2, 50–74.9% occurrence; 3, 25–49.9% occurrence; 4, < 25% occurrence.

Species		% frequency of occurrence	Level of frequency
Chromis chromis	Damselfish	100.0	1
Diplodus vulgaris	Common two banded bream	100.0	1
Serranus cabrilla	Comber	100.0	1
Diplodus annularis	Annular sea bream	97.1	1
Diplodus sargus	White sea bream	95.6	1
Scorpaena spp.	Scorpion fish	92.6	1
Parrablennius rouxi	Blenny	82.0	1
Gobius cruentatus	Red mouthed goby	75.0	1
Symphodus spp.	Wrasse	75.0	1
Spondyliosoma catharus	Black sea bream	69.1	2
Spicara spp.	Picarel	57.3	2
Oblada melanura	Saddled sea bream	55.9	2
Mullus surmuletus	Red mullet	54.4	2
Apogon imberbis	Cardinal fish	51.5	2
Conger conger	Conger eel	27.9	3
Coris julis	Rainbow wrasse	27.9	3
Diplodus puntazzo	Sharpsnout sea bream	23.6	4
Dentex dentex	Common dentex	11.8	4
Labrus bimaculatus	Cuckoo wrasse	5.9	4
Mugil sp.	Grey mullet	5.9	4
Phycis phycis	Forkbeard	5.9	4
Sciaena umbra	Brown meagre	4.4	4
Anthias anthias	Swallowtail sea perch	2.9	4
Mycteroperca rubra	Comb grouper	2.9	4
Seriola dumerili	Greater amberjack	2.9	4
Boops boops	Bogue	1.5	4
Dicentrachus labrax	Sea bass	1.5	4
Engraulis encrasicolus	European anchovy	1.5	4
Gobius niger	Black goby	1.5	4
Parablennius incognitus	Blenny	1.5	4
Sparus aurata	Gilthead bream	1.5	4

Benthic faunal assemblage development also affected the hollow passageways through the pyramid where denser knots of hydroids, bryozoans, sponges and bivalves had developed, making these passages excellent shelters and possible grazing zones for both adult and young individuals of many species (Colour Plate 6). Fish censuses carried out between 1994 and 1998 showed a further change in fish communities (Relini *et al.*, 1998b). The goby (*G. cruentatus*) disappeared whilst other species increased in abundance and the following species were recorded for the first time: two blennids (*Parablennius gattoruggine*, Brünnich); *P. ocellaris*, L., the wrasse (*Symphodus doderleni*, Jordan) and the Mediterranean moray eel (*Muraena helena*, L.).

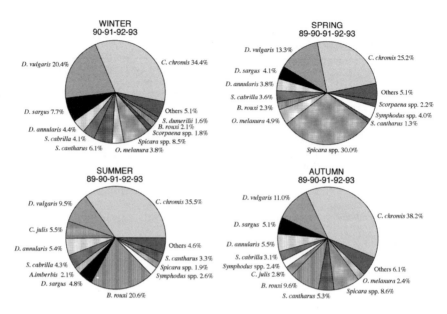

Figure 6.   Seasonal changes in fish population composed censused by divers. Mean value of 5 years observation.

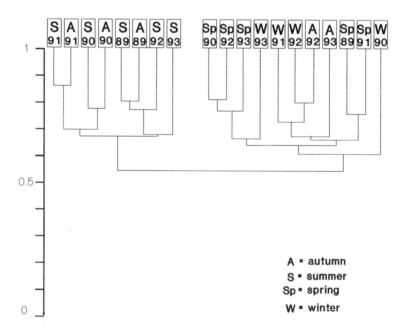

Figure 7.   Cluster analysis of seasonal assemblages computed using the Kluczynski Index (see Relini *et al.*, 1994b).

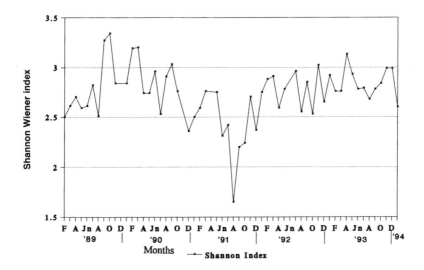

Figure 8. Shannon-Wiener index computed each month for fish community.

Recruitment

During the censuses the presence of young was often recorded, indicating that the artificial structures acted as a nursery. Three size classes, juveniles, medium, large were defined *a priori* according to the maximum recorded total length of each species in the literature. For example, juveniles of *Serranus cabrilla* were defined as individuals with a total length of < 9 cm, medium from 9 cm to 18 cm and large > 18 cm (Fig. 9). In the case of the *Coris julis* (Fig. 10) only juvenile fish were ever recorded, and these presented exclusively in the summer and autumn months. Recruitment was greatly reduced in the summer of 1992. *Serranus cabrilla* juveniles were mainly recorded from July to December (Fig. 9). Even for *C. chromis* and *Symphodus* spp. recruitment occurred exclusively in the late summer and autumn months. (For further information see Relini *et al.*, 1997a).

Biomass

The biomass trend as a monthly average of the censuses is shown in Fig. 11 (study area 280 m², 1026 m³ of water volume, encompassing 80 m³ of the artificial reef, i.e. an area around two pyramids with a connecting 'corridor'). The values are a rough estimate of biomass because of the methodology used. The biomass range was between 117 g m⁻² (33 g m⁻³ taking water volume into account), and 424 g m⁻³ (in relation to the volume of the two pyramids) for February 1992 and 9.59 g m⁻² (2.69 g m⁻³, 34 g m⁻³) for February 1993. The averages of the 62 censuses were 31.68 g m⁻² (8.89 g m⁻³, 114.06 g m⁻³). These are high biomass values compared to natural areas and other studies carried out on Mediterranean artificial reefs (Relini *et al.*, 1995a) and, in particular, compared to muddy/sandy seabed areas around the Loano artificial reef where only low levels of biomass were observed (unpublished observations). Recent work has shown that the presence of a FAD within the reef complex can increase the number of fish, and so the biomass, by a multiple of nine (Relini *et al.,* 1995c).

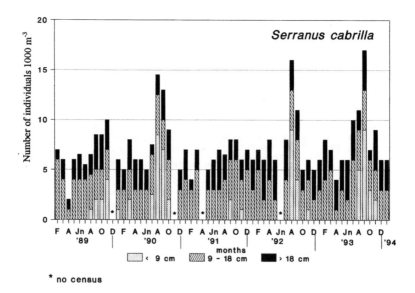

Figure 9.   Counts of the comber *Serranus cabrilla* (three different size classes from February 1989 to January 1994).

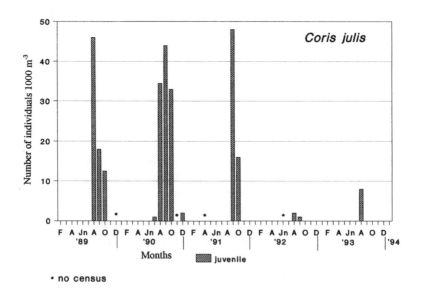

Figure 10.   Numbers of juveniles of the rainbow wrasse *Coris julis* in each month.

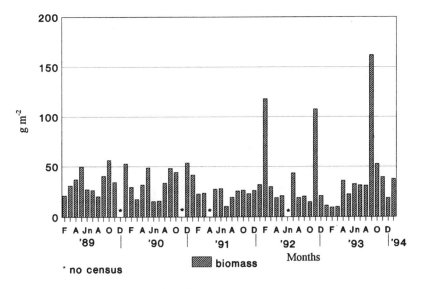

Figure 11. Monthly biomass values (g m⁻²).

Trammel Fishing

Between January and September 1989 some members of the local fishing co-operative filled out specially written questionnaires detailing a total of 191 catches, 33 of which were made using long-lines and 158 using trammel nets. These provided both qualitative and quantitative data (species, number of individuals, weight and size) with regard to the catches outside the pyramids (Relini *et al.,* 1990a). The maximum catches recorded were 1.9 kg 100 m⁻¹ of trammel net and 15.5 kg per 100 hooks for long-lines. From May 1991 to August 1992, 12 trammel catch surveys were carried out by researchers to improve knowledge of the catchable macrofauna present in the area. The results obtained, numbers of individuals and weight are shown in Fig. 12. The catch per 100 m of trammel ranged from 4.61 kg to 1.01 kg, with an average of 2.32 kg. The trammel yield (average 2.32 kg 100 m⁻¹) was clearly higher than obtained in ecologically comparable seas, such as Sicily, with regard to both natural areas (0.64 kg 100 m⁻¹; Toccaceli and Levi, 1990) and artificial reefs (0.31 kg 100 m⁻¹; Arculeo *et al.,* 1990). In Loano the catches were similar to those obtained on artificial reefs 1 year after deployment in the highly productive environment of the Adriatic Sea (2.80–1.71 kg 100 m⁻¹; Bombace *et al.,* 1990).

Conclusions

The fish population around the Loano artificial reef shows a high species richness, high biomass and good catch rates with trammel nets; 33.3% of species censused by the visual method can be considered as belonging to the resident population.

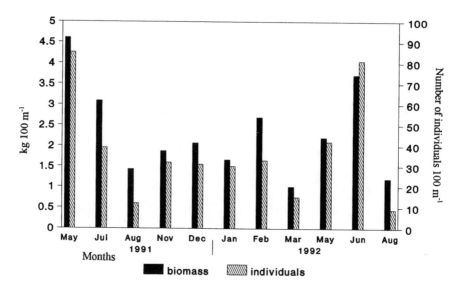

Figure 12.   Catches by trammel net in weight of fishes and number of individuals.

By comparing the Loano species with that of fish censused in the natural rocky area of Port-Cros (Marseilles, France), where 43 species were recorded (Harmelin, 1987), it can be seen that the two environments have certain characteristics in common. Twenty-seven species are common to both biotopes and *Chromis chromis* is numerically dominant in both communities. The family represented by the most species, both at Port-Cros and Loano, is the Sparidae, followed by the Labridae and Serranidae. The numerical predominance of species belonging to these first two families seems to be one of the general characteristics of rocky sea-beds in the north-west Mediterranean (Harmelin, 1984, 1987).

Many of the species noted at Loano are the same as those caught or observed in *Posidonia* meadows (Bell and Harmelin-Vivien, 1982; Harmelin-Vivien, 1982, 1983; Harmelin-Vivien and Francour, 1992). These authors also state that the majority of those species that frequent *Posidonia* meadows are not exclusive to this biocoenosis. Further evidence of the importance of artificial reefs is given by the exchanges which occur between the two habitats, i.e. the rocky (though artificial) substrata and the *Posidonia* meadow. Differences found between the fish communities at the Loano and Lavagna (a small reef built in 1993) artificial reefs are considered to be related to the respective ages of each structure and the presence or absence of *Posidonia* (Relini et al., 1997b).

General Conclusions

The results presented show that artificial reefs made from concrete blocks act in the same way as a natural rocky seabed and increase both species diversity and fish biomass. It seems likely, therefore, that artificial reefs could be used as a successful tool in the management of coastal waters, providing areas of food, shelter, and also nurseries for many species.

It is interesting to point out that although the benthic biomass settled on blocks varies between different artificial reefs in Italian waters, the fish communities are very similar (Relini *et al.*, 1994b) especially in terms of biomass.

The Loano artificial reef may be considered similar to a natural rocky habitat on which exchanges occur mainly with surrounding *Posidonia* beds, wrecks, community of brown alga *Cystoseira*, and skiophilous assemblages such as coralligenous and pre-coralligenous (Peres and Picard, 1964; in Peres, 1982). The Loano artificial reef, with its central protected area, can also be considered as a centre of reproduction, exporting macrobenthic organisms and fishes (in the adult and/or larval phase) to adjacent habitats.

The Loano artificial reef provides an instrument for diversifying the environment and providing organisms with their various required habitats. In this way it contributes to maintaining a high level of biodiversity, and has created new ecological niches which lead to more complicated food webs.

Finally the communities settled on artificial substrata can be used as a transplant to degraded zones in need of assistance to recover from ecological damage.

The main goals of the Loano artificial reef project were completely reached as confirmed from the results obtained from macrobenthos and fishes studied over the 10-year investigation.

References

Arculeo, M., G. Bombace, G. D'Anna and S. Riggio. 1990. Evaluation of fishing yields in a protected and an unprotected coastal area in N/W Sicily. FAO Fisheries Reports. **428**: 70–83.

Ardizzone, G.D., M.F. Gravina and A. Belluscio. 1989. Temporal development of epibenthic communities on artificial reefs in the central Mediterranean Sea. *Bulletin of Marine Science.* **44**(2): 592–608.

Badalamenti, F., G. Gioccone, M. Gristina and S. Riggio. 1985. An eighteen month survey of the artificial reef off Terrasini (Sicily): the algal settlement. *Oebalia.* **11**: 417–425.

Balduzzi, A., S. Belloni, F. Boero, R. Cattaneo, M. Pansini and R. Pronzato. 1982. Prime osservazioni sulla barriera artificiale della Riserva sottomarina di Monaco. *Naturalista Siciliano.* **3**: 601–605.

Balduzzi, A., F. Boero, R. Cattaneo-Vietti, M. Pansini and R. Pronzato. 1985. Etude du benthos sur les structures immergées dans la Réserve de Monaco. *Bulletin de l'Institut Océanographique,* Monaco. **4**: 163–165.

Balduzzi, A., F. Boero, R. Cattaneo-Vietti, M. Pansini and R. Pronzato. 1986. Long-term photographic records from the artificial reefs of Monaco (1980–1985). *Rapport de la Commission International de la Mer Méditerranée.* **30**: 264.

Bell, J.D. and M.L. Harmelin-Vivien. 1982. Fish fauna of French Mediterranean *Posidonia oceanica* seagrass meadows. I-Community structure. *Tethys.* **10**(4): 337–347.

Bombace, G. 1977. Aspetti teorici e sperimentali concernenti le barriere artificiali. Atti IX Congresso SIBM, Ischia. 29–41.

Bombace, G. 1981. Note sur les expériences de création de récifs artificiels en Italie. *Studies and Reviews (GFCM-FAO), Rome,* **58**: 321–337.

148   G. Relini

Bombace, G. 1989. Artificial reefs in the Mediterranean sea. *Bulletin of Marine Sciences.* **44**(2): 1023–1032.

Bombace, G., G. Fabi and L. Fiorentini. 1990. Preliminary analysis of catch data on artificial reefs in Central Adriatic. FAO Fisheries Reports. **428**: 86–98.

Cecere, E., A. Petrocelli, G. Relini and O. Saracino. 1993. Phytobenthic communities on fouling panels on Loano (Savona, Ligurian Sea). *Oebalia.* **19**: 163–172.

Falace, A. and G. Bressan. 1990. Dinamica della colonizzazione algale di una barriera artificiale sommersa nel Golfo di Trieste: macrofouling. *Hydrores.* **7**(8): 5–27.

Harmelin, J.G. 1984. Suivi des peuplements ichtyologiques du parc national de Port-Cros (Méditerranée, France). Mise en place d'un inventaire périodique. *Travaux Scientifiques du Parc National Port-Cros, France.* **10**: 165–168.

Harmelin, J.G. 1987. Structure et variabilité de l'ichtyofaune d'une zone rocheuse protégée en Méditerranée (Parc national de Port-Cros, France). *Marine Ecology.* **8**(3): 263–284.

Harmelin-Vivien, M.L. 1982. Ichtyofaune des herberies de Posidonies du parc national de Port-Cros. I: composition et variation spatio-temporelles. *Travaux Scientifiques du Parc National Port-Cros, France.* **8**: 69–92.

Harmelin-Vivien, M.L. 1983. Ichtyofaune des herberies de Posidonies des cotes provençales francaises. *Rapport de la Commission International de la Mer Méditerranée.* **28**(3): 161–163.

Harmelin-Vivien, M.L. and P. Francour. 1992. Trawling or visual censuses? Methodological bias in the assessment of fish population in seagrass beds. *Marine Ecology.* **13**(1): 41–51.

Montanari, M. and G. Relini. 1973. Variazioni stagionali del fouling di pannelli immersi a diverse profondità nell'Avamporto di Genova. Atti V Congresso Soc. It. Biol. Marina Ed. Salentina, Nardò. 305–326.

Peres, J.M. 1982. Major benthic assemblages. *Marine Ecology.* **5**(1): 373–521.

Relini, G. 1979. Ricerche in corso in Liguria sulle barriere artificiali. (Artificial reef investigation in progress in the Ligurian Sea). Atti 8° Simposio Nazionale sulla Conservazione della Natura, Bari, Aprile 1979: 79–87.

Relini, G. 1982a. La barriera artificiale nel Golfo Marconi (Mar Ligure). (The artificial reef in Marconi Gulf (Ligurian Sea)). *Naturalista Siciliano.* **3**: 593–599.

Relini, G. 1982b. Esperienze di barriere artificiale in Liguria. (Investigation on artificial reefs in the Ligurian Sea). Atti del Convegno delle Unità Operative afferenti ai sottoprogetti Risorse Biologiche e inquinamento Marino. (Roma, 10–11 Nov. (1981)). 155–164.

Relini, G. 1983a. Twelve years of experiments on artificial reefs in the Gulf of Genoa (Italy). *Journée d'études sur les aspects scientifiques concernant les récifs artificiels et la mariculture suspendue.* CIESM Cannes (1982): 73–75.

Relini, G. 1983b. Esperienze di barriere artificiali in Mar Ligure. (Artificial reef experiments in the Ligurian Sea). Atti Convegno Gestione e Valorizzazione della fascia costiera, Ancona 4–5 giugno (1982). *Gazzettino della Pesca.* **5**: 20–23.

Relini, G. 1984. Gestione della fascia costiera Ligure: problemi ed iniziative. (Management of coastal water in the Ligurian Sea: problems and plans). Atti del Convegno Nazionale 'Difesa del Mare', Genova 19 maggio (1984): 81–90.

Relini, G. and P. Cormagi. 1990. Colonisation patterns of hard substrata in the Loano artificial reef (Western Ligurian Sea). FAO Fisheries Reports. **428**: 108–113.

Relini, G. and S. Moretti. 1986. Artificial reef and *Posidonia* bed protection off Loano (Western Ligurian Riviera). FAO Fisheries Reports. **357**: 104–108.

Relini, G. and L. Orsi Relini. 1971. Affondamento in mare di carcasse di automobili ed inquinamento. (Car bodies in the sea and pollution). *Quaderni Civica Stazione Idrobiologia Milano.* **3/4**: 31–43.

Relini, G. and L. Orsi Relini. 1989. The artificial reefs in the Ligurian Sea (N-W Mediterranean): Aims and results. *Bulletin of Marine Science.* **44**(2): 743–751.

Relini, M. and G. Torchia. 1993. Distribuzione spaziale dei pesci sulla barriera artificiale di Loano (SV). *Biologia Marina Suppl. al Notiziario SIBM*. **1**: 195–200.

Relini, G. and M. Wurtz. 1977. La scogliera artificiale di Varazze (Mar Ligure) a sei anni dall'immersione. (The artificial reef at Varazze (Ligurian Sea) after six year immersion). Atti IX Congresso S.I.B.M., Ischia, Maggio 1977: 363–371.

Relini, G., A. Peirano, L. Tunesi and L. Orsi Relini. 1986. The artificial reef in the Marconi Gulf (Eastern Ligurian Riviera). FAO Fisheries Reports. **357**: 95–103.

Relini, G., M. Relini and G. Torchia. 1990a. Fishes of the Loano artificial reef (Western Ligurian Sea). FAO Fisheries Reports. **428**: 120–127.

Relini, G., N. Zamboni and F. Sonmezer. 1990b. Development of macrobenthos community in the Loano artificial reef. *Rapport de la Commission International de la Mer Méditerranée*. **32**(1): 27.

Relini, M., N. Zamboni, F. Tixi and G. Torchia. 1994a. Patterns of sessile macrobenthos community development in an artificial reef in the Gulf of Genoa (NW Mediterranean). *Bulletin of Marine Science*. **55**(2): 747–773.

Relini, M., G. Relini and G. Torchia. 1994b. Seasonal variation of fish assemblages in the Loano artificial reef (Ligurian Sea, NW Mediterranean). *Bulletin of Marine Science*. **55**(2): 401–417.

Relini, M., G. Relini and G. Torchia. 1995a. Fish population pattern in a coastal artificial habitat in the N-W Mediterranean. In Eleftheriou, A., A. Ansell and A. Smith. (eds.) *Biology and Ecology of Shallow Coastal Water*. Proceedings of the 28th EMBS, Crete. Olsen and Olsen, DK, pp. 359–368.

Relini, M., G. Torchia and G. Relini. 1995b. La barriera artificiale di Loano. *Biologia Marina Mediterranea*. **2**(1): 21–64.

Relini, M., G. Torchia and G. Relini. 1995c. The role of a FAD in the variation of fish assemblages on the Loano Artificial Reef (Ligurian Sea, N-W Mediterranean). *Proceedings of ECOSET'95*. Published by Japan International Marine Science and Technology Federation. **1**: 1–5.

Relini, G., M. Relini and G. Torchia. 1997a. Reclutamento delle specie ittiche su alcune barriere artificiali della Liguria (Fish recruitment on artificial reefs in the Ligurian sea) *Biologia Marina Mediterranea*. **4**(1): 269–276.

Relini, M., G. Torchia and G. Relini. 1997b. Fish assemblages in the Ligurian Artificial Reefs (N-W Mediterranean). In Hawkins, L.E., and S. Hutchinson with A.C. Jensen, M. Sheader and J.A. Williams. (eds.). *The Responses of Marine Organisms to Their Environment*. Proceedings of 30th EMBS, Southampton Oceanography Centre. pp. 337–343.

Relini, G., S.E. Merello and G. Torchia. 1998a. Influenza della sedimentazione sulla colonizzazione e successione ecologica di substrati duri (Influence of sedimentation on colonization and ecological succession on hard substrata). *Biologia Marina Mediterranea*. **5**(1): 154–163.

Relini, G., F. Torchia and S.E. Merello. 1998b. Studi ecologici nelle barriere artificiali italiane: osservazioni in Liguria (Ecological studies on Italian artificial reefs: observations in Liguria). *Biologia Marina Mediterranea*. **5**(3): 1822–1834.

Riggio, S., G. Badalamenti, R. Chemello and M. Gristina. 1986. Zoobenthic colonisation of a small artificial reef in the Southern Tyrrhenian: results of a three-year survey. FAO Fisheries Reports. **357**: 109–116.

Riggio, S., F. Badalamenti, R. Chemello and M. Gristina. 1990. Zoobenthic colonisation of a small artificial reef in the Southern Tyrrhenian Sea: results of a three-year survey. FAO Fisheries Reports. **428**: 138–153.

Toccaceli, M. and D. Levi. 1990. Preliminary data on an experimental trammel net survey designed to estimate the potential of a planned artificial reef near Mazara del Vallo (Italy). FAO Fisheries Reports. **428**: 154–158.

9. Artificial Reefs in the Principality of Monaco: Protection and Enhancement of Coastal Zones

DENIS ALLEMAND[1], EUGENE DEBERNARDI[2] and WILLIAM SEAMAN, Jr.[3]

[1]*Conseiller Scientifique de l'Association Monégasque pour la Protection de la Nature, Professeur à l'Université de Nice Sophia-Antipoli, Directeur de Recherche à l'Observatoire Océanologique Européen du Centre Scientifique de Monaco, Avenue Saint Martin, MC-98000 Monaco.* [2]*Président de l'Association Monégasque pour la Protection de la Nature. Secrétariat 7, rue de la Colle, MC-98000 Monaco – Principauté de Monaco.* [3]*Associate Professor, Department of Fisheries and Aquatic Sciences, and Associate Director, Florida Sea Grant College Program, University of Florida, Gainesville, Florida, USA.*

Introduction

Some of the earliest uses of modern artificial habitats in coastal waters of Europe took place in Monaco. Not only were these deployments of fabricated structures forerunners of current marine habitat alteration in many other areas, but also their focus on habitat restoration and protection instead of the traditional goal of fishery enhancement was unique. This chapter describes a variety of designed reef structures to protect nearshore seagrass bed communities and to enhance red coral colonization. Deployment of reefs in underwater reserves was a novel concept in resource management at the time (1976).

Background

The Principality of Monaco, city and state, is located in the south-eastern corner of the Alpes Maritimes Department (France) between Menton and Nice on the Mediterranean Sea (Fig. 1). An independent state (population 30 000) that measures 195 ha with a linear coast line of approximately 2 km, Monaco is situated between the foot of the Alps and the Ligurian Sea. It enjoys a very mild microclimate (average temperature of 16°C) due to this location. Since 1976 a co-operative programme between governmental and civic interests has used artificial reefs in management of coastal resources.

The proposal for deploying artificial reefs followed the establishment of two protected zones, or reserves, along the coast of Monaco, an initiative of S.A.S. Prince Rainier III. Subsequently, the project was organized and managed by the Association Monégasque pour la Protection de la Nature (AMPN). Each reserve deployed different design of reef module consistent with the goals of restoration and protection of seagrass communities in the Undersea Reserve of Monaco, and enhancement of red coral growth in the Coral Reserve (Fig. 1).

A.C. Jensen et al. (eds.), Artificial Reefs in European Seas, 151–166

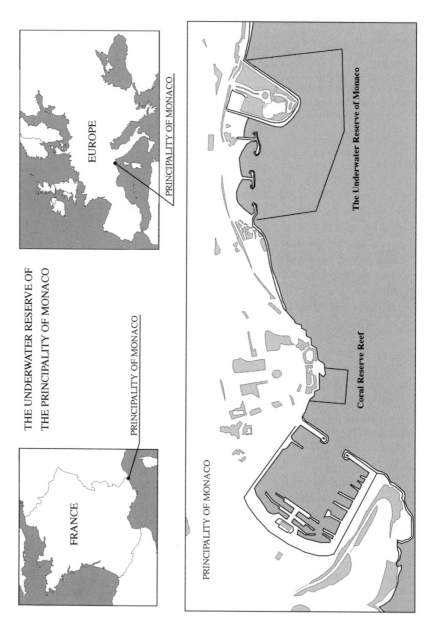

Figure 1. Location of the Principality of Monaco. Location of two underwater reserves along the shoreline of Monaco.

Undersea Reserve of Monaco

The reserve is located on the eastern side of the Monaco coast (Fig. 1). It follows the beaches of the Larvotto quarter, after which it was named originally. The trapezoid shape of the reserve has the coast as its base, which encompasses two sea walls made of stone blocks, down to 12 m in depth. A central breakwater protects the beaches of Larvotto.

The reserve has a surface area of approximately 50 ha and a depth ranging from 0 to 40 m. The seabed is sand, mud and gravel. A *Posidonia oceanica* bed of approximately 14 ha stretches from the west at a depth of 8–25 m (Fig. 2).

The surface perimeter of 2.2 km is marked by 42 buoys to denote the boundary for boats. The buoy's moorings are 120 kg anchors. Eight metres from the surface an intermediate buoy ensures that tension is maintained on the surface buoys. The surface buoy is linked by a loop which deadens the effect of swells. All metallic elements were eliminated to avoid electrolytic corrosion of components. Accurate mapping of the seabed was conducted by the Marine Nationale Française.

The Undersea Reserve of Monaco was created by Order of the Sovereign (No. 5851, 11 August 1976; completed by Order of the Sovereign No. 6256, 25 April 1978, and 10426, 9 January 1992). Legal arrangements specify a ban on fishing and spear fishing of all types, a ban of the use of motorized boats and a ban on laying

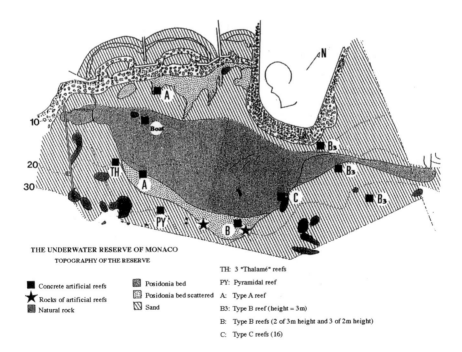

THE UNDERWATER RESERVE OF MONACO
TOPOGRAPHY OF THE RESERVE

- ■ Concrete artificial reefs
- ★ Rocks of artificial reefs
- ▨ Natural rock

- ▨ Posidonia bed
- ▦ Posidonia bed scattered
- ▨ Sand

TH: 3 "Thalamé" reefs
PY: Pyramidal reef
A: Type A reef
B3: Type B reef (height = 3m)
B: Type B reefs (2 of 3m height and 3 of 2m height)
C: Type C reefs (16)

Figure 2. Map of the undersea reserve of Monaco and siting of artificial reefs.

anchors. The purpose is to ensure maximum protection of marine biological diversity, both flora and fauna.

Previously, this area was greatly impoverished by intensive fishing. For example, the *Posidonia* beds were trawled daily by fishermen for over 30 years, using 'ganguis', trawl nets that have a 15 m opening with a 10–15 mm mesh. These were dragged through the beds of *Posidonia oceanica* to capture any animals seeking shelter there, seriously damaging the beds and destroying vast stretches. This type of fishing is now banned.

Coral Reserve

This reserve was created to protect the only coral slope in the territorial waters of the Principality. The reserve was created by Order of the Sovereign No. 8681, 9 August 1986 (completed by order No. 10426, 9 January 1992). The reserve is located at the foot of Loew's coral slope ('Pointe Focignana'), within an area of one hectare and a maximum depth of 38 m (Fig. 1).

The quantity of red coral (*Corallium rubrum*) gathered in the Mediterranean is estimated to be 70 tonnes per year. Although this species is not endangered, its high economic value and attractiveness to divers have left much of the shallow areas of the Mediterranean coast devoid of coral. This has caused concern about how to manage this resource and stimulated research on a species previously not well studied (Allemand, 1993).

L'Association Monégasque pour la Protection de la Nature proposed an *in situ* study of transplantation of red coral and a study of its growth and development under controlled conditions. The main experimental objectives were a study of adaptation and development of corals transplanted from their natural substratum on to an artificial one, under natural conditions at the same depth, monitoring growth and development of the colonies transplanted into artificial caves, and determining future reproduction of these transplanted corals.

Artificial Habitat Design and Placement

The earliest attempt to build an artificial reef in Monaco was unsuccessful. In 1977, 300 tonnes of rocks ranging from 30 to 400 kg were deployed from a barge as three reefs. However the depth (28–30 m) and layout of materials proved unattractive to fish. Since then, stable habitats have been deployed in the two reserves described previously.

Undersea Reserve Reefs

From 1979 to 1992, 31 objects were deployed in the Undersea Reserve of Monaco (Fig. 2). Of these, 30 were fabricated specifically as five different designs for aquatic habitat structures. The principal ones are described below. Funding of the reserve is assured by the AMPN, from membership fees, donations and the personal participation of S.A.S. Prince Rainier III, with technical assistance services provided by the Government of the Principality.

Cast Concrete Reef Units

The first specifically designed reefs in Monaco were built using roughcast alveolar concrete of the type used in the construction of floors in buildings. These were easier to transport than rocks and boulders, having the great advantage of being constructed on land. They were cemented together to make an elevated structure with many cavities which facilitated the settlement of benthic organisms. The individual blocks weighed 30 kg in air and had dimensions of 50 × 20 × 16 cm. The larger reefs were constructed on concrete slab bases, then taken to the site by a barge equipped with a crane.

Helicopters were used to deploy 16 smaller reefs, each weighing 500 kg; deployment took 15 min. Reefs were placed within a few metres of each other at a depth of 25 m. Unhooking of each reef was controlled by a diver equipped with communications equipment. The speed of execution and relatively low cost of this operation, due to the proximity of the heliport, make this solution of particular interest.

Three designs (reef types A, B and C) were constructed (Figs 3–5); principal physical characteristics are given in Table 1 (see also Colour Plates 7, 8 and 9). Each included 20 cm diameter Thor steel rings for lifting. To build each Type A reef 200 roughcast blocks were cemented randomly on a concrete slab. Each block in longitudinal section exposes three cavities in the shape of a section through a pyramid (Fig. 3). Two Type A reefs were immersed in June 1979, one at 8 m on the border of the seagrass bed, and the other on sand at 30 m. Their relative lightness provides the advantage that they do not sink into the soft seabed.

To improve attraction and colonization on the reefs, a second design (Type B) was constructed in an octagonal shape (Fig. 4). Reefs of two different weights

Figure 3. Principal reefs built from alveolar concrete roughcasting: Type A.

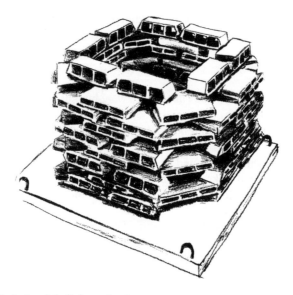

Figure 4. Principal reefs built from alveolar concrete roughcasting: Type B.

Figure 5. Principal reefs built from alveolar concrete roughcasting: Type C.

were built (Table 1). This shape allowed for more natural lighting, while maintaining a zone of shade in the central shaft of 1 m² located in the centre of the reef. These reefs resemble a hollow tower. The top row of roughcast blocks contained four steel rods (diameter 30 mm; height 1.80 m), to hold pierced roughcast blocks which could be moved for study. Such removal of blocks was impossible with the Type A reefs, because of the cementing. Five units of type B reefs were grouped together while three other units were deployed along a line in the east part of the

Table 1. Physical characteristics of three types of concrete-fabricated artificial reefs deployed in Monaco

Type of reef	Dimensions of slab (metres)	Number of roughcast blocks on slab	Height on slab (metres)	Unit weight (tonnes)	Number deployed	Depth (metres)
Type A	3 × 2.5 × 0.15	200	1.5–2.0	7	2	8.3
Type B	3 × 2.5 × 0.15	ca. 400	2.0	10	3	14–22–34
		ca. 600	3.0	12	5	32
Type C	1 × 1 × 0.1	ca. 18	ca. 0.8	0.5	16	25

reserve (Fig. 2). The final reef design, Type C, used roughcast blocks on four levels, with an average height of 0.8 m and a central shaft of 0.5 × 0.5 m (Fig. 5). These reefs (16 units) were grouped together at a depth of 25 m.

Miscellaneous Structures

Three other reefs have been deployed in this reserve (Fig. 2). In December 1986 a pyramidal reef, made by stacking about 30 alveolar caissons (2 × 1 × 0.8 m) of reinforced concrete, was assembled at a depth of 28 m on sand (Fig. 6). The caissons have the shape of a parallelepiped rectangle open at one end. They are in the form of a tube of reinforced concrete 2 m in height and cross-section (1 × 0.8 m). They were piled on top of one another, leaving an open space of 1 m (between the end of each caisson lengthways) and 0.8 m along the side of each caisson which created a labyrinth for the fauna. This system also offered the cavities formed by the caissons themselves (a volume of 11.50 m³). This reef has overall dimensions of length 14 m, width 10 m, height 3 m. It was immediately colonized by wrasse (*Labrus turdus*), bream (*Oblada melanura*), goatfish (*Mullus barbatus*), saupes (*Boops salpa*) and moray eel (*Murena helena*; see Colour Plate 10).

In August 1989 three artificial shelter reefs, called 'Thalamé', patented by the Assainissement Entretien Environnement Ecologie (AEEE) company of Nice, France were submerged on the border of the *Posidonia* beds at 20 m (Anonymous, 1990; Fig. 2). This reef is in the shape of a turtle carapace (about 3.30 m in diameter; height 1 m) with three openings level with the seabed, which creates a shelter for coastal fish (Fig. 7). A few observational dives confirmed the presence of bream both inside and outside the shelter of these reefs, to the exclusion of conger and moray eels.

There has been only one reef made from materials that came to hand deployed in Monaco coastal waters. In June 1989 a teak boat was donated to be submerged in the reserve. This 12 m boat, 4 m wide, with a keel height of 3.5 m, was stripped of its engine, windows and all metal parts, and thoroughly cleaned to remove grease and oil. It now rests at 20 m depth (Fig. 2) on a sandy seabed, where it was colonized exceptionally quickly by macroalgae, invertebrates and fishes (bream, painted comber, damsel fish).

Figure 6.   Pyramid reef.

Coral Reserve Reefs

Given that the red coral (*Corallium rubrum*) prefers shaded regions (caves, slopes shielded from natural light), it was important that the experiment to enhance its abundance be conducted under conditions mimicking its natural habitat. It was decided that artificial caves would be used. Their dimensions, which took into account the need to have divers working inside, were 3 × 2 × 2.2 m. The thickness of walls was 0.10 m (Fig. 8; Debernardi, 1992). Each cave weighed 4 tonnes.

In December 1988 the Entreprise des Grands Travaux Monégasque moulded the four concrete caves. The company financed most of the construction costs. With the support of the technical services division of the public works department, the caves were submerged on 21 December 1988. Initially the caves were placed in the port to 'clean' the concrete. Two were later deployed in the coral reserve, at the foot of Loew's slope, 300 m east of the port at 30 m depth, and the others placed at the southern limit on the main reserve (Larvotto quarter), about 600 m from the first two, at 27 m (Fig. 2). The caves in these two groups were arranged perpendicular to one another to study any affects of marine currents.

To 'seed' the caves coral colonies of non-commercial size (base diameter: 8 mm; height about 3 cm) were collected on Loew's slope (Cattaneo-Vietti and Bavestrello, 1994). The colonies were first transplanted by screwing them to polypropylene panels. Each panel (60 × 19 × 3 cm) received six branches of coral. Some panels were fixed vertically (the coral was horizontal), and others placed horizontally (the coral was upside down). After 6 months, more than 60% of the colonies were lost. This was due either to necrosis of the tissue, or breakage of the calcareous axis.

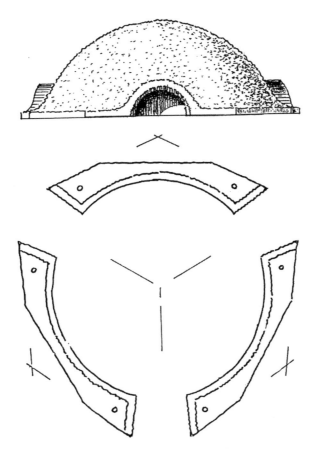

Figure 7. 'Thalamé' reef unit (Patented by AEEE company). The diameter of the base is about 3.30 m, and the total height is about 1 m.

Results were, however, still considered promising, because the remaining cultures were active and healthy.

A second attachment procedure was developed with the help of scientists at the aquarium of the Oceanographic Museum in Monaco. This time transplanted corals were fixed with a special underwater epoxy resin (UW paste, Devcon Ltd., Ireland). In this case, the colonies appeared to withstand transplantation, and the percentage of active cultures approached 100% 4 years afterwards (see Colour Plate 11). In addition to polypropylene supports, porphyry supports were tried.

Research Results and Discussion

When the project to establish artificial reefs in Monaco was started, the concept of creating underwater reserves (e.g., Bohnsack, 1993) was not well established.

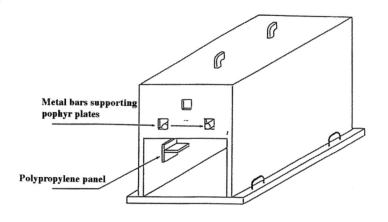

Metal bars supporting pophyr plates

Polypropylene panel

LONGITUDINAL SECTION   CROSS SECTION

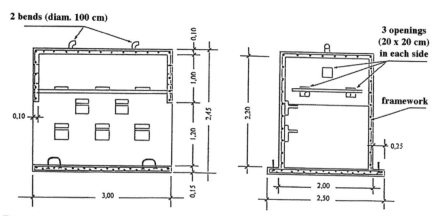

2 bends (diam. 100 cm)

3 openings
(20 x 20 cm)
in each side

framework

Figure 8.   Coral caves used in coral transplantation (From Derbernardi, 1992).

Further, using reefs for purposes other than the traditional goal of enhancing fishing (Stone *et al.*, 1991) was not a common practice.

Therefore deployment of fabricated structures (instead of surplus materials) in the two reserves along the coast of Monaco for purposes of ecosystem restoration and protection represented somewhat of a new approach in artificial reef practices in Europe. Assessment of the performance of the various artificial habitats has addressed biological issues and includes both experimental and monitoring approaches to acquire data.

Undersea Reserve Protection and Restoration Reefs

Prior to establishment of the undersea reserve, depletion of fauna by intensive fishing had led to the disappearance of species such as sea urchins, rock lobster and fish including the brown meagre (*Corvina nigra*). Coupled with a subsequent ban on fishing, the artificial reefs have increased biological diversity in this area. Species such as *Palinurus vulgaris, Corvina nigra, Sparus aurata* and *Scorpaena scrofa* have reappeared.

The evolution of the benthic population was followed by a group from the Istituto di Zoologia dell'Università di Genova, Italy (Balduzzi *et al.*, 1986; Boero, 1982; Cattaneo, 1982; Pansini, 1982). At least 36 species of sponges, 31 species of hydroids, 41 species of bryozoans and 12 species of opistobranch mollusc have been identified on the artificial structures (Balduzzi *et al.*, 1986). A 3-D computer reconstruction of the evolution of algal and benthic fauna colonization of the reef has been developed (de Vaugelas *et al.*, 1992).

Given the relative high sedimentation rate outside the *Posidonia* beds, and the mostly horizontal, rocky natural substrata, a fauna that resists high levels of sedimentation is apparent. The artificial reefs accumulate less sediment due to their mostly vertical orientation. These reefs have created areas that have greater diversity and are therefore qualitatively richer than the natural communities. These artificial habitats have allowed the establishment of benthic communities analogous to those found on rocky vertical walls (Balduzzi *et al.*, 1986). Hydroids and bryozoans established themselves rapidly as early colonizers, but their numbers diminished as a more complex community developed. The encrusting bryozoans were replaced after five years by species with an erect form. Attachment of sponge larvae has been favoured by the irregular surface of the roughcasting. These structures, however, inhibit the colonization of burrowing species.

Barnabé and Chauvet (1992) evaluated the ichthyofauna in the reserve (Table 2; Figs 9, 10). The majority of species populating the artificial reefs were characteristic of rocky coastal zones (Labridae and Sparidae) (Harmelin, 1987). In part because of their volume, heterogeneous structure and varying depth, the breakwaters constituted the richest zones in terms of numbers of species as well as individuals. Reefs composed of large natural stone blocks at a depth of 30 m also contain impressive numbers of fish. This area was particularly rich in two-banded bream (*Diplodus vulgaris*) of all sizes. All the size classes were present. The absence of fishing seemed to have led to an abundance of juveniles (Barnabé and Chauvet, 1992). The most abundant fishes were the damsels (Pomacentridae) and picarels (Centracanthidae). The alveolar reefs (Type C) represented oases, attracting numerous and varied fauna, dominated by damsel fish (*Chromis chromis*).

It was noted that these reefs were sparsely populated when isolated, containing on average two rockfish, two combers and a few damsels. Reefs of this type, when grouped, became 'super-reefs', which were far more attractive to the fauna (Barnabé and Chauvet, 1992). The breakwaters served as nurseries because they were in shallow water areas where larval fish assembled. The deeper reefs (more than 10 m below the surface) could not fulfil this function, but instead attracted a fixed population of species (Barnabé and Chauvet, 1992). Invertebrates were introduced into the reserve in the hope of restoring the stocks decimated by overfishing. To this end 25 000 sea urchins (*Paracentrotus lividus*) were transplanted

Table 2.   Fish species observed at the Undersea Reserve (adapted from Barnabé and Chauvet, 1992; English names from Fischer *et al.*, 1987).

Family	Species	Common name	
		French	English
Sparidae	*Diplodus annularis*	sparaillon	annular bream
	Diplodus sargus	sargue	white bream
	Diplodus vulgaris	sar	two-banded bream
	Diplodus puntazzo	becofino	sheepshead bream
	Boops boops	bogue	bogue
	Boops salpa	saupe	saupe (salema)
	Dentex dentex	denti	common dentex
	Oblada oblada	oblade	saddled bream
	Cantharus lineatus	canthare	black bream
	Sparus aurata	dorade royale	gilthead sea bream
Centracanthidae	*Spicara maena*	mendole	blotched picarel
Mugilidae	*Chelon labrosus*	muge à grandes lèvres	thick-lipped grey mullet
	Liza aurata	muge doré	golden-grey mullet
Serranidae	*Dicentrarchus labrax*	loup	sea bass
	Mullus surmuletus	rouget	striped red mullet
	Serranus scriba	serran écriture	painted comber
	Serranus cabrilla	serran chèvre	comber
Pomacentridae	*Chromis chromis*	castagnole	damsel fish
Scorpaenidae	*Scorpaena porcus*	rascasse brune	black scorpion fish
	Scorpaena scrofa	grande rascasse rouge	red scorpion fish
Blennidae	*Blennius gattorugine*	blennie	tompot blenny
	Blennius rouxi	petite blennie	blenny
Carangidae	*Seriola dumerilii*	sériole	greater amberjack
Sciaenidae	*Corvina nigra*	corb	brown meagre
Labridae	*Coris julis*	girelle	rainbow wrasse
	Symphodus cinereus	crénilabre cendré	long striped wrasse
	Symphodus doderleini	labre de Doderlein	five-spotted wrasse
	Symphodus roissali	crénilabre à 5 taches	axillary wrasse
	Symphodus mediterraneus	labre melops	peacock wrasse
	Symphodus tinca	crénilabre paon	corkwing wrasse
	Symphodus melops	merle	brown wrasse
	Labrus viridis	grand labre vert	green wrasse
Congridae	*Conger conger*	congre	conger eel
Muraenidae	*Muraena helena*	murène	moray eel

into the reserve (Gras, 1988). Some *Pinna nobilis* were also transplanted in two phases. Their development is being studied by scientists from Marseilles, who have recorded several juveniles at the site (de Gaulejac, 1992) since the study started.

Coral Reserve Enhancement Reefs

The colonies of red coral adapted well to the conditions within the artificial caves. A year after the beginning of this experiment, numerous organisms (sponges, hydroids, zooanthids, ascidians) proliferated in the caves without disturbing the

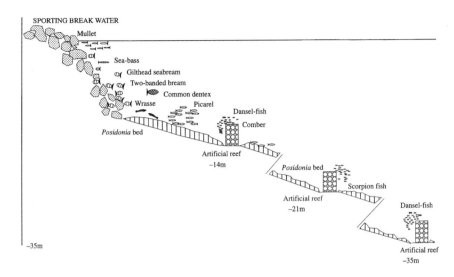

Figure 9. Schematic representation of fish population along a line from the Sporting breakwater to the artificial reef at 35 m (not drawn to scale) (Adapted from Barnabé and Charvet, 1992).

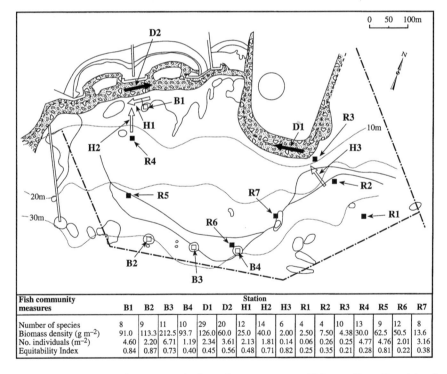

Fish community measures	Station															
	B1	B2	B3	B4	D1	D2	H1	H2	H3	R1	R2	R3	R4	R5	R6	R7
Number of species	8	9	11	10	29	20	12	14	6	4	10	13	9	12	8	
Biomass density (g m⁻²)	91.0	113.3	212.5	93.7	126.0	60.0	25.0	40.0	2.00	2.50	7.50	4.38	30.0	62.5	50.5	13.6
No. individuals (m⁻²)	4.60	2.20	6.71	1.19	2.34	3.61	2.13	1.81	0.14	0.06	0.26	0.25	4.77	4.76	2.01	3.16
Equitability Index	0.84	0.87	0.73	0.40	0.45	0.56	0.48	0.71	0.82	0.25	0.35	0.21	0.28	0.81	0.22	0.38

Figure 10. Fish community measures in the underwater reserve of Monaco (from Barnabé and Charvet, 1992). Fish were studied on artificial reefs (R), (B), *Posidonia* beds (H) and artificial breakwaters (D).

coral (Cattaneo-Vietti *et al.*, 1992). The coral itself inhibited the growth of these organisms around its base.

In April 1991, two years after the experiment started, many small coral colonies (10–15 m^{-2}) were observed on the ceiling of the caves. By January 1993, these colonies had grown well, some being over 30 mm high. Many of these colonies were quite squat. These growth rates were much faster than observed in the laboratory, where no growth was detected six months after transplant (Chessa *et al.*, 1992). This shows that the parameters for successful growth *in vitro* of red coral culture are not yet well understood but that the artificial caves provide appropriate conditions for growth.

Other small coral colonies also attached themselves on the Devcon resin, polypropylene panels and the walls of the cave. In February 1992, 150 young colonies were counted on the inside back wall of the vault of the cave placed on the western side of Loew's slope. A few, barely visible, colonies were counted on the two main side walls. In September 1992 several dozen larger colonies were counted and photographed on the same walls. These observations seem to indicate that the substratum was acceptable to the coral, and allows attachment and development.

In conclusion, it is too early to suggest that this type of experiment could be used in attempts at recolonization. Nevertheless the good results relating to the manipulation of the colonies and the good fixation of the corals in the short term gives us hope for future use of these caves as 'diffusers' of larvae for experimental reintroduction of red coral in zones where it has disappeared.

Set-up of a New Type of Artificial Cave for Rearing Red Coral

The concrete caves set up within the first phase of the programme of coraliculture had some disadvantages such as weight, the low levels of water movement within the caves and the difficulty of observation without disturbing the coral colonies. In order to overcome these problems, a glass fibre cave prototype was constructed in collaboration with the biologists from the aquarium of the Oceanograpic Museum of Monaco (see details in Allemand *et al.*, 1995). This prototype is light (about 160 kg; approximate dimensions: $2.5 \times 1.7 \times 2$ m) and can be easily transported and immersed. Its new design improved hydrodynamic conditions allowing the establishment of water current inside the cave and around the coral colonies. The accessibility for divers to coral colonies was made easier thanks to removable side panels which allowed observations without any difficulties.

This cave was immersed in the Monaco coral reserve in July 1993 at 39 m depth. Eighteen coral colonies and 36 sections (basal or apical part of coral colonies) were transplanted to study coral propagation. First results showed that transplanted colonies and sections survived in this cave and were able to produce larvae. This experiment also demonstrated the significant ability of red coral to reproduce asexually.

The good results of this new experiment and the successful attachment of the coral larvae gives us hope for future use of these kind of caves, not only as 'diffusers' of larvae but also for red coral propagation from mother colonies by vegetative reproduction, in experimental reintroduction of red coral in zones where it has disappeared.

References

Allemand, D. 1993. The biology and skeletogenesis of the Mediterranean red coral. A review. *Precious Corals and Octocorals Research.* **2**: 19–39.

Allemand, D., E. Debernardi, P. Gilles, N. Ounais, D. Théron and T. Thévenin. 1995. La réserve à corail rouge. In *XX ans au service de la nature*. AMPN, Monaco: pp. 121–130.

Anonymous. 1990. Immersion de 3 récifs-abri artificiels 'Thalamé' dans la réserve sous-marine de Monaco. Association Monégasque pour la Protection de la Nature. Compte-rendu des activités 1988–1989; p. 7.

Balduzzi, A., F. Boero, R. Cattaneo-Vietti, M. Pansini and R. Pronzato. 1986. La colonisation des structures artificielles immergés dans la réserve sous-marine de Monaco. Association Monégasque pour la Protection de la Nature. Compte-rendu des activités 1984–1985: pp. 19–33.

Barnabé, G. and C. Chauvet. 1992. Evaluation de la faune ichtyologique dans la réserve sous-marine de Monaco. Association Monégasque pour la Protection de la Nature. Compte-rendu des activités 1990–1991: pp. 51–59.

Boero, F. 1982. The benthic populations of the submarine reserve of Monaco: 2. Hydroids. *Journée d'études sur les aspects scientifiques concernant les récifs artificiels et la mariculture suspendue*. Cannes. C.I.E.S.M.: pp. 85–86.

Bohnsack, J.A. 1993. Marine reserves: they enhance fisheries, reduce conflicts and protect resources. *Oceanus.* **36**: 63–71.

Cattaneo, R. 1982. The benthic populations of the submarine reserve of Monaco: 3. Opisthobranch molluscs. *Journée d'études sur les aspects scientifiques concernant les récifs artificiels et la mariculture suspendue*. Cannes. C.I.E.SM.: pp. 87–88.

Cattaneo-Vietti, R. and G. Bavestrello. 1994. Four years rearing experiments on the Mediterranean red coral. *Biologia Marina Mediterranea.* **1**: 413–420.

Cattaneo-Vietti, R., G. Bavestrello, M. Berbieri and L. Senes. 1992. Premières expériences d'élevage de corail rouge dans la réserve sous-marine de Monaco. Association Monégasque pour la Protection de la Nature. Compte-rendu des activités 1990–1991: pp. 35–41.

Chessa, L.A., M.C. Grillo, A. Pais and L. Vitale. 1992. Natural and artificial settlements of red coral, *Corallium rubrum* (L.): preliminary observations. *Rapport de la Commission International de la Mer Méditerranée.* **33**: 32.

Debernardi, E. 1987. Conception et construction des récifs artificiels. Paper given at the Fourth International Conference on Artificial Habitats for Fisheries, University of Florida, November 2–6, 1987.

Debernardi, E. 1992. Expérience de coralliculture dans les eaux territoriales de la Principauté de Monaco. Association Monégasque pour la Protection de la Nature. Compte-rendu des activités 1990–1991: pp. 32–33.

de Gaulejac, B. 1992. Implantation de *Pinna nobilis* dans la réserve sous-marine de Monaco afin d'augmenter le nombre de géniteurs. Association Monégasque pour la Protection de la Nature. Compte-rendu des activités 1990–1991: pp. 44–45.

de Vaugelas, J., F. Loquês and G. Obolenski. 1992. Description des récifs artificiels de la réserve de Monaco en images de synthèse 3D. Association Monégasque pour la Protection de la Nature. Compte-rendu des activités 1990–1991: pp. 47–49.

Fischer, W., M. Schneider and M.-L. Bauchot. 1987. Fiches FAO d'identification des espèces pour les besoins de la pêche (révision 1). Méditerranée et Mer Noirc. Zones de pêche. 37. **Vol II**. Vertébre's. Rome FAO: pp. 761–1530.

Gras, G. 1988. Tentative de repeuplement en oursins comestibles de la réserve sous-marine de Monaco: transfert massif de *Paracentrotus lividus* en provenance du Golfe de Marseille. Association Monégasque pour la Protection de la Nature. Compte-rendu des activités 1986–1987: pp. 26–29.

Harmelin, J.G. 1987. Structure et variabilité de l'ichtyofaune d'une zone rocheuse protégée en Méditerranée (Parc National de Port-Cros, France). *Mar. Ecol. Pubbl. Staz. Zool. Napoli.* **8**: 263–284.

Pansini, M. 1982. Les peuplements benthiques de la réserve sous-marine de Monaco: I. Spongiaires. *Journée d'études sur les aspects scientifiques concernant les récifs artificiels et la mariculture suspendue.* Cannes. C.I.E.S.M.: pp. 83–84.

Stone, R.B., J.M. McGurrin, L.M. Sprague and W. Seaman Jr. 1991. Artificial habitats of the world: synopsis and major trends. In Seaman, W. Jr and L.M. Sprague (eds.) *Artificial Habitats for Marine and Freshwater Fisheries*, Academic Press, San Diego, California, pp. 31–60.

10. Artificial Reefs in France: Analysis, Assessments and Prospects

GILBERT BARNABÉ[1], ERIC CHARBONNEL[2], JEAN-YVES MARINARO[3], DENIS ODY[4] and PATRICE FRANCOUR[5]

[1]Université Montpellier II, Station Méditerranéenne de l'Environnement Littoral, 1 Quai de la Daurade, 34200 Sète, France. [2]Gis Posidonie, Faculté des Sciences de Luminy, 13288 Marseille, Cédex 9, France. [3]Université de Perpignan, Laboratoire de Biologie Marine, 52 Av. de Villeneuve, 66860 Perpignan, France. [4]890, carraire de Salins, 13090 Aix-en-Provence, France. [5]Gis Posidonie, Laboratoire Environnement Marin Littoral, Faculté des Sciences, Université Nice-Sophia Antipolis, Parc Valrose, 06108 Nice, Cédex 8, France.

Introduction

French artificial reef experiments started in 1968, with pilot reefs made from waste materials being placed in local initiatives. It was only at the beginning of the 1980s that a concerted programme was developed, reef shape and materials being specially designed and specified to withstand marine conditions. Almost 40 000 m^3 of concrete reefs have since been deployed along the French Mediterranean coast, with two complementary objectives:

(1) preservation of traditional and artisanal coastal fisheries, by creating a barrier against illegal trawling within the 3 nautical mile limit;
(2) increasing biological production in impoverished marine areas, and promoting recovery of the natural environment, degraded by human activities (e.g. coastal development and destruction of seagrass beds).

Artificial Reefs on the French Atlantic Coasts

French Atlantic coasts are diverse in terms of their morphology but have two features in common: the presence offshore of a substantial, gently sloping continental shelf substratum of mobile silty sand, and exposure to westerly winds which generate long swells. These factors in combination create conditions which are potentially deleterious to artificial reefs and their fauna.

Three examples of Atlantic reefs, all constructed from waste materials, are worthy of mention:

(1) Pilat. The oldest reef, consisting of 26 old car bodies, was placed in water 12–18 m deep at Pilat (Basin of Arcachon) in 1971. The car bodies rapidly filled with sand and corroded.
(2) Langrune sur Mer. Two separate reef installations, based on ballasted old tyres, were deployed in 12 m of water, the first in 1975 and the second, of 170 tyres, in 1983. The French national agency for waste (Agence Nationale pour

A.C. Jensen et al. (eds.), Artificial Reefs in European Seas, 167–184
© 2000 Kluwer Academic Publishers. Printed in Great Britain.

la Récupération et l'Elimination des Déchets, ANRED) participated in this project. Tyres are generally considered to be non-polluting in sea water and have been widely used elsewhere in the world for reef construction. Nevertheless their use in reef building remains a somewhat controversial issue.

(3) Gulf of Gascony. Several reefs, also composed of tyres, have been established at Mimizan, on the coast of Landes, since 1988, with the support of a recreational divers' association which seeks to restore and enhance the marine fauna of the area. The artificial reef sites are located on a sandy seabed at 12–25 m depth. Because of the heavy swell to which the area is exposed secure moorings are necessary. Study of the reefs by diving is also rendered difficult by the swell. Nevertheless it has been possible to identify 50 species (of which 16 are fish) associated with the reef whereas the peripheral benthic fauna is impoverished in terms of both numbers and diversity (de Casamajor, 1992). A new reef project is in progress near Cap Breton, deploying 800 concrete modules stacked into chaotic heaps at a cost of FF 1 000 000 (G. Fourneau, personal communication).

In summary, on the basis of the three examples given, Atlantic reefs suffer various problems associated with siltation, damage caused by the action of swell and difficulties associated with observation by SCUBA divers. In consequence there have been few artificial reef deployments on the French Atlantic coasts in comparison with the Mediterranean.

Recently, a new project in the Gulf of Gascogne proposed the use of artificial reefs for aquaculture (sea ranching). Juvenile sea bass (*Dicentrarchus labrax*), coming from hatcheries, will be released in the open sea near artificial reefs. Fish will be pre-conditioned to an acoustic signal in an attempt to persuade them to stay around the reefs, which will provide food and shelter (Hydro M, personal communication).

Artificial Reefs on the French Mediterranean Coasts: Location and Installation

The Geographic and Hydrological Context

Broadly speaking, the French Mediterranean coasts can be divided into two parts. In the east the coast shows a high, indented profile falling abruptly into deep water. The coastal sub-littoral band of the east coast (Région Provence-Alpes-Côte d'Azur) is narrow, owing to the almost total absence of a continental shelf. It is characterized by a nearly continuous belt of the seagrass *Posidonia oceanica* which grows in shallow water down to a depth between 30 and 38 m. The area is bathed by the westward-flowing Liguro-Provençal current. By contrast, in the west, isolated rocky outcrops separate large sandy beaches which themselves represent the limits of an immense continental shelf extending from Cap Creus to Cap Sicié.

Conditions west of the mouth of the Rhône river (Golfe du Lion-Région Languedoc-Roussillon) are quite different, as the 'Mediterranean' character of the coast is modified by the western Ekman deflection of the Rhône waters. A resultant feature of the Golfe du Lion is that the waters are more productive than those to the east. This is a point which must be remembered, not only in relation to the colonization of reefs in the two regions, but also when considering the socio-economic reasons for establishing specific types of artificial reefs.

For ease of subsequent discussion, assume that the west coast corresponds to the sub-littoral of the Languedoc-Roussillon (LR) region and the east coast incorporates the Provence-Alpes-Côte d'Azur (PACA) region. The locations of artificial reefs in these two areas are indicated in Fig. 1 and Table 1.

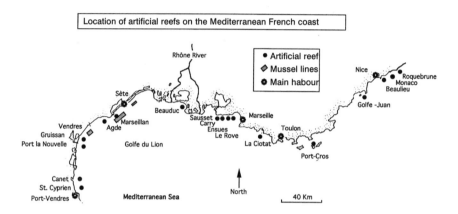

Figure 1. Location of artificial reef areas on the French Mediterranean coasts.

Table 1. Artificial reefs along the French Mediterranean coasts.

Deployment site	Volume placed (m³)
East Coast (PACA)	
Roquebrune	4300
Beaulieu	2880
Golfe Juan	7970
Port-Cros	35
La Ciotat	749
Marine Park Côte Bleue	3906
Total east coast	19840
West Coast (LR)	
Beauduc	640
Marseillan	510
Agde	3000
Gruissan	6500
Port-La-Nouvelle	4216
Canet	2116
Saint-Cyprien	2244
Total west coast	19226
Total	39066

Aims, Materials and Participants

The aims of reef deployment tend to differ between the Languedoc-Roussillon and Provence-Alpes-Côte d'Azur regions. In Languedoc-Roussillon, artisanal coastal fishermen have sought the support of local authorities to install obstacles (stakes and artificial reefs) to protect long lines and fixed nets from being damaged by trawlers making illegal fishing incursions within the 3 nautical mile limit. While many reefs in the area primarily serve this protective purpose, a few have been introduced to increase biological production by generating new habitats similar to natural rocky areas.

By contrast, in the Provence-Alpes-Côte d'Azur region and at Monaco, the reefs have largely been introduced to assess the potential for the restoration of fauna and *Posidonia oceanica* seagrass beds damaged by coastal developments. Such activity has impacted irreversibly on 15% of the natural habitat between the surface and 10 m (Meinesz *et al.*, 1993).

In France the licensing conditions for artificial reefs vary from region to region. Typically a region will appoint an administrative body which is charged with the responsibility of assembling relevant data and evaluating the input from all interested parties (e.g. fishermen, yachtsmen, fish-farmers, ecologists, tourist bodies, environmentalists, scientific organizations, territorial and administrative groups). Funding bodies need to be approached and the proposal has to be authorized by the French state. The overall procedure allows coherence, but with input from many participants.

The first reef deployment in the French Mediterranean, in 1968/69, consisted of old car bodies. Altogether, a total volume of some 400 m^3 was deposited in 20 m of water on a sandy seabed off the bathing resort of Palavas (Languedoc-Roussillon). This reef was deployed by the Compagnie Générale Transatlantique, with support from public funds.

Tyre-based reefs of 3000 m^3 and 3500 m^3 were established in 1980 off Port-la-Nouvelle (Languedoc-Roussillon) in 35–39 m of water, and between 1979 and 1983 in Golfe Juan (Provence-Alpes-Côte d'Azur) at a depth of 27–30 m, respectively. Apart from these, all other artificial reefs in the French Mediterranean, amounting in total to some 39 000 m^3, have been constructed from concrete and specially designed. Most of these reefs were deployed between 1985 and 1989; they are roughly equally divided between the Languedoc-Roussillon and Provence-Alpes-Côte d'Azur regions (Table 1).

Generally, the decision to place artificial reefs involved a consensus between parties with differing concerns and interests: administrators anxious to find alternatives to imposing unpopular fishing restrictions, professional fishermen concerned at the loss of fishing grounds but recognizing the inevitability of the imposition of other restrictions if they did not agree to reef deployment and scientists eager to explore new means of potentially enhancing and/or preserving fish stocks and protecting the environment. With the consensus established, initial scientific studies started to prepare for reef deployment.

Artificial Reefs on the East Coast (Provence-Alpes-Côte d'Azur)

The initial Golfe Juan reef construction programme ran from 1979 to 1983 and involved the deployment of 27 000 used tyres assembled in batches or as single units. These, together with 937 m³ of bricks and breeze blocks, formed a reef 'field' of some 3500 m³. In addition, further alveolar reefs (Fig. 2) were established off the French Riviera coast (80 × 8 m³ modules) and at Carry (9 × 13 m³ modules plus 27 × 4 m³ modules).

Initial reef deployment involved a somewhat arbitrary arrangement of the materials. From 1985 however the programme was extended using industrially made concrete reefs, specially designed to optimize assembly *in situ* into predeter-

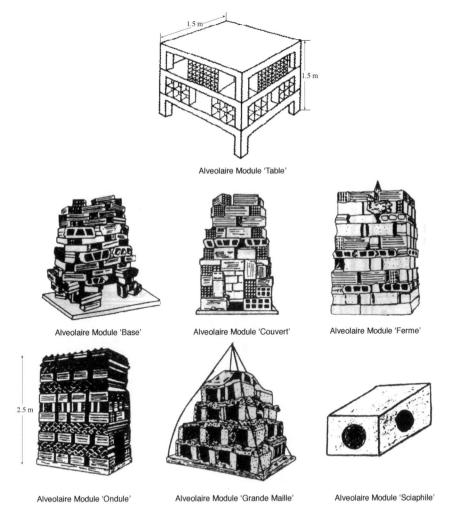

Figure 2. Example of alveolar reef units, made of breeze blocks and small bricks. The volumes of each unit vary from 1 to 8 m³ (from Charbonnel, 1989).

mined configurations. A variety of artificial reef module types have been used. Small cubic modules of 1–2 m³ (Fig. 3) have been deployed in chaotic heaps, ranging in volume from 20–500 m³ (Fig. 4). Additionally, large-volume modules (158 m³) have been placed as single units (Fig. 5). Anti-trawling reefs have been placed offshore of the Côte Bleue Marine Park near Marseilles. Sited off the cities of Martigues, Sausset, Carry, Ensuès and Le Rove, these anti-trawling reefs are

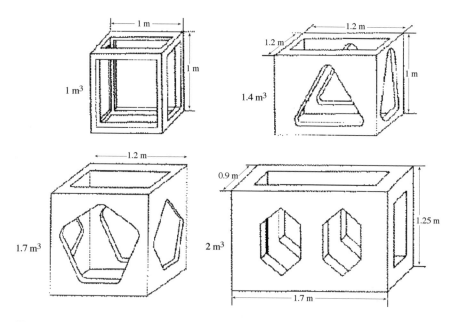

Figure 3. Small cubic reef modules of 1, 1.4, 1.7 and 2 m³ (from Charbonnel and Francour, 1994).

Figure 4. Small cubic modules, deployed in chaotic piles with a volume of 20–500 m³ (from Charbonnel and Francour, 1994).

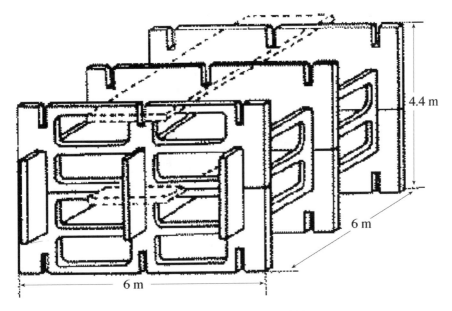

4.4 m

6 m

6 m

Figure 5. Large-volume unit (158 m³) (from Duval-Mellon, 1987).

made from 83 (12 tonnes each) rock blocks, 100 'sea rock' reef modules (a truncated concrete pyramid 2 × 2 m at the base and 0.5 × 0.5 m at the top), each having a volume of 1.6 m³ (Fig. 6) and 91 modules made with concrete telegraph poles (each module weighing 10 tonnes and having a volume of 13 m³).

Overall, artificial reefs with a total volume of more than 19 000 m³ have been deployed in the Provence-Alpes-Côte d'Azur region (Table 1): Martigues (1331 m³), Sausset (1321 m³), Carry (256 m³), Ensuès (592 m³), Le Rove (335 m³), La Ciotat (749 m³), Port-Cros (35 m³), Golfe Juan (7970 m³), Beaulieu (2880 m³) and Roquebrune (4300 m³). Typically the reefs have been located 15–50 m deep on a sandy seabed, below the limit of *Posidonia oceanica* meadow. The global cost of the Provence-Alpes-Côte d'Azur programme of reef deployment from 1985 to 1989 was of the order of FF 7.5 million, of which the greater part was provided by the EU (FEOGA), the French State, the Provence-Alpes-Côte d'Azur Region and local cities. The artificial reef deployment programme ceased in 1989, though a new reef was created in 1997 in the Côte Bleue Marine Park which has used concrete telegraph poles as reefs (Charbonnel and Francour, 1994a), at a cost of FF 1 100 000.

Artificial Reefs on the West Coast (Languedoc-Roussillon)

Following the 1980 trial programme at Port la Nouvelle, a consortium representing administrative, scientific and fishery interests planned, and in 1985 installed, a series of substantial reefs in the sub-littoral zones of five cities (Saint-Cyprien, Canet, Port-la-Nouvelle, Gruissan and Agde; Table 1) at depths ranging

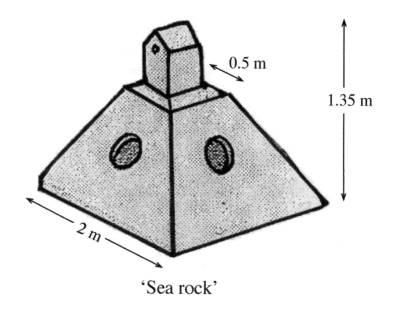

'Sea rock'

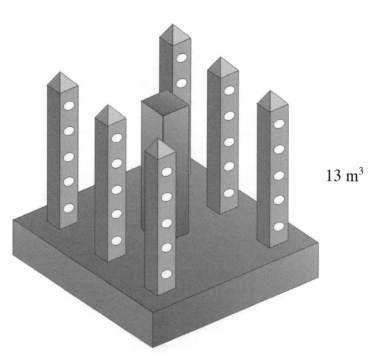

'Telegraph pole unit'

Figure 6. Anti-trawling 'sea rock' reef unit (1.6 m³) and module made from concrete telegraph poles (13 m³) (Gis posidonie, unpublished).

from 18 to 30 m. These artificial reefs were constructed from two components: large-volume units of 158 m³ (Fig. 5) and 10.4 m³ hexagonal units (Fig. 7). Total volume of these reefs was some 18 000 m³. The cost of this exercise was more than FF 7 million, of which a substantial part was met by EU funding, and about one-quarter was allocated to research. This research funding was largely allocated to preliminary geological studies at Saint-Cyprien and Canet, pre-installation study of the hydrology at Gruissan, module technology development, fishery impact assessment at Saint-Cyprien and Gruissan and bibliographic research. Little remained to support wider research into the ecological impact of these artificial reefs.

Since 1988 the emphasis has shifted further towards a policy of fishery management through the installation of anti-trawling reefs. In that year, a private agency (CEM-Marseille) took the initiative in placing a series of reefs to form an obstacle to trawling in the Golfe de Beauduc (near the mouth of the Rhône river, Fig. 1). Altogether, this exercise involved the deployment of 410 modules of the 'sea rock' type (total volume 640 m³). The project cost exceeded FF 1.5 million, again jointly funded by the EU (50%), the French State and regional groups (Tocci, 1996).

More recently, finance from the Languedoc-Roussillon region allowed the initiation of a project designed to protect both the molluscan culture zone off Marseillan-Agde (Fig. 1) and the fish stocks in the 10–20 m depth range. The reefs placed were made from units which comprised two concrete pipes of 1 m and 1.9 m diameter respectively, placed one inside the other (Fig. 8). Each unit was 2.5 m long and weighed 8.5 tonnes. In total, some 60 units were laid off Marseillan in 1992 and 200 units off Agde in 1995; each unit is about 200 m away from its neighbours. The cost was FF 450 000 for Marseillan and FF 1.8 million for Agde. Reef deployment is still in progress off Marseillan and Agde, and comparable protection for the long-line culture of mussels is under consideration in the cities of Frontignan and Aigues-Mortes.

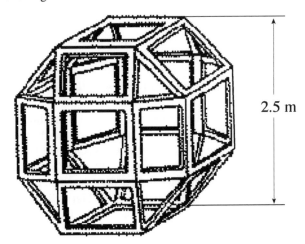

2.5 m

Figure 7. Hexagonal reef unit (10.4 m³) (from Duval-Mellon, 1987).

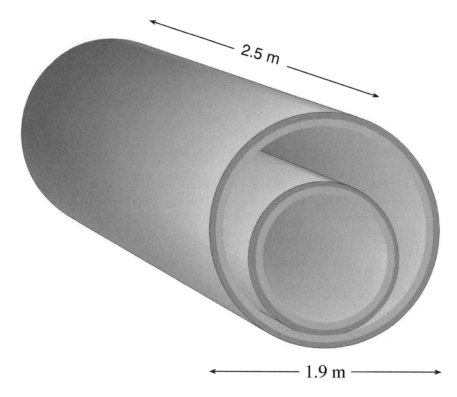

Figure 8.   Pipe unit reef (8.5 m³) used to prevent illegal trawling (from Collart and Charbonnel, 1998).

Analysis of the Effects of Artificial Reefs

Significant funding has been allocated in France for the construction and deploy-ment of artificial reefs. Nevertheless, impartial evaluation of the impact of reefs has been limited and some divergence of opinions has arisen which is worthy of further analysis.

Research on the Reefs of the East Coast (Provence-Alpes-Côte d'Azur)

Studies on fish assemblages, undertaken in parallel with the reef deployments, have been conducted in the Côte Bleue Marine Park (Charbonnel and Francour, 1994b; Ody, 1987); the National Park of Port-Cros (Ody and Harmelin, 1994); the French Riviera/Alpes-Maritimes (Charbonnel, 1989, 1990) and Monaco (Barnabé and Chauvet, 1991). The sampling method used has been visual census using SCUBA (Harmelin-Vivien and Harmelin, 1975; Harmelin-Vivien *et al.*, 1985), adapted to the specific architecture of each artificial reef, using a sampling stratification (Charbonnel *et al.*, 1995) technique.

The criteria used in biological characterization (summarized in Table 2) include number of species, abundance of individuals, population density and biomass. A

Table 2. Comparison of the available biological data from artificial reefs from the east coast (Provence-Alpes-Côte d'Azur region) of the French Mediterranean, for fish populations (after Charbonnel and Francour, 1994a). Certain species were deliberately excluded by the authors (microphages, cryptic species, *Coris julis*). These data are not therefore precisely identical and orders of magnitude only should be considered. 'Rugosity' is defined as being the structural complexity of the reef, on a rising scale of 1 to 4. – = no data available.

Authors	Location	Kind of reefs	Deployment date	Rugosity	Species richness	Number species/census	Density (ind. m⁻³)	Biomass (g m⁻³)
Ody (1987)	Carry,	Alveolar reef	1983	4	35–37	18.1–18.9	2.8–3	144–179
	Sausset	2 m³ reef units	1985	3	22–28	10.8–11.4	0.7–1	82–116
		Large reef unit 158 m³	1985	1	15–24	7–7.3	0.1–0.2	5–17
Charbonnel and Francour (1994)	Sausset	1.7 m³ reef units	1989	3	26–28	13.7–16.2	2–4.4	277–421
		Large reef unit 158 m³	1985	1	18	7.9–11	1.1–1.8	24–94
		1.7 m³ +158 m³ reef units	1985/89	2	22–25	13.4–16.2	1.1–3.2	106–464
Charbonnel (1989)	Golfe Juan	Alveolar reef	1983	4	19	10.1	2.7	556
	Beaulieu	Assembled tyres	1982	3	34	17.1	1.5	191
	Roquebrune	1 m³ reef units	1985	2	28–35	14–19.7	1.2–2	345–384
		2 m³ reef units	1987	3	20–27	8–11.4	0.4–1.3	55–279
		1 or 2 m³ + alveolar reef units	1983/85	3	24–32	10.7–15.8	0.5–1.2	189–279
		Large reef unit 158 m³	1986	1	9–18	4.2–8.2	0.08–0.2	7–27
Ody and Harmelin (1994)	Port-Cros	1 m³ alveolar reef units	1985/88	3	32–38	11.5–13.5	2.6–3.7	–
Barnabé and Chauvet (1991)	Monaco	Alveolar reefs	1977/83	4	4–13	–	–	2–62
		Rocky blocks	1977/83	3	9–12	–	2.9–7.9	94–212

comparison of these data shows that the character of biological development varies from one to another, depending essentially on reef design criteria (Charbonnel and Francour, 1994a).

One of the more striking features associated with the deployment of a reef is the appearance of new species and the increase in the size of fish assemblages in the area relative to those observed in the area prior to installation (increase of a factor of 150 for the abundance of fish and 500 for the biomass). Typically, fish assemblages on artificial reefs are similar to those on natural rocky areas; the differences can be attributed to the architecture of the reef. The three-dimensional structure of the reef determines the composition of the fish assemblage and reefs which increase the spatial heterogeneity are good fish attractants. The effectiveness of a module increases with its surface-to-volume ratio (Table 3), which varies from 1 to 30, according to Charbonnel (1989) and Ody (1987). In this context, artificial reefs constructed with small modules (1–2 m³) placed in haphazard chaotic heaps of around 100 m³ are particularly effective (Charbonnel, 1989, 1990). Aggregations of commercial species, such as the sparids, are often associated with such structures. By contrast, large-volume modules (158 m³), inspired by Japanese technology, have proved rather disappointing in the Mediterranean context, because of their very simple structure (a very large single chamber; Charbonnel, 1989, 1990) which apparently has too large a void space to be useful to fish. The time scale required to achieve the full potential of a given structure as a fish attractant seems to be of the order of two years (Ody, 1987).

Studies on the reefs in the marine reserve at Monaco have extended the range of ecological aspects examined. Here differences in colonization which can be related to both depth and reef volume have been reported (Barnabé and Chauvet, 1991), corroborating the findings of Charbonnel (1989, 1990) on reefs in the neighbouring French Riviera (Roquebrune, Beaulieu and Golfe Juan areas). Coastal protection works consisting of rocky blocks extending from 0 to 12 m depth in the Monaco reserve may also be regarded as shallow water artificial reefs, which play an important nursery role (judging by the richness in numbers of juvenile and post-larval forms present).

Table 3.   Surface-to-volume ratio of reef modules used on the French Mediterranean coasts.

Type of module	Free volume in m³	Total surface area in m³	Surface area/volume
Bonna large reef unit	158	175	1
Comin hexagonal medium reef unit	10.4	22	2
Doizon small cubic reef unit	1	4	4
Sabla small reef unit	1.4	8.6	6
Covered small reef unit	1.7	11	6.5
Covered medium reef unit	2	13	6.5
Pipe	8.5	53.9	6.5
Sea-rock	1.8	14.2	8
Alveolar reef unit	1 to 8	10 to 240	10 to 30

Research on the Reefs of the West Coast (Languedoc-Roussillon)

Various studies have followed the installation of artificial reefs as anti-trawling barriers in the Languedoc-Roussillon region. Some of these, as for example those of the reefs at Agde, Beauduc and Marseillan, relate primarily to the fauna of the reefs themselves whilst those at Gruissan have had wider implications.

Agde

Installed on the 3 mile limit in 16 m water depth, the reef consists of three piles of 5–15 hexagonal units. A short survey of the reef in 1988 (Magnan and Vray, 1989) found the surface of the structures to be extensively colonized by oysters and mussels with an estimated harvest by fishermen of 53 tonnes. The authors suggest that a limited exploitation would be possible, because these reefs support a biomass of mussels and conger eels of some 250 kg m^{-2} whereas the neighbouring soft seabed was only sparsely populated.

Beauduc

Tocci (1996) reported the effectiveness of the reefs at Beauduc in discouraging trawling. He also noted that these reefs, like those at Agde, were covered in oysters and mussels and, in addition, have a diverse associated fauna including sea bass, octopus and lobster.

Marseillan

Barnabé (1995) indicated that the pipe units used as anti-trawling reefs serve as a habitat for conger eels, sea bass and especially lobsters. Some relatively uncommon species (*Spirographis spallanzani, Echinus acutus*) were found but the main colonizers of the surfaces were mussels, forming a near-continuous carpet on the cylinders, with flat oysters also present. The presence of this 'living carpet' provides a habitat for a classic interstitial fauna. Also noteworthy is the abundance of juvenile fish in and around the modules. Sparids have been observed using the pipes as a refuge. These fish typically swim above the modules until danger threatened, when they enter the reef unit. Sea bass also occur around the modules whilst conger eels occupy the spaces between the two tubes. In addition to the preliminary study of Barnabé (1995), a subsequent one year biological survey was undertaken on the reefs off Marseillan and Agde (Collart and Charbonnel, 1998).

Saint-Cyprien and Gruissan

A fishing survey using trawl-net, gillnet and questionnaire, conducted by the government research organization IFREMER, found that the range of catch weights of fish from these reefs was 20% greater than that for fish taken in the neighbouring zone (Duval-Mellon, 1987). In a further IFREMER study on the

Gruissan artificial reef, Duclerc and Bertrand (1993) concluded that the reef had no enhancing effect on fish catches. In the light of the apparent success of artificial reefs on the west Mediterranean coast it is of interest and importance to analyse these findings further. Such analysis invites discussion as to the environmental and structural factors which lead to success or failure of an artificial reef in a particular role. It is important to gain such understanding in the sense that overall policy with respect to artificial reefs should not be unduly influenced by a single study, however carefully executed, without consideration to the wider context. Several points are worthy of further attention.

The Type of Artificial Reef

The Gruissan artificial reef was deployed with large-volume units (158 m³; Fig. 5). Studies of such large modules in the Provence-Alpes-Côte d'Azur region have shown them to be less effective as fish attractants than chaotic piles of smaller units. Supporting this conclusion are the findings of Ody (1987) and Charbonnel (1989, 1990) that the structural heterogeneity of a reef determines its biological complexity. In a similar vein, Marinaro (1995) stresses the importance of the surface of volume relationship; a feature which also disadvantages the reefs at Gruissan (Table 3). We may ask therefore whether the end result of Gruissan might not have been different if another type of structure, more appropriate to the role of the reef, had been used (Ody, 1990).

The Depth of the Reef

In their study at Monaco, Barnabé and Chauvet (1991) draw attention to the importance of structures in shallow water to populations of juveniles. The abundance of juveniles at such sites clearly implies the potential importance of artificial reefs in providing a resource and protection for the young which may, in the longer term, influence overall production by increasing juvenile survival. No such significant role in the relation to juveniles would have been expected at Gruissan because the water depth, at 28 m, was too great for juvenile fish.

The Type of Evidence Used to Make the Assessment

As befits a fishery-related survey, the study by Duclerc and Bertrand (1993) placed special emphasis on fishery returns and reports of catches from fishermen for the period immediately before, during and after reef installation (1984/86) followed by a comparison with data for the period 1987/89. They noted an apparent overall improvement in the catch by the local fishing fleet by 2283 kg (more than 15%) over the period. The average annual catch per boat in the fleet also appeared to have risen over the period by between 14 and 27% according to the type of net used (trammel-net or gillnet). Of individual species, the catches of flatfish (*Scopthalmus maximus*) and of sea bass were reported as increasing. Nevertheless, interpreting the data conservatively, Duclerc and Bertrand (1993) considered that such differences could be accounted for by natural variation and concluded that the study of professional fishing catches did not provide any evidence of significant variation in catches that could be attributed to the reefs. Given the care needed in interpreting inquiries from fishermen, cautious interpretation of these data from the Gruissan reef was justified in the light of the limitations of the survey techniques used. By the same token, however, these findings do not provide the desirable degree of certainty necessary for the making of policy decisions affecting proposals involving other types of reef construction designed for other purposes at different locations. However, IFREMER's policy decision was not to support research on artificial reefs, other than FADs (Fish Attracting Devices). This may indeed seem to represent a situation where a generalization based on a particular example has influenced much of the French effort in reef building programmes. The question one may well ask is, 'Would current IFREMER policy have been more favourably disposed towards reefs had their research been extended more widely to other reef systems?' However, in 1998 IFREMER reconsidered artificial reef deployment as a multi-purpose tool for global coastal management, and a report is due in the year 2000.

Mussel Culture by Long Lines and Aquaculture Cages

Less controversy surrounds the introduction of long lines for mollusc culture. To the purist these can scarcely be defined as artificial reefs but our justification in including them is that they were in the terms of reference of the FAO working party on artificial reefs, as were fish-attracting devices (FADs) (FAO, 1990). Long-line systems for mollusc culture have been introduced at three areas in the Languedoc-Roussillon region (Frontignan, Sète-Marseillan and Vendres, Fig. 1). Altogether, some 2000 ha are devoted to this purpose. This area is subdivided into 3–9 ha lots which range in depth from 15 to 35 m. Two lines, each 200 m long, supported by floats, are located in each 3 ha lot; together these are capable of supporting 20 tonnes of mussels (Barnabé, 1990).

Preliminary skin-diving studies revealed that these long lines attracted a secondary fauna, a component of which, particularly where the ropes were weighed down towards the seabed, were sea bass and sea bream. In some respects therefore these rope systems reflect the ability of artificial reefs to act as a base for organisms utilizing plankton as a food resource and in turn provide a niche for a secondary fauna. There can be no question, however, of long lines serving all the functions of artificial reefs. First they lack structural stability and are buffeted during storms; second they lack substance and cannot serve as obstacles to illegal fishing; and third they lack temporal stability, since the emplacement of the young mussels on the ropes is dependent upon human activity and the mussel stock is not itself self-replicating. No permanent ecosystem is established. Moreover, the expected life of a long-line system is about 5 years, compared with an estimated life of an artificial reef of about 40 years.

Floating aquaculture cages in open water can, like the long-line systems, act as attractants to some types of fish. Work in progress at Monaco (Barnabé, unpublished) on surface cages anchored above 90 m depth has shown that they are surrounded by many fish species, such as horse mackerel (*Trachurus trachurus*) and bogues.

Conclusions and Future Prospects

The research on French artificial reef systems indicates that such structures can be effective in various roles though exceptions can occur. Noteworthy amongst the exceptions is the study reported by IFREMER (Duclerc and Bertrand, 1993) where a variety of factors, including reef size and architecture, led to the conclusion that any favourable effects on post-deployment fisheries catch could not be distinguished from natural variation during the period of sampling. This study apart, it seems that suitably located reefs of appropriate structural form can have positive effects. Naturally however, the environmental effect of reefs will vary according to location, the nature of the surrounding substrata and whether the waters in the vicinity are eutrophic or oligotrophic.

Traditionally, artificial reefs have been associated with fishery activities. Currently, as we have seen, the roles played by artificial reefs are multiplying. The function as a fish attractant remains but now additional purposes are being fulfilled: for example, shallow water artificial reefs are frequented by post-larval and juvenile fish and so play the role of a nursery. Moreover, artificial reefs have been used to create obstructions to prevent illegal trawling, so protecting seagrass beds and sensitive inshore fisheries. Artificial reefs have also been used in some areas with the aim of enhancing faunal diversity. Coupled with such management of coastal resources and ecological aspects, an aesthetic aim, linked to tourism, could be also promoted, that of using shipwrecks for diving activities. An example of the latter was the sinking of an old 30 m boat near Gruissan in March 1994 to create a new sport-diving location. Many new projects aiming to use shipwrecks for diving activities are in progress, despite the French law which includes shipwrecks in the legal definition of waste material.

The diversity of roles is a testament to the multi-functional potential of artificial reefs. Nevertheless, much remains to be done in ascertaining the reef design (size, structural complexity and architecture) and materials best suited to the fulfilment of specific roles in particular locations. Substantially more knowledge is needed on the feature of reefs which attract target species and of the means by which enlarged stocks of favoured species can be sustained without unbalancing the neighbouring ecosystem. Such knowledge requires *in situ* studies of fish behaviour, a field which still remains poorly developed in France.

Marine protected areas represent an ideal location for the study of the efficiency of reefs in relation to the support of unexploited stocks. Another necessary, and as yet under-examined, research area is the socio-economic aspects of reefs where fish stocks are exploited. Here studies undertaken by scientists working in collaboration with the professional fishermen should prove invaluable. As adult fish stocks become depleted by fishing pressure, the importance of improving reproductive efficiency through larval and post-larval protection is increased. We know that artificial reefs can serve as nursery areas in qualitative terms but there is an urgent need for data to quantify their effectiveness.

The cost of reef deployment is a major obstacle to their development. The appropriate use of non-toxic industrial waste in reef construction could alleviate this issue, one such example being the use of old concrete telegraph poles as artificial reefs (Charbonnel and Francour, 1994a). Old tyres have been widely utilized as reefs in south-east Asia and in USA but, though the issue is somewhat controversial, there seems to be a contra-indication to their use, because of the fear of possible leaching of pollutants from tyres. Much other waste material has the potential for recycling as reefs if cleared of all toxic components (e.g. shipwrecks).

Artisanal coastal fisheries, in spite of their decline, represent, on a world scale, an effectiveness 20 times superior to industrial fisheries. The deployment of more reefs might well assist in the preservation of these localized fishing activities which are an important element in regional economies. Fishermen have not always had the resources in the past to establish fish-attractant structures on any scale. With the increasing recognition of the potential benefits of artificial reefs, interest in further deployment is gathering pace as is the pressure from those with other aims in relation to environmental management and recreational activities. The challenge for the future is not whether or not to deploy artificial reefs but rather how to meld the disparate reef-related aims and activities so as to optimize the potential for the common good.

These prospects come within the scope of an ecological development and management of the coastal zone which is in no sense utopian and can drastically change mankind's relationship with this fragile and coveted space. However, artificial reefs represent only one part of the overall management of coastal resources and it is salutary to recall in this context that currently only 0.5% of the French Mediterranean maritime zone is protected.

References

Barnabé, G. 1990. Open sea culture of molluscs in the Mediterranean. *Aquaculture*, E. Horwood, publ. New York, pp. 429–442.

Barnabé, G. 1995. Suivi de la zone de pêche protégée de Marseillan. 1er rapport. Univ. Montpellier II, Station Méditerranéenne de l'Environnement Littoral. pp. 1–10.

Barnabé, G. and C. Chauvet. 1991. Evaluation de la faune ichtyologique dans la réserve sous-marine de Monaco. *C.R. Activ Assoc Mon Protect Nat.* 1990–1991: 51–59.

de Casamajor, M.N. 1992. Suivi biologique du récif artificiel de Porto. Report Mémoire de Maitrise des Sciences et Techniques, Univ. Bordeaux III, Talence, pp. 1–146.

Charbonnel, E. 1989. Evaluation des peuplements ichtyologiques des récifs artificiels dans les établissements de pêche des Alpes-Maritimes. Rapport final. Report Conseil Général des Alpes-Maritimes, Conseil Régional PACA, CEE-FEOGA et Parc National de Port-Cros. pp. 1–96.

Charbonnel, E. 1990. Les peuplements ichtyologiques des récifs artificiels dans le département des Alpes-Maritimes (France). *Bulletin Société Zoologie Française.* **115**: 123–136.

Charbonnel, E. and P. Francour. 1994a. Etude sur les possibilités d'utilisation des poteaux électriques comme récifs artificiels. Report GIS Posidonie publ., Marseille, pp. 1–43.

Charbonnel, E. and P. Francour. 1994b. Etude de l'ichtyofaune des récifs artificiels du

Parc Régional Marin de la Côte Bleue en 1993. Report GIS Posidonie publ., Marseille, pp. 1–66.

Charbonnel, E., P. Francour, J.G. Harmelin and D. Ody. 1995. Les problèmes d'échantillonnage et de recensement du peuplement ichtyologique dans les récifs artificiels. *Biologia Marina Méditerranea.* **2**: 85–90.

Collart, D. and E. Charbonnel. 1998. Impact des récifs artificiels de Marseillan et d'Agde sur le milieu marin et la pêche professionnelle. Bilan du suivi 1996/97. Contrat Conseil Régional Languedoc-Roussillon & Conseil Général de l'Hérault. CEGEL & GIS Posidonie publ., pp. 1–168.

Duclerc, J. and J. Bertrand. 1993. Variabilité spatiale et temporelle d'une pêche au filet dans le Golfe du Lion. Essai d'évaluation de l'impact d'un récif artificiel. Report IFREMER, DRV 93.003/RH/Sète. pp. 1–42.

Duval-Mellon, C. 1987. Impact halieutique des récifs artificiels du Languedoc-Roussillon. Report IFREMER, DRV. 87.016/RH/Sète. pp. 1–196.

F.A.O. 1990. Les récifs artificiels et la mariculture. Rapport de la première session du groupe de travail, CGPM, FAO Rome, Rapp. Pêches 428. pp. 1–162.

Harmelin-Vivien, M. and J.G. Harmelin. 1975. Présentation d'une méthode d'évaluation *in situ* de la fauna ichtyologique. *Travaux Scientifiques du Parc National Port-Cros, France.* **1**: 47–52.

Harmelin-Vivien, M., J.G. Harmelin, C. Chauvet, C. Duval, R. Galzin, P. Lejeune, G. Barnabé, F. Blanc, R. Chevalier, J. Duclerc and G. Lassere. 1985. Evaluation visuelle des peuplements et populations de poissons: problèmes et méthodes. *Revue Ecologie (Terre Vie).* **40**: 467–539.

Magnan, N. and F. Vray. 1989. Stratégie d'exploitation du récif artificiel du Grau d'Agde Report ISIM, STE, Univ Sci Tech Languedoc, Montpellier. pp. 1–36.

Marinaro, J.Y. 1995. Artificial reefs in the French Mediterranean: a critical assessment of previous experiments and a proposition in favour of a new reef-planning policy. *Biologia Marina Méditerranea.* **2**: 65–76.

Meinesz, A., J.M. Astier, E. Bellone, J.R. Lefèvre, I. Genot, B. Hesse and P. Vitiello. 1993. Impact des aménagements gagnés sur la mer le long des côtes françaises de la Méditerranée (Régions Provence-Alpes-Côte d'Azur et Corse). *Bulletin Ecology.* **24**: 100–103.

Ody, D. 1987. Les peuplements ichtyologiques des récifs artificiels de Provence (France, Méditerranée Nord-Occidentale). Thèse 3ème cycle. Univ Aix-Marseille II. pp. 1–183.

Ody, D. 1990. Les récifs artificiels en France. Bilan; analyse; perspectives. *Bulletin Société Zoologie Française.* **114**: 49–55.

Ody, D. and J.G. Harmelin. 1994. Influence de l'architecture et de la localisation de récifs artificiels sur leurs peuplements de poissons en Méditerranée. *Cybium.* **18**: 57–70.

Tocci, C. 1996. Champs récifaux formant obstacle aux arts traînants. *Journal Recherche Océanographique.* **21**(1–2): 42–44.

11. Artificial Reefs in Spain: The Regulatory Framework

SILVIA REVENGA, FRANSISCO FERNÁNDEZ, JOSE LUIS GONZÁLEZ
and ELADIO SANTAELLA

Secretaría General de Pesca Marítima, Ministerio de Agricultura, Pesca y Alimentación, C/ Ortega y Gasset, 57, 28006 Madrid, Spain

Introduction

Spain is a leading exponent of the deployment and use of artificial reefs in Europe. Spain is also the premier consumer of fish within the European Union. These two facts are not unrelated. Demand for fish is high: the average annual Spanish fish consumption in 1994 was 30 kg per person (MAPA, 1995). In recent years the tightening of international fishing regulations has restricted the activity of the Spanish 'distant water' fishing fleet while there has been conflict within the domestic fleet, each element competing for fish from the same fishing grounds but using different gear. To mitigate these problems the Spanish government has promoted a policy of deploying artificial reefs since 1987 (SGPM, 1987, 1991; Fig. 1) with the support of public funds from the European Union and from national and regional Spanish budgets.

Aims of Artificial Reefs in Spain

An artificial reef is a tool of Spanish fisheries policy that may be directed towards:

(1)   protection and enhancement of fish stocks;
(2)   impeding illegal trawling;
(3)   protecting marine habitats (mainly seagrass, i.e. *Posidonia* beds);
(4)   reducing conflicts between different commercial fisheries.

Surveys conducted among artisanal fishermen operating in artificial reef fishing grounds prove that the deployment of artificial reefs are appreciated because of their social benefits. Fishermen, aware of the benefits that artificial reefs can provide, are not only asking for reef deployment, but actively promoting reef creation and contributing to part of the cost. There is a similarity here with the Japanese, for example, where reefs are being built to focus the efforts of the commercial fishing fleet, with designs based on those found to be successful previously.

Artificial reefs appear to be a tool of coastal management that helps to share coastal zone resources among users; in the Spanish case these are primarily professional fishermen. This is a very interesting aspect of reef deployment because conflicts between different groups of fishermen are growing as fish catches decrease. Reefs may provide a way to diffuse such conflict by effectively partitioning resources between trawl fishermen and local, more artisanal, fisheries.

A.C. Jensen et al. (eds.), Artificial Reefs in European Seas, 185–194
© *2000 Kluwer Academic Publishers. Printed in Great Britain.*

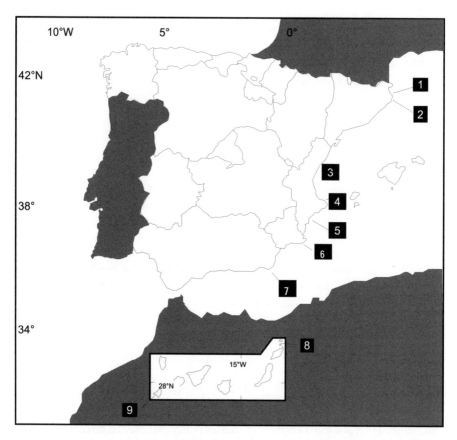

Figure 1.   Marine reserves of Spain. 1. Islas Medas (established 1990); 2. Ses Negres (1993); 3. Islas Columbretes (1990); 4. Cabo de San Antonio (1993); 5. Isla Tabarca (1986); 6. Cabo de Palos e Islas Hormigas (1995); 7. Cabo de Gata-Nijar (1995); 8. Isla Graciosa e islotes del norte de Lanzarote (1995); 9. Punta Restinga-Mar de las Calmas (1996).

Artificial reefs are also being placed in or near marine reserves (Fig. 1). Spain currently has nine marine reserves. In most of these marine reserves some kind of artificial reef has been placed:

(1)   Marine Reserve of Islas Medas: tetrapods have been tested.
(2)   Marine Reserve of Cap Negre: small ceramic units are being tested with a view to enhancing crustacean colonization.
(3)   Marine Reserve of Cabo de San Antonio: the Secretariat of Marine Fisheries has placed an artificial reef around the protected area but beyond its limits (S.G.P.M. and Tecnologia Ambiental S.A., 1995a).
(4)   Marine Reserve of Isla Tabarca: artificial reefs have been built inside the reserve (production units made out of concrete as well as wooden hulls).
(5)   Marine Reserve of Cabo de Palos e Islas Hormigas: an artificial reef (similar to that at Cabo de San Antonio) has been deployed to protect the beds of *Posidonia* (S.G.P.M. and Tecnologia Ambiental S.A., 1995b).

(6) Marine Reserve of Cabo de Gata: an artificial reef with production nuclei and protection belts has been constructed around the reserve.

The Use of Artificial Reefs in Spain

The first legal and policy regulation of Spanish artificial reefs dates back to May 1982 and focused on marine re-stocking, on both natural and artificial reefs. Artificial reefs to enhance fisheries were encouraged. No public funding was committed to support reef deployment.

Interest in artificial reefs was stimulated by Spain's entry into the European Economic Community (EEC) in 1986 (Santaella and Revenga, 1995). EEC regulations allowed grants for artificial reef deployment under the Multi Annual Guidance Programmes (MAGP). In the 1987–1991 MAGP, Spain applied for financial support for reef creation, considering reefs as a tool to reduce conflicts between fishing interests in Spanish waters, and to experiment with their usefulness in habitat protection and fishery enhancement. Regulations in force during 1987 required promoters to be public organizations, fishermen's associations or a government organization. All fishing had to be prohibited on and within 200 m of artificial reefs for 3 years following deployment.

Artificial reef projects were implemented between 1988 and 1993 off the Spanish coasts of the Mediterranean Sea and the Cantabric Sea (northern Spain), and around the Canary Islands. These were shallow artificial reefs, less than 50 m deep, placed in areas where trawling was also prohibited. The monitoring studies associated with these artificial reefs revealed that fishing bans were often ignored, as nets, traps or other fishing gear were found entangled on the reefs.

During the same period, the regional government of Asturias (northern Spain) placed six deep artificial reefs in waters shallower than 100 m. Spanish regulations ban trawlers from fishing in water shallower than 50 m in the Mediterranean Sea and off the SW Atlantic coast, and 100 m in the Cantabric Sea, so trawling around these reefs was prohibited (SGPM and MAPA, 1994). There were no additional, reef-specific fishing bans associated with these deployments. Four reefs were constructed from 88, 265, 150 and 150 concrete reef units, respectively, each unit weighing 3.5 tonnes. The other two reefs shared 230 units: 150 designed as 'protection' units (3.5 tonne solid units with sharp edges intended to deter and prevent trawling activity) and 80 modules as 'production' nuclei (5.5 tonne concrete structures containing holes). One hundred and seventy square kilometres of seabed were protected from trawling activity by these six artificial reefs.

From 1994 onwards, there were changes in regulations associated with artificial reef deployment both in Spain and in the wider European Union (SGPM, 1994). The physical presence of the reefs was considered sufficient to regulate the type of fishing activity possible in their vicinity, and no additional regulations were associated with the structures. Artificial reefs were regarded as tools which promoted protection (from trawling), enhancement and development of fisheries. Many reefs are now located in designated protection zones, previously studied by the Instituto Español de Oceanografía (Fig. 2). However, artificial reefs can also be placed in areas not previously designated as protected zones, as long as the aims of the projects are to protect, enhance and develop fisheries. In this case the presence of an artificial reef automatically creates a no-trawling, protected zone in the waters above the structure and within 200 m of the reef margin. Other fishing methods are allowed within this zone.

Legal Aspects

Responsibilities: General and Regional Administrations

Marine fisheries are managed in Spain by both national and regional administrations. The national Ministerio de Agricultura, Pesca y Alimentación (MAPA), (Ministry of Agriculture, Fisheries and Food) regulates fishery activity beyond the baylines (lines drawn across the mouth of a bay between two headlands) or beyond the low-water tide mark up to 12 nautical miles offshore (i.e. Spanish territorial waters). Fishery activity between the shore and the baylines is controlled by regional administrations (Ministerio de Obras Publicias y Transportes (MOPT), 1992).

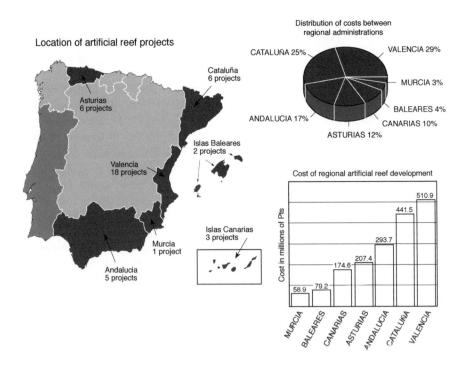

Figure 2.   Costs and location of Spanish artificial reef deployment (1986–1994).

Before an artificial reef can be authorized, the developers must seek permission to use the seabed for reef deployment. In Spain use of the seabed must be for public benefit, and the Ministry of Environment is responsible for monitoring the use of the seabed and awarding permits to place artificial reef modules on the sea floor.

There are two kinds of permit issued to allow the seabed to be used for reef deployment; the type and application procedure depend on whether the promoter is a Ministry (MAPA for instance in the case of artificial reefs), or another institution. Permits are in both cases issued under the 1988 Spanish 'law of coasts'.

Ministry Applications

A ministry obtains a 'reservation for a precise use', through a direct request to the government. The permit application procedure required of the ministries is different, but no less rigorous in its requirements than that undertaken by other applicants. In both cases the deeds go through the Ministerio de Medio Ambiente (Ministry of Environment).

The 'reservation for a precise use' tool was first used in 1994–1995 when the Secretaría General de Pesca Marítima (SGPM; Secretariat of Marine Fisheries) decided to implement its own artificial-reef initiatives, and not just fund them as before. Previously, only regional and local administrations had deployed artificial reefs with SGPM providing support by sharing costs and obtaining European Commission funding.

Other Applications

Regional government, local administration or others apply to the Ministerio de Medio Ambiente to lease an area of seabed on which they wish to deploy a reef. Such leases are for 30 years and are issued under the Spanish 'law of coasts', which was created to reduce environmental damage caused by development in the coastal zone. This law has hardened the requirements that proposals have to meet. Within this legal framework any application to create an artificial reef has to fulfil the following criteria:

(1) Artificial reefs must be constructed from environmentally benign materials. No scrap materials are permitted.
(2) At least 15 m of water must exist between the top of the reef and the sea surface for navigation purposes.
(3) Artificial reefs designed to discourage illegal fishing practices must seek to maximize their efficiency in this respect.
(4) If the artificial reef or barrier is to be made from wooden fishing vessel hulls, these must be clean: no engines, oil tanks or any polluting substances may be present. Hulls must be holed and ballasted and should contain sharp structures, or beams, in order to improve the performance of the barrier against illegal trawling activity by ripping nets.

(5) The exact geographical limits of the reef must be drawn on a nautical chart, as must the exact location of the reef modules and the extent of the protected area.
(6) Each reef module must be fully described, and the module distribution pattern specified.
(7) Any site-specific conditions the promoter must meet when the artificial reef is installed should be explained.
(8) The prohibitions and special conditions relating to fishing on the reef must be fully detailed.

It should be emphasized that the Spanish policy on artificial reefs is that of deployment for the public good, provided that the authorization to install gives no specific group preferential rights to fish on the reef.

Co-ordination of Artificial Reef Deployment

The system for authorizing reef replacement, which falls within the jurisdiction of up to 10 regional agencies and the MAPA is guided by the 'JACUMAR' (Junta Nacional Asesora de Cultivos Marinos) committee. This is a joint committee that brings together Spanish marine fisheries administrators. An 'artificial reef' working group of the JACUMAR committee provides a focal point for reef promoters to share experiences and discuss reef-related topics (optimum shapes, weights, methods to deploy modules and guidelines for follow-up studies) in order to provide a consensus of opinion and develop guidance in a number of key areas. Four meetings have already taken place, the conclusions of which are leading to the development of a 'field manual' providing guidelines on reef development based on practical experience (SGPM, 1992; SGPM Grupo JACUMAR, 1995).

Technical Conclusions of the Spanish Working Group on Artificial Reefs

Types of Artificial Reefs and Modules

In Spain three designs of artificial reef modules have been found to work satisfactorily. Concrete is the material of choice for all purpose-built modules and to minimize cost, modules are based on items already commercially available, such as tunnel sections (Fig. 3). The three designs are:

(1) Protection modules: simple structures intended to prevent trawling activity, made out of concrete where shape (square, cylindrical or other) is not as important as weight. Original designs varied between 3 and 5 tonnes, more recently heavier modules have been used weighing up to 8 tonnes. Normally protection modules incorporate projecting and pointed beams (usually sections of old railway track). These are intended to snag and rip trawl nets.

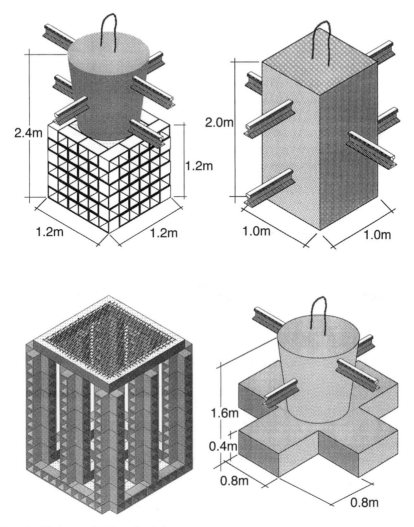

Figure 3. Various artificial reef unit designs.

(2) Production modules: bigger and heavier structures than those used for protection, these stand proud of the seabed and have holes and sheltered spaces in them.

(3) Mixed modules: protection units incorporating a certain structural complexity.

There are two basic types of artificial reef:

(1) Protection artificial reefs: these are intended to interfere with illegal fishing practices, usually trawling in waters shallower than 50 m in the Mediterranean and off the SW Atlantic coast, and 100 m in the Cantabric Sea. Artificial reefs must comprise large numbers of protection modules. The most effective pattern for protecting an area from trawling is still the subject of current

research (see Ramos-Esplá *et al.*, chapter 12, this volume) but the number of modules should be the minimum required to achieve the desired result, and they should be placed in more than one row: 2–4 rows are recommended.

(2)   Production or concentration artificial reefs: these are deployed to provide an increase in seabed habitat diversity. Modules are bigger, heavier and more complex than protection units. Surfaces are rough to encourage epibiotic colonization and the structure itself should promote fish recruitment. Fish appear to be attracted to spatial heterogeneity, which could lead to over-fishing without some protection from illegal trawling. Old wooden fishing vessel hulls can be used in this role but are in themselves attractive to fish so do not provide an absolute solution.

Based on experience of deploying and monitoring both kinds of artificial reefs, Spanish structures designed as protection units but with some habitat complexity added (a mixture of protection and production modules) seem to be working well. Surfaces have been colonized by epibenthos, including mussels, and experimental fishing off Asturias (a typical protection reef) shows an increase in finfish catches and evidence of fish recruitment to the reefs (Sánchez, 1995). This apparent success has led to the most recent artificial reefs in the Mediterranean being designed with several production modules (forming a nucleus) surrounded by protection struc-tures. Artificial reefs placed solely for production are still considered as pilot experiments in the Spanish context.

Artificial reef deployment is designed to protect fish stocks. In general, reefs tend to be deeper in the Cantabric Sea than in the Mediterranean Sea and off the SW Atlantic coast, reflecting the inshore seabed bathymetry and prohibition of trawling in waters shallower than 100 and 50 m, respectively. Basic differences between the Cantabric and Mediterranean fisheries have a direct influence on the characteristics of the artificial reefs used. Cantabric Sea artificial reefs can use as many as 300 protection units (3–4 tonnes with sharp edges) seeded in large areas (maximum 1800 ha) in depths between 40 and 100 m. In the Mediterranean Sea artificial reefs were first deployed as small pilot experiments made out of complex building blocks standing high (3–5 m) on the seabed in water less than 50 m deep. In recent years artificial reefs have been designed as mixed structures: protection barriers surrounding production nuclei. The main objective in deploying such reefs is that of protection but they also allow the function of production units to be studied.

Methods Involved in the Deployment of Artificial Reefs

Deployment of artificial reefs requires vessels of 3000–5000 tonnes dead weight, with a lifting capability of 10 tonnes. Differential GPS is required for accurate positioning of the modules and the most effective deployment technique is to rig the crane with a 'shooting hook' which can be released remotely, removing the need to use divers for deployment, increasing operational speed and safety whilst reducing the cost of reef deployment.

Standardized monitoring of reefs is needed to allow their efficacy to be judged against initial target criteria, so enabling the implementation of better-planned reef layouts in the future. Baseline studies must be undertaken to provide data, in support of the project proposal, and a pre-reef deployment baseline, against which results from the compulsory follow-up studies (lasting for 3 years for reefs deployed prior to 1994, or 5 years if placed after 1994) can be compared. Required data include aspects relating to the physical environment, biological communities and fishing fleet operations.

Preliminary results from annual monitoring of artificial reefs in the Cantabric Sea show improved recruitment of both juvenile and high-priced fish. Before the deployment of artificial reefs off Asturias (1992), there had been a decline in the catch of hake (*Merluccius merluccius*). After reef deployment, experimental catches in 1994 and 1995 showed two important results which differed from the 1991 and 1992 baseline catches: higher yields (kg ha^{-1}) and a change in the dominant species caught. Although high-priced species (e.g. hake) reappeared, lower-value species, such as sharks (*Scyliorhinus canicula*) provided high yields. This effect could be a result of changing the topography of the seabed; muddy bottoms seeded with artificial reefs may function as a sort of newly developed rocky sea bed (Sánchez, 1995).

Funding of, and Statistics Relating to Artificial Reefs in Spain

Funding of artificial reefs has been achieved under Spanish and EEC (now European Union: EU) regulations.

Between 1986 and 1993, there were two MAGPs (1987–91 and 1991–93). During these periods artificial reefs were actively promoted by regional and local administrations. Such programmes were financed by the sum of contributions from the EEC, Spanish general budgets and the promoter itself (regional or local authority). During the 1987–91 MAGP, 35 reef projects were implemented, 19 receiving national and EEC funding, and 16 supported solely by regional government. In the second MAGP a further 22 reef deployments were authorized, 17 receiving EEC support, and the remainder being funded by regional administrations (Fig. 2).

Since 1994, with the Spanish Fisheries Sector Planning Programme in force, artificial reefs placed to promote stock enhancement have been classified as public investments without private participation. In this case the cost is financed by the EU and the promoter itself, usually a public administration. In 1996, the Secretaría General de Pesca Marítima started to promote artificial reefs. Public funds available for artificial reef development amount to more than 1500 million pesetas.

Conclusions

Spain has been involved in deploying artificial reefs for more than 10 years, long enough to have a considered opinion about their uses. A lot is known about the technical aspects, costs, civil works, and gaining permissions, but there is still much research to be done on how artificial reefs function in the marine environment.

Statistics relating to the number and type of module used in artificial reefs reveal certain trends.

Artificial reefs in the north of Spain tend to be based on protection structures (high number of units, low weights) and seldom production units. In recent years experimental fishery campaigns within areas with artificial reefs show a significant growth in fish recruitment.

Off the SW Atlantic coast and in the Mediterranean Sea the trend is towards mixed units, placed as fishery and habitat protection structures, although the first artificial reefs (those of Comunidad Valenciana) were small production pilot projects.

Artificial reefs are being requested by fishermen because they act as tools to reduce conflict between different fishing groups. In addition, the improved environmental education of Spanish fishermen is making them realize the scarcity of marine resources and how important it is to optimize catches with minimum effect on fisheries stocks and on other stock users.

Future advance requires development on two fronts: education and resource protection (artificial reefs, marine reserves) and the results to date suggest that there is reason to be optimistic about the future of Spanish coastal fisheries.

References

MAPA 1995. La alimentación en España 1994.

MOPT 1992. Las aguas interiores en la ordenación del litoral. Unpublished report.

Sánchez, F. 1995. Informes sobre Pescas Experimentales en las Playas de Arrecifes Artificiales de Cudillero y Llanes durante las Campañas Demersales 1994 y 1995. Unpublished report by the Instituto Español de Oceanografía.

Santaella, E. and S. Revenga. 1995. Les récifs artificiels dans la politique de structures de la pêche en Espagne. *Biologia Marina Mediterranea.* **2**: 95–98.

S.G.P.M. 1987. Programa de Orientación Plurianual 1987–1991. Sector Arrecifes Artificiales. Unpublished report: 44 pp.

S.G.P.M. 1991. Programa de Orientación Plurianual 1992–1996. Sector Zonas Marinas Protegidas. Unpublished report: 24 pp.

S.G.P.M. 1992. Conclusiones del Grupo de Trabajo sobre Arrecifes Artificiales y Reservas Marinas. Alicante. 4–5 de Noviembre 1991.

S.G.P.M. 1994. Plan Sectorial de Pesca 1994–1999. Ambito de actuación de zonas marinas costeras.

S.G.P.M. and MAPA. 1994. Legislación Pesquera, 1994.

S.G.P.M. Grup JACUMAR. 1995. Jornadas de Arrecifes Artificiales. Cartagena 11–12 de Julio.

S.G.P.M. and Tecnología Ambiental S.A. 1995a. Proyecto Técnico de Arrecifes Artificiales para la Protección de Especies Marinas de Interés Pesquero, en Cabo de Palos (Murcia).

S.G.P.M. and Tecnología Ambiental S.A. 1995b. Proyecto Técnico de Arrecifes Artificiales para la Protección de Especies Marinas de Interés Pesquero, en Cabo de San Antonio (Alicante).

12. Artificial Anti-trawling Reefs off Alicante, South-Eastern Iberian Peninsula: Evolution of Reef Block and Set Designs

[1]ALFONSO A. RAMOS-ESPLÁ, [2]JUAN E. GUILLÉN, [1]JUST. T. BAYLE and [1]PABLO SÁNCHEZ-JÉREZ

[1]Dept. Ciencias Ambientales (Laboratorio de Biologia Marina), Universidad de Alicante, E-03080 Alicante, Spain. [2]Instituto de Ecología Litoral, ctra. Benimagrell-5, E-03560 Campello (Alicante), Spain

Introduction

Many authors and organizations believe that to achieve effective regulation and management of the Mediterranean coastal zone, not only do physical and biological factors have to be considered but also the socio-economic circumstances of the different communities in the area. Doumenge (1981) pointed out that one of the main objectives of such regulation is to preserve fish stocks and so ensure that long-established coastal fishing communities are able to enjoy a sustainable level of catches: "…it is becoming increasingly difficult to maintain natural fish stocks at a sufficiently high level to ensure a stable base for their future exploitation… these living resources are now in danger of being over-exploited because of an uncontrolled increase in catching-performance as a result of the increase in the number of boats being used, many of which have enormous catch-capacities". He also pointed out that some effective measures to ensure a future sustainable development of the coastal zone are: "…statutory and physical controls to protect the coastal seabed from illegal bottom-trawling which is an environmentally destructive activity comparable to deforestation on land".

History and Necessity

This concern about protecting the seabed is not new. The government of what is now the United Kingdom has received reports from fishermen since 1350 (Anonymous, in De Groot, 1984) when they informed the House that catches were deteriorating and that bottom-trawling (it is referred to as trynk nets and wondryrchoun) was destroying the 'living slime' and the plant growing on the seabed. Later, in the 18th century the use of bottom-trawl gear was denounced (Cornide and Sáñez-Reguart, in Urteaga, 1987) for two reasons: "…it caught and killed immature fish in nets with too small mesh, even though they had no more use than for fertilising fields; and the effect of trawling and sweeping the seabed destroyed the previously optimum environmental conditions…".

Even when pair trawlers were only sailing craft there was concern that trawling allowed a few to profit from large catches, while artisanal fisheries shared the

A.C. Jensen et al. (eds.), Artificial Reefs in European Seas, 195–218 benefit more widely among coastal communities. Today, coastal trawlers have powerful engines of more than 250 H.P. and trawling gear has increased significantly in size and weight from that of the 1700s. Rough and uneven seabeds offer little or no resistance to the sweep of the net. Trawling is the main fishing method used in the Western Mediterranean, contributing 60% of the total catch (Oliver, 1983).

Otter-trawling and, to a lesser extent, beam-trawling are the methods that do most damage to benthic ecosystems (de Groot, 1984; Hutchings, 1990; Jones, 1992). The dragging of heavy structures (otter-boards, butterflies, sweeps, ground-ropes), that can weigh more than 2 tonnes, across coastal seabeds seriously degrades both the topography and associated habitats, decreasing spatial hetero-geneity. Trawls are a non-selective fishing method that can negatively affect coastal fish resources.

The Impact of Bottom-Trawling on Coastal Mediterranean Biocoenoses

One of the most important marine communities in the Mediterranean is *Posidonia oceanica* meadows. These seagrass beds have an ecological and economic importance (Boudouresque *et al.,* 1994) providing:

(1) a climax community of soft, coastal seabeds;
(2) a high spatial heterogeneity, with different ecological niches;
(3) spawning and nursery areas for commercial species (fishes, decapod crustaceans and cephalopods);
(4) food and shelter for juveniles and adults of living coastal resources.

In recent years the area of seabed covered by meadows has started to decline. Bottom-trawling is the principal cause of this observed regression (the impact starts at a depth of 10 m) of the deep *Posidonia oceanica* meadows together with surface-dwelling, calcareous, long-lived algal seabeds (which form maerl beds and coralligenous blocks with important fauna, such as *Lophogorgia ceratopyta*) (Ardizzone and Pelusi, 1984; Bombace, 1989; Guillén *et al.,* 1994; Relini and Moretti, 1986; Riggio, 1990; Sánchez-Lizaso *et al.,* 1990). Repeated heavy physical damage is causing a serious regression of seagrass meadows and their associated biota (Sánchez-Jérez and Ramos-Esplá, 1996) in almost all of the coastal Mediterranean. This loss can be reflected in the decrease of shelter for juveniles and prey species (Bell and Westoby, 1986).

To protect these important biological communities and to prevent conflicts between selective small-scale (trammel, long-lines) and non-selective (trawl, seine) fisheries, Mediterranean European Union countries forbid the use of trawls at a depth of less than 50 m (CE Regulation 1626/94, for the preservation of the Mediterranean fishery resource). Such legislation is also used to protect seagrass beds (France, regions of Catalonia and Valencia). This law is supported by the European Union Directive 92/43 (habitats of priority interest for conservation) and Regulation 1626/94 (to forbid trawling and seine netting on the marine phanerogram beds).

The Necessity of Artificial Anti-trawling Reefs (AARs)

Despite control and surveillance actions carried out by the marine police and fisheries authorities, illegal fishing still takes place. Consequently, effective measures are required to prevent such fishing activities. Polovina (1991) pointed out that artificial reefs should be considered as part of an overall fishery management plan and may be useful in closing off areas to trawling, thus protecting juveniles in shallow nursery grounds. In this way, deterrent structures can operate as production reefs which protect recruitment and favour the growth of juveniles while causing the fishing mortality to decline.

This natural re-population (or recovery) effect is reported in the Order (11/05/82) of the Spanish Ministry of Agriculture and Fisheries, which points out that artificial reefs should constitute an efficient system of re-population, capable of preventing trawling in forbidden areas. In this respect, the Spanish Working Group Studying Artificial Reefs and Marine Reserves (SGPM 1992, 1995) has as an objective the combination of reef deployment with other measures (such as the sinking of wooden hulls, marine reserves, close seasons and selection of fishing methods) to manage coastal living resources. The aim is to achieve sustainable exploitation which will facilitate the survival of traditional fisheries.

Lack of knowledge about the impact of AARs on the fish population and structure of seagrass meadows is an important issue in the discussion on their use in fishery management. Almost all evidence about the effects of artificial reef modules on fish assemblages has been obtained from sandy seabeds (D'Anna *et al.*, 1995; Johnson *et al.*, 1994) and the efficacy of artificial reefs in more productive benthic systems is still unclear. AARs introduce new artificial habitat into a natural ecosystem, providing an excellent tool for research into the ecological relationship between habitat structure and fish assemblage.

Background and Aims of the AARS in the SE Iberian Peninsula

The need to protect inshore, high-diversity communities (mainly seagrass beds of *Posidonia* and *Cymodocea*, and calcareous benthic algae) and consequently the nursery areas of fish (Scorpaenidae, Serranidae, Sparidae and Mullidae) and spawning grounds of cephalopods (cuttlefish and squid) motivated the empirical development of shallow water AARs for sub-littoral conservation and the possible enhancement of coastal living resources. These structures were intended to protect selective small-scale fisheries as opposed to those using non-selective fishing methods (trawl, seine).

Aims of the AAR Projects

The objectives of the AAR projects can be summarized following the themes developed by Bohnsack and Sutherland (1985):

(1) To protect coastal zone marine biocoenoses and species from the mechanical impact of trawling. This is especially important for high-diversity communities such as seagrass beds (*Posidonia, Cymodocea*) and associated fauna of biological interest (e.g. pen shell *Pinna nobilis*) together with calcareous algae aggregations (coralligenous and coastal detritic communities and associated fauna, e.g. gorgonian *Lophogorgia ceratophyta*).

(2) To protect juvenile fish and nursery areas (especially those of red mullet *Mullus surmuletus*; Scorpaenidae, *Scorpaena* spp. and Sparidae, *Pagellus acarne* and *Pagrus pagrus*), and cephalopod spawning areas (*Sepia officinalis* and *Loligo vulgaris*).

(3) As a management tool to prevent conflicts between large-scale trawling and purse-seining in small-scale fisheries.

(4) As a means of controlling the CPUE (catch per unit effort), by promoting the use of more selective fishing gear (long-line, troll-line, trammel) as opposed to non-selective methods (trawl, seine).

(5) To increase the spatial heterogeneity and variety of substrata (hard substratum) on soft seabeds.

(6) As a natural laboratory to investigate the structure, insularity, ecological succession and habitat complexity of certain communities (not impacted by trawling); as well as monitoring areas. A reef represents a developing experimental system which, with the passage of time, comes to resemble neighbouring environments (Bellan-Santini, 1992).

(7) To assess other types of artificial reef using experimental units (e.g. attraction/ concentration benthic modules, and fish attraction devices).

Stages of AAR Building

Surveys of the broad south-eastern Iberian coastal sector and subsequent bionomical mapping (Aranda *et al.*, 1992a, 1992b; Martínez *et al.*, 1990; Ramos, 1984, 1985; Ramos *et al.*, 1993) showed that wide areas of *Posidonia* meadows had been damaged by the mechanical impact of trawling. In some meadows up to 48% of the seagrass area had been damaged (Sánchez-Lizaso *et al.*, 1990). An extensive AAR, with units placed over a wide area, was considered to be better than an intensive reef in providing protection for the seagrass beds.

The first AAR studies in the south-eastern Iberian Peninsula were carried out by the Maritime Fisheries Institute in Alicante. This work focused on protecting a sector of the Marine Reserve of Tabarca against trawling (Ramos and Trapote, 1987). Subsequently, the University of Alicante and the Instituto de Ecología Litoral deployed the AAR off Campello (Martínez *et al.*, 1990) and the later projects off Benidorm, La Vila Joiosa and Altea. Concrete was the chosen material for all of these reefs.

Legal Requirements, Permission and Funding

According to Spanish law (Order of 11/05/82), a reef installation proposal must be justified by preliminary studies which produce a bionomic chart and local fishery

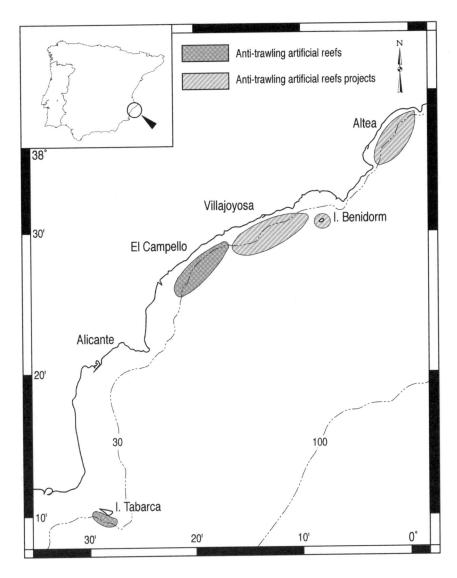

Figure 1. Location of artificial reef projects in Alicante.

data. The project proposal must be examined by the Instituto Español de Oceanografía (research organization linked to the Ministry of Agriculture and Fisheries), the Ministry of Defence, the Merchant Navy Administration and the National Federation of Fishermen Associations (*Cofradias*). If the project proposal is accepted, permission must be obtained from the Ministry of Public Works, Transport and Environment (Coastal Administration) to make use of an area of territorial waters. In addition, if the area comes under the jurisdiction of a regional government (within baylines) the corresponding Regional Fishery Department must also give their permission.

The reef projects at Tabarca and Campello were financed by the following organization: 50% EEC (D.G. XIV); 35% Ministry of Agriculture, Fisheries and Food; 15% Regional Council of Agriculture and Fisheries (Tabarca project), and 15% Municipality of Campello (Campello project). The preliminary studies of all the projects mentioned were financed by the respective town councils of Alicante, Campello, La Vila Joiosa and Altea.

Methodology

Preliminary and On-going Studies

These studies were done before and after the deployment of the anti-trawling artificial reefs.

Physical, Chemical and Biological Characteristics

Data were obtained on the following aspects of the environment:

(1) Water column. Seasonal samples provided information on temperature, salinity, transparency (using a Secchi disk), oxygen, nutrients, surge and currents (at sea surface and seabed), chlorophyll a concentration and seston (methodologies after Strickland and Parsons, 1972).
(2) Substratum. Data were collected to describe the nature of the seabed (hard/soft); texture (Buchanan, 1984), stability and distribution of the sediment and the location of the trawling tracks. Possible sedimentary inputs (such as rivers or creeks) were also assessed.
(3) Benthic communities. Bionomical mapping followed techniques described by Meinesz et al. (1983). Data collection (between depths of 0 and 35 m) was undertaken by SCUBA divers being towed on a hydroplane along a line perpendicular to the shore (Ramos, 1984). Depth was simultaneously recorded by echo-sounder. The depths at which the sea floor was affected by the impact of trawling were established.
(4) Posidonia beds. Quality was evaluated using certain meadow descriptors such as shoot density, cover and maximum summer biomass (Sánchez-Lizaso, 1993).
(5) Ichthyofauna. Fish counts were obtained by means of SCUBA visual census (Harmelin-Vivien et al., 1985; Bortone and Kimmel, 1991) and beam-trawl (Harmelin-Vivien, 1984).
(6) Macrobenthos. Seasonal sampling of vagile epifauna was achieved by beam trawl, epibenthic dredge (Guillen and Pérez-Ruzafa, 1993) and a diver operated hand-net (Sánchez-Jérez and Ramos-Esplá, 1994) for small mobile fauna (mainly crustaceans). Removable fouling plates were also installed on the blocks to study macrobenthic colonization.

The general outline of the AAR research is summarized in Figure 2.

Local Fisheries

(1) Local trawling fisheries. After evaluating the damage caused by otter-trawling to coastal biological communities, the characteristics of the local trawler fleet (number of units, engine power, type of trawl gear) were assessed. This information was used subsequently, when designing the AAR.
(2) Study of artisanal fisheries. Mainly small-scale fisheries were found to be active in the zone proposed for the reef (Martínez-Hernández, 1993).
(3) Experimental fishing used trammel and long-line around the AAR, mainly focused on some target species (e.g. rocky red mullet *Mullus surmuletus*).

Characteristics of the AAR

In order to achieve maximum effectiveness at minimum cost, the blocks were designed (size, weight and arrangement) taking into account the characteristics of the trawl gear (Fig. 3), the power of the vessels, the impacted area, depth and substrata (see calculation appendix). Apart from theoretical considerations, the work has a strong empirical component based upon trial-and-error experience.

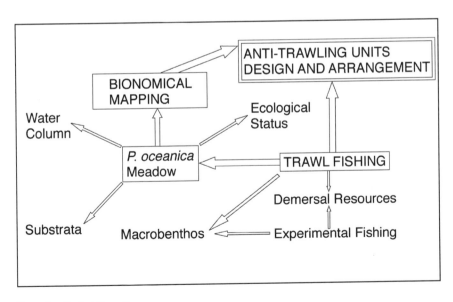

Figure 2. Project flow diagram.

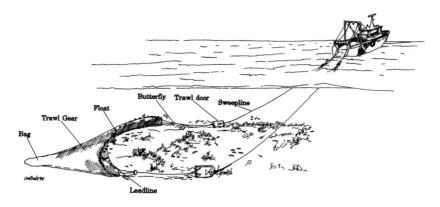

Figure 3.   Main elements of the bottom otter-trawl.

Anti-trawling Block Design

Characteristics of reef blocks and units were defined as follows (Grove *et al.*, 1991):

(1) Durability. Blocks should last underwater for 40 years. Reinforced concrete was the material of choice.
(2) Safety. The block must be free of toxic substances.
(3) Mobility. It must be possible to re-float the reef blocks so that their arrangement could be changed in response to the activity of the trawling fleet (e.g. to close possible trawling corridors).
(4) Functional character. The blocks must be in the form of a cube and heavy enough to stop trawlers with 800 HP engines. Reef blocks should also have protruding metal beams to increase the deterrent effect by ripping nets.
(5) Economy. The structure should be as cost-effective as possible, protecting the maximum area at the lowest possible cost. Reef units should also provide the greatest habitat complexity.
(6) Size and weight of the block. Using data from the trawling fleet survey the optimum reef block weight as a function of engine power was calculated (see Appendix). As an example, in order to obstruct the passage of 700–800 HP trawlers, blocks with a weight in air of 7.20 and 8.77 tonnes respectively, would be needed (Fig. 4). The first blocks made weighed 8.32 tonnes (massive cubes with 1.5 m sides) and were deployed in the Tabarca AAR. Later blocks weighing 7.90 tonnes (hollow cubes, 2 m sides) were used in the Campello AAR.
(7) Shape and complexity of the block. Experience of the fishermen suggested that a cubic shape should be the most effective shape to deter trawling, as it would be hardest to remove. At first solid blocks were used (Tabarca AAR, Fig. 5). However, after this experience and taking account the work carried by Bohnsack *et al.* (1991) it was decided to construct hollow blocks with holes

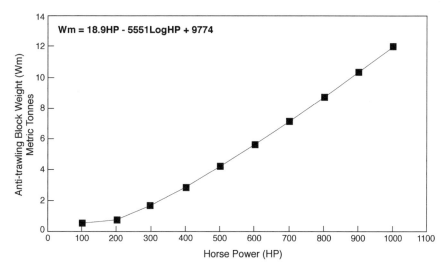

Figure 4. Optimum anti-trawling reef block weight (W$_m$) as a function of trawler engine power (HP).

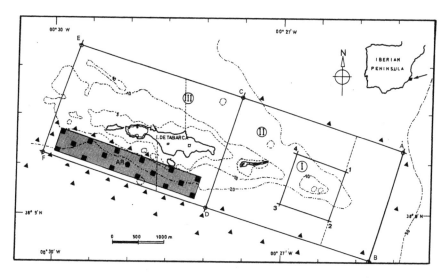

Figure 5. Anti-trawling artificial reef of Tabarca marine reserve (shaded area).

(2–3 holes/face) and fill the core with pipes and bricks. These reef units were deployed in the Campello AAR (Fig. 5). It was hoped that the openness and spatial complexity of the interior of the block would help to protect juvenile fish as well as decapod crustaceans.

(8) Deterrent elements. To enhance the deterrent effect of the reef units on trawling activity, projecting iron beams (5, each 0.5 m long) were set into the

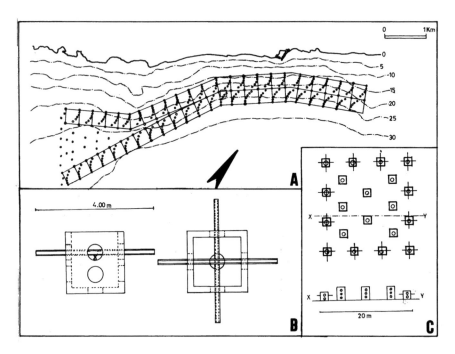

Figure 6. Anti-trawling artificial reef of El Campello: a) arrangement of blocks; b) block characteristics; c) attraction/concentration set.

faces of the Tabarca reef blocks; and 2 lengths of railway track (each 4 m long), welded in a cross at the centre, protruded from the holes in the sides of the Campello reef blocks (Fig. 6; Colour Plate 12). These were designed to rip nets.

Spatial Arrangement and Location

The spatial distribution of reef units had to be carefully planned to ensure that trawlers were prevented from trawling regardless of their course. Minimum distance between blocks was determined as a function of trawl parameters (coastal otter-trawl width and length) and a basic anti-trawling set (BAS) or pattern of unit distribution was established. When steaming parallel to the coast trawl gear working in 20 m depth has a total length of 300 m (net, sweepline, warp) and a distance between otter-boards of 40–80 m. These parameters have served to establish, respectively, the basic parallel and perpendicular distance (to the shore-line) between the blocks.

Tabarca
The BAS of Tabarca (Fig. 5) was a rectangle (400 × 200 m) with five blocks, one in each corner (vertices) with the fifth in the centre. This represents a protected

area/block ratio of 1.6 ha/block. The location of the AAR was to the south of the Marine Reserve of Tabarca, and was intended to protect an area of 80 ha (2000 × 400 m) with 25 blocks. The seabed occupied by the blocks was only 0.014% of the total protected area.

Campello

After the experience of Tabarca AAR, the blocks were arranged in accordance with a model of the area impacted (Martínez *et al.*, 1990). The area was represented as a network of hexagons. The circles circumscribing the hexagons generated a series of curves which protected the area against trawling in all directions. Since most trawling was done in the same direction (parallel to the coast), the model was simplified leaving a series of curves in this direction only. Consequently, the maximum distance between curves was found to be 300 m, which gave the distance between the blocks in the direction parallel to the shore-line. The distance between blocks perpendicular to the coast was 50 m (see Appendix 1).

The arrangement shown in Fig. 6 was designed for a BAS composed of nine blocks in a quadrat set of 9 ha (300 × 300 m), four blocks on the vertices and five on the curved diagonal. This represents a protection ratio of 1 ha/block. The total area protected was over 540 ha with 358 blocks, the units occupying 0.027% of the surface area.

Other Projects

Experience gained from reefs at Tabarca and Campello allowed the development of projects (off Benidorm, La Vila Joiosa and Altea) with a greater distance between blocks (500 m parallel to shore and 100 m at right angles to the shore). This represents a quadrat BAS of 500 m side using eight blocks, giving a protection ratio of 3.125 ha/block. Other BAS with a smaller number of blocks (five units in 25 ha, protected area 5 ha/block) have been considered for areas with little illegal trawling activity.

The evolution of the characteristics of the anti-trawling blocks and arrangement and the areas protected by the different projects are shown in Table 1.

Location, Transportation and Sinking

All the reefs were located parallel to the shore-line on *Posidonia oceanica* and *Cymodocea nodosa* meadows and calcareous algae beds degraded to a greater or lesser extent by trawling, at depths between 15 and 30 m.

The blocks were transported and deployed with lifting bags in the marine reserve of Tabarca, whereas a barge and crane (10 tonnes) were used off Campello. The latter method was the most cost-effective.

Monitoring of Fish Colonization

Fish colonization was monitored at Campello for 3 years after installation of the Campello AAR. An experimental set of reef units was placed between 15 and 17 m water depth. Four quadrats of 10 × 10 m area on *Posidonia* meadow were

Table 1.   Main characteristics of artificial anti-trawling reefs deployed.

	Tabarca	Campello	Benidorm	Villajoyosa	Altea
Block characteristics	Solid cubes	Hollow cube with holes in walls	Hollow cube with holes in walls	Hollow cube with holes in walls	Hollow cube with holes in walls
Holes/dia (m)	–	9/0.5	9/0.5	9/0.5	9/0.5
Side length (m)	1.5	2.0	2.0	2.0	2.0
Surface (m^2)	13.7	34.0	34.0	34.0	34.0
Anti-trawling block weight (kg)	8370	7900	7900	7900	7900
Anti-trawling basic set (m)/No. units	400 × 200/5	300 × 300/10	300 × 300/5	350 × 350/10 500 × 700/8	500 × 500/9 500 × 500/5
Protected area (ha)	80	540	180	813	804
No. of blocks	25	358	80	215	118
Density (units/ha)	0.31	0.66	0.44	0.26	0.14
Block costs (ECU)	2368	1124	1124	1124	1124
Protected ha costs (ECU)	734	742	520	292	157

randomly chosen and two reef units placed in each one. Another four quadrats (50 × 2 m) on *Posidonia* meadow were randomly chosen, located away from any block (more than 100 m). Visual census techniques were used to estimate population densities of fish in each of the quadrats with artificial habitat (two blocks) and the control quadrats (Sánchez-Jérez and Ramos-Esplá, 1995). The control quadrat provided data on fish assemblage trends in the natural habitat. All the surveys were done around midday. Two replicate censuses taken 1 h apart, each of 15 min duration were achieved during each survey. Individual fish size was estimated with a ruler attached to the end of a meter stick (Bohnsack and Bannerot, 1986). Total number of fish and numbers of the most important species, distinguishing between juveniles and adults, per 100 m^2 seabed area were calculated.

The differences between the fish populations of meadow quadrats with reef blocks and the control were tested by permutations of the rank of similarity matrix (analysis of similarities ANOSIM; number of permutations = 5000), test which is a distribution-free analogue of one way ANOVA. The *a priori* hypothesis was that there was no difference in the overall fish assemblage between quadrats with blocks and the control. The fish assemblage from each treatment was graphically represented by two-dimensional display of multi-dimensional scaling (MDS; see Clarke, 1993). Temporal patterns of juvenile and adult abundance for some numerically plentiful and/or important species (*Diplodus vulgaris, D. annularis, Chromis chromis, Apogon imberbis, Symphodus tinca* and *Mullus surmuletus*) were represented graphically in respect to both treatments.

Results

Environmental Factors

Water Column

The water masses in the area were found to be warm-temperate and oligotrophic (Zoffmann *et al.,* 1985). The surface temperature fluctuated between 12.5 (February) and 28°C (August). The surface salinity varied between 36.9 (October) and 38 ppt (April). The water was well oxygenated (5.5–8.6 mg oxygen dm^{-3}) and had a low concentration of chlorophyll a (0.1–0.7 mg m^{-3}).

The currents at the minimum depth of the installation (15 m) reached a maximum of about 0.5 knots. The maximum surge of the waves at this depth was 5.5 m but this height was only reached in 3% of the cases.

Seabed Substratum

The substratum was composed of living or dead *Posidonia* rhizomes and soft sediment. The mean composition of the sediment (at a depth of 17 m) in non-degraded cum degraded beds in Campello was: coarse sand 15–5%, fine sand 50–35%, very fine sand and mud 35–60%. This represented a stable sediment (seagrass rhizomes, muddy sand, detritus, both coarse and fine biogenic sediments (shells and tests), and mud). The seabed profile was flat (average slope of 3.25% at Tabarca and 1.6% at Campello).

Survey of Some Target Species

The study was mainly aimed at the target species (*Posidonia, Pinna, Mullus*) and other fishes. It was considered important to know the habitat requirements of the target species and the factors that restrict the size of the population in order to effectively design and locate future AARs.

Posidonia Beds and Macrobenthos

Posidonia oceanica
The initial study of the *Posidonia* beds revealed that 25–45% of the deep meadows had been altered or degraded by trawling (Ramos, 1984; Sánchez-Lizaso *et al.,* 1990; Aranda *et al.,* 1992a, 1992b). Six years after the installation of the AAR of Tabarca, the *P. oceanica* meadow had recovered significantly. Before installation of the reef (1988), the mean shoot density was 10 shoots m^{-2}, by 1992 it had increased to 60 shoots m^{-2} (Sánchez-Lizaso, 1993).

Pinna nobilis
A similar effect was observed with the great sea pen-shells (*Pinna nobilis*), a species vulnerable to illegal trawling, and considered to need protection (Vicente

and Moreteau, 1991). Prior to reef deployment *Pinna nobilis* was absent, 18 months after the reef installation of Campello (1992) a population density of 5 juvenile specimens 250 m⁻² was observed, an important recovery for this species.

Small vagile macrobenthos
This fraction of epibenthos comprised mainly the crustaceans Peracarida (Amphipoda, Isopoda, Mysidacea) and Eucarida (Decapoda). These represent an important food source to demersal resources, mainly for the juvenile stages of fish and cephalopods. After a canonical correlation analysis, the community structure of invertebrates was seen to be strongly influenced by *Posidonia oceanica* features such as shoot density and accumulated detritus (unpublished data).

The density of the different crustacean populations were influenced, but not necessary increased, by high shoot density. There were correlations among community structure and seagrass features (values of ponderate correlation between -0.54 and 0.64).

Influence of the Blocks on the Ichthyofauna Populations

Fish Assemblages and Colonization of the Blocks
In the period between the deployment of the AAR until the last survey in August 1995, 41 species were found in meadow quadrats with reef blocks. Only 29 species were found in the control meadow quadrat. This general trend was confirmed by the analysis of similarities (ANOSIM). Fish assemblages sampled around blocks and those in seagrass meadows (without blocks) were significantly different between seasons and within seasons. The statistic global R of ANOSIM had values between 0.478 and 0.813 with a significant difference (p 0.01) in all cases. In the two-dimensional representation of MDS, the results showed a strong grouping by habitat from the first samples (stress value $= 0.05$). The structure of the fish assemblages was more influenced by the existence of artificial habitat than by the seasonal or interannual changes.

Initially most species present in the quadrats were species common around *Posidonia* which made use of both the artificial and the seagrass habitat. *Chromis chromis* was the first species which sheltered in the blocks, followed by *Symphodus* spp. and some serranids (*Serranus cabrilla* and *S. scriba*). An important group of species with a preference for rocky habitat colonized the artificial reef units, increasing the species richness significantly. Six months after reef block deployment, species such as *Diplodus puntazzo, D. cervinus, Epinephelus marginatus, Apogon imberbis, Sciaena umbra* and *Phycis phycis* were relatively common in the reef units. Fish belonging to the same taxonomic family showed similar behaviour in relation to the artificial habitat. The sparids, centracantids, *Balistes carolinensis* and carangids were vagrants that were frequent short-term visitors to the reef units. Both *Apogon imberbis* and *Chromis chromis* recruited to blocks, but the adults of *A. imberbis* stayed within the units and the adults of *Chromis chromis* emigrated from the reef to the *Posidonia*. Families such as Labridae did not reveal any special preference for artificial habitat, sheltering in the blocks sporadically.

In the experimental AAR of Campello, distance between blocks was an important feature for determining fish assemblage structure (Sánchez-Jérez and Ramos-Esplá, 1995).

Influence of the Reefs on Fish Species

The sparid *Diplodus vulgaris* showed a rapid population increase in the presence of artificial habitat. Five weeks after installation, adults began to occupy the blocks showing, 3 months later, a population abundance 3–5 times greater around the reef blocks than in the *Posidonia* meadow. In the meadow, the density of this species was low, less than 1 individual per 100 m^{-2}. The adults did not show any seasonal pattern of abundance, exhibiting vagrant behaviour in relation to the reef blocks. Groups of juveniles (< 40 mm length) were attracted by blocks and some had been recruited to the artificial habitat. Average total density of *D. vulgaris* was 6 ± 4.12 individuals 100 m^{-2}.

The other sparid, *Diplodus annularis*, showed no significant increase in population after the AR blocks were placed. The first individuals around the blocks were observed 5 months after deployment. This species made little use of the blocks and was more plentiful in the surrounding natural habitat. Only transient groups of juveniles were found around the blocks in some surveys, for example the summer of 1993. Irregularity in adult presence and a substantial recruitment characterised the *D. annularis* population in *Posidonia* meadows, with a total average density of 2.58 ± 2 individuals 100 m^{-2}.

Chromis chromis showed a clear preference for the blocks, occupying these just the day after deployment. *C. chromis* recruited to the blocks, adults were present all year but with strong seasonal numeric variations. The recruitment of juveniles was completed during the summer, increasing the density of adults around the artificial habitat in winter. In comparison with *Posidonia* meadow, *C. chromis* was more abundant around the reefs, 11.36 ± 7.06 in quadrats with reef blocks against 1.32 ± 2 individuals 100 m^{-2} in *Posidonia*.

Species such as the labrid *Symphodus tinca* did not show any clear preference for artificial substrata. In the first few weeks after placement some individuals were found swimming around and within the blocks. Adults were present in very similar densities on both natural and artificial habitats but juveniles maintained a constant presence in the meadow, and were not seen around the blocks. However, another labrid population, *Coris julis* increased in size because of the existençe of artificial habitat. In the autumn and spring of 1994 there was a difference in the abundance of juveniles relative to the number of adults present around the blocks (spring 2 juveniles per 6 adults 100 m^{-2}, autumn 4 juveniles per 7 adults 100 m^{-2}).

A species with a clear preference for the artificial habitat was *Apogon imberbis* which was first recorded 8 months after reef deployment and continuously thereafter. The population appeared to be stable with a relatively constant number of fish associated with each experimental block (4.2 ± 2.7 individuals 100 m^{-2}). Recruitment occurred during the summer, when there was an increase of juveniles without a significant reduction in adult density.

In the case of *Mullus surmuletus*, a species with great economic importance, the blocks did not influence population abundance. The density of *M. surmuletus* in and around the blocks during the surveys was < 0.5 individuals 100 m⁻². In the meadow this species had a low abundance but was continuously present. Groups of juveniles were found in *Posidonia* during the summers of 1993–1995, increasing their abundance slightly during the last two years.

Local Fishery Trends
Artisanal fishery studies (Martínez-Hernández *et al.,* 1996; personal communication) noted that commercial landings data suggested an increase in the catch of some fish such as red mullet (*Mullus surmuletus*) and sea scorpions (*Scorpaena porcus, S. notata*). Red mullet catches have increased by about the 12% since the deployment of the Campello AAR.

Discussion and Conclusions

The deterrent effect on trawling of the heavy cubic blocks (7–8 tonnes) and arrangement designed for the basic sets of the Tabarca and Campello AARs have proved to be effective and trawlers now avoid the reef zone. On the other hand, the small-scale local fishery (trammel, long-line) shows signs of recovery. However, existing data are not sufficient to enable us to confirm a definitive improvement in the local artisanal fishery.

The use of AAR with dispersal blocks spread over a wide area not only protects coastal communities from damage caused by trawling but also can avoid the possible fish aggregation and concentration effect that could increase the risk of overfishing (Polovina, 1989). Importantly, an extensively arranged AAR allows fishing by artisanal methods and angling, thereby helping to decrease the populations of adult predators, so favouring the natural nursery role of the meadows.

The results of the monitoring programme show that the instalment of new, artificial habitat affected the structure of the fish assemblage by increasing the total fish abundance and the number of species, and changing the densities and spatial distribution of some fish populations. The results showed that the effects were similar among seasons during the last two years of monitoring (R global of ANOSIM with $p < 0.01$), and the composition of samples was relatively comparable, depending on the existence of artificial habitats within seasons.

After three years, the more common biological objectives for the use of AAR for fisheries management, fish attraction, fishery yield enhancement and stock diversification (Milon, 1989) were, generally, achieved. The existence of blocks in the meadow increased the recruitment and concentration of fish above that of previous *Posidonia* meadow stock, e.g. some sparids such as *Diplodus annularis, Chromis chromis*, labrids (*Symphodus* spp. and *Labrus* spp.) or serranids (*Serranus scriba, S. cabrilla*) were added. The blocks facilitated an increase in the number (S = 12) and density of habitat-limited species by providing increases in food resources, reproductive habitat and protection from predators. *Sciaena umbra*,

Epinephelus marginatus, some sparids (such as *Diplodus sargus*) and *Balistes carolinensis* were species with habitat-limited recruitment (because of their preference for rocky seabed) that were attracted. New habitat seemed to both attract fish from the surrounding environment and increase fish biomass by increasing the recruitment of some species.

Rapid colonization of a newly deployed AAR before food resources could have developed indicated that food availability on the reef may be less important to fish than shelter, at least initially (Sale, 1980). Some fish rapidly colonized the blocks, for instance *Chromis chromis, Diplodus vulgaris* and/or some *Symphodus* spp. Some of these were later seen feeding on the fouling community of the artificial habitat (*Coris julis, Symphodus tinca* and *Diplodus* spp. (personal observation)).

Initial fish colonization happened in two ways: individuals moved from the *Posidonia* meadow to the blocks, and adults came to the reef from natural rocky habitat. Consequently, the immediate effect of increasing seabed habitat complexity with an AAR was the redistribution of fish already in the area. Secondary recruitment by obligate reef-dwelling species (not present in the seagrass habitat) such as *A. imberbis* and territorial species such as *C. chromis* presumably will have increased the overall abundance of these species.

Although artificial reefs are classically thought to increase critical, limiting resources which attract and benefit fish, some fish populations may not be limited by shelter or food availability (Bohnsack, 1991). This appeared to be the case with *Mullus surmuletus,* a species of great economic importance, whose adult population appeared to be unaffected by the presence of reef blocks. A slight increase in the numbers of juvenile red mullet was detected in *Posidonia* meadow, suggesting a recovery to pre-trawling levels.

AARs in seagrass meadows could be utilized to raise fish abundance or biomass, and species richness, to reduce user conflicts or just to protect *Posidonia oceanica* meadows from physical damage. The response of the more economically important and biologically interesting species to the reef was related, in varying degrees, to both behavioural preferences and habitat limitation. These must be considered when assessing the probability of an artificial reef producing new fish biomass or attracting/aggregating fish.

The Future

According to Bohnsack (1991), artificial reef studies should not just contribute information about the reef location (geographical area and isolation) but also about the surrounding habitat, reef size and the seasonality of certain factors. These additional data are possibly more important than just describing the habitat complexity when it comes to determining the population density of fish on a reef.

Controlled experimental studies on the building of reefs should aim at improving recruitment, the growth of juvenile fish and spawning success, in particular protection of cephalopod breeding sites. Constructing reefs to improve larval recruitment may prove more economical and effective in increasing the spatial heterogeneity than attraction-concentration reefs (Bohnsack and Sutherland, 1985). If this is the case such larval recruitment reefs should increase available shelter and food resources (e.g. small crustaceans) and subsequently improve recruitment to, and catch rates in, neighbouring areas. As a follow-up to the studies described, some open questions and possible research trends for the future are set out below:

(1) To improve and optimize the AAR design. The block (weight, size, shape) and the arrangement (location, density and spatial configuration) designs of the AAR can be optimized, based on our empirical experience. More effective deterrent blocks, basic sets and large scale programmes could be designed and the cost-effectiveness of the structures improved.

(2) Habitat complexity. Studies are inconclusive regarding the effect of increasing the habitat complexity of small units (incorporating pipes, bricks, and other building materials inside the block). According to Bohnsack *et al.* (1991) it is necessary to understand the role played by habitat complexity (size, shape, holes, relief, roughness, filling material) to improve the protection of juveniles.

(3) Recruitment. Apart from the prevention of trawling, the importance of predation and shelter on early recruitment should be evaluated. Fish populations of single dispersed reef units, which prevent the presence of permanent populations of predators, may favour the protection of juveniles and should be studied. The distance between reef blocks of similar design in a habitat is an extremely important aspect; structures with the same complexity have different fish assemblages depending on how far they are each from the other (Bohnsack, 1991).

(4) Standardization of methods of monitoring. The need to standardize the methodology used was pointed out by the SGPM (1992, 1995). Such standardization is necessary to determine the behaviour, life cycles and food of ecologically and economically important target species (e.g. red mullet, sea bream, cuttlefish and pen shell).

(5) Small-scale fisheries and anglers. Can the AAR reduce the conflicts between professional and/or sport fishermen? The use of deterrent blocks can provide a means of improving fishery resource division, but to reach this end large-scale programmes need to be established.

The installation of artificial reefs should be based on scientific studies which determine the objectives and characteristics of reefs. Decisions are often based on political expediency, absolute cost, navigational and pollution considerations without taking into account the biological and socio-economic effects. To adopt artificial reef models designed for other situations and install them without the necessary scientific support is a misuse of large quantities of money and shows a frivolous attitude to the management of resources (Bohnsack and Sutherland, 1985; Pereira, 1993).

Appendix

To design an anti-trawling artificial reef the weight of each block unit and the arrangement of these must be considered in relation to the characteristics of the vessel's engine (power) and fishing gear.

Block Weight

To calculate the maximum block weight to deter a trawler, the trawl power, the total trawl resistance and the gear resistance must be considered.

(a) Trawl power

According to De la Cueva (1974), the available power (AP) of the trawl vessel can be obtained from the nominal power (HP):

$AP = HP.Ks.Kp.Keu.$ (1)

where: Ks = coefficient of sea state (approx.: calm sea = 1.0; sea 2–3 = 0.9; sea 3–4 = 0.8; sea 4–6 = 0.7); Kp = coefficient of propulsion (< 300 r.p.m. = 0.24–0.28; 300 r.p.m. = 0.22; > 300 r.p.m. = 0.20); Keu = coefficient of engine utility (approx. 0.8).

Taking into account these coefficients (calm sea and r.p.m. = 300) and replacing in (1):

$AP = 0.176HP$ (2)

(b) Total trawl resistance

The required power (RP) to trawl a resistance (R), from fishing gear, catch, objects, etc., with a determined speed (s in metres per second) is expressed as:

$RP = R.s/75$, then $R = RP.75/s$ (kp units) (3)

Considering the total resistance (TR) in relation to available power (AP), with a trawling speed of 3 knots (s = 1.5 m seg^{-1}), and replacing in (2) and (3) will be:

$TR = AP \times 75/1.5 = 8.8HP$ (4)

(c) Gear resistance

The otter resistance of the trawl represents 20–28% of the total gear resistance (De la Cueva, 1974).

Otter-trawl resistance (OR) can be obtained from the formula:

$OR = Kt.d.s^2.S$ (5)

where:

Kt = coefficient of trawling or resistance (Kt with a speed of 3 knots = 0.79), d = Mediterranean sea water density (international system = 104.85 kg seg^{-2} m^{-3}), s = trawl speed in m seg^{-1} (we considered an average speed of 3 knots = 1.5 m sec^{-1}), S = otter boards trawl surface.

Gear resistance (GR) can be determined from the otter-trawl resistance. If percentage otter-trawl resistance is 25%, then GR = 40R, and replacing from (5): $GR = 745.48 \times S$ (6)

With reference to local fishery data and the study of De la Cueva (1974), taking the logarithmic function between otter-boards surface (S) and trawl power (HP)

$S = 3.47 \times \log HP - 6.11$

Replacing S in (6) we get:

$GR = 2586.8\log.HP - 4554.9$ (7)

(d) Block weight

To calculate the block weight as a function of trawl power, factors favourable to trawling (calm sea, maximum power, minimum depth and a speed of 3 knots) are taken into consideration. Block resistance (BR) is considered to be equal to, or greater than, the total trawl resistance less the gear resistance (BR $\geq$ TR $-$GR).

The block resistance depends on the block weight in water (BW) and with a coefficient of trawling $(Kt) = 0.79$, we obtain:

$BR = 0.79BW$;

where, $BW = BR/0.79$.

The weight of the block in air (BA) will be, considering a reinforced concrete density = 2.7 kg dm^{-3} and the density of sea water = 1.027 kg dm^{-3}.

$BA \geq 1.67(TR-GR)/0.79 = 2.11(TR-GR)$

Replacing TR and GR in formulas (4) and (7), we can get:

$BA \geq 18.9HP - 5551\log HP + 9774$ (Fig. 4)

Block Arrangement

(a) Arrangement at right angles to the coast line

This depends on the width of the gear opening (otter-boards opening). The knowledge of horizontal net opening (NH), net length (NL), and sweepline lengths (SL) allow the distance between otter-boards (OD) to be estimated (Fig. 3). Using the similar triangles theory:

$OD = NH(SL + NL)/NL$ (8)

A small coastal trawler often presents the following characters: NH = 10 m; NL = 30 m; SL = 100 m. By the application of formula (8), the theoretical distance between otter-boards would be 43.5 m. According to these data, a minimum distance of 50 m between blocks in the direction at right angles to the coastal line was considered a feasible option.

(b) Arrangement parallel to the shore line

To find the minimum distance between blocks in the direction parallel to the coast line, the total length of the gear (LG) is taken into account. This is equal to the sum of the warp (WL), sweepline (SL) and net (NL) lengths:

$LG = WL + SL + NL$ (9)

The minimum warp length (WL) at 15 m depth is 70 m, using the formula (De la Cueva, 1974) WL = 3z + 25 (z = depth in m); sweepline length = 100 m; and net length = 30 m. Therefore, and replacing in (9), the fishermen usually release a minimum of 200 m of gear length.

(c) Basic anti-trawling set (BAS)
With the right-angle and parallel minimum distances a rectangular anti-trawling basic set of 200 × 50 m (1 ha/4 blocks = (0.25 protected ha/block) with the blocks at the vertex seemed to be an effective option.

The sea area could be increased to 200 × 100 m with a block in the middle of rectangle (2 ha/5 blocks = 0.4 protected ha/block). However, this minimum BAS can be increased, mostly based on the experience and research to achieve benefit for minimum cost at Tabarca and Campello.

References

Aranda, A., J.E. Guillén, L. Martínez, A.A. Ramos and P. Sánchez-Jérez. 1992a. Estudio científico-técnico para la protección de los fondos de la Vila Joiosa. Proyecto de arrecifes artificiales, Informe Técnico, Istituto de Ecologia Litoral – Ayuntamiento de la Vila Joiosa.

Aranda, A., J.E. Guillén, L. Martínez, A.A. Ramos and P. Sánchez-Jérez. 1992b. Estudio científico-técnico para la protección de los fondos de la Bahia de Altea. Proyecto de arrecifes artificiales, Informe Técnico, Istituto de Ecologia Litoral – Ayuntamiento de Altea.

Ardizzone, G.D. and P. Pelusi. 1984. Yield and damage of bottom trawling on *Posidonia* meadows. In Boudouresque, C.F., A.J. de Grissac and J. Olivier. (eds). *International Workshop on* Posidonia oceanica *Beds*. GIS Posidonia. **1**: 63–72.

Bell, J.D. and M. Westoby. 1986. Abundance of macrofauna in dense seagrass is due to habitat preference, not predation. *Oecologia*. **68**: 205–209.

Bellan-Santini, D. 1992. Biodiversité dans les récifs artificiels. Rapports et Communications. CIESM. **33**: 376.

Bohnsack, J.A. 1991. Habitat structure and the design of artificial reefs. In: Bell, S.S., E.D. McCoy and H.R. Mushinsky. (eds.) *Habitat Structure. The Physical Arrangement of Objects in Space*. Chapman and Hall, London. pp. 412–426.

Bohnsack, J.A. and S.P. Bannerot. 1986. A stationary visual censual technique for quantitatively assessing community structure of coral reef fishes. NOAA Technical Report NMFS. **41**: 1–15.

Bohnsack, J.A. and D.L. Sutherland. 1985. Artificial reef research: a review with recommendations for future priorities. *Bulletin of Marine Sciences*. **37**: 11–39.

Bohnsack, J.A., D.L. Johnson and R.F. Ambrose. 1991. Ecology of Artificial Reef Habitats and Fishes. In Seaman, W. and L.M. Sprague. (eds.) *Artificial Habitats for Marine and Freshwater Fisheries*. Academic Press, London, pp. 61–107.

Bombace, G. 1989. Artificial reefs in the Mediterranean sea. *Bulletin of Marine Sciences*. **44**: 1023–1032.

Bortone, S.A. and J.J. Kimmel. 1991. Environmental Assessment and Monitoring of Artificial Habitats. In Seaman, W. and L.M. Sprague. (eds.), *Artificial Habitats for Marine and Freshwater Fisheries*. Academic Press, London, pp. 177–236.

Boudouresque, C.F., A. Meinesz, M. Ledoyer and P. Vitiello. 1994. Les herbiers à phanérogames marines. In Bellan-Santini, D., J.C. Lacaze and C. Poizat. (eds.) *Les Biocénosis Marines et Littorales de la Méditerranée: Synthèse, Menaces et Perspectives*. Collection Patrimoines Naturels, Múseum National d'Histoire Naturelle de Paris. **19**: 98–118.

Buchanan, J.B. 1984. Sediment Analysis. In Holme, N.A. and A.D. McIntyre. (eds.) *Methods for the Study of Marine Benthos*, 2nd edn., Blackwell Scientific Publications, Oxford, pp. 41–65.

Clarke, K.R. 1993. Non-parametric multivariate analysis of changes in community structure to environmental variables. *Australian Journal of Ecology*. **18**: 117–143.

D'Anna, G., F. Badalamenti, R. Lipari, A. Cuttitta and C. Pipitone. 1995. Fish assemblage analysis by means of a visual census survey on an artificial reef and on natural areas in the Gulf of Castellammare (NW Sicily). *Proceedings of the Sixth International Conference of Aquatic Habitat Enhancement, ECOSET'95*, Tokyo, **1**: 221–226.

De Groot, S.J. 1984. The impact of bottom trawling on benthic fauna of the Northern Sea. *Ocean Management*. **9**: 177–190.

De la Cueva, M. 1974. Artes y aparejos. Tecnología pesquera. Subsecretaría de la Marina Mercante, 258 pp.

Doumenge, F. 1981. Problèmes de l'aménagement integré du littoral Méditerranéen. Studies and Reviews, CGPM, FAO. **58**: 329–350.

Grove, R.S., C.J. Sonu and M. Nakamura. 1991. Design and Engineering of Manufactured habitats for fisheries. In Seaman, W. and L.M. Sprague. (eds.) *Artificial Habitats for Marine and Freshwater Fisheries*. Academic Press, London, pp. 109–152.

Guillén, J.E. and A. Pérez-Ruzafa. 1993. Composición, estructura y dinámica de crustáceos decápodos asociados a las comunidades arenosas del SE Ibérico. Publicaciones Especiales del Instituto Español de Oceanografía. **11**: 175–183.

Guillén, J.E., A.A. Ramos, L. Martínez and J.L. Sánchez-Lizaso. 1994. Antitrawling reefs and the protection of *Posidonia oceanica* (L.) delile meadows in the western Mediterranean sea: demand and aims. *Bulletin of Marine Science*. **55**(2–3): 645–650.

Harmelin-Vivien, M. 1984. Description d'un petit chalut à perche pour récolter la faune vagil d'herbiers de posidonies. Rapports et Procès-Verbaux, CIESM. **27**: 199–200.

Harmelin-Vivien, M, J.G. Harmelin, C. Chauvet, C. Duval, R. Galzin, R. Lejeune, G. Barnabé, F. Blanc, R. Chevalier, J. Duclerc and G. Lasserre. 1985. Evaluation visuelle des peuplements et populations des poissons: Méthodes et problèmes. *Revue d'Ecologie (Terre et Vie)*. **40**: 467–539.

Hutchings, P. 1990. Review of the effects of trawling on macrobenthic epifaunal communities. *Australian Journal of Marine and Freshwater Research*. **41**: 111–120.

Johnson, T.D., A.M. Barnett, E.E. De Martini, L.L. Craft, R.F. Ambrose L.J. Purcell. 1994. Fish production and habitat utilization on a Southern California artificial reef. *Bulletin of Marine Science*. **55**(2–3): 709–723.

Jones, J.B. 1992. Environmental impact of trawling on the seabed. A review. *New Zealand Journal Marine Freshwater Research*. **26**: 59–67.

Martínez, L., A.A. Ramos, J.L. Sánchez-Lizaso and J.E. Guillén. 1990. El proyecto de arrecife artificial en el litoral marine de El Campello (Alicante). *O.P. El litoral II*. **18**: 72–81.

Martínez-Hernández, M. 1993. Datos preliminares sobre la pesquería artesanal de El Campello (Alicante) en relación a las especies demersales. Publicaciones Especiales del Instituto Español de Oceanografía. **11**: 375–381.

Martínez-Hernández, M., J.E. Guillén and P. Sánchez-Jérez. 1996. Evolución de la pesca artesanal en al área protegida por el arrecife artificial de El Campello (SE Ibérico). *IX Simposio Ibérico de Estudios del Bentos Marino*, Alcalá de Henares, pp. 100–101.

Meinesz, A., C.F. Boudouresque, C. Falconetti, J.M. Astier, D. Bay, J.J. Blanc, M. Bourcier, F. Cinelli, S. Cirik, G. Cristiani, I. Di Geronimo, G. Giaccone, J.G. Harmelin, L. Laubier, A.Z. Lovric, R. Molinier, J. Ssyer and C. Vamvakas. 1983. Normalisation des symboles réprésentation et la cartographie des biocénoses benthiques littorales de la Méditerranée. *Annales de l'Institut Océanographique, Paris*. **59**: 155–172.

Milon, J.W. 1989. Economic evaluation of artificial habitat for fisheries: progress and challenges. *Bulletin of Marine Science*. **44**(2): 831–843.

Oliver, P. 1983. Les resources halieutiques de la Méditerranée. Première partic: Méditerranée occidentale. Studies and Reviews, CGPM, FAO. **59**: 1–139.

Pereira, F. 1993. Arrecifes artificiales. In Castelló Orvay, F. (ed.) *Acuicultura Marina: Fundamentos Biológicos y Tecnología de la Producción*. Universitat de Barcelona, Barcelona, pp. 691–702.

Polovina, J.J. 1989. Artificial reefs: nothing more than benthic fish aggregators. *CalCOFI*. **30**: 37–39.

Polovina, J.J. 1991. Fisheries applications and biological impacts of artificial habitats. In Seaman, W. and L.M. Sprague. (eds.) *Artificial Habitats for Marine and Freshwater Fisheries*. Academic Press, London, pp. 153–176.

Ramos, A.A. 1984. Cartografía de la pradera superficial de *Posidonia oceanica* en la Bahía de Alicante. In Boudouresque, C.F., A.J. de Grissac and J. Oliver. (eds.) *International Workshop on* Posidonia oceanica *Beds*, GIS Posidonie. **1**: 57–71.

Ramos, A.A. 1985. Contribución al conocimiento de las biocenosis bentónicas litorales de la Isla Plana o Nueva Tabarca (Alicante). Universidad de Alicante – Ayuntamiento de Alicante, pp. 111–147.

Ramos, A.A. and A. Trapote. 1987. Proyecto de arrecife artificial (antiarrastre y de atracción/concentración) en la Reserva marina de la Isla Plana o Nueva Tabarca (Alicante). Informe Técnico, Instituto Marítimo-Pesquero – Consellería de Agricultura i Pesca, Generalitat Valenciana.

Ramos, A.A., J.L. Sánchez-Lizaso, A. Aranda and J.E. Guillén. 1993. Estudio bionómico de los fondos de la isla de Benidorm (SE Ibérico). Publicaciones Especiales del Instituto Español de Oceanografía. **11**: 431–439.

Relini, G. and S. Moretti. 1986. Artificial reef and *Posidonia* bed protection of Loano (Western Ligurian Riviera). FAO Fisheries Report. **357**: 104–109.

Riggio, S. 1990. A short review of artificial reefs in Sicily. FAO Fisheries Report. **428**: 128–138.

Sale, P.F. 1980. The ecology of fishes on coral reef. *Oceanography and Marine Biology Annual Review*. **18**: 367–421.

Sánchez-Jérez, P. and A.A. Ramos-Esplá. 1994. Influencia de la estructura de la pradera de Posidonia oceanica (L.). Delile en la comunidad animal asociada: peces y anfípodos. VII Simposio Ibérico de Estudios del Bentos Marino. Girona.

Sánchez-Jérez, P. and A.A. Ramos-Esplá. 1995. Influence of spatial arrangement of artificial reefs on *Posidonia oceanica* fish assemblage in the West Mediterranean sea: importance of distance among blocks. *Proceedings of the Sixth International Conference on Ecological System Enhancement Technology for Aquatic Environments, ECOSET'95*, Tokyo. **2**: 646–651.

Sánchez-Jérez, P. and A.A. Ramos-Esplá. 1996. Detection of environmental impacts by bottom trawling on *Posidonia oceanica* (L.) Delile meadows: sensivity of fish and macroinvertebrate communities. *Journal of Aquatic Ecosystem Health*. **5**: 239–253.

Sánchez-Lizaso, J.L. 1993. Estudio de la pradera de *Posidonia oceanica* (L.) Delile de la Reserva marina de Tabarca (Alicante): Fenologia y producción primaria, Tesis Doctoral, Universidad de Alicante (unpublished).

Sánchez-Lizaso, J.L., J.E. Guillén and A.A. Ramos. 1990. The regression of *Posidonia oceanica* meadows in El Campello (SE Iberian Peninsula). Rapports et Communications, CIESM. **32**: 7.

Seaman, W. and L.M. Sprague. 1991. Artificial habitat practices in aquatic systems. In Seaman, W. and L.M. Sprague. (eds.) *Artificial Habitats for Marine and Freshwater Fisheries*. Academic Press, London, pp. 1–29.

SGPM. 1992. Programa de orientación plurianual 1987–91 y 1992–96, Grupo de trabajo sobre arrecifes artificiales y reservas marinas, Secretaría General de Pesca Marítima, Ministerio de Agricultura, Pesca y Alimentación, 17 pp.

SGPM. 1995. Jornadas de arrecifes artificiales, Grupo de JACUMAR, Secretaría General de Pesca Marítima, Ministerio de Agricultura, Pesca y Alimentación, 92 pp.

Strickland, J.D.H. and T.R. Parsons. 1972. *A Practical Handbook of Seawater Analysis*. Fisheries Research Board of Canada, Ottawa.

Urteaga, L. 1987. *La tierra esquilmada*. Serval/CSIC, Barcelona.
Vicente, N. and J.C. Moreteau. 1991. Statut de *Pinna nobilis* L. en Méditerranée (Mollusque Eulamellibranche). In Boudouresque, C.F., M. Avon and V. Gravez. *Les espèces marines à protéger en Méditerranée*. pp. 159–168.
Zoffmann, C., A.A. Ramos and F. Rodriguez-Varela. 1986. Datos preliminares oceanograficos y de contaminación marina en la Isla Plana o Nueva Tabarca (Alicante). In Ramos, A.A. (ed.) La Reserva Marina de la Isla Plana o Nueva Tabarca (Alicante). Ayuntamiento, Universidad de Alicante, pp. 95–110.

13. Artificial Reef Programme in the Balearic Islands: Western Mediterranean Sea

ISABEL MORENO

Marine Biology Laboratory, Universitat de les Illes Balears, 07071 Palma, Spain

Introduction

Although the coasts of the Balearic Islands offer great topographic variety, in general each island has a rocky, steep northern coast whilst the southern coast slopes gently onto a sedimentary seabed. The waters off the Balearics are oligotrophic, clear and warm (average temperature in the first 30 m is 18.5°C), with a high salinity (37–38 ppt). In this low-nutrient, high-water-transparency environment *Posidonia oceanica* meadow is the most important benthic community, especially along the southern coasts. The seagrass meadows trap sediment, protect the coast from wave damage, provide a habitat for many animals and other plants and act as a nursery for larvae and juveniles of many species. Maintaining seagrass meadows in good condition is most important for the health of the whole water mass. Many factors, but especially illegal trawling, sand dredging and coastal construction, have led to a large stretch of the *Posidonia* meadow being damaged and destroyed.

Faced with such problems in Balearic coastal waters and after considering research carried out on artificial reefs in different parts of the Mediterranean (e.g. Ardizone *et al.*, 1982; Augris *et al.*, 1984; Badalamenti *et al.*, 1985; Bombace 1981, 1982, 1989; Debernardi, 1980, 1984; IFREMER, 1987; Lefevre *et al.*, 1983; Relini, 1979; Relini *et al.*, 1983; Relini and Moreti, 1986; Riggio *et al.*, 1985, 1986) and in other countries (e.g. CIESM, 1982; Colunga and Stone, 1974; FAO, 1983) as well as in Japan (e.g. Asada *et al.*, 1986) and the United States of America (Aska, 1978, 1981; Bockstael *et al.*, 1985; Bockstael and Caldwell, 1986; Bohnsack, 1983; Bohnsack and Sutherland, 1985; D'Itri, 1985) an artificial reef programme for the Balearic Archipelago (Fig. 1) was launched by the local Government in 1988. The aim of the programme was to prevent further damage to the seabed, especially from trawlers, enhance its natural regeneration and increase biomass and biodiversity in the area.

The first step was to survey the coastal zone and develop criteria to establish suitable reef sites. This survey (Moreno *et al.*, 1989) took a special note of seabed topography, type of sediment, established biological community, state of conservation, pollution levels and human impact on the seabed. This led to the establishment of artificial reefs off the four main islands of the archipelago (Table 1; Fig. 2) on soft seabed with *Posidonia oceanica* meadows and bare sandy patches at a depth range of 25–30 m.

A.C. Jensen et al. (eds.), Artificial Reefs in European Seas, 219–233
© 2000 *Kluwer Academic Publishers. Printed in Great Britain.*

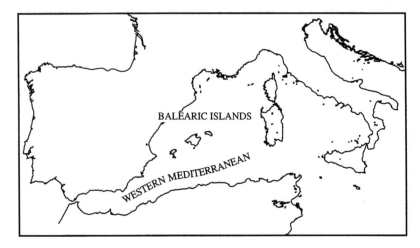

Figure 1.   Location of the Balearic Islands in the western Mediterranean.

Artificial Reef Placement and Type

Three types of benthic artificial reef modules have been deployed in the area (Table 1) for environmental protection and for fisheries production. The former has compact concrete cylindrical units (Fig. 3a), 1.69 m high with a diameter of 1.00 m at one end and 1.18 m at the other. Two iron spikes (0.25 × 0.25 m cross-section), protrude 0.5 m from the concrete on each side. They are spread extensively on the seabed, always less than 50 m deep. The units catch and tear nets and their objective is to prevent illegal fishing in the area. This type of reef has been deployed around Majorca in parts of Palma Bay, Santa Ponça Bay in the SW of the island (Fig. 2) and Dragonera Strait, on the SW coast between the islands of Majorca and Dragonera.

The second type of artificial reef is for fishery production. It consists of a hollow alveolar unit (Fig. 3b) with a base of 2.5 × 2.5 m and 1.76 m high. It has perforated 0.5 m thick walls with tunnel-like cavities leading to the hollow interior. An average of 50 units are placed in an area of approximately 4 ha. As well as being a physical obstacle to illegal fishing they are designed to aggregate many nektonic organisms, providing shelter and spawning grounds, as well as acting as a substratum for epibiota. The units can be arranged in groups following an outline rectangular pattern, such as in Palma and Santa Eulalia Bays or arranged following lines or in six clusters of eight modules, such as off Tramontana and Migjorn respectively.

A third multipurpose unit has been developed: this is a variation of the production reef type (unit volume 4.29 m³ and weighing 3.7 tonnes; Fig. 3c) with surface cavities and devices to hook nets. These multipurpose reef units have been used in Minorcan waters.

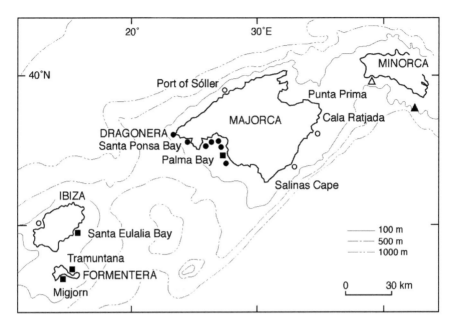

Figure 2. Positions of the different types of artificial reefs (circles: protection reefs; squares: production reefs and triangles: mixed reefs, planned in white and already deployed in black).

Legislation and Funding

The legislation on which this activity is based is an order of the Spanish Ministry of Agriculture and Fisheries in 1982 and a Regulation of the EEC dated 1986. Once the Coastal Authorities have approved a reef structure the Balearic Government then grants the order for reef deployment and forbids all types of fishing in the area for three years.

The structure, the work of deploying the modules and the subsequent monitoring have been sponsored by the EEC, Spanish funds for fisheries, the Balearic Government and the foundation Ramón Areces of Madrid.

Monitoring

Environmental monitoring focused on the four fisheries production reefs placed off the islands of Majorca, Ibiza and Formentera between 1991 and 1993. The reefs were visited every 4 months, resulting in a total of nine surveys. Temperature, salinity, oxygen, nutrients, chlorophyll and zooplankton were measured. The study of the nekton was carried out by visual census (Harmelin-Vivien *et al.*, 1985) following the replica method used by Ambrose and Swarbrick (1989), Bregliano and Ody (1985) and Fowler (1987). A photographic survey monitored percentage cover of epibiota on the reef surfaces. Collection of benthic specimens and small samples of unit material, tiles previously placed with the reef with later sampling

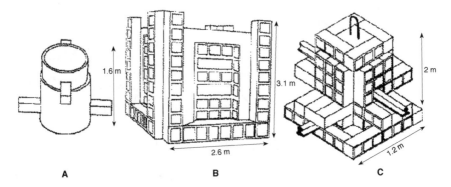

Figure 3. Types of artificial reef module/unit placed. (a) Protection module; (b) production modules, (c) mixed type.

Table 1. Characteristics of the artificial reef off the Balearic coastline. Locations are given as central points of the artificial reef area.

Island	Area	Location	Date placed	Number of modules	Type	Comment
Majorca	Palma Bay	39°27.01N 02°42.05E	July '89	50	Production	
	Palma Bay	39°27.01N 02°42.05E	Jan. '91	495	Protection	Dispersed
	Santa Ponça Bay	39°31.00N 02°27.00E	May '91	55	Protection	Dispersed
	Dragonera Strait	39°36.00N 02°25.50E	May '91	200	Protection	Dispersed
Ibiza	Santa Eulalia Bay	38°58.05N 01°32.44E	July '90	50	Production	
Formentera	Tramuntana	38°41.75N 01°30.00E	July '90	50	Production	
	Migjorn	38°48.05N 01°27.50E	July '90	50	Production	
Minorca	Punta Prima	39°48.27N 04°15.75E	July '91	35	Multipurpose	

in mind, also took place. The reef unit tiles were studied under the microscope to identify the biota to species level. In order to have a qualitative appraisal of the reef impact on the surrounding area a survey with a remotely operated vehicle (ROV) was carried out on the Palma reef at the end of the 3-year monitoring period.

Results of the Monitoring

The results of the monitoring showed that the environmental factors measured did not change because of the artificial reef's presence and that there was no noticeable disturbance of the water conditions at microscale levels.

In the reef areas 52 nektonic species were recorded (Table 2), all of them common in the area. Labridae was the most represented family. The colonization of reefs at 30 m by characteristic rocky seabed and *Posidonia* meadow species took place during the first year, and the species composition of the fish community established remained fairly stable during the study period. Samples from different years are not separated in the qualitative affinity dendrogram produced from the Jaccard index.

Although the total number of species present in the area changed very little with time, the average number of species per unit and the density of nektobenthic fish belonging to categories 3 (e.g. sparids), 4 (e.g. *Mullus surmuletus*), 5 (e.g. Labridae, Serranidae and *Sciaena umbra*) and 6 (e.g. Blennidae, Gobiidae, Tripterygidae and Scorpaenidae) (Harmelin, 1987) increased significantly (R = 0.60; $P < 0.001$ and R = 0.48; $P < 0.001$, respectively).

Different types of relationship between fish species and the reef units were observed. This behaviour can be summarized into three categories:

(1) The residents which lived within the units and nested inside (*Serranus cabrilla, Epinephelus guaza, Symphodus tinca* and different species of *Gobius* and *Scorpaena*; see Colour Plate 13);
(2) The visitors which were attracted by the units, but left them when disturbed, although remaining in the area (*Spondyliosoma cantharus* and several species of *Diplodus*);
(3) The transitory species, which, although living in the surroundings and being more abundant close to the reef than elsewhere, did not show any special relationship with the units themselves (*Mullus surmuletus, Spicara smaris, S. maena* and *Seriola dumerili*).

The fish community established in and around the four monitored reefs was very similar. However, the type of seabed made a significant difference, not only to the time the colonization took place, but also to the abundance and variety of species. The units placed on a healthy and well-developed *Posidonia* meadow in a good state of conservation populated more rapidly and with a higher density (Fig. 4), number (Fig. 5) and diversity of species (Fig. 6) than reefs deployed in poor-condition meadow or on a bare sandy or muddy seabed.

The variations in fish size frequency distribution observed around the four reefs was attributed to seabed type. The reefs deployed on *Posidonia* meadow showed a higher proportion of small individuals than those placed on mud, where medium- and large-sized individuals were more abundant.

The principal role of an artificial reef, as far as fish are concerned, seems to be the attraction to the reef of juveniles or adults from nearby areas; any increase of biomass in the whole area is not yet proven. This concentration of fish, when fishing is permitted, may lead to an overfishing of the area (Bohnsack, 1989; Polovina, 1989).

The colonization of the reef units by benthic species (Table 3) was also rapid (see Colour Plates 14 and 15). The epibiota population stabilized 18 months after deployment. A sequence in the number of individuals, species and groups of

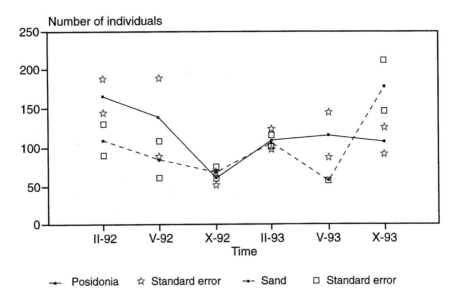

Figure 4.   Average fish density per reef on *Posidonia* meadow and sand in Santa Eulalia artificial reef.

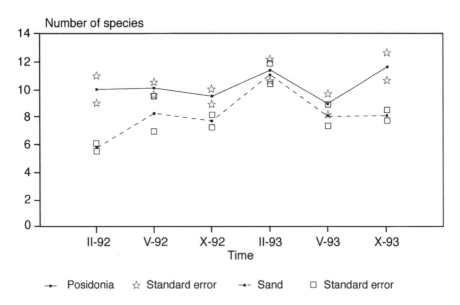

Figure 5.   Average number of fish species per reef module on *Posidonia* meadow and sand in Santa Eulalia artificial reef.

organisms was seen. Six months after deployment, the reef substratum was mainly colonized by blue-green algae and colonial animals with stolonial growth forms such as small hydroids (*Clytia hemisphaerica, C. linearis, Obelia dichotoma* and

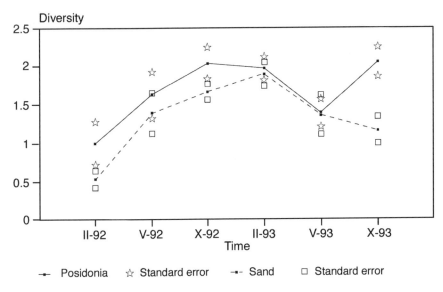

Figure 6. Average fish diversity (Shannon) per reef unit on *Posidonia* meadow and sand in Santa Eulalia artificial reef.

Campanalecium medusiferum) and bryozoans (*Aetea truncata* and *Nolella gigantea*), as well as serpulid polychaetes (*Serpula vermicularis* and *Spirorbis pagenstecheri*).

Ten months after placement, an important increase in epifaunal abundance, although not in number of species, was observed. Fourteen months after deployment, animals with vertical growth appeared, mainly hydroids with taller colonies than the previous colonizers (*Eudendrium racemosum* and *Sertularella gaudichaudi*) and bryozoans (*Caberea boyri*, *Scrupocellaria delilli* and *Bugula gautieri*). At this stage, small calcareous sponges (*Sycon* and *Grantia*) began to appear. Eighteen months after deployment colonial forms with laminar growth had appeared: sponges, both calcareous (*Clathrina*) and Demospongiae (*Hemimycale columella* and *Spirastrella cunctatrix*), encrusting bryozoans (*Calpensia* and *Schizomavella*) and Ascidiacea (*Didemnium maculosum*). At the same time, the serpulids increased in number and size. Competition for space was then observed and some cases of epiphytism and epizoism appeared (the hydroid *Obelia dichotoma* and the bryozoans *Caberea boryi* and *Nolella gigantea*). From this stage onwards the coverage of surfaces was complete but with some species, especially sponges, growing over other organisms to form big assemblies that concealed the former shape of the structures (Fig. 7).

No significant differences were detected between colonization of the four sides of the artificial reef units at three of the sites. Only at the Santa Eulalia (Ibiza) reef was slight siltation observed on the northern side of some units, resulting in a small difference in the epibiota population. In this same reef area the units placed on sand were colonized by many specimens of the sea urchin *Arbacia lixula*, whose

Table 2.   Fish species recorded in the four artificial reefs studied. *indicates presence at site.

Species	Artificial reef sites			
	Palma Bay	Santa Eulalia	Tramontana	Migjorn
Muraena helena		*	*	*
Conger conger		*		*
Zeus faber				*
Phycis phycis	*			
Epinephelus alexandrinus	*	*	*	*
Epinephelus caninus		*	*	*
Epinephelus guaza	*	*	*	*
Serranus cabrilla	*	*	*	*
Serranus scriba	*	*	*	*
Apogon imberbis	*	*	*	*
Trachurus sp.				*
Seriola dumerili		*	*	*
Sciaena umbra	*	*	*	*
Mullus surmuletus	*	*	*	*
Boops boops		*		*
Sarpa salpa		*		*
Diplodus annularis	*	*	*	*
Diplodus puntazzo	*	*	*	*
Diplodus sargus	*	*	*	*
Diplodus vulgaris	*	*	*	*
Oblada melanura	*	*		*
Spondyliosoma cantharus	*	*	*	*
Spicara maena	*	*	*	*
Spicara smaris	*	*	*	*
Chromis chromis	*	*	*	*
Coris julis	*	*	*	*
Labrus bimaculatus			*	
Labrus merula	*	*	*	*
Labrus viridis	*	*	*	*
Symphodus cinereus	*			*
Symphodus doderleini	*	*	*	*
Symphodus mediterraneus	*	*	*	*
Symphodus melanocercus	*	*	*	*
Symphodus ocellatus	*	*	*	*
Symphodus rostratus	*	*	*	*
Symphodus tinca	*	*	*	*
Thallassoma pavo		*	*	
Gobius auratus			*	*
Gobius cruentatus	*		*	*
Gobius buchichi		*		*
Gobius geniporus	*	*	*	*
Gobius vittatus			*	*
Gobius sp.	*	*	*	*
Blennius rouxi		*	*	*
Blennius sp.		*		
Tripterygion delaisi	*	*	*	*
Tripterygion melanurus		*		
Tripterygion sp.	*	*	*	
Scorpaena notata	*			*
Scorpaena porcus	*	*	*	*
Scorpaena scrofa	*	*	*	*
Balistes carolinensis		*		*

Table 3. Benthic species recorded in the four artificial reef areas studied. *implies presence at site.

Artificial reef sites	Palma Bay	Santa Eulalia	Tramontana	Migjorn
Phycophyta				
Chlorophyceae				
Bryopsis plumosa		*		
Acetabularia mediterranea	*	*	*	*
Codium bursa	*			
Derbesia tenuissima	*	*	*	
Derbesia *lamourouxi*		*		
Phaeophyceae				
Feldmania sp.		*		
Halopteris filicina	*	*	*	*
Halopteris scoparia		*		
Aglaozonia parvula		*		
Dictyota dichotoma	*	*	*	*
Padina pavonica	*	*	*	*
Arthrocladia villosa		*		
Sphacellaria plumula	*			
Zanardinia prototypus	*			
Rhodophyceae				
Peyssonnelia rubra	*		*	
Peyssonnelia squamaria	*	*	*	*
Mesophyllum sp.	*			
Lithothamnium sp.	*	*	*	*
Lithophyllum sp.	*		*	
Polysiphonia sublifera	*			
Rhizopoda				
Foraminifera				
Miniacina miniacea	*		*	
Porifera				
Calciospongidae				
Clathrina clathrus	*	*	*	*
Clathrina coriacea	*		*	*
Grantia compressa		*	*	
Sycon ciliatum	*	*	*	*
Demospongiae				
Anchinoe tenacior	*			*
Oscarella lobularis	*	*		
Spirastrella cunctatrix	*	*	*	*
Verongia cavernicola		*		
Hemimycale columella	*		*	
Mycale sp.	*			
Cnidaria				
Hydrozoa				
Halocordyle disticha		*		
Eudendrium racemosum	*	*	*	*
Halecium sp.	*	*		*
Campalecium medusiferum	*	*	*	*
Cuspidella humilis	*	*		
Sertularella gaudichaudi		*	*	*
Clytia hemisphaerica	*	*	*	*
Clytia linearis	*	*	*	*
Clytia paulensis	*		*	

Table 3 continued on next page

Table 3.   continued

Artificial reef sites	Palma Bay	Santa Eulalia	Tramontana	Migjorn
Obelia dichotoma	*	*	*	*
Obelia bidentata				*
Antenella secundaria	*			*
Filellum serpens	*			
Endoprocta				
Barentsia discreta	*			
Mollusca				
Gastropoda				
Astraea rugosa			*	*
Hinia reticulata	*			
Thais hemaestoma	*			
Coryphella pedata	*			
Flabellina affinis	*			
Platodoris argo			*	
Bivalvia				
Pinna nobilis			*	
Lithophaga lithophaga	*	*		*
Arca noae	*			
Chlamys varia	*			
Hiatella arctica		*		
Ostrea edulis	*	*		*
Anomia ephipium	*			
Cephalopoda				
Octopus vulgaris				*
Annelida				
Polychaeta				
Serpula vermicularis	*		*	*
Pomatoceros triqueter	*	*		
Spirorbis pagenstecheri	*	*	*	*
Filograna implexa	*			
Protula sp.	*			
Spirographis spallanzani	*			
Arthropoda				
Crustacea				
Balanus sp.	*			*
Palinurus elephas		*		*
Tentaculata				
Bryozoa				
Aetea truncata	*	*	*	*
Bugula gaitieri	*		*	
Beania mirabilis	*			
Scrupocellaria delilli	*		*	*
Caberea boryi	*	*	*	*
Sertella mediterranea	*	*	*	*
Crisia ramosa	*	*	*	*
Lichenopora sp.	*	*	*	*
Nolella gigantea	*	*	*	*
Valkeria uva	*	*	*	
Savygniella lafontii	*			
Schizobrachiella sp.	*			
Schizomavella sp.		*	*	*
Plagioecia sp.			*	

Table 3 continued on next page

Table 3. continued

Artificial reef sites	Palma Bay	Santa Eulalia	Tramontana	Migjorn
Calpensia nobilis	*	*		*
Entalophoroecia deflexa	*	*		
Echinodermata				
Crinoidea				
Antedon mediterranea	*			*
Leptometra phalangium		*		*
Holothuroidea				
Holothuria tubulosa	*		*	
Asteroidea				
Marthasterias glacialis	*		*	
Echinaster sepositus	*	*	*	
Echinoidea				
Arbacia lixula		*		*
Paracentrotus lividus	*			*
Sphaerechinus granularis	*	*	*	*
Tunicata				
Ascidiacea				
Aplidium proliferum				*
Aplidium conicum	*			*
Amaroucium conicum	*			
Didemnum maculosum	*	*	*	*
Diplosoma sp.	*			
Phallusia mammilata	*			
Halocynthia papillosa	*	*	*	*
Microcosmus sabateri	*	*		

Figure 7. Biological colonization of the reef modules in Palma Bay.

grazing resulted in a decrease of epibiota, opening up new surfaces for colonization. This followed the same pattern as seen previously after unit deployment.

The results from the Balearic reefs, when compared with the work of other authors in the Mediterranean, essentially agree with those of Balduzzi *et al.* (1986), Relini and Cormaggio (1989) and Riggio *et al.* (1986), all describing Italian reefs. However, they differ from most of the reefs on the Italian coast in that mussels are absent. This may be due to the depth of the reefs (Relini *et al.*, 1986) or to the oligotrophic character of Balearic waters.

Results of the Artificial Reef Deployment Programme

To discuss the results of the programme as a whole it is necessary to separate the results of the fishery production reefs from those of the environmental protection (anti-trawling) reefs.

Fishery Production Reefs

Using fishery production reefs to provide a physical obstacle for trawling gear has only been partly successful. The reef units are big and, being clustered on the seabed, are easily detected with echo-sounders, making it relatively easy for a trawler to evade. Several reef modules have been severely damaged by fishing activity and some nets have been tangled in the modules and torn. Information on how often trawlers fished in the area with no net damage is not available, but it can be supposed that at least one trawler will not repeat the experience.

The role of the units as new substrata to be colonized has been successful. Epibiotic colonization has reached over 100% cover and appears stable (Fig. 7a,b).

Comparing the ROV survey of Palma Bay 4 years after deployment with the images taken before the deployment, the situation does not seem very different. Inside the reef area the *Posidonia* meadows tend to be in better condition than in the area outside, just as it was before. It may be concluded that these structures have stopped further degradation of the seabed ecology but a true 'regeneration' has not taken place. It is important to point out that during the study period, beach replenishment works were carried out in Palma Bay. This clouds the results somewhat, as it is not possible to distinguish the impact of reef deployment from that of beach replenishment.

Environmental Protection Reef Units

The smaller environmental protection reef units have not been monitored, but because of their smaller size and random distribution are more difficult for trawlers to evade. With their iron stakes they should prove more damaging to nets than fishery production units. It seems that this type of unit was more effective in stopping illegal fishing and the seagrass meadows were better protected. Since they are smaller than the fishery production units and do not have cavities, their ability to attract fish is lower. This means that when fishing is resumed after the 3 years of prohibition the danger of overfishing the area is less.

Summary

Analysing the Balearic artificial reef programme so far, it is interesting that biological success seems to be related to the state of conservation of the seabed and the water column, rather than to fisheries production. It may be more interesting to focus on a different solution to the problem. Perhaps instead of deploying artificial structures it would be a better long-term investment to restore the natural environment and its balance directly by taking steps to stop environmental damage and avoid future problems. As the Balearic Islands are small in area and human impact is recent and not of an industrial nature this may be more feasible here than in other European coastal areas. The strengthening of fishing regulations (both commercial and recreational) and their thorough implementation may be more effective and naturally balanced than deploying artificial reefs to damage illegal fishing gear, something which has not proved to be totally effective. Regardless of the effectiveness of reefs, measures for the correct management of the coastal zone (Bohnsack, 1989) are necessary in the Balearics.

The recent interest in using artificial reefs in many coastal areas and the problems and scepticism aroused (Cognetti, 1989; Ody, 1990) stress the need to have an accurate overview of the role of these devices on different seabeds. Experiments deploying reef units that have different aims, size, shape, composition and surface textures, on various types of seabed would provide this. Such information would provide a scientific database describing the impact of reef deployment and the relationship of these structures within various marine ecosystems as a whole, which could be used to model ideas for reef placement prior to executing extensive plans. Data relating to the use and catch on fishing grounds and related areas before and after reef deployment are also required. This information should include the amount and composition of the catches with the different gear used, as well as the size frequency distribution of the species caught. The knowledge gained from these experiments could be of great help to the developers or authorities involved in reef programmes.

Future of the Programme

Additional artificial reef placements are planned during the next five years (Fig. 2): in the surroundings of San Antonio Bay in the west of Ibiza Island, in Majorca near the port of Sóller in the north and between Salinas Cape and Cala Ratjada in the east. An artificial reef is planned for the south-west of Minorca Island. Building on previous experience, the reefs off Majorca and Ibiza will be environmental protection reefs, while the reef planned for Minorcan waters will have mixed multipurpose reef units.

Acknowledgements

The work on the artificial reefs in Balearic waters has been sponsored by the Foundation Ramón Areces of Madrid and by the Fisheries and Mariculture Authority of the Balearic Government, to whom we express our gratitude. Also I want to thank very specially all the components of the Artificial Reef Research Team of the University of the Balearic Islands: Dra. I. Roca, O. Reñones, J. Coll, J. Moranta, G. Morey, F. Mir, M. Salamanca and J. Brotons as well as Dr. B. Morales for reading and discussing the manuscript.

References

Ambrose, R.F. and S.L. Swarbrick. 1989. Comparison of fish assemblages on artificial and natural reefs off the coast of southern California. *Bulletin of Marine Science.* **44**(2): 718–733.

Ardizone, G.D., C. Chimenz and A. Belluscio. 1982. Benthic communities on the artificial reef of Fregene. CIESM. 55–57.

Asada, Y., Y. Hirasawa and F. Nagasaki. 1983. L'aménagement des pêches au Japon. FAO Doc. Tech. Pêches. **258**: 1–25.

Aska, D.Y. (ed.) 1978. Artificial reefs in Florida: proceedings of a Conference held in June 10 and 11 at the Bayboro Campus. University of Florida. Florida Sea Grant College Reports. **24**: 1–69.

Aska, D.Y. (ed.) 1981. Artificial Reefs Conference Proceedings. Florida Sea Grant Reports. **41**: 1–227.

Augris, C., P. Cochonat and S. Gullaume. 1984. Récifs artificiels à Languedoc-Roussillon. Reports IFREMER DERO/GM, 1–35.

Badalamenti, F., G. Giaccone, M. Gristina and S. Riggio. 1985. An eighteen month survey of the artificial reef off Terrasini (NW Sicily): the algal settlement. *Oebalia.* **XI**: 417–425.

Balduzzi, A., F. Boero, R. Cattaneo, M. Pansini and R. Pronzato. 1986. The colonization of the artificial reef of the Monaco Natural Reserve. Association Monégasque par la Protection de la Nature: 25–28.

Bockstael, M.E. and L. Caldwell. 1986. Economic analysis of artificial reefs: a pilot study of selected methodologies. Artificial Reef Development Center Report. **6**: 1–90.

Bockstael, M.E., A. Graeffe and I. Strand. 1985. Economic analysis of artificial reefs: an assessment of issue and methods. Artificial Reef Development Center Report. **5**: 1–94.

Bohnsack, J.A. 1983. Species turnover and the order versus chaos controversy concerning reef fish community structure. *Coral Reefs.* **1**: 223–228.

Bohnsack, J.A. 1989. Are high densities of fishes at artificial reefs the result of habitat limitation or behavioral preference? *Bulletin of Marine Science.* **44**(2): 631–645.

Bohnsack, J.A. and D.L. Sutherland. 1985. Artificial Reef Research: a review with recommendation for future priorities. *Bulletin of Marine Science.* **37**(1): 11–39.

Bombace, G. 1981. Notes on experiments in artificial reefs in Italy in management of living resources in the Mediterranean coastal areas. *Stud. Rev. Gen. Fish. Council.* **58**: 309–324.

Bombace, G. 1982. Il punto sulle barriere artificiali: Problemi e prospetive. *Naturalista Siciliano.* **IV** Suppl.: 573–591.

Bombace, G. 1989. Artificial Reefs in the Mediterranean Sea. *Bulletin of Marine Science.* **44**: 1023–1032.

Bregliano, P. and D. Ody. 1985. Structure du peuplement ichtyologique de substrat dur à travers le suivi des récifs artificiels et d'une zone témoin. Colloques françaises-japonaises. *Océanographie Marseille.* **6**: 101–112.

CIESM 1982. Journée d'études sur les aspects scientifiques concernant les récifs artificiels et la mariculture suspendue Mónaco. 1–128.

Cognett, G. 1989. Artificial Reefs. Why? *Marine Pollution Bulletin.* **20**: 20–21.

Colunga, L. and R. Stone. (eds.) 1974. *Proceedings of the first International conference on Artificial Reefs.* College Station Center for Marine Resources, Texas A and M University 1–151.

Debernardi, E. 1980. Underwater marine reserve of Monaco. Notes on Artificial Reefs. Stud. Rev. G.F.M.FAO. **58**: 339–341.

Debernardi, E. 1984. Contribution d'une revenue dans l'aménagement des pêches cotières. XVII éme Session de C.G.P.M.

FAO 1986. Consultation Technique sur la conchyliculture en mer et les récifs artificiels. Fish Reports. **357**: 1–175.

Fowler, A.J. 1987. The development of sampling strategies for population studies of coral reef fishes. A case study. *Coral Reefs.* **6**: 49–58.

Harmelin, J.G. 1987. Structure et variabilité de l'ichtyofaune d'une zone protegée en Mediterranée (parc National de Port-Cros, France) Pub. della Stazione Zoologia di Napoli 1: *Marine Ecology.* **8**(3): 263–284.

Harmelin-Vivien, M.L., J.G. Harmelin, C. Chauvet, C. Duval, R. Galzin, P. Lejeune, G. Barnabé, R. Blanc, R. Chevalier, J. Duclerc and G. Lasserre. 1985. Evaluation des peuplements et populations de poissons. Méthodes et problèmes. *Revue Ecologie (Terre Vie).* **40**: 467–539.

IFREMER. 1987. Les récifs artificiels en Mediterranée française Doc DRY-RH Sète 1–16.

Lefevre, J.R., J. Duclerc, A. Meinsz and M. Ragazzi. 1983. Les récifs artificiels des établissementes de pêches du Golfe Juan et de Beaulieu-sur-mer, Alpes Maritims. CIESEM 1982.

Moreno, I., I. Roca, I. Barceló, I. Massutí, S. Puigserver and M. Salamanca. 1989. Estudio ecológico de la plataforma costera Balear y valoración de las zonas potencialmente aptas para la instalación de arrecifes artificiales. Dirección General de Pesca y Cultivos Marinos, Govern Balear, pp. 1–60.

Ody, D. 1990. Les récifs artificiels en France: bilan, analyse, perspectives. *Bulletin Société Zoologique Française.* **114**(4): 49–55.

Polovina, J.J. 1989. Artificial reefs: nothing more than benthic fish aggregations. *CalCOFI Report.* **30**: 37–39.

Relini, J. 1979. Richerche in corso in Liguria sulle barriere artificiali Atti 8° Simposio Nazionale sulla conservazione della Natura, Bari. pp. 79–87.

Relini, G. and P. Cormaggio. 1989. Colonization patterns of hard substrate in the Loano artificial reef. FAO Fisheries Report. **429**: 108–113.

Relini, G. and E. Moretti. 1986. Artificial reef and *Posidonia* bed protection off Loano (Western Ligurian Riviera). FAO Fisheries Report. **357**: 104–108.

Relini, G., C.N. Bianchi, G. Matricardi and E. Pisano. 1983. Research in Progress on colonization of hard substrata in the Ligurian Sea. CIESM. 1982. pp. 77–78.

Relini, G., A. Peirano, L. Tunesi and L. Orsi-Relini. 1986. The artificial reef in the Marconi Gulf (Eastern Ligurian Riviera). FAO Fisheries Report. **357**: 95–103.

Riggio, S., G. Giaccone, F. Badalamenti and M. Gristina. 1985. Further notes on the development of benthic communities on the artificial reef off Terrasini (NW Sicily). CIESM. **29**: 321–323.

Riggio, S., F. Badalamenti, R. Chemello and M. Gristina. 1986. Zoobenthic colonization of a small artificial reef in southern Tyrrhenian: Results of a three year survey. FAO Fisheries Report. **357**: 109–119.

14. Artificial Reefs of the Canary Islands

RICARDO HAROUN and ROGELIO HERRERA

Littoral Research Unit, Dpto. de Biología, Universidad de Las Palmas de Gran Canaria, 35017 Las Palmas, Canary Islands, Spain

Background

The Canarian Archipelago is a chain of volcanic islands located in the eastern Atlantic Ocean, near the north-west African coast (Fig. 1). Most of the islands lack a coastal shelf and the seabed drops quickly to great depth.

Since the 1960s local fishermen of the Canarian Archipelago have sunk obsolete boats and other materials in coastal areas to concentrate fish. In 1981 an artificial reef was created with debris from the Spanish Air Force Base in Gando Bay (Gran Canaria Island) (Castillo *et al.*, 1991). Later collaboration between Canarian and American researchers resulted (during 1990) in the first formal fish attraction devices (FAD) project in the Canarian Archipelago. These FAD were located off the southern coast of Gran Canaria.

In the early 1990s the Canarian Autonomous Government promoted deployment of four artificial reefs in shallow coastal areas off Lanzarote (two artificial reefs), Gran Canaria (one reef) and La Palma (one reef) islands. These artificial reefs were placed with the objective of increasing coastal fishery yield. The four reefs received financial support from the Spanish Government and the European Union (EU), under the Spanish Multi-annual Guidance Programmes (MAGPs 1987–1991 and 1992–1996) (Anonymous, 1992; Gómez-Buckley and Haroun, 1994). Other artificial reefs are planned in other islands (i.e., one artificial reef off the southern coast of Fuerteventura island).

Description

In the summer of 1990, 10 FADs were placed in 20 m of water off the south-western coast of Gran Canaria island (near to Puerto Rico Marina). Each FAD was made of polypropylene mesh, 10 m long, 1 m wide which was anchored to the sea floor by five cinder blocks and suspended in the water column by three polystyrene buoys. Each $50 \times 20 \times 25$ cm cinder block was made of concrete mixed with volcanic gravel and weighed 25 kg in air. In addition to the FAD, five 100-cinder-block piles, each about 3 m high, were deployed 30 m distant from the FAD.

The four artificial reefs off Lanzarote, Gran Canaria and La Palma were placed on sandy/rocky seabed between 18 and 26 m deep (Fig. 1). Each was built using concrete modules of different design (Tables 1 and 2). The artificial reefs located off La Palma and Gran Canaria were constructed in July 1991 and November 1991 respectively, with five different types of concrete modules (Fig. 2; 1.2–8.5 tonnes)

A.C. Jensen et al. (eds.), Artificial Reefs in European Seas, 235–247

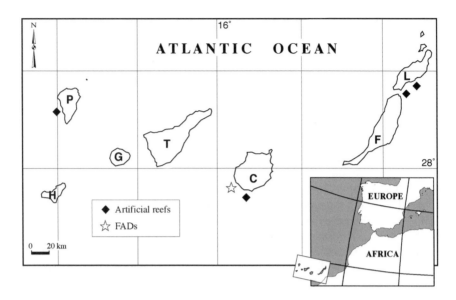

Figure 1. Location of artificial reefs and FADs in the Canary Islands; L, Lanzarote; F, Fuerteventura; C, Gran Canaria; T, Tenerife; G. Gomera; P, La Palma and H, Hierro.

Table 1. Dimensions of each module used in La Palma, Gran Canaria and Lanzarote artificial reefs.

Name	Shape	Material	Dimensions (m)	Weight (t)	Volume (m³)	Comments
C1	cube	concrete	$2 \times 2 \times 2$	6.2	2.2	Stacked bricks in central part
TC	rectangular base	concrete	$4 \times 2 \times 0.25$	6.2	4.7	Many stacked ceramic 'boxes' $20 \times 30 \times 40$ cm
T6	rectangular prism	concrete	$5 \times 3 \times 2$	8.3	13.3	6 tubes joined by two concrete rings
C3	rectangular	concrete	$3.7 \times 6 \times 1.6$	8.5	9.3	semi-closed walls made of small bricks
Alveolar	cube	concrete	$3.1 \times 2.6 \times 2.6$	9.1	19.2	reef blocks pierced with holes
Anti-trawling	block	concrete	$1 \times 0.8 \times 0.8$	0.8	1.9	small concrete blocks with 4 protruding arms of $0.5 \times 0.25 \times 0.25$ m

on 24 000 m² (84 modules) and 15 300 m² (52 modules) of seabed, respectively. The two reefs constructed off Lanzarote in June 1993 (north and south) have the same layout, 9.1 tonnes, alveolar, concrete modules (Fig. 3), 35 in one reef and 34 in the other. Each reef site used 9800 m² of seabed. All reefs were placed under a

Table 2. Summary of reef construction in Canary Islands.

Reef site	Date	Seabed area m^2	No. of modules	Type of modules
La Palma	November 1990	15,000	52	C1, TC, T6, C3, anti-trawling
Gran Canaria	November 1991	24,000	84	C1, TC, T6, C3, anti-trawling
Lanzarote, North	July 1993	9,800	35	Alveolar
Lanzarote, South	July 1993	9,800	34	Alveolar

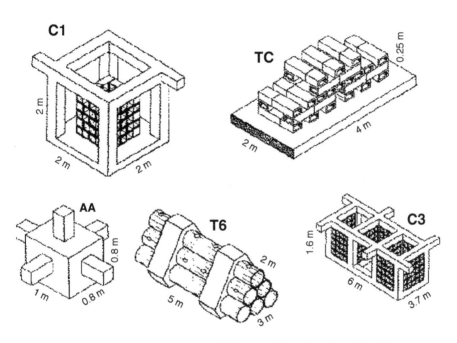

Figure 2. Various module types from the artificial reefs of Gran Canaria and La Palma.

licence granted by the Spanish Central Government, after fulfilling legal requirements laid down by the Defence Ministry and the Navigation and Environment Agencies (see Revenga *et al.*, chapter 11, this volume).

The reef units of La Palma and Gran Canaria were placed by professional divers using air-filled lifting bags to support the units in the water as they were towed from a nearby fishing harbour (Tazacorte and Arguineguín respectively). Slow release of the air from the bags controlled the descent and final positioning of the units on the seabed.

The reef units placed off Lanzarote were deployed in their respective sites (north and south) by crane from the deck of a large barge; all units were taken to the deployment site on the barge, swung over the side of the barge and released, dropping onto the seabed.

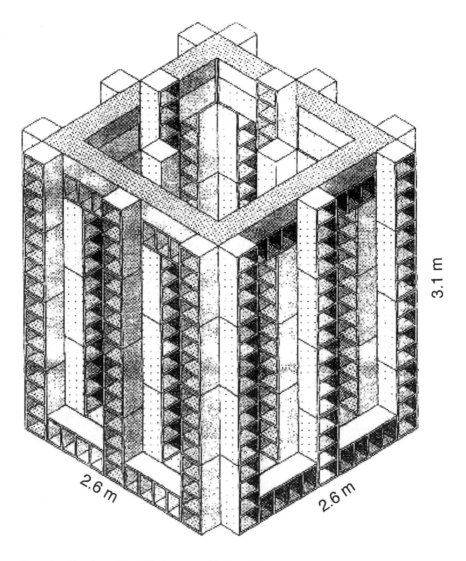

Figure 3.   Alveolar unit used in the two artificial reefs off Lanzarote.

The first method (using divers) allowed precise positioning of reef units in several reef groups, while the second (deployment with a crane) was a more cost-effective method of deploying reef units in a random fashion.

The four artificial reefs, together with a 200 m wide band around each site, were designated as protected areas; fishing was prohibited for the first three years following deployment of the concrete modules.

Research Protocols and Main Results

Baseline Studies

Gran Canaria FAD

To assess the fish assemblage associated with the structure, the FAD area was surveyed within one month of deployment and again one year later, using a visual census technique. As a result of winter storms, the polypropylene nets of the FAD were almost totally destroyed. Because of this damage the 12-month survey in July 1991 assessed fish presence around the remaining cinder blocks or surrounding sandy seabed.

La Palma Artificial Reef

Before the placing of the concrete blocks, SCUBA divers completed a baseline biological assessment of the benthic biota and fish on and around the selected reef site in October 1990. Visual census techniques were used within the artificial reef area and on a nearby natural reef. The seabed granulometry was determined.

Gran Canaria Artificial Reef

A baseline environmental appraisal of the selected area was carried out between November 1990 and March 1991. The study sampled the artificial reef area, coastal rocky seabed (0–6 m deep) and a nearby natural reef (17 m deep). Substratum type (FURUNO echo sounder), inorganic nutrient concentrations and plankton analysis (water samples collected with Niskin bottles) and light penetration (LICOR underwater spherical sensor) were assessed/measured and a SCUBA diver survey described the biota. The combined results provided a bionomical (ecological) chart which showed the reef site as a sandy platform in 17–21 m of water, with isolated natural reefs (16–20 m deep) nearby. The seabed sediment was organic white sand, rich in detritus of mollusc shells, echinoderm tests and coralline calcareous boulders. The topographic profile remained almost unchanged until, 9 km from the coast, an underwater cliff defined the end of the Island's coastal platform. The waters were oligotrophic, nutrient levels being below detection limits of the instruments used. Benthic survey results showed a rich community of invertebrates typical of biogenic sandy seabed and several different species of green algae (Haroun *et al.*, 1994a).

Lanzarote Artificial Reefs

No true baseline environmental surveys were carried out before the deposition of the concrete units. A compromise baseline study was achieved by investigating nearby sandy seabed during the first survey of the reef structure in December 1993. Around the northern reef, elasmobranch fish (sharks and rays) were the numerically dominant group. In the southern reef there were some patchy seagrass beds in which numbers of benthic macroinvertebrates and epiphytic macroalgae were found.

Scientific Monitoring Programme

Following artificial reef placement a 3-year monitoring study of the benthic communities associated with each artificial reef was started. These ran between November 1990 to December 1993 (La Palma), November 1991 to February 1995 (Gran Canaria) and December 1993 to December 1995 (Lanzarote).

    Field work sample and data collection was primarily undertaken by scientific divers using SCUBA. The physical stability of the artificial reef units, and their additional components (bricks, ceramic tiles, etc.) was assessed. The benthic communities present on horizontal and vertical surfaces of the reef units were sampled and later identified. The presence and percentage cover of macroalgae was recorded using visual census techniques and any unidentified specimens collected for further taxonomic analysis. Invertebrates communities were observed on a reef unit basis, presence/absence, class sizes and behaviour (i.e. spawning activity of cephalopods) all being recorded. The fish communities associated with the modules were surveyed using a point-count visual census technique. This sampling method recorded the number of individuals and their class size for each species during a 5-minute survey within a seabed area of 100 m^2 (Bortone *et al.*, 1986, 1989). Each survey area included a reef module. Qualitative data of fish and invertebrates presence were also obtained with underwater video and pictures (see Colour Plates 16, 17 and 18). Analysis of the raw data was done on a PC using CSS Statistics Package software.

Main Results

Gando Bay

The structure of the first (1981) artificial reef in Gando Bay (aeroplane debris and other metallic compounds) has been almost completely flattened. At present this site does not provide an effective attracting point for any coastal fish community.

Gran Canaria FAD

The number of fish species and individuals increased in the study area after the FAD were deployed. It was difficult to assess if FAD per se were the feature of the reef design responsible for the increase in fish numbers or if the FAD and associated cinder blocks had achieved their full potential of attracting fish to the area. There were some concerns, although no quantitative data to support or deny, that disturbance at the site from recreational activities (motor boats, water skiing), winter storms or overfishing may have reduced the fish numbers in the surveyed area from the maximum obtainable.

La Palma Artificial Reef

On the La Palma artificial reef the structure of benthic communities was dominated by crustose macroalgae and bryozoa, with some hydroid species becoming more abundant along the edges of the reef units which were exposed to the strongest water movement. The long-spined sea urchin *Diadema antillarum* was present in high densities (>100 individuals/unit) on the modules placed on natural rocky substrata, whereas on those units located on a sandy seabed the numbers of this amphiatlantic species were much lower.

The structure of the fish communities was similar to those encountered around any Canarian rocky seabed occupied by *D. antillarum*. However, the expected natural reef fish population was impoverished. Very few rocky seabed species, such as *Serranus atricauda*, *Diplodus* spp. or *Sparisoma cretense*, were observed. This was thought to be the combined effect of illegal fishing activity in and around the reef area (Wildpret *et al.*, 1994), strong swell action during winter storms and a high rate of siltation on the reef module surfaces, the latter two preventing the development of more complex epibenthic communities.

With hindsight the choice of site was poor (too close to the coastline and with high siltation rate) and this was felt to be responsible for the reef not realizing its full potential.

Gran Canaria Artificial Reef

After reef deployment, fish originally resident on nearby natural reefs are thought to have colonized the new structure (such as *S. atricauda*, *S. cretense*, *Diplodus* spp., *Scorpaena maderensis*, *Balistes carolinensis*) over the next 2 years. In addition, some juvenile fish thought to have recruited directly to the reef units were recorded (mainly *S. cretense*, *Apogon imberbis*, *Thalassoma pavo*) together with medium-size and large pelagic predatory fish (such as *Pseudocaranx dentex*, *Seriola* spp.). These predators have a diverse range of fish prey and appeared to use reef area as a feeding ground.

Two years after placement the artificial reef supported diverse epibenthic communities. At this time there were very few *D. antillarum* in the reef area. Six months later, a rapid increase in numbers of this omnivorous species resulted in a drastic impoverishment of epibenthic species, demonstrated by the decrease in macroalgae cover (Fig. 4) and in the number of macroinvertebrate species (e.g. hydroids, bryozoans and mollusca) (Fig. 5).

Following the rapid increase in the *D. antillarum* populations, very few juvenile fish (e.g. *Coris julis*), were observed around reef units (Fig. 6) (Haroun *et al.*, 1995). However, over time there appeared to be a slow decrease in total fish biomass. The triggering factors leading to the rapid increase of the *D. antillarum* population have not been established, but appear to be related to high fishing pressure in the reef area, especially during the second year after deployment.

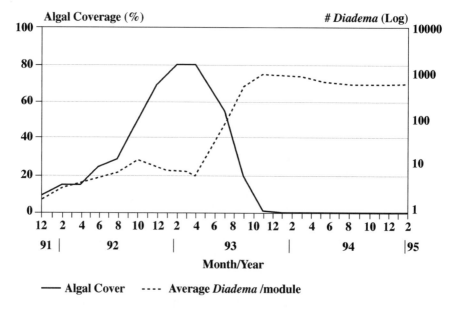

Figure 4. Variation of macroalgal cover and mean number of the sea urchin *Diadema antillarum* during the study period on Gran Canaria artificial reef.

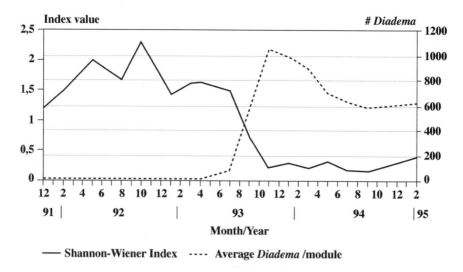

Figure 5. Variation in the Shannon-Wiener diversity index for macroinvertebrates and mean number of sea urchin *D. antillarum* during the study period on Gran Canaria artificial reef.

Colour Plate 1. Squirrel fish (Lessipian migrants) sheltering within a tyre reef off the Mediterranean coast of Israel (S. Breitstein).

Colour Plate 2. Piddocks, *Pholas dactylus*, in coal ash blocks on the Senigalia articial reef (Gianna Fabi).

Colour Plate 3. Small red scorpion fish (*Scorpaena rotata*) settled among the algae on the Loano reef.

Colour Plate 4. Common two-banded sea bream (*Diplodus vulgaris*) grazing algae settled on the Loano artificial reef (Marco Relini).

Colour Plate 5. A young spiny lobster (*Palinurus elephas*) in a hole within a concrete block on the Loano reef (Marco Relini).

Colour Plate 6. A colony of the bryozoan *Pentapora fascialis* partly covering the entrance of a hole. The fish, *Apogon imberbis*, is common in shaded areas of the Loano reef (Marco Relini).

Colour Plate 7. Type A artificial reef off Monaco (Denis Allemand).

Colour Plate 8. Type B artificial reef off Monaco (Dennis Allemand).

Colour Plate 9. Type C artificial reef off Monaco (Denis Allemand).

Colour Plate 10. Moray eel and Conger eel with tunicates in a Monégasque reef crevice (Denis Allemand).

Colour Plate 11. Red coral on surfaces with artificial caves off Monaco (Denis Allemand).

Colour Plate 12. Spanish 'anti-trawling' reef modules being deployed (Alphonso Ramos Espla).

Colour Plate 13. Sea scorpion using the reef habitat for shelter in a Balearic reef (Isabel Moreno).

Colour Plate 14. Starfish grazing the epibiota on a Balearic artificial reef (Isabel Moreno).

Colour Plate 15. The sea urchin *Diadema* grazing on a Balearic reef (Isabel Moreno).

Colour Plate 16. Cryptic blenny resting on algae within a Canary Islands' reef (Fernando Espino).

Colour Plate 17. Fish shoaling close to a reef off the Canary Islands (Fernando Espino).

Colour Plate 18. Fish shoals within and close to a reef module off the Canary Islands (Fernando Espino).

Colour Plate 19.   Squid eggs within a pipe reef off the Canary Islands (Daniel Montero).

Colour Plate 20.   Epibiota on the Poole Bay artificial reef (Antony Jensen).

Colour Plate 21.   Diver inspecting tagged lobster on the Poole Bay artificial reef
(Antony Jensen).

Colour Plate 22.   Edible crab, *Cancer pagurus*, on the quarry rock Noordwijk artificial reef shortly after deployment (Rob Leewis).

Colour Plate 23.   Plumose anemones on the Noodwijk artificial reef (Rob Leewis).

Colour Plate 24.   Polish pipe reef supporting mussels acting as biofilters to improve water quality (Juliusz Chojnacki).

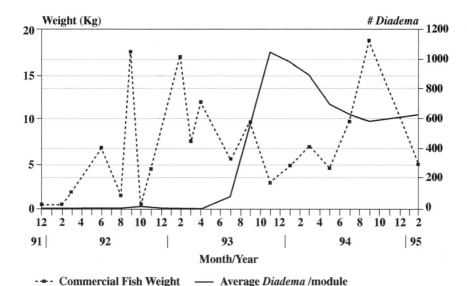

Figure 6. Relationship between commercial fish biomass and mean number of sea urchin *D. antillarum* along the study period in Gran Canaria artificial reef.

Lanzarote Artificial Reef

Lanzarote artificial reef data analysis is still preliminary (Haroun *et al.*, 1994b), but it is possible to draw some comparisons with other artificial reefs. The epibenthic communities were less developed than seen elsewhere, probably due to a greater amount of sand abrasion and the homogeneity of the modules, which provided only one type of microhabitat and few sheltered areas. Large predatory fishes (mainly *Seriola* spp.) were regularly observed in the areas around the reefs. A very low level of illegal fishing was observed. Seasonal variations in the composition of the fish community were seen, with higher number of species and greater biomass during the summer months, compared to winter months.

In all the reefs, especially that of Gran Canaria, different species of cephalopods (such as *Octopus* and *Sepia*) used the small holes and crevices of the units to lay eggs (Fig. 7; Colour Plate 19), mainly in late spring and summer. The increased numbers of cephalopods seen in the reef areas are thought to be a consequence of this utilization of reef habitat.

Discussion

Comparison of fish populations before and after reef construction shows that the artificial reefs successfully provided new, hard habitat on previously sandy seabeds. Prior to reef deployment, fish populations were typical of flat, sandy habitats; after placement fish populations were dominated by rocky seabed species

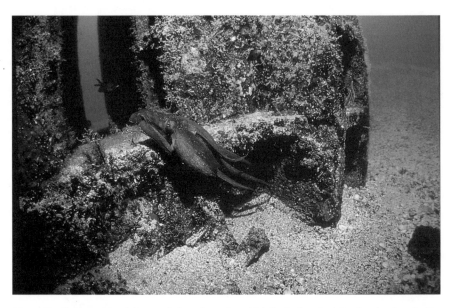

Figure 7.   Octopus vulgaris is a cephalopod species which finds abundant food and refuge in the modules (R. Herrera).

(Bombace *et al.*, 1994). In successive surveys of the Gran Canaria artificial reef, the species number of boney fish (osteichthyes) increased over time as the cartilagenous fish (elasmobranch) species declined from the numbers found during the preliminary study. The change in the structure of the fish and epibenthic communities after the 'bloom' of the *D. antillarum* population mainly affected the epibenthic species, with a smaller impact on the commercial fish biomass, although total recruitment of juvenile fish has been reduced since that time (Haroun *et al.*, 1994b, 1995).

Site selection appears to be the prime factor influencing the success of Canarian artificial reef projects; the results from the first FAD project off Gran Canaria Island (Bortone *et al.*, 1994a) and the La Palma artificial reef are clear cautionary examples of poor site selection. The nets of the FAD in the south-western coast of Gran Canaria were destroyed by the final stages of two tropical storms that sometimes reach the Canarian Archipelago during the winter months. The selection of the artificial reef area off La Palma was mainly a political decision, made without any scientific consultation.

The presence of large pelagic predatory fish, such as *Seriola* spp., seems to be related to the profile of reef units. This appeared to be especially true for the Lanzarote reefs, which have taller units than those of Gran Canaria and La Palma (Bohnsack and Sutherland, 1985; Thierry, 1988). In contrast the reef units used off Gran Canaria and La Palma were occupied by a larger number of benthic species (from crustacea to cephalopods) because there were different microhabitats of various sizes available. Apart from the resident species, a small group of pelagic fishes (*Seriola* spp. and *Boops boops*) aggregated around the reef area and occupied the water column.

Conclusions

The increase in the number of fish species after reef deposition was related to the introduction of a hard substratum that provided a habitat for species that prefer rocky seabed. Several typical sandy-seabed species were found close to the modules and used the reef areas as feeding places. Some rays, such as *Taeniura grabata* (Fig. 8), used the modules as a daytime refuge, also taking advantage of the presence of the 'cleaner shrimp' *Hippolysmata grabhami*.

Significant differences in the use of each module by various fish species were observed. The module design and size appeared to determine the colonization patterns of a given species. As reported by Bortone *et al.* (1994b), the module size and the variety of designs are considered primary parameters that seemed to determine the abundance and number of fish species from the start of the colonization process. Although the assessment of the Lanzarote reefs is not finished it seems that they attracted a higher number of predatory pelagic fish than the other Canarian reefs.

Compared with natural reef areas, Canarian artificial reefs did not hold significantly higher densities of sparid fish and only small serranids were observed, probably as consequence of space limitations on the artificial reefs. Only one type

Figure 8. The black stingray (*Taeniura grabata*) is a species that frequently uses the wide open spaces of the modules to rest (F. Espino).

of module (the TC unit) had enough openings to be considered to be a fish refuge. Some resident species (such as *Chromis limbatus, Apogon imberbis* and *Canthigaster rostrata*) and other non-resident species (e.g. *Coris julis*) used the modules as a nursery area. During February and April some predators such as *Seriola* spp. aggregated in the reef areas, probably to feed on the juvenile stages of coastal pelagic species, such as *Atherina presbyter*, which found refuge in the modules.

Reef site selection is a key factor in the later development of associated organisms; a more diverse benthic community was recorded from the Gran Canaria artificial reef than the La Palma reef probably because of the higher sedimentation rates at the latter reef. At almost all the Canarian reefs there was a constant disturbance by illegal fishing in and around the reef domain which meant that fish population assessments were always influenced by harvesting, preventing a complete appreciation of the true fish diversity and density of a reef. In the future, closer collaboration between fishermen, politicians and technical personnel will be needed if scientists are to obtain a clearer and more quantitative understanding of the artificial reefs in the Canarian waters.

Acknowledgements

This research is mainly based on the Multi-Annual Artificial Reef Programme run by the Dirección General de Pesca, Gobierno de Canarias, to whom we want to express our gratitude. Special thanks are extended to other members of the Littoral Research Unit, such as Dra. M.A. Viera-Rodríguez, E. Soler-Onís, N. Pavón-Salas, A. Casañas-Barrios, D. Montero-Vítores, T. Moreno-Moreno, for their collaboration in data collection and analysis. The suggestions of Dr A.C. Jensen, which have improved the writing style of this contribution, are highly appreciated.

References

Anonymous 1992. Jornados sobre Arrecifes Artificiales y Reservas Marinas. Minist. Agricultura, Pesca y Alimentación, Alicante (Spain), 4–5 Noviembre 1992: 5 pp.

Bombace, G., G. Fabi, L. Florentini and S. Spernaza. 1994. Analysis of the efficacy of artificial reefs located in five different areas of the Adriatic Sea. *Bulletin of Marine Science.* **55**(2–3): 559–580.

Bortone, S.A., R.W. Hastings and J.L. Oglesby. 1986. Quantification of reef fish assemblages: a comparison of several *in situ* methods. *Northeast Gulf Science.* **8**(1): 1–22.

Bortone, S.A., J.J. Kimmel and C.M. Bundrick. 1989. A comparison of three methods for visually assessing reef fish communities: time and area compensated. *Northeast Gulf Science.* **10**(2): 85–96.

Bortone, S.A., J. Van Tassel, A. Brito, J.M. Falcón, J. Mena and C.M. Bundrick. 1994a. Enhancement of the nearshore fish assemblage in the Canary Islands with artificial habitats. *Bulletin of Marine Science.* **55**(2–3): 602–608.

Bortone, S.A., T. Martin and C.M. Bundrick. 1994b. Factors affecting fish assemblage development on a modular artificial reef in a northern Gulf of Mexico estuary. *Bulletin of Marine Science.* **55**(2–3): 319–332.

Bohnsack, J.A. and D.L. Sutherland. 1985. Artificial Reef Research: A review with recommendations for future priorities. *Bulletin of Marine Science*. **37**: 11–39.

Castillo, R., J.A. Gómez and P. Guzmán. 1991. Experiência de un biotopo artificial en la Bahía de Gando, Gran Canaria. In C. Munic (ed.) *Jornadas Atlânticas do Proteção do Médio Ambiente Açores Madeira Canarias e Cabo Verde Angra do Heroismo, Portugal*, pp. 367–379.

Gómez-Buckley, M.C. and R.J. Haroun. 1994. Artificial reefs in the Spanish coastal zone. *Bulletin of Marine Science*. **55**(2–3): 1021–1028.

Haroun, R.J., M. Gómez, J.J. Hernández, R. Herrera, D. Montero, T. Moreno, E. Portillo, M.E. Torres and E. Soler. 1994a. Environmental description of an artificial reef site in Gran Canaria (Canary Islands, Spain) prior to reef placement. *Bulletin of Marine Science*. **55**(2–3): 932–938.

Haroun, R.J., R. Herrera, T. Moreno, A. Casañas, N. Pavón and H. Haack. 1994b. Análisis estructural de las comunidades bentónicas y demersales en los arrecifes artificiales de Lanzarote. Consejería de Pesca y Transportes, Gobierno de Canarias. 132 pp.

Haroun, R.J., R. Herrera, T. Moreno, A. Casañas, F. Espino, N. Pavón, L. Medina, M.E. Torres and E. Soler. 1995. Seguimiento Científico del Arrecife de Arguineguín, Gran Canaria. Cuarto Informe Periodo: Enero-Diciembre 1994. (Consejería de Pesca y Transportes, Gobierno de Canarias). 108 pp.

Thierry, J.M. 1988. Artificial reef in Japan – a general outline. *Aquacultural Engineering*. **7**: 321–348.

Wildpret, W., A. Brito and E. Barquín. 1994. Tercer Informe del Proyecto de Seguimiento Científico del Arrecife Artificial de Tazacorte (Consejería de Pesca y Transporte, Gobierno de Canarias). 55 pp.

15. Portuguese Artificial Reefs

CARLOS COSTA MONTEIRO and MIGUEL NEVES SANTOS

Instituto de Investigação das Pescas e do Mar, Centro Regional de Investigação Pesqueira do Sul, Av. 5 de Outubro, s/n 8700 Olhão, Portugal

Introduction

The use of artificial reefs for marine habitat enhancement is a recent development in Portuguese coastal waters. To date, four reefs have been constructed, off the south coast of the Algarve. The first two were established in 1990 as a pilot project supported by the Integrated Plan for Regional Development (PIDR). The third and fourth were deployed in 1998 and were co-supported by Instrumento Financeiro de Orientação das Pescas (IFOP, Financial Tool for Fisheries Orientation). All of these programmes were proposed by the Fisheries and Marine Research Institute (IPIMAR) and have been implemented by its Regional Centre in the Algarve (CRIPSul), with the first author responsible for the coordination.

Most of the information included in this chapter relates to the pilot project, the results of which provided the justification for the development of the most recent artificial reef programmes in Algarve coastal waters.

The specific environmental reasons for initiating the programme and for locating the experiment on the south coast included:

- the presence of a number of highly productive lagoon and estuarine systems in the region;
- the relative paucity of natural rock outcrops (natural reefs);
- the high fishing intensity offshore;
- the need to provide a means of compensating for the effect of the fishing in order to sustain fishery potential.

The Lagoon Environment as a Nursery

The estuarine and coastal lagoon systems, and particularly the Ria Formosa Lagoon (Fig. 1), have an important function as nursery grounds for a number of ichthyological populations that cyclically migrate to the littoral region from offshore (Monteiro *et al.,* 1987, 1990). They thus play a major role in the maintenance of fish stocks.

The Paucity of Natural Rocky Systems

Where present elsewhere, sublittoral rock outcrops not only supply a potential food resource via their associated fauna but also provide shelter for juvenile fish from natural predators.

A.C. Jensen et al. (eds.), Artificial Reefs in European Seas, 249–261
© 2000 Kluwer Academic Publishers. Printed in Great Britain.

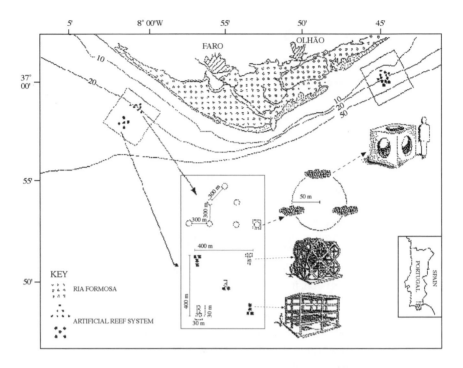

Figure 1.   Location of Faro and Olhão artificial reefs, with insert indicating Ria Formosa. Structural organization of artificial reef systems and component concrete blocks.

Fishing Pressure

The ease of accessibility of the inshore waters and high commercial value of the catch, together with the by-catch of juvenile fish, have depleted some coastal fish stocks. This in turn has generated a need to find alternative or complementary means to compensate for the effects of fishing.

With this as a background the two first reef systems (pilot project) were initiated to assess the applicability of deployment of artificial reefs in Portuguese waters. The location of the reef systems off the Ria Formosa facilitated the examination of the effects of the interaction between lagoon and coastal waters in relation to recruitment to fish stocks.

Objectives

The underlying objective of Portuguese artificial reef development is to enhance fisheries. In the present context however there were also subsidiary aims including:

- the testing of three different types of module and reef structural organization;
- the evaluation of the impact of artificial reefs at both ecological and fishery levels;

- establishment of the means by which artificial reefs may be used as an instrument for the management of coastal fisheries;
- development of multi-disciplinary knowledge in a domain of growing public interest.

Deployment of the Artificial Reefs

Legal Aspects

The Portuguese authorities have recently introduced specific laws concerning the placing of material in the sea; these laws comply with the spirit of the international maritime law relating to pollution control and safety of navigation. Accordingly, permission to deploy the reefs had to be sought from the appropriate authority, namely, the DGPNTN (Department of Ports, Navigation and Marine Transports) within the Ministry of Agriculture, Rural Development and Fisheries.

Preliminary Study

Pilot studies of the area were undertaken prior to installation of the reefs. Physical aspects considered included currents, slopes, seabed sediment, etc. These determinations, together with the use of scale models of the reefs were used to assess the potential stability of the reef modules (ETERMAR, 1989a, 1989b). Biological features examined included benthic and ichthyological fauna (Guerra and Gaudêncio, in press; Martins *et al.*, 1992; Monteiro, 1989; Monteiro and Carvalho, 1989; Santos, 1990). Other aspects (Monteiro and Carvalho, 1989), such as distance from the lagoon inlet (fish migration), distance from ports (fish landing), depth (between 15 and 40 m) and interference by fishing gear (i.e. bivalve dredging), were also taken in account.

Modules and Reef Organization

To test two particular environments (differing in terms of distance from the coast, substratum type, currents, etc.), identical reef systems were installed, one to the east and the other to the west of the main opening of the Ria Formosa (Fig. 1). The artificial reef systems consist of a protection reef and an exploitation reef. The aim of the protection reef is to create a shelter for the juvenile populations that cyclically migrate from the lagoon and the continental shelf. The protection reef consists of 735 small concrete cubic units (Fig. 2a) with a total volume of 2017 m³, distributed in 21 groups at depths that range from 15 m to 22 m; overall it occupies a total area of 39 ha.

The exploitation reef consists of 20 large reinforced concrete blocks, with a total volume of 3036 m³, comprising two different shapes (Fig. 2b,c), distributed in five groups at depths ranging from 25 m to 40 m. Altogether these occupy a total area of 21 ha. The distance between the protection and the exploitation reefs varies from 0.9 km in the Olhão reef system to 2.7 km in the Faro reef system.

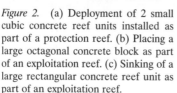

Figure 2. (a) Deployment of 2 small cubic concrete reef units installed as part of a protection reef. (b) Placing a large octagonal concrete block as part of an exploitation reef. (c) Sinking of a large rectangular concrete reef unit as part of an exploitation reef.

Method

To assess the impact of the project a multi-disciplinary study was undertaken between 1990 and 1995. The components of this study ranged from chemical assay to assessment of benthic and ichthyological colonization.

Chemical Studies

Samples of seabed sediment and particulate material in suspension were collected every 2 months for evaluation of environmental changes (water column and sediment) and comparison of the chemical/physical characteristics of the water column and sediment both in the reef zone and in its neighbourhood. The parameters analysed in the water column were temperature, salinity, dissolved oxygen, pH, nutrients (UV and visual spectrophotometric methods), chlorophyll (fluorimetric methods) and concentration and composition of dissolved material and primary production. Analyses of seabed sediment and suspended particles in the water column (collected by sediment trap) included measurement of granulometry, pH, total inorganic and organic carbon, nitrogen, phosphorus, calcium, silica, magnesium, zinc and copper.

Macro-algal Studies

Macro-algal colonization was assessed from sampling undertaken every 2 months. Direct under-water observation, photo and video records were used to follow the spatio-temporal variability of species, and species succession. For biomass evaluation, samples of macro-algae were taken seasonally.

Benthic Studies

In order to determine benthic spatio-temporal variability, colonization ratio and species succession, seabed sediment was collected and some wooden fouling panels (20 × 20 cm) attached to the reef were recovered every 2 months.

Ichthyological Studies

The methodologies used in reef ichthyological sampling included: observation using a remotely operated vehicle (ROV), visual census by scientific SCUBA divers and photo and video records, following the methods described by Bortone *et al.* (1986), Harmelin-Vivien *et al.* (1985) and Helfman (1983). Several types of fishing gear (gillnets, long-lines and traps) were also used as indirect methods of evaluation.

The studies sought:

- an understanding of reef functioning related to fish communities, and mechanisms of reef colonization;
- the evaluation of the ichthyological fauna available on the reef, at both qualitative and quantitative levels;
- the definition of a reef exploitation strategy.

The following biological and population parameters were determined monthly: species richness, diversity, fish and biomass density, spatio-temporal variability, strategy and chronology of reef colonization, dynamics of populations with commercial interest and selectivity of fishing gears.

Results

The impact study has finished, and results have been reported (Guerra and Gaudêncio, in press; IPIMAR, 1994; Monteiro *et al.*, 1991; Monteiro *et al.*, in press; Santos, 1997; Santos *et al.*, 1996, 1997; Santos and Monteiro, 1997, 1998). The following account summarizes the principal findings.

Chemical

The chemical composition in the water column near the two reef systems, coastal water and the adjacent lagoon, has been compared. The results from August 1992 (Fig. 3) showed that the concentration of ammonia, silicates, organic nitrogen and

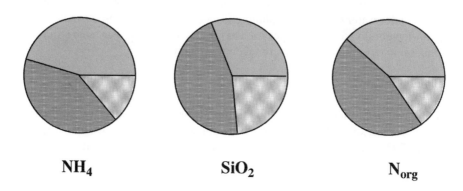

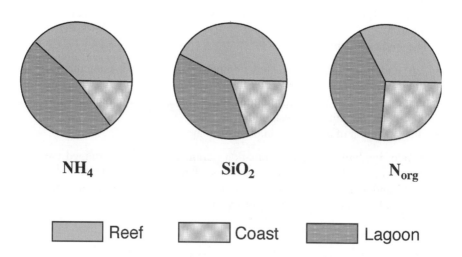

Figure 3.   Mean values of NH_4, SiO_2 and N_{org} (mM) in the water around the artificial reef systems, in the lagoon and off the coast.

organic phosphorus was similar in the lagoon and in the vicinity of the reefs (cluster analysis; Steinhaus coefficient). These concentrations are approximately 2–3 times higher than that recorded in coastal waters.

During the summer period a layer of higher salinity and raised nutrient concentration was also observed in the reef area at a depth of 10 m. This was considered to be associated with the export of nutrient-rich water from the lagoon during the ebb tide.

Macro-algae

Colonization by macro-algae was most intense between 15 m and 20 m depth. A total of 36 species were recorded (Fig. 4); four Cyanophycea, six Chlorophycea, five Phaeophycea and 21 Rhodophycea. Of these, 25 species were present in the Olhão reef system (Olhão) and 21 in the Faro reef system (Faro). The number of species of Rodophycea did not differ significantly during spring and summer, but fell in winter. In quantitative terms *Codium* spp. were well represented (maximum wet weight 364 g 0.1 m^{-2}), as were *Dictyota dichotoma* and *Ulva lactuca*. All others showed low values of biomass (wet weight <1 g 0.1 m^{-2}).

Benthos

Rapid colonization rates were recorded. The first macro-organism species to settle on both reefs was *Balanus crenatus*, but the two reef systems subsequently showed different patterns of colonization.

Faro Reef System

After the first month the most abundant species was *B. crenatus*, covering 60% of the reef area; the rest was occupied by some hydroids (e.g. *A. dichotoma, E. ramosum, S. ellisi*), amphipods and polychaetes. After the second month the hydroids occupied almost 100% of the reef surface, but some bryozoans and polychaetes were also recorded.

Figure 4. Macro-algae colonization of a reef.

Olhão Reef System

After the first month *B. crenatus* covered 100% of the reef surface. During the second month a few polychaetes, hydroids, bryozoans and the bivalve *Anomia ephippium* began to colonize the reef. This reef showed a higher density of colonization than the Faro reef system, due to the abundance of *B. crenatus*.

A similar, site-related, difference in macro-benthos density was also observed in the seabed sediment, before reef deployment.

Ichthyofauna

During the 39 months following reef deployment, 84 species were recorded, 69 in the Faro reef system and 71 in the Olhão reef system. Fifty-six species were common to both reefs, and 42 common to the reefs and the adjacent lagoon (Ria Formosa). The Sparidae (e.g. *D. bellottii, D. vulgaris, P. acarne*) and Soleidae (e.g. *D. cuneata, M. azevia*) were well represented with both juvenile and adults present since reef deployment (Fig. 5). Statistical analysis (Wilcoxon matched pairs test and ANOVA) showed no significant differences in terms of total and mean species richness ($P > 0.05$), within the reef modules, nor were there significant differences between the Faro and Olhão reef systems (Santos *et al.*, 1997a). However, these indices were always higher at the artificial reefs than at the control sites, particularly in the Olhão area.

Reef colonization rate (in terms of the accumulated species richness) was faster in the Faro reef system than in the Olhão reef system. The value of 80% coverage was reached 10 months earlier on the Faro reef (Fig. 6).

The effects of artificial reef deployment were particularly noticeable with regard to fishing yield. Yields were always higher at the reefs than at the respective control sites. This increase varied between 30% and 100%, respectively, at Olhão and Faro reefs (Santos *et al.*, 1995) (Fig. 7).

Conclusion

The artificial reef systems of the south coast of Portugal are still very young, and so their colonization is still progressing. Nevertheless it is already possible to conclude from the results that the main objectives of the pilot programme have been, or will be, achieved. The various types of reef modules used were found to be suited to the oceanographic conditions around the coast. Four years after reef deployment, there had been no observable silting up or sinking nor had there been displacement of modules. The concrete material used for constructing the modules displayed the expected capacity to act as a substratum for fouling organisms. Reef colonization was particularly fast during the first 3 months, and followed the usual pattern of colonization (Fig. 8) as described by several authors (Bohnsack and Sunderland, 1985; Buckley and Hueckel, 1985; Goran, 1979). Below 20 m depth, macro-algae colonization was very poor, due to low penetration of light and silting.

Figure 5. Fish around an exploitation reef.

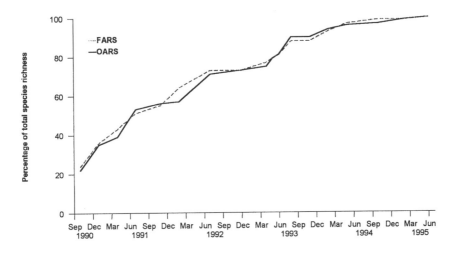

Figure 6. Reef fish colonization: percentage of accumulated species richness for the two artificial reef systems.

This biological development in the reef areas was matched by local differences at the chemical level. These seemed to be the conjugate results of:

(1) the accumulation of organic material at the seabed, with subsequent benthic re-mineralization and nutrient release;

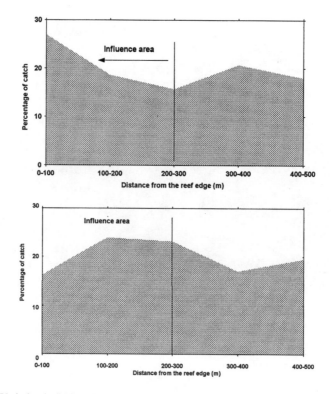

Figure 7.   Variation in fishing yield with distance from the reef edge.

(2)   the impact of the water following from the lagoon; this served to accelerate and increase the productivity of the reef systems;
(3)   other ocean effects on shallow waters.

We concluded that, for fish species:

(1)   the Ria Formosa lagoon made a strong contribution to reef colonization;
(2)   the equilibrium between the different fish species was not disturbed by the reef deployment, although the two reefs show individual differences;
(3)   the reefs played an important role protecting juvenile communities;
(4)   the reef systems proved to be an interesting tool for coastal fisheries management under the geo-ecological conditions of the Algarve, diversifying catches and increasing the fishing yields.

Perspectives

Based on the confirmed positive effects of the pilot programme, at the ecological level, fisheries and coastal resources management will continue until 2001. By the beginning of the new millennium five large artificial reef systems will be constructed (Fig. 9), which will cover a total area of more than 30 km^2. Associated with these new reefs, off-shore aquaculture experiments will be developed in order to evaluate the potential of this association for an integrated coastal resources management strategy.

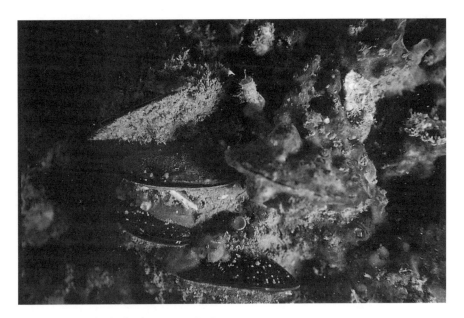

Figure 8. Mussel colonization on a reef unit.

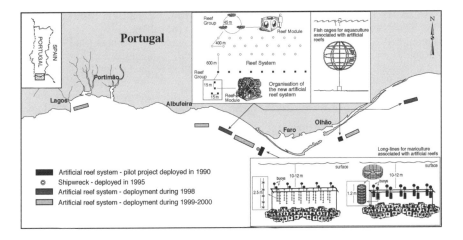

Figure 9. Proposed reef deployment along the southern Portuguese coast.

Acknowledgements

We thank all the members of the team who have been involved in this project, José Carlos Oliveira, Manuela Falco, Carlos Vale, Miriam Guerra and Maria José Gaudêncio for collecting and analyzing these data.

References

Bohnsack, J.A. and D.L. Sutherland. 1985. Artificial reef research: a review with recommendations for future priorities. *Bulletin of Marine Science.* **37**(1): 11–39.
Bortone, S.A., R.W. Hastings and J. Oglesty. 1986. Quantification of reef fish assemblages: a comparison of several in situ methods. *Northeast Gulf Science.* **8**(1): 503–508.
Buckley, R.M. and O.J. Hueckel. 1985. Biological processes and ecological development on an artificial reef in Puget Sound, Washington. *Bulletin of Marine Science.* **37**(1): 50–59.
ETERMAR 1989a. Instalação de Recifes Artificiais na Costa Algarvia. Projecto Piloto. Vol. 1, Estudo Preliminar: 30 pp.
ETERMAR 1989b. Instalação de Recifes Artificiais na Costa Algarvia. Projecto Piloto. Vol. Final, Proj. Estab.: 31 pp.
Fager, E.W. 1971. Pattern in the development of marine community. *Limnology and Oceanography.* **16**(2): 241–253.
Goren, M. 1979. Succession of benthic community on artificial substratum at Elat (Red Sea). *Journal of Experimental Marine Biology and Ecology.* **38**: 1949.
Guerra, M.T. and M.J. Gaudêncio. In press. Macrozoobentos da costa algarvia (zonas da Barra de S. Luis e Ilha da Armona). Bol. UCTRA.
Harmelin-Vivien, M.L., J. Harmelin, C. Chauvet, C. Duval, R. Galzin, P. Lejeune, G. Barnabé, F. Blanc, R. Chevalier, J. Duclerc and G. Lasserre. 1985. Evaluation visuelle des peuplements et populations des poissons: méthodes et problèmes. *Revue Ecologie (Terre et Vie).* **40**: 467–539.
Helfman, G.S. 1983. Underwater methods. In Bielsen, L.A. and D.L. Johnson. (eds.) *Fisheries Techniques.* American Fisheries Society, Bethesda, Maryland, pp. 349–369.
IPIMAR 1994. Relatório progresso Programa STRIDE. 16 pp.
Martins, R., M.N. Santos, C.C. Monteiro and M.L.P. Franca. 1992. Contribuição para o estudo da selectividade das redes de emalhar de um pano fundeadas na costa portuguesa no biéno 1990–1991. Relat Tecn Cient INIP, Lisboa (**62**), 27 pp.
Monteiro, C.C. 1989. La Faune Ichtyologique de la Lagune Ria Formosa (Sud Portugal). Répartition et Organisation Spatio-Temporelle des Communautés: Application à l'Aménagement des Ressources. Thèse Icoct Univ Scien Techn. Languedoc Montpellier: 219 pp.
Monteiro, C.C. and M.P. Carvalho. 1989. Os Recifes Artificiais como Contributo Fundamental para o Ordenamento das Pescarias Litorais Algarvias. Rel Teen Cien INIP., no. **1**, 16 pp.
Monteiro, C.C., M.M. Falcão and M.N. Santos. 1994. The artificial reef of the south coast of Portugal. *Bulletin of Marine Science.* **55**(2–3): 1346.
Monteiro, C.C., T. Lam Moi and G. Lasserre. 1987. Distribution chronologique des poissons dans deux stations de la lagune Ria Formosa (Portugal). *Oceanologica Acta.* **10**(3): 359–371.
Monteiro, C.C., G. Lasserre and T. Lam Hoi. 1990. Organisation spatiale des communautés ichtyologiques de la Lagune Ria Formosa (Portugal). *Oceanologica Acta.* **13**(1): 79–96.
Monteiro, C.C., M.N. Santos, M.M. Falção and C. Vale. In press. Os Recifes Artificiais da Costa Algarvia e a sua Interrelaçao com a Ria Formosa. Bol. UCTRA.

Santos, M.N. 1990. Ictiofauna das Zonas de Implantação dos Recifes Artificiais da Costa Algarvia – Caracterização Preliminar. Selectividade da Rede de Emalhar Fundeada para o Besugo – Pagellus acarne (Risso, 826). Rel. Estág. Licenc. University of Algarve: 47 pp.

Santos, M.N. 1997. Ichthyofauna of the Artificial Reefs of the Algarve coast. Exploitation Strategies and Management of local Fisheries. PhD Thesis, University of Algarve: 267 pp.

Santos, M.N. and C.C. Monteiro. 1997. The Olhão artificial reef system (south Portugal): fish assemblages and fishing yield. *Fisheries Research*. **30**: 33–41.

Santos, M.N. and C.C. Monteiro. 1998. Comparison of the catch and fishing yield from an artificial reef system and neighbouring areas off Faro (Algarve, south Portugal). *Fisheries Research*. **39**: 55–65.

Santos, M.N., C.C. Monteiro and G. Lassèrre. 1996. Faune ichtyologique comparée de deux récifs du littorale la Ria Formosa (lagune, Portugal): résultats préliminaires. *Oceanologica Acta*. **19**(1): 89–97.

Santos, M.N., C.C. Monteiro and G. Lassèrre. 1997. A four year overview of the fish assemblages and yield on two artificial reef systems off Algarve (south Portugal). In Hawkins, L.E., S. Hutchinson with A.C. Jensen, M. Sheader and J.A. Williams. (eds.) *Responses of Marine Organisms to their Environments*. Proceedings 30th European Marine Biology Symposium, Southampton, pp. 345–352.

16. The Poole Bay Artificial Reef Project

ANTONY JENSEN, KEN COLLINS and PHILIP SMITH

School of Ocean and Earth Science, University of Southampton, Southampton Oceanography Centre, European Way, Southampton, SO14 3ZH, UK

Introduction

Artificial reef research is relatively new in the UK. The Poole Bay artificial reef was the second licensed reef in the UK, the first being a little-investigated quarry rock reef off the east coast of Scotland (Todd *et al.*, 1992). The Poole Bay reef is unique in Northern Europe, being the first artificial reef made from cement-stabilized pulverized fuel ash (PFA) and flue gas desulphurization (FGD) gypsum derived from power stations. The reef was designed to assess the environmental acceptability of this novel material and was deployed in Poole Bay, off the central south coast of England, in June 1989 (Fig. 1).

The project arose from concerns during the late 1980s that, as most of the UK's electricity was generated by coal-fired power stations, a steady increase in the amount of PFA (a by-product of coal combustion) produced was likely. Around 50% of the PFA was taken by the construction industry, the remainder being consigned to landfill disposal sites with some (from Blythe power station, NE England) being dumped at sea. Sea disposal was due to be phased out by 1995 and landfill costs were expected to rise, making disposal of a low-toxicity bulk material in landfill an expensive option. In addition, a European Community directive to reduce sulphur dioxide emissions to 60% of the 1982 level by the year 2003 prompted the proposal to fit FGD plant to major coal-fired power stations. The favoured option was the use of limestone slurry sprayed through the flue gases, producing gypsum as an end-product. With the adoption of this process, the estimated volume of gypsum that would be produced far exceeded the requirements of the UK construction industry, providing a further disposal problem. One possible option, and one with a potential positive benefit, was the development of artificial reefs from cement-stabilized PFA and gypsum blocks. The environmental concern was that heavy metals, naturally present in coal, become concentrated in the ash formed during combustion, and so may leach into the environment from an artificial reef. To counter this, it was suggested that stabilizing PFA with cement would immobilize heavy metals (or other components) and provide a hard substratum for the attachment of organisms. The work in the UK built on the pioneering coal-waste artificial reef programme (CWARP) in the USA (Woodhead *et al.*, 1985, 1986).

The Poole Bay artificial reef was constructed using three different mixtures of PFA, FGD gypsum, cement and gravel made up into $40 \times 20 \times 20$ cm blocks at a commercial block-making plant (Collins *et al.*, 1990, 1991; Collins and Jensen, 1992a). Fifty tonnes of blocks were formed into eight conical units each 1 m high

A.C. Jensen et al. (eds.), Artificial Reefs in European Seas, 263–287
© 2000 Kluwer Academic Publishers. Printed in Great Britain.

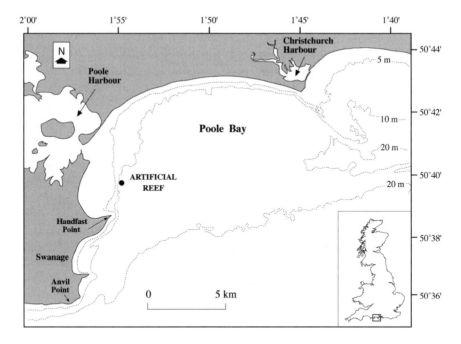

Figure 1.   Location of the Poole Bay artificial reef.

by 4 m diameter. The eight units replicate three different PFA/gypsum mixtures and one concrete control (Fig. 2). The reef structure is 10 m below MLWS (tidal range 2 m) on a flat sandy seabed.

Elsewhere, in an environmentally positive approach to the re-use of bulk waste material, commonly disposed of by dumping in landfill sites or at sea, groups in the United States, Europe, Taiwan and Japan are also examining cement stabilization technology. This work has utilized oil ash (Metz and Trefry, 1988; Nelson *et al.,*1988) and incinerator ashes (Breslin *et al.,* 1988) and ash from coal-fired power stations (Kuo *et al.,* 1995; Relini *et al.,* 1995, chapter 21, this volume; Suzuki, 1985, 1995) in experimental reef structures.

The initial Poole Bay artificial reef monitoring programme was designed to study:

(1) the physical integrity of the structures (Collins *et al.,* 1991);
(2) the environmental acceptability of the reef in terms of the heavy metal content (derived from the coal) of the blocks and the attached epifauna (Collins *et al.,* 1994a);
(3) differences in the patterns of biological colonization between the PFA/gypsum reefs and between PFA/gypsum reefs and the control reefs;
(4) changes in the infauna close to the reef;
(5) the fisheries potential of the structures.

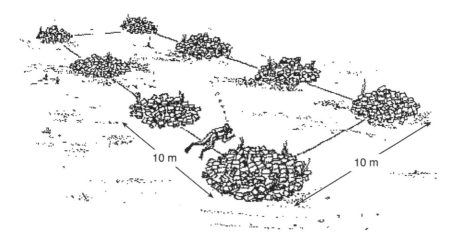

Figure 2.   Detail of the Poole Bay artificial reef.

As the study progressed the fisheries aspect of the reef became more prominent and further work encompassed fish behaviour (gadoid and labrid) and lobster (*Homarus gammarus*) and crab (*Cancer pagurus*) behaviour on artificial and natural reefs. These two latter aspects were facilitated by the in-house development of an electromagnetic telemetry system (Collins *et al.*, 1996b, in press).

Methodology

General

The artificial reef has been studied intensively by scientific divers using SCUBA. The frequency of diving increased from monthly, in winter, to weekly during the summer months. The majority of the techniques used to gather data or samples from the reef involved the use of scientific divers.

Chemical Monitoring

The Ministry of Agriculture, Fisheries and Food (MAFF) is the UK regulatory authority for marine disposal/construction. Methods similar to those recommended by MAFF (Harper *et al.*, 1989) were used to determine the heavy metal content blocks and biological samples.

Initially blocks were sampled by divers removing a block corner using a hammer and chisel underwater and the sample was then analysed for heavy metals in the laboratory. As the programme developed, a technique was adopted to provide more information on the movement of chemicals within the blocks. Reef blocks from each of the three mixtures were raised, 3 and 4 years after deployment. A $5 \times 5 \times 10$ cm deep core was cut from the centre of each block using a diamond saw. The cores were sectioned 7, 20, 33, 60 and 100 mm from the surface. In all cases, for corners and sections, samples were dried in an oven at 90°C for several days to constant weight and any epifauna was carefully scraped off. The block section samples were placed inside a polyethylene bag and crushed with a hammer, then passed through a 500 mm sieve to remove the inert gravel fraction.

Both the concrete and ash reef blocks were rapidly colonized by a wide variety of epibiota (Fig. 3), fish and crustacea (Collins et al., 1991; Collins and Jensen, 1992b; Jensen et al., 1994). A range of epibiota and reef-associated fauna (algae, hydroids, bryozoans, sponges, ascidians, molluscs and crustacea) were removed by divers from each control site and reef unit. Fish were caught in small traps set on each reef unit. The principal species or genera included within the taxonomic groups used are shown in Table 1.

After collection, organisms were frozen and stored in numbered polyethylene bags. Any adhering reef block material on the epibiota samples was removed by scraping/brushing and finally rinsing the organisms in a small amount of distilled water. Larger species (molluscs, crabs and fish) were dissected into different tissue types. The samples and tissues were placed in plastic weighing boats and dried in the same way as the block samples.

All block and biological samples were analysed for cadmium, chromium, copper, lead, manganese and zinc. In addition the concentrations of calcium and magnesium were measured in the block section samples. Samples were digested at 80°C for 24 h in concentrated nitric acid before flame atomic absorption spectrophotometry using a Pye Unicam SP9 AAS. BDH 'Spectrosol' standards were used. Additionally, standard reference material (TORT-1 lobster Hepatopancreas, National Research Council, Canada, Marine Analytical Chemistry Standard) was analysed to check the accuracy of the standards. Results for the range of elements described in this paper were within one standard deviation of the certified values. One blank was included for every 10 samples. All samples from immersed blocks were analysed alongside pre-deployment block samples for reference and standardization. The standard deviation of determinations was within ±1% of the mean, although the reproducibility of sampling due to the non-uniformity of block material increased this to ±5%.

The nitric acid digestion of PFA does not completely mobilize the heavy metals. To determine the fraction remaining after nitric acid digestion, representative block samples were also totally digested in boiling hydrofluoric acid. After evaporation to dryness, addition of perchloric acid and evaporation to dryness, distilled water was added and the digests were analysed as above.

Table 1.   Principal species sampled.

Red algae	Calliblepharis ciliata, Bonnemaisonia sp., Chondria sp.
Brown algae	Dictyota dichotoma, Dictyopteris membranacea, Desmarestia spp.
Hydroids	Halecium halecinum, Sertularella sp., Tubularia sp. Aglaophenia sp., Plumularia sp.
Bryozoans	Bugula spp., Vesicularia spinosa, Bicellariella sp.
Sponges	Scypha ciliata, Amphilectus sp., Halichondria spp.
Ascidians	Ascidia mentula, Styela clava

Figure 3.   Diver and epibiota on reef.

Photographic Monitoring

Monitoring stations were set up on all eight reef units. Individually marked blocks were photographed *in situ* on their vertical and horizontal surfaces, recording a 20 × 15 cm area of block. In this way a total area of some 2 m² was examined every 1–2 months (frequency was lower during the winter months). Images were generally clear and detailed enough to allow identification and quantification using estimations of percentage cover and random point analysis (e.g. Bohnsack, 1979).

The principal groups identified were: red algae (*Bonnemaisonia, Calliblepharis, Chondria*), hydroids (*Halecium, Aglaophenia, Plumularia*), erect bryozoans (*Bugula, Crisia, Bicellariella*), encrusting bryozoans (*Escharoides*), barnacles (*Balanus*), ascidians (*Ascidia, Aplidium, didemids*), tube worms (*Pomatoceros*) and sponges (*Amphilectus, Scypha*).

These groups are distinctive and could be distinguished in photographs. In many cases a considerable proportion of the block photograph was covered in an indistinct 'turf' usually composed of small hydroids and bryozoans.

Infaunal Studies

Seabed samples were collected from a transect running north-west from the reef at 1 m, 5 m, 10 m, 20 m and 150 m distance from the reef and also in the centre of the reef complex.

Fish Studies

Pouting (*Trisopterus luscus*) numbers and growth rates were quantified from still photographs and video footage taken around the reef. To support these data pouting were caught in fish traps and by diver-operated nets. Body length was measured and gut contents examined. Divers quantified fish numbers by swimming two transects, one over and the second around the perimeter of each reef unit.

Fish behaviour in relation to current strength and direction and light intensity was recorded over 46 h on the margin of a reef unit using a Sony V-90 camcorder with a wide-angle (×0.5) lens, plane port housing and custom-built time-lapse controller. Video footage was recorded for 5 s every 2.5 minutes (a 50 W lamp was switched on at night for the duration of filming sequence). Light intensity on the seabed was measured with an Alec MDS-L ultra-miniature light intensity recorder. Tidal current, speed and direction were measured using a Valeport BFM208 self-recording current meter. A number of fish were tagged with streamer tags to identify individuals captured from the reef under observation (Fowler *et al.*, 1999).

To categorize school position the static field of view recorded by the video camera was subjectively divided into two: 'vertical height in the water column above seabed' and 'horizontal distance from reef unit edge'. Two further categories were created to record the number of pouting in the field of view and their orientation relative to current direction.

The dominant position (horizontal distance and vertical height), numbers of individuals and orientation of the pouting were recorded every 15 minutes. Current speed, direction, water depth and light intensity were also noted.

R × c contingency table analysis (Fowler and Cohen, 1990) was used to establish whether pouting position, numbers and orientation were significantly related to current speed, direction or water depth.

Corkwing wrasse (*Crenilabrus melops*) behaviour around seaweed nests was filmed with a Sony Hi8 V900 video camera in an Amphibico housing with 100 W ac filming lamp mounted on a tripod 0.5–1 m from the nest recording continuously for up to 1 h. Longer-term studies have been made with a time-lapse system comprising a Sony V90 video camera in custom housing containing controlling circuitry and power supply for a 50 W lamp which is switched on when filming during the night. The camera was positioned 60 cm away from the nest with a field of view approximately 60 cm across. This recorded 5 s of film every 2.5 minutes over a period of 22 h (Collins *et al.*, 1996).

Lobster Studies

Conventional tagging of lobsters and crabs began in the summer of 1990 (Collins *et al.*, 1991). Lobster pots (parlour pots), as used in the local fishery, placed by each of the eight reef units were baited and emptied regularly by divers. Lobsters and crabs from the pots were measured and tagged before being returned to the reef unit adjacent to the pot of capture. Lobsters were marked with three tags: one each of T-bar, streamer and claw tags (Hallprint-Pty, Australia). The claw tag assisted in diver observations, since lobsters tend to sit in the shelter entrance with chelae foremost, but was expected to be lost during the moult. It was hoped that the T-bar and streamer tags, anchored in the dorsal musculature, would be retained through ecdysis. These two types of intramuscular tags were used in order to test their relative effectiveness.

To obtain information on within-reef location two lobsters were acoustically tagged. Divers were able to locate these animals precisely within reef units using a hydrophone on a cable linked to a receiver in an underwater housing. The acoustically tagged lobsters were recaptured after 1 month (near the end of the transmitter battery life) and the tags replaced, enabling tracking for a further month.

Acoustic telemetry of lobster movements has been superseded by the in-house development of an electromagnetic telemetry system, designed to track a large number of lobsters automatically and record environmental information (water temperature, light intensity, current direction and velocity, tidal changes and wave action) on the same time scale (Collins *et al.*, in press). Since electromagnetic signals pass through rock, lobsters can be detected when they are within crevices in the reef. On the other hand, electromagnetic telemetry is a short-range system and it was not possible to monitor lobsters that left the artificial reef. The telemetry system recorded movements between each of the eight reef units and activity indicated by a tilt switch incorporated into the transmitting tag attached to lobsters. Tag signals were detected with an array of loop aerials, one around each reef unit, connected to a central receiver and data logger (Fig. 4). In total, 41 lobsters were tagged with electromagnetic transmitters during a 2-year study period; one animal was monitored continuously for over 11 months and up to 11 individuals were monitored simultaneously. A 6-week electromagnetic telemetry study using identical equipment to that used on the artificial reef was conducted on a natural reef in Poole Bay in June and July 1997. Ten lobsters were tracked for over 1 week, seven for over 2 weeks and up to seven lobsters were tracked simultaneously.

Results

Physical Integrity

There has been no visible deterioration in undisturbed ash block integrity. Only those toppled by winter storms and displaced by dredging have shown some evidence of damage. Trials were conducted in parallel by the Building Research Establishment (Dunster and Collins, 1990) utilizing blocks with a range of mixes

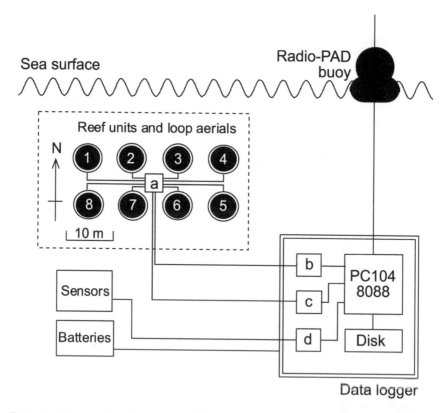

Figure 4. Diagram of the Poole Bay artificial reef electromagnetic telemetry system. a. analog aerial selector switch, b. receiver, c. shift register, d. analog-to-digital corrector.

immersed in seawater. Results suggest that block compression strength increases with time. The results of studies in the USA with stabilized coal ash products suggest long-term stability (Carleton and Muratore, 1985; Labotka *et al.*, 1985; Roethel and Oakley, 1985).

Analysis of Block Heavy Metal Content

The initial routine analysis of block corners was intended to give an indication of changes occurring in the surface of the reef blocks. The averaged (over all the PFA/gypsum mixtures) results from pre-deployment samples when compared to those after 2 years' immersion showed that only in the case of cadmium was there an indication of leaching. Analysis of block core profiles showed this leaching to be restricted to the outer 2–4 cm, and to have halted within the first 18 months of submergence (Collins *et al.*, 1991). The total leaching, from an initial block concentration of 5 ppm, was a fraction of a percent of the total block cadmium content. Later work, 3 and 4 years after deployment, provided the results presented in Fig. 5 showing average concentration of heavy metals against depth within the ash reef blocks (two of each of three mixtures) after 4 years' immersion. There is no significant change (beyond one standard deviation) in any of the metal concentrations through the block section.

Changes in the concentrations of calcium and magnesium were evident, showing loss of calcium and increase in magnesium at the surface. Sectioning the blocks reveals a lighter-coloured outer zone approximately 10 mm deep, which presumably represents the main area of interaction with the seawater. The depth of this visible zone has remained constant in sections cut from blocks sampled after 1, 2, 3 and 4 years immersion, suggesting that changes occurred rapidly after deployment and slowed subsequently. Increasing compressive strength of the ash blocks with time immersed has already been reported (Collins *et al.*, 1992a). This has also been noted by Carleton and Muratore (1985), Roethel and Oakley (1985) and Suzuki (1985, 1995).

Roethel and Oakley (1985) reported the loss of calcium, largely due to the solution of gypsum, from coal waste/FGD sludge blocks during the initial stages of the CWARP. Replacement of calcium by magnesium in the same materials was noted by Labotka *et al.* (1985). Hockley and Van der Sloot (1991) examined mineralogical and chemical changes in a CWARP block (made from coal ash and FGD gypsum) after 8 years' immersion. The depth of the region of change in composition was between 1 and 2 cm from the surface. The most notable feature of this zone was the exchange of magnesium for calcium. Minerals derived from the FGD gypsum (gypsum and calcium sulphite) and cement/PFA (ettringite and portlandite) were replaced by calcite.

In a study of leaching from coal ash and coal-ash products, Van der Sloot *et al.* (1985) demonstrated considerably reduced availability of metals from cement-stabilized products. The metal concentration profiles of a coal-waste block from the CWARP experiment, immersed for 1.5 years, were also examined. They showed no significant uptake or loss of iron, cobalt, chromium or vanadium, a slight loss

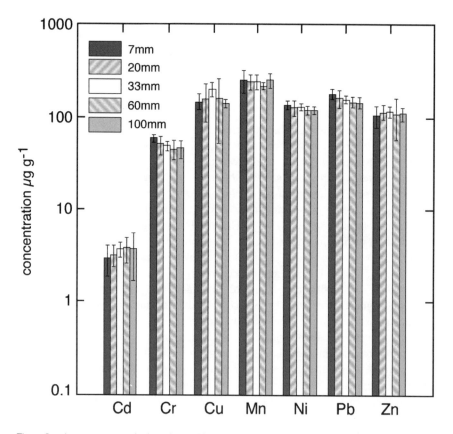

Figure 5.   Average concentrations (±1 s.d.) of heavy metals at different depths within an ash reef block section after 4 years' immersion.

of antimony, zinc or copper and, because of large variations within the data set, no clear pattern for arsenic and cadmium. Molybdenum and lead were identified as being significantly lost from the surface layer. Some uptake of manganese was found and an inward flux rate was estimated. Roethel and Oakley (1985), working with similar blocks after 500 days' immersion, also note an apparent enrichment by manganese and proposed a number of mechanisms for this. Scavenging of trace metals by coal-ash blocks in tank studies has been reported by Seligman and Duedall (1979) and Heaton *et al.* (1982). Hockley and Van der Sloot (1991), monitoring CWARP blocks immersed for 8 years, reported no loss of zinc, copper or manganese.

Analysis of Epibiota and Predators' Metal Content

The results of analyses from samples of organisms taken in 1993 (4 years after deployment) are shown in Fig. 6. In each case at least 10 individuals or colonies of each organism from ash and control reefs were analysed. Both sets of data show

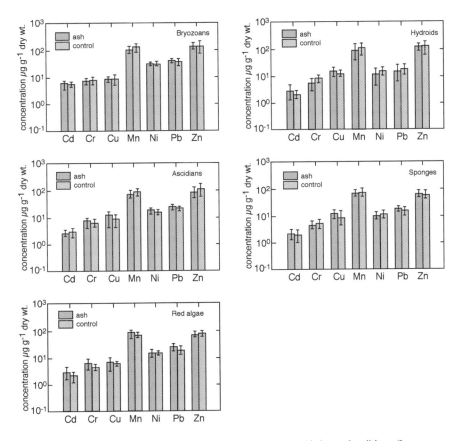

Figure 6. Comparison of heavy metal concentrations (±1 s.d.) in reef epibiota (bryozoans, hydroids, ascidians, sponges and red algae) sampled 4 years after deployment.

no significant differences between the heavy metal content of organisms growing on the ash reef blocks and those on the control surfaces.

The possibility of transfer and concentration of excess heavy metals by predatory fauna higher in the food chain has been considered. Detailed examination of the reef by divers through the life of the reef has shown that some species are resident and actively browse on the reef epibiota. The concentrations of heavy metals in the predatory gastropod mollusc, *Oceanebra erinacea*, have been reported (Collins *et al.*, 1994a). Results for another predatory gastropod mollusc (*Buccinum undatum*), crabs (*Necora puber*) and fish (*Labridae*) from the ash and control reefs were presented by Collins and Jensen (1995). There are larger, commercially important crustacea resident on the artificial reef: lobsters (*Homarus gammarus*) and crabs (*Cancer pagurus*) (Collins *et al.*, 1991). However, these are generally nocturnal and have not been observed feeding. By contrast the smaller crab, *Necora puber*, both has been seen feeding and is present in larger numbers allowing more representative sampling. The most numerous fish on the reef are pouting (*Trisopterus luscus*) which form shoals of several hundred around each unit during the day and disperse at night. Stomach content analysis indicates that this species feeds on small crustacea from the surrounding seabed. Wrasse species (Labridae; corkwing, *Crenilabrus melops* and goldsinny, *Ctenolabrus rupestris*) are resident and territorial, staying close to and within a specific reef unit. It is therefore assumed that the diet must be largely derived from that reef unit.

Chemical studies (Collins *et al.*, 1991, 1992a; Collins and Jensen, 1995) and colonization studies (Jensen *et al.*, 1994) show no appreciable difference between epibiota from the different PFA/gypsum mixtures and the control surfaces. Workers on the CWARP study (Woodhead *et al.*, 1985, 1986) and Italian reef studies (Relini *et al.*, 1995) found no evidence for trace metal transfer from coal-waste block to epifauna. Price *et al.* (1988) demonstrated no significant difference between settlement, growth and heavy metal (Cu, Fe, Zn) content of oysters growing on oyster shell and coal-ash discs.

Colonization by Epibiota and Seasonal Patterns of Recruitment

The colonization pattern of the reef blocks appeared, initially, to be following two cycles: a seasonal fluctuation overlying the natural succession from primary, opportunistic species to those which recruited later in the reef's history and successfully competed for space (Jensen *et al.*, 1994). A difference could also be seen between colonization on vertical and horizontal surfaces, the former being dominated by faunal species whilst the latter, receiving more illumination, supported more algae.

Cluster analysis was used to examine the 30-month data set. There appeared to be little difference in colonization between the different PFA/gypsum mixes (Jensen *et al.*, 1994). The concrete control blocks are represented in most of the major clusters indicating no overall difference between the colonization of the three PFA/gypsum mixes and the control.

Continuation of photographic monitoring and a repeat of the analysis on data collected between 1989 and 1996 (Beaumont, 1997) showed no significant difference in species composition, diversity and total percentage cover on horizontal surfaces of ash and concrete control reef blocks. Differences were detected between species composition, total percentage cover (but not diversity) on vertical surfaces on ash and concrete control reef blocks. These differences were not considered to be the effect of chemical leaching (diversity should be different in the presence of toxic leachate) but to a variation in physical (e.g. light, current, siltation) or biological (predation, previous settlement history) parameters between the block's vertical surfaces.

As the reef matured, convergence with local natural reefs at a similar depth became more apparent. Typically there was increasing similarity in the range of species present but differences in the proportion of surfaces covered by various groups. Sponges, particularly encrusting species, form an important component of natural reef epifauna (see Colour Plate 20). These have been slow to establish on the artificial reef but are now increasing in number. Five sponge species, new to the reef, were identified between 1989 and 1994.

Increasing density of epifauna and flora provided greater opportunity for grazing by mobile species. Corkwing wrasse and swimming crabs (*Necora puber*) have been observed browsing on barnacles attached to block surfaces. The crabs, in turn, are themselves preyed on by cuttlefish, *Sepia officinalis*, itself the subject of a local commercial fishery.

Lobster Studies

Lobsters (*Homarus gammarus*) and crabs (*Cancer pagarus*), both commercially valuable species, were recorded on the reef within 3 weeks after deployment (Jensen *et al.,* 1994). Lobsters dominated the pot catches numerically and the size distribution (Fig. 7) was typical of an exploited inshore fishery (R.C.A. Bannister, personal communication), with a mode just below the minimum landing size (85 mm carapace length) and few larger. This suggests that initial colonization of the artificial reef by lobsters was by movement of adults from the local fishery area (rather than by post-larval lobsters settling from the plankton). The fact that the nearest known natural lobster habitat is at least 2 km from the site suggests that lobsters travelled a considerable distance over seabed with little shelter. Scarratt (1968) noted similarly rapid colonization of an artificial reef in eastern Canada by American lobsters (*Homarus americanus*), although his site was closer to natural lobster ground. On the Poole Bay artificial reef, adult lobsters were evenly distributed among the eight reef units (pile of blocks), with no detectable bias in relation to block material. The standing stock of adult lobsters was estimated to be between two and three individuals per reef unit. Subsequently, juvenile lobsters as small as 25 mm carapace length have been found on the reef; they may have settled from the plankton.

The conventional T-bar and streamer tags have performed in a similar and satisfactory manner, both being retained through the moulting phase (Colour Plate 21). These tags have allowed collection of information relating to reef residence time and movements of lobsters away from the reef. Individuals have been found at the site more than 4 years after first being tagged and released there. Repeated capture, sightings and acoustic tracking results indicate a range of 'on-reef' behaviour, from prolonged residence of one reef unit, to great mobility between them (Collins and Jensen, 1992b). In addition, several animals have moulted successfully during their stay on the reef and several ovigerous females have been seen. These findings suggest that the reef units meet the habitat requirements of at least some lobsters. However, adult lobsters on the reef have moved in from elsewhere and some have left the site again. Animals tagged on the reef have been recaptured 3–4 km to the north of the reef, 6 km to the south, 12 km to the south-east and some 15 km to the south-west. It therefore seems that at any one time, the reef may contain resident and itinerant lobsters, although it is possible that individuals assume either status at different times. What prompts change from one mode of behaviour to another is unknown.

In addition to the tagging programme at the artificial reef, lobsters below the minimum landing size were captured, tagged and released in the Poole Bay lobster

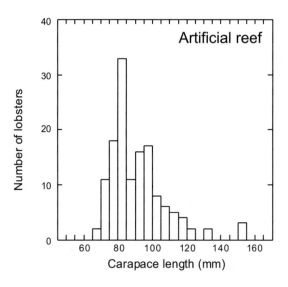

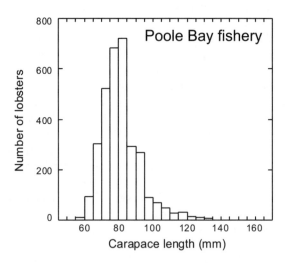

Figure 7   Size frequency distribution of lobsters caught in Poole Bay.

fishery. A substantial proportion of these animals (37%) were recaptured, many of them more than once. Over a period of 2 years, most recaptures were within 2 km of the release position. Although some lobsters moved over 20 km from release, few individuals moved outside the Poole Bay fishery area. For example, only two lobsters tagged in Poole Bay were recaptured in the neighbouring fishery of Christchurch Ledge, 5 km away, during the programme (Jensen *et al.,* 1992).

Electromagnetic telemetry indicated that over periods of several weeks lobsters inhabited particular units of the artificial reef, with excursions to other areas of the reef (Smith *et al.,* 1998). The greatest degree of movement occurred in summer, declining during autumn until few inter-reef unit movements were recorded from late December until early May. These changes in degree of movement were related to water temperature, but not to wave height, mid-day illumination or tidal range (which would reflect maximum current speeds). Activity indicated by the tilt switch also varied seasonally in relation to water temperature and, taking that effect into account, activity was negatively correlated with mid-day illumination.

Throughout the spring and summer, inter-reef unit movements were predominantly nocturnal, with a peak 2–3 h after sunset. In October and November, a time when daytime illumination at the seabed was declining, a greater proportion of movements was recorded during the day. From March until October, there was a marked diel pattern in activity resembling that of inter-reef unit movements, with a peak in the early part of the night. In January and February, activity was low and did not vary significantly in relation to time of the day. A smaller-scale telemetry study on a natural reef in mid-summer revealed similar diel patterns of movement and activity to those seen on the artificial reef at the same time of the year. It therefore seems that the diel patterns seen on the artificial reef were not an artefact of the unnatural characteristics of the site. In addition, there was some evidence from the natural reef for site fidelity shown by lobsters caught and released by a commercial fisherman, consistent with long periods of occupancy of particular reef units by lobsters on the artificial reef.

This general pattern of nocturnal activity and seasonal variation in its expression has been noted in intensive diving studies of American lobsters, *Homarus americanus*, (Cooper and Uzmann, 1980; Ennis, 1984; Karnofsky *et al.,* 1989). However, electromagnetic telemetry provides the potential to assess seasonal changes in activity in male and female lobsters of various sizes and the extent to which these seasonal changes are influenced by environmental factors.

Fish Studies

Fish recruitment was rapid. Shoals of pouting (*Trisopterus luscus*) were observed around the reef structure immediately after deployment. Wrasse followed. In 1994 the wrasse population (corkwing, *Crenilabrus melops* (L.); ballan, *Labrus bergylta* Ascanius and goldsinny, *Ctenolabrus rupestris* (L.)) was assessed by visual census (Collins *et al.,* 1996). Independent assessment showed an average population of 2–3 individuals of each species per reef unit. Differences in behaviour were noted; corkwing and goldsinny stayed close to or within the reef units whilst ballan wrasse were more mobile and patrolled the margins of the units. Fish were observed feeding on reef epifauna and goldsinny were observed 'cleaning' red mullet (*Mullus surmuletus*) which rested on the base of the reef units.

Shoal Size

The most numerous shoaling fish congregating around the reef units has been pouting, *Trisopterus luscus* (Fig. 8), a small gadoid. This species appears to use the reef units for shelter from tidal currents (up to 0.75 m s^{-1}), swimming close to the blocks during peak flow and dispersing upward and outward during slack water. Shoal size was normally around 200 individuals per reef unit in summer (Pennington, 1991). Using a length–weight relationship for pouting (Evans *et al.*, 1983) the biomass of the reef population was estimated for July/August of 1989 and 1990. Each reef unit appeared to have around it a dry weight of pouting of 207–1572 kg ha^{-1}. These values are similar in range to that on tropical reefs; e.g. 180 kg ha^{-1} (Bardach, 1959); 175–1950 kg ha^{-1} (Goldman and Talbot, 1976); 920–2373 kg ha^{-1} (Williams and Hatcher, 1983).

Diurnal Migrations of Pouting and Daytime Relocation Patterns

All pouting, tagged and untagged, had left the camcorder field of view within 15–30 min of light intensity falling below 344 lux (minimum detectable by the light meter), and did not return until 45–60 minutes before light intensity exceeded 344 lux at dawn. Diver observation of a pouting school around 1 reef unit through dusk noted the following phases:

(1) Dusk -30 min; school dispersed around reef unit.
(2) Dusk -10 min; school congregated on one side of reef unit.
(3) Dusk $+10$ min; compact school 10 m away from reef.
(4) Dusk $+30$ min; school had left reef unit.

Thirty-five pouting, (0- and 1-group) fish were tagged. Two were recaptured 2 days after release, with divers sighting a further two tagged pouting, one 30 minutes, and the second 5 days, after liberation. These results suggest that a proportion of the tagged pouting returned to the reef at dawn after 'off-reef' movement at dusk.

    None of the tagged pouting, recaptured or sighted by divers, was on the reef unit of their release, or on immediately adjacent reef units.

Tide-related Behaviour of Pouting

The vertical height of the school in the water column and horizontal distance from the reef unit were related to current speed (Fig. 9). R × c contingency table analysis (Fowler and Cohen, 1990) of the complete data set (46 h) confirmed the significance ($P < 0.01$) of the influence of water speed on pouting behaviour.

Figure 8.   Shoal of pouting *Trisopterus luscus* on the artificial reef.

Current speeds greater than 0.3 m s^{-1} were significantly related to the pouting moving close to the reef unit ($P < 0.001$) and positioning themselves low in the water column ($P < 0.0005$). Current speeds of less than 0.1 m s^{-1} saw the pouting increasing their height in the water column and moving away from the reef unit.

Corkwing Wrasse Nesting

One year after deployment, brightly coloured male corkwing wrasse were observed building seaweed nests between the reef blocks. The fish were observed collecting algae drifting past the reef, presumably in response to limited algal growth on the reef blocks. Nesting activity was seen in following years with a sequence of nests being built and abandoned on each reef unit, a single nest per unit being completed and used. During the peak of the nest building period (May to July) the fish became very defensive of their nests, chasing away other fish, and were not intimidated by the presence of divers. Activity of the males during the day appeared to be continuous, rearranging the nest material, patrolling within about 2 m of the nest, presumably feeding and collecting nest material (Fig. 10), usually seaweed of various types (Collins *et al.,* 1996). Video time-lapse photography showed that excursions from the nest decreased in frequency towards dusk and ceased after nightfall. During the darkness the fish lay on their sides, at the base of the nest, occasionally changing position. As dawn approached, the wrasse started to move and resumed excursions from the reef as light levels increased.

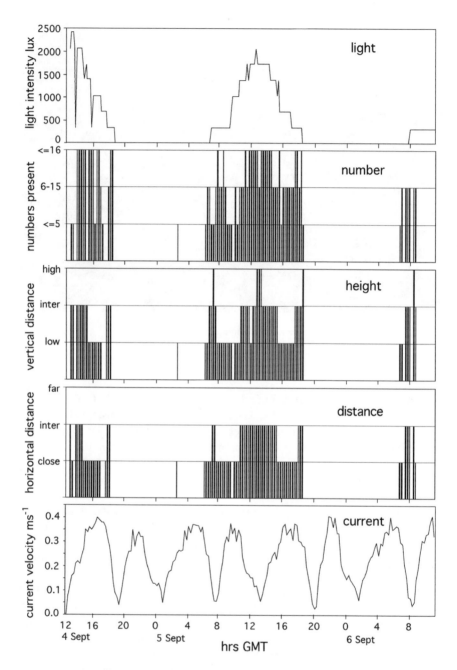

Figure 9.   Position of pouting in relation to water speed (after Fowler *et al.*, 1999).

Reproduction

Observations show the positive role of the reef in the reproduction of a number of species. Male corkwing wrasse (*Crenilabrus melops*) have constructed and defended nests of seaweed built between reef blocks every year since 1990. Lobster residence and reproductive state are mentioned above. A moulting aggregation of 50 spiny spider crabs (*Maja squinado*) occurred at the base of one reef unit in July 1990. Such gatherings are thought to be associated with mating (Carlisle, 1957; Jones and Hartnoll, 1997; Stevcic, 1971). In August 1990, velvet swimming crabs (*Necora puber*) were seen paired and in March 1991 there were large numbers of paired hermit crabs, *Pagarus bernhardus*. A number of molluscs have laid eggs on the reef blocks. Over each winter since deployment large numbers of whelks

Figure 10.   Corkwing wrasse with seaweed for nest.

(*Buccinum undatum*) and *Archidoris pseudoargus*, a large nudibranch, have congregated on the reef to mate and lay eggs. Both cuttlefish and squid have laid eggs on lines and lobster pots within the reef site.

Benthic Infauna

Cluster analysis revealed little between-site variation in the infauna along the north-west transect (Jensen *et al.*, 1994). The variations observed are similar to other annual polychaete population fluctuations seen previously which are considered to be part of natural change within the Bay (Jensen *et al.* unpublished data).

The data set suggests that the infaunal community in the vicinity of the reef was not overtly affected by the presence of the reef and that a 'feeding halo' had not been formed by grazing pressure of animals (such as fish) from the reef. These data support the view (Ambrose and Anderson, 1990; W. Nelson, personal communication, and some of the results of Davis *et al.*, 1982) that the presence of an artificial reef does not necessarily create such a degree of predation pressure on the infauna that a 'halo' of low diversity and abundance can be detected around a reef.

Sediment Granulometry

Generally sediments were dominated by the -1, 2 and 3 Phi fractions (granules and above, fine sand and very fine sand respectively). The silt clay fraction (> 4 Phi) was generally below 5%. This distribution of particles is typical for the area (Jensen *et al.*, unpublished data).

Data analysis showed little long-term change in the sediment granulometry. This supports the inference from the infaunal data that the reef has not substantially altered the character of the nearby seabed.

Discussion

Results from the Poole Bay artificial reef suggest that material such as cement-stabilized PFA can provide an environmentally acceptable structure in the marine environment. The reef surfaces have been rapidly and completely covered in epibiota with associated mobile fauna.

The Poole Bay reef, estimated to hold between 16 and 24 lobsters at any one time, has shown that artificial structures can provide good lobster habitat. Although the experimental site is small, it demonstrates that potential exists for linking a lobster hatchery and rearing reef in a ranching operation (see Jensen *et al.*, chapter 23, this volume). Economic analysis suggests that at the 'public good' level (Whitmarsh, 1997) such a concept is feasible at 1996 prices. The use of environmentally acceptable recycled materials may help to lower costs and so promote the lobster reef concept (Jensen and Collins, 1997).

In the UK at present 50% of PFA production is sold to the construction industry. A significant proportion of the remainder is dumped in landfill sites. However, since 1989 the privatized electricity generators in the UK have switched from coal- to gas-fired power stations on cost and environmental grounds. This has reduced the tonnage of PFA produced, the related disposal/re-use problems and the need for generating companies to consider novel methods for re-use of PFA.

Construction of marine structures such as artificial reefs is licensed by MAFF in England and Wales. MAFF's current policy position is that it will not license reef construction purely for fisheries reasons: there must be some well-founded reason for reef construction in the first instance, such as coastal protection or environmental rehabilitation, before it will consider approval. If approval is probable, MAFF will then look to maximize the benefits to fisheries from individual proposals. This policy could be turned to the benefit of the shellfish industry.

In the UK the most likely reason for building a marine structure is for coastal defence, especially in the south and east of England. The UK Climate Change Impacts Review Group (1991) suggest that by 2030 higher temperatures (1.5–2.1°C), warmer winters than at present, a higher incidence of hot summers and increased precipitation (5% higher during the winter of 2030) will result in a sea level rise of some 20 cm (to be adjusted for vertical land movements). An increased incidence of storms is predicted, so increasing the possibility of enhanced wave heights and generation of storm surge. In addition to these predicted climatic effects, the south-east of England experiences isostatic adjustment subsidence at the rate of 1 mm year^{-1} (Huntley, 1980).

Modern coastal defence philosophy is turning away from the 'hard' defences typified by concrete sea-walls. There are moves towards a 'soft engineering' approach, absorbing wave energy before it impacts vulnerable beaches and cliffs. For example the Happisburg to Winterton scheme (Gardener *et al.*, 1997; Hamer *et al.*, chapter 24, this volume) consists of sixteen 200 m long reefs, 300 m below high water spring tide, parallel to the shore along some 8 km of East Anglian coastline. Reefs such as this, if built with a multi-functional role, could enhance or create a fishery in an area with limited hard substrata, if placed in the sublittoral zone (Jensen *et al.*, 1998; Reeve *et al.* 1998).

In the UK, several areas need to be addressed if effective use of artificial reefs is to become a reality. Work must be continued to establish the exact function of an artificial reef in terms of biomass productivity. Research into optimal PFA block mixes to maximize block strength must be published so that engineers have a comparison of prime material and stabilized waste strengths. Block design must also be optimized to provide the required mechanical strengths and maximize the habitat potential for targeted species, such as the lobster. A greater knowledge of recruitment levels, stocking densities and behaviour of lobsters is also essential if artificial reefs are to be successfully managed in the future. It is hoped that the research in Poole Bay will help to answer some of these questions and assist in the design of larger-scale reefs in the UK and Europe.

Acknowledgements

National Power and PowerGen Joint Environmental Programme supported the Poole Bay artificial reef project (1989–1991) and MAFF have supported the lobster and crab tagging programme (1990–1997). Our thanks go to all members of the School of Ocean and Earth Science diving team for assistance throughout the programme and to the fishermen of Poole, Mudeford and Swanage who have supported and assisted with the programme.

References

Ambrose, R.F. and T.W. Anderson. 1990. Influence of an artificial reef on the surrounding infaunal community. *Marine Biology*. **107**: 41–52.

Bardach, J.E. 1959. The summer standing crop of fish on a shallow Bermuda reef. *Limnology and Oceanography*. **4**(1): 77–85.

Beaumont, J. 1997. An analysis of epibiotic colonisation on concrete and waste material blocks of an artificial reef situated in Poole Bay, UK. Unpublished honours degree project thesis, Department of Oceanography, University of Southampton.

Bohnsack, J.A. 1979. Photographic quantitative sampling of hard bottom benthic communities. *Bulletin of Marine Science*. **29**(2): 242–252.

Breslin, V.T., F.J. Roethel and V.P. Schaeperkoetter. 1988. Physical and chemical interactions of stabilised incineration residue with the marine environment. *Marine Pollution Bulletin*. **19**(11B): 628–632.

Carleton, H.R. and J. Muratore. 1985. Effect of exposure on the physical properties of coal-waste blocks in the ocean. In Duedall, I.W., D.R. Kester and P.K. Park. (eds.) *Wastes in the Oceans Vol. 4: Energy Wastes in the Ocean*. John Wiley & Sons, New York, pp. 668–690.

Carlisle, B.D. 1957. On the hormonal inhibition of moulting in decapod Crustacea. II. The terminal anecdysis in crabs. *Journal of the Marine Biological Association of the UK*. **36**: 291–307.

Collins, K.J. and A.C. Jensen. 1992a. Stability of a coal waste artificial reef. *Chemistry and Ecology*. **6**: 79–93.

Collins, K.J. and A.C. Jensen. 1992b. Acoustic tagging of lobsters on the Poole Bay artificial reef. In Priede, I.G. and S.M. Swift. (eds.) *Wildlife Telemetry. Remote Monitoring and Tracking of Animals*. Ellis Horwood, London, pp. 354–358.

Collins, K.J. and A.C. Jensen. 1995. Stabilised coal ash reef studies. *Chemistry and Ecology*. **10**: 193–203.

Collins, K.J., A.C. Jensen and A.P.M. Lockwood. 1990. Fishery enhancement reef building exercise. *Chemistry and Ecology*. **4**: 179–187.

Collins, K.J., A.C. Jensen and A.P.M. Lockwood. 1991. Artificial Reefs: using coal fired power station wastes constructively for fishery enhancement. *Oceanologica Acta*. **11**: 225–229.

Collins, K.J., A.C. Jensen, A.P.M. Lockwood and A.H. Turnpenny. 1994a. Evaluation of stabilised coal-fired power station waste for artificial reef construction. *Bulletin of Marine Science*. **55**(2–3): 1251–1262.

Collins, K.J., J. French and A.C. Jensen. 1994b. Electromagnetic tracking of lobsters on an artificial reef. Proceedings of the 6th International Conference Electronic Engineering in Oceanography, 19–21 July 1994, Churchill College, Cambridge. Institute of Electrical Engineers, **394**: pp. 1–5.

Collins, K.J., A.C. Jensen and J.J. Mallinson. 1996. Observations of wrasse on an artificial reef. In Sayer, M.D.J., Treasurer, J.W. and Costello, M.J. (eds.) *Wrasse: Biology and Use in Aquaculture*. Oxford: Blackwell Scientific Ltd., pp. 47–54.

Collins, K.J., I.P. Smith and A.C. Jensen. in press. Lobster (*Homarus gammarus*) behaviour studies using electromagnetic telemetry. In Fifth European Conference on Wildlife Telemetry, Strasbourg, 25–30 August 1996.

Cooper, R.A. and J.R. Uzmann. 1980. Ecology of juvenile and adult *Homarus*. In Cobb, J.S. and B.F. Phillips. (eds.) *The Biology and Management of Lobsters, Vol. 2. Ecology and Management*. Academic Press, New York pp. 97–142.

Davis, N., G.R. VanBlaricom and P.K. Dayton. 1982. Man-made structures on marine-sediments: effects on adjacent benthic communities. *Marine Biology*. **70**: 295–303.

Dunster, A.M. and R.J. Collins. 1990. Flue gas desulphurisation gypsum/PFA cement stabilised blocks – mix design and marine durability studies. Building Research Establishment Client Report to the Central Electricity Research Laboratories.

Ennis, G.P. 1984. Territorial behavior of the American lobster *Homarus americanus*. *Transactions of the American Fisheries Society*. **113**: 330–335.

Evans, N.A., R.N. Whitfield, R.N. Bamber and P.M. Espin. 1983. *Lernaeocera lusci* (Copepoda: pennellidae) on bib (*Trisopterus luscus*) from Southampton Water. *Parasitology*. **86**(1): 161–174.

Fowler, J. and L. Cohen. 1990. *Practical Statistics for Field Biology*. Open University Press, Milton Keynes.

Fowler, A.J., A.C. Jensen and K.J. Collins. 1999. Age structure and diel activity of pouting on the Poole Bay artificial reef. *Journal of Fish Biology*. **54**: 944–954.

Gardner, J., B.A. Hamer and R. Runcie. 1997. Physical protection of the seabed and coasts by artificial reefs. In Jensen, A.C. (ed.) *European Artificial Reef Research*. Proceedings of the first EARRN conference, March 1996, Ancona, Italy. Southampton Oceanography Centre, pp. 17–38.

Goldman, B. and F.H. Talbot. 1976. Aspects of the ecology of coral reef fishes. In Jones, O.A. and R. Endean. (eds.) *Biology and Geology of Coral Reefs*. Volume 4 Biology II. Academic Press, New York, pp. 125–154.

Harper, D.J., C.F. Fileman, P.V. May and J.E. Portmann. 1989. Aquatic Environmental Protection Analytical Methods No. 3. Methods of Analysis for Trace Metals in Marine and Other Samples. MAFF, Lowestoft.

Heaton, M.G., J.S. Buyer, J.P. Hershey, I.W. Duedall and R.D. Giaque. 1982. Elemental analysis of raw and stabilised coal waste materials. *Environmental Technology Letters*. **3**: 529–540.

Hockley, D.E. and H. Van der Sloot. 1991. Long term processes in a stabilised coal-waste block exposed to seawater. *Environmental Science and Technology*. **25**: 1408–1414.

Huntley, D.A. 1980. Tides of the north-east European Continental Shelf. In Banner, F.T. (ed.) *The Northwest European Shelf Seas. II. Physical and Chemical Oceanography, and Physical Resources*. Elsevier Oceanography series 24b. Chapter 9.

Jensen, A.C. and K.J. Collins. 1997. The use of artificial reefs in crustacean fisheries enhancement. In Jensen, A.C. (ed.) *European Artificial Reef Research*. Proceedings of the first EARRN conference, March 1996, Ancona, Italy. Southampton Oceanography Centre, pp. 115–122.

Jensen, A.C., K.J. Collins and E. Free. 1992. Poole Bay lobster tagging programme: July to September 1991. Report to MAFF. SUDO/TEC/92 1C.

Jensen, A.C., K.J. Collins, A.P.M. Lockwood, J.J. Mallinson and A.H. Turnpenny. 1994. Colonisation and fishery potential of a coal waste artificial reef, Poole Bay, United Kingdom. *Bulletin of Marine Science*. **55**(2–3): 1263–1276.

Jensen, A.C., B. Hamer and J. Wickins. 1998. Ecological implications of the construction of coastal defences. In Allsop, W. (ed.) *Coastlines, structures and breakwaters*. Institution of Civil Engineers, London, 19–20 March 1998. Thomas Telford, London.

Jones, D.R. and R.G. Hartnoll. 1997. Mate selection and mating behaviour in spider crabs. *Estuarine Coastal and Shelf Science*. **44**: 185–193.

Karnofsky, E.B., J. Atema and R.H. Elgin. 1989. Field observations of the social behaviour, shelter use, and foraging in the lobster, *Homarus americanus*. *Biological Bulletin*. **176**: 239–246.

Kuo, S-T., T-C. Hsu and K-T. Shao. 1995. Experiences of coal ash artificial reefs in Taiwan. *Chemistry and Ecology*. **10**: 233–247.

Labotka, A.L., I.W. Duedall, P.J. Harder and P.J. Schlotter. 1985. Geochemical processes occurring in coal-waste blocks in the ocean. In Duedall, I.W., D.R. Kester and P.K. Park. (eds.) *Wastes in the Oceans Vol. 4: Energy Wastes in the Ocean*. John Wiley & Sons, New York, pp. 718–739.

Metz, S. and J.H. Trefry. 1988. Trace metal considerations in experimental oil ash reefs. *Marine Pollution Bulletin*. **19**: 633–636.

Nelson, W.G., P.M. Navratil, D.M. Savercool and F.E. Vose. 1988. Short-term effects of stabilised oil ash reefs on the marine benthos. *Marine Pollution Bulletin*. **19**: 623–627.

Pennington, D. 1991. Assessment of the reef fish populations of the Poole Bay Artificial Reef including analysis of basic trophic relationships. Unpublished honours degree project report, Department of Oceanography, Southampton University.

Price, K.S., K. Mueller, J. Rosenfeld and T. Warren. 1988. Stabilised coal ash as a substratum for larval oyster settlement: a pilot field study. In *Advances in Chesapeake Bay Research*, Proceedings of a Conference, 29–31 March 1988, Baltimore, Maryland. Chesapeake Research Consortium Publication 129. CBP/TRS 24/88 pp. 29–31.

Reeve, D.E., K.J. Collins, J.F. Wickins, B.A. Hamer, A.C. Jensen and I.P. Smith. 1998. *Coastal engineering and fisheries*. Proceedings of the June 1998 MAFF conference of River and Coastal Engineers, Keele, 1–3 July 1998.

Relini, G., G. Dinelli and A. Sampaolo. 1995. Stabilised coal ash studies in Italy. *Chemistry and Ecology*. **10**: 217–232.

Roethel, F.J. and S.A. Oakley. 1985. Effects of seawater on the mineralogical and chemical composition of coal waste-blocks. In Duedall, I.W., Kester, D.R. and Park P.K. (eds.) *Wastes in the Oceans Vol. 4: Energy Wastes in the Ocean*. John Wiley & Sons, New York, p. 691–715.

Scarratt, D.J. 1968. An artificial reef for lobsters (*Homarus americanus*). *Journal of the Fisheries Research Board Canada*. **25**: 2683–2690.

Seligman, J.D. and I.W. Duedall. 1979. Chemical and physical behaviour of stabilized scrubber sludge and fly ash in seawater. *Environmental Science and Technology*. **13**: 1082–1087.

Smith, I.P., K.J. Collins and A.C. Jensen. 1998. Electromagnetic telemetry of lobster (*Homarus gammarus* (L.)) movements and activity: preliminary results. *Hydrobiologia*. **371/372**: 133–141.

Smith, I.P., K.J. Collins and A.C. Jensen. 1998. Movement and activity patterns of the European lobster, *Homarus gammarus* (L.), revealed by electromagnetic telemetry. *Marine Biology*. **132**: 611–623.

Stevcic, Z. 1971. Laboratory observations on the aggregations of the spiny spider crab (*Maja squinado* Herbst). *Animal Behaviour*. **19**: 18–25.

Suzuki, T. 1985. A concept of large artificial ridges using a new hardened product made from coal ash. In Kato, W. (ed.) *Ocean Space Utilisation '85*. Springer Verlag, Tokyo, pp. 611–618.

Suzuki, T. 1995. Application of high volume fly ash concrete to marine structures. *Chemistry and Ecology*. **10**: 249–258.

Thierry, J-M. 1988. Artificial reefs in Japan – a general outline. *Aquacultural Engineering*. **7**: 321–348.

Todd, C.D., M.G. Bently and J. Kinnear. 1992. Torness artificial reef project. In Baine, M.S.P. (ed.) *Proceedings of 1st British Conference on Artificial Reefs and Restocking*. September 1992, Orkney, ICIT, Stromness, pp. 15–21.

U.K. Climate Change Impacts Review Group. 1991. The potential effects of climate change in the United Kingdom. 1st report to the Department of the Environment, June 1991, HMSO, London, 117 pp.

Van der Sloot, H.A., J. Wijkstra, C.A. van Stigt and D. Hoede. 1985. Leaching of trace elements from coal-ash products. In Duedall, I.W., D.R. Kester and P.K. Park. (eds.) *Wastes in the Oceans Vol. 4: Energy Wastes in the Ocean*. John Wiley & Sons, New York, pp. 408–495.

Whitmarsh, D. 1997. Cost benefit analysis of artificial reefs. In Jensen, A.C. (ed.) *European Artificial Reef Research*. Proceedings of the first EARRN conference, March 1996 Ancona, Italy. Southampton Oceanography Centre, pp. 175–194.

Williams, D.M. and A.I. Hatcher. 1983. Structure of fish communities on outer slopes of inshore, mid-shelf and outer shelf reefs of the Great Barrier Reef. *Marine Ecology Progress Series*. **10**: 239–250.

Woodhead, P.M.J., J.H. Parker, H.R. Carleton and I.W. Duedall. 1985. Coal Waste Artificial Reef Program, Final Report, CS-3936. March 1985, EPRI, California.

Woodhead, P.M.J., J.H. Parker and I.W. Duedall. 1986. The use of by-products from coal combustion for artificial reef construction. In D'Itri, F.M. (ed.) *Artificial Reefs – Marine and Freshwater Applications*. Lewis Productions, Chelsea, Michigan, pp. 265–292.

17. An Artificial Reef Experiment off the Dutch Coast

ROB LEEWIS[1] and FRANK HALLIE[2]

[1]National Institute for Public Health and Environment, P.O. Box 1, 3720 BA Bilthoven, The Netherlands. [2]Rijkswaterstaat, North Sea Directorate, P.O. Box 5807, 2800 HV Rijswijk, The Netherlands

Introduction

The Dutch working group on Alternative Materials in Aquatic Engineering was the first to suggest placing an artificial reef off the Dutch coast in 1986. Although some initial research into the idea was commissioned, it did not lead to a field experiment (see Intron B.V., 1986; Leewis, 1986).

In 1989, the North Sea Directorate of Rijkswaterstaat incorporated the idea in a management policy paper, later accepted by the Dutch government. As a consequence, a field experiment was set up with the following aims:

(1) To investigate the feasibility of constructing an artificial reef off the Dutch coast.
(2) To study biological colonization and succession on such a reef and describe the results in terms of species diversity, abundance and biomass.
(3) To establish the environmental impact of the reef on the surrounding seabed by monitoring changes in infaunal species composition and biomass as well as any alterations in seabed morphology and granulometry.

The hypothesis that an artificial reef could greatly enhance (local) biodiversity and biomass, was based on 12 years of research experience by the Tidal Waters Division of Rijkswaterstaat (now the National Institute for Coastal and Marine Management; RIKZ) on the species composition and biomass of epibiota settled on the dikes in the south-west of The Netherlands (De Kluyver and Leewis, 1994; Leewis and ter Kuile, 1985; Leewis and Waardenburg, 1989, 1990; Leewis et al., 1984, 1989, 1994; Waardenburg et al., 1984), as well as a number of shipwrecks on the Dutch Continental Shelf (DCS) (Leewis and Waardenburg, 1991; Leewis et al., 1997; chapter 25, this volume).

Two extensive literature studies were carried out, one addressing the technological and environmental aspects of artificial reefs (FUGRO, 1991), the other describing biological aspects (Van Moorsel, 1991). An engineering study (De Looff, 1993) showed that the planned reef was technically feasible.

On September 10th, 1992, after a baseline assessment of seabed morphology and granulometry, the benthic communities (epibenthos and infauna) and the fish fauna in an area up to 1 km from the selected deployment site (Van Moorsel, 1992), the reef was placed 8.5 km off the coast near Noordwijk, at a depth of 18 m. The geographical position of the reef is: 52° 15′ 00″ N and 04° 17′ 30″ E (see Fig. 1).

A.C. Jensen et al. (eds.), Artificial Reefs in European Seas, 289–305
© 2000 Kluwer Academic Publishers. Printed in Great Britain.

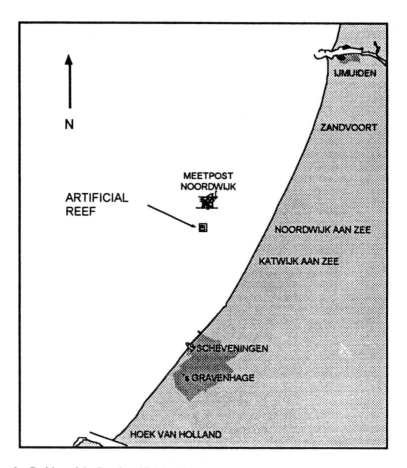

Figure 1.   Position of the Dutch artificial reef (arrow).

Materials and Methods

The material used to build the artificial reef was rough basalt blocks of 60–300 kg, or 40–70 cm in diameter, although in practice a considerable proportion of the blocks were smaller than specified (Van Moorsel, 1993) Material was deployed from a side-dumping vessel, positioned by four anchors.

Four oval, 125 tonne reef units, each about 12 m diameter and 1.5 m high were formed. The reef units were placed in a straight line, at right angles to the direction of the prevailing current (which is approximately parallel to the coast), dispelling concerns that reef units might shelter each other from the current and so facilitate inter-reef migration of animals within the current shadow.

Baseline Assessments

Seabed sediment

Seabed core samples were taken in July 1992, at stations along two transects perpendicular to the line of reef units. The corer was a round boxcorer of the Reineck/Wuttke type, with an inside diameter of 29.5 cm (surface area 0.06835 m^2). Each transect contained 30 stations, 10, 20, 40, 100, 200, 400 and 1000 m distance from the line of reef units.

Each core sample was screened through a 1 mm sieve to obtain the macrofauna, which were stored in 4% formaldehyde. In the laboratory, these samples were further fractioned and the fauna identified to species where possible and counted.

The National Geological Service (RGD) took three sediment cores to a maximum depth of 3.6 m in July 1992. These were analysed for median grain size and calcareous substances (MTZM, 1993a). Side scan sonar images, as well as other physical measurements of the area were also taken.

Fishing

Fish and other mobile fauna were sampled using a 3 m beam trawl, equipped with a standard shrimp net (mesh size 10 × 10 mm, decreasing to 5 × 5 mm at the cod-end), in July 1992. A trawl run was 150 m long, parallel to the line of reef units and conducted at 3.3 knots (MTZM, 1993a; Van Moorsel, 1992), at distances of 20, 40, 100, 200, 400 and 1000 m from the reef on both upstream and downstream sides.

Monitoring

After deployment of the reef, the following aspects were monitored monthly (Table 1), except for the winter months December through April, when weather conditions prevent working: hard substratum macrofauna (epifauna); fish 'on' the reef units; fish and mobile epifauna from the area around the reef. Once a year reef shape was assessed. Following the sampling routine shown in Table 1 monitoring continued until the summer of 1995.

Hard substrata macrofauna (epifauna)

Given the turbidity of the southern North Sea, the reef was too deep to support algal growth. Percentage cover and/or presence was noted down according to an adapted Braun-Blanquet scale (Braun-Blanquet, 1964) (Table 2); categories r to 5 were used for the attached organisms and some mobile organisms like sea stars. Fish were scored according to 'present' or 'abundant'. Some organisms were collected for identification in the laboratory. Small reef blocks were temporarily turned over to study colonization on undersides not in contact with the seabed.

Several aspects of the reef were photographed and video sequences were obtained.

Table 1.   Temporal distribution of the various monitoring activities.

Dates	Dives	Box cores	Fishing
27/07/92–30/07/92 (pre assessment)		*	*
10/09/92 (deployment)			
21/09/92–22/09/92	*		
08/10/92	*		*
05/11/92	*		
04/05/93–10/05/93	*	*	*
08/06/93–10/06/93	*		
22/07/93	*	*	*
19/08/93–20/08/93	*		
01/09/93–02/09/93	*		
06/10/93–08/10/93	*	*	*

Table 2.   Braun-Blanquet (1964) scores for hard-substrata macrofauna (epifauna).

Code	Cover (%)	No. of individuals per 1, 10 or 100 m^2*
r	<0.05	<5.00
+	0.05–0.50	5.00–50.00
1	0.50–5.00	50.00–500.00
2	5.00–25.00	
3	25.00–50.00	
4	50.00–75.00	
5	75.00–100.00	
°		Present
∞		Abundant

*Where possible the percentage cover was estimated; if not, then the code would be determined by means of a number of organisms, dependent on the surface area per individual: individual surface = 1 cm^2 (e.g. porcelain crab) per m^2; 10 cm^2 (e.g. swimming crab) per 10 m^2; 100 cm^2 (e.g. North Sea crab) per 100 m^2.

Biomass
In 1992 it proved impossible to take representative biomass samples, but in 1993 samples were taken during 15 dives (MTZM 1993b). Sampling was equally successful in 1994 and 1995. Reef block areas of 12.5 cm^2 were scraped off as completely as possible, faunal samples being caught in fine mesh bags. In the laboratory the fauna were separated into seven taxonomic groups, then dried at 80°C for 4 days and ashed at 560°C to provide ash-free dry weight (AFDW) data.

Stomach contents
Specimens of the most numerically abundant fish species near the reef were caught and their stomach contents analysed to provide an indication of their diet.

Seasonal epifaunal colonization
To allow an appreciation of the seasonal availability of sedentary epifauna 'fouling panels' in the form of bricks with holes were placed on a frame or on poles hammered into the sea bed near to the reef. During each sampling cruise a number of

these 'fouling bricks' were collected and others left in their place, so that a picture of the organisms available to colonize reef surfaces in each season could be developed. This experiment was not totally successful; several times the frame and bricks were damaged (probably by fishing activities) destroying the data sequence.

Legal Aspects

The deployment of the four reef units was licensed as is necessary for the placing of any kind of structure or material into the sea. In Holland the licence is granted by the Director General of the Directorate-General for Public Works and Water Management (Rijkswaterstaat). The small scale of this pilot project simplified the licensing procedure. Particularly important components of the 5-year licence were the restrictions placed on the construction materials that could be used. In the Netherlands it is strictly forbidden to put anything into the sea that can be considered to be chemical waste. In the case of the reef even the use of waste concrete rubble was prohibited, because the exact origin of such material could not be established.

Other licence requirements included the assurance of reef removal and restoration of the seabed, if requested by the authorities, the marking of the reef site with illuminated navigation buoys for the first 2 years after deployment and informing all interested parties of the reef's location. Removal was considered to be technically possible but expensive; navigation buoys were put in place and the reef's location was advertised in a Dutch marine periodical (similar to the UK's 'Notice to Mariners') and given to the operational centre of the Dutch coastguard.

Results

Baseline Assessment

The area around the reef was found to be flat and sandy (median grain size 220–250 μm; the sand fraction appeared to range from 105 to 240 μm; very fine to very coarse). Water depth ranged from 17.3 to 18.2 m. There were no obstacles worth noting within an area of 1 km^2 around the reef.

Trawling showed a dominance of brown shrimp (*Crangon crangon*) in the catch. Fourteen fish species were recorded. The numerically important species were *Trachurus trachurus* (only small individuals), *Callionymus lyra, Pomatoschistus minutus* and *Limanda limanda*. The latter was the most important fish species in terms of biomass. The various trawl runs showed only small differences in species composition and numbers of individuals between catches. Only the southernmost runs contained relatively large numbers of shrimps and other organisms.

The infaunal species composition found in the core samples was typical for the near-coast ecosystem off the Dutch coast (Groenewold and Van Scheppingen, 1988, 1989). The most prominent species was *Spiophanes bombyx* (a tube-dwelling polychaete), found in all samples, in numbers ranging from 88 to 2707 individuals m^{-2}. *Lanice conchilega* was found in 83% of the samples, with a maximum number of 454 individuals m^{-2}. Nephtidae were important in biomass. Other species present were *Echinocardium cordatum, Bathyporeia* sp. *Urothoe poseidonis,* and *Montacuta ferruginosa,* a commensal species on *Echinocardium.*

Some of the larger organisms dominated the biomass (e.g. *Ensis arcuatus* formed about 50% of the total biomass m^{-2}).

The cores from the southern most stations contained the highest numbers of organisms. Together with the high numbers of shrimp this suggests that this is a relatively important area for predators like fish.

Monitoring

Monitoring started 2 weeks after deployment of the reef.

Shape of the reef units

The shape of each reef unit was more or less elliptical, a consequence of the method used to deploy the blocks which was partly chosen as being cost effective. As the orientation of the long axis of the ellipses was oblique to the prevailing current direction, each reef unit was influenced in an asymmetrical way by water movement. This is undesirable from an experimental replicate point of view, and the deployment method would be altered to avoid this effect if the experiment was to be repeated.

Each reef unit covered a seabed area of about 14 × 8 m, unit height varying between 1 and 1.8 m. Around the base of each ellipse unit, loose blocks were spread in a band 1–3 m wide, and a fan of small stones and gravel was found spreading out to the south-west, the direction of the current at the moment of dumping (Fig. 2).

By the summer of 1993 the shape of the reef units had changed little, indicating that the stones had settled into a rather firm 'construction'. The fans of smaller stones on the south-west side of the units had disappeared under a layer of sand. In a few cases small stones were found in the box core samples.

The quarry rock used appeared to be 'polluted' by some phosphorous slags; a granite block was also observed. Some of the basalt blocks appeared to be painted in several colours – a consequence of an earlier experiment. It is not known whether this has influenced the results. On some of the reef blocks, yellow lichens were present. These started to peel off after about one month in the sea, but their presence will certainly have had a negative influence on epifaunal colonization during the first few months. As these were autumn and winter months, however, this is not considered a serious problem.

Morphology of the surrounding seabed

No scour gullies were observed beside the reef units. There were, however, small scour gullies close to some of the reef units and beside individual stones, the deepest being 0.3 m. These were temporary features. In 1993 silt accumulations

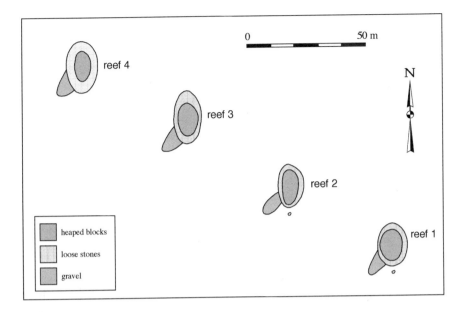

Figure 2.   Size, shape and mutual position of the four reef units.

were found on the NNE and SSW sides of the reef units, with a thickness of 10–15 cm and 5 cm, respectively.

Colonization and percentage surface cover of the reef unit

There were three monitoring cruises in 1992, six in 1993, four in 1994 and three in 1995.

Twelve days after deployment of the reef the first attached organisms were found, a colony of the hydroid *Tubularia indivisa*, 15 mm in length. Swimming crabs (*Liocarcinus holsatus*), sea stars and hermit crabs were common; they occurred on the reef in the same abundance as on the surrounding sandy seabed. A few individuals of *Macropodia rostrata* and *M. parva*, as well as one North Sea crab (*Cancer pagurus*; see Colour Plate 22), were also observed. *Trisopterus luscus* (pouting) had been present from the eleventh day, and two specimens of the eel *Anguilla anguilla* were seen. This is remarkable, because this species was not observed in 5 years of shipwreck monitoring in the southern North Sea (Leewis *et al.*, chapter 25, this volume).

One month after deployment, the number of hydroid species present had increased. The first nudibranchs (*Tergipes tergipes* and *Facelina bostoniensis*) had followed their prey (the hydroids) and were already producing egg masses. Some large plumose anemones (*Metridium senile*; see Colour Plate 23) had been transported to the reef units by currents and successfully re-attached. Barnacles were present on the exposed parts of the reef in numbers of 15–20 m^{-2}. The school of *Trisopterus luscus* held about 50 specimens.

By the beginning of November 1992 the number of species had decreased slightly, but percentage surface cover had increased from (roughly) 30% to 50%. The percentage of surface area covered by barnacles had increased, mainly because the individuals had increased in size. The bryozoan *Electra pilosa* had increased its total surface cover. Some epizooic hydroids were recorded for the first time. Loose stones around the reef units were by this time catching some drifting hydroids with their attached fauna, loosened elsewhere by the first autumn storms.

The first sampling in May 1993 showed about 7 species on the reef, and 100% epifaunal cover. Hydroids dominated (mainly Campanularidae) the reef surfaces, this changing in the following months to a dominance of bryozoa such as *Electra pilosa* and *Bowerbankia* cf *gracilis*. Tube-dwelling Amphipoda and Gammaridea were present on quite large surfaces, characterized by a kind of 'turf' collecting a lot of sediment. *Sagartia troglodytes* was present during early summer (cover 5–25% on the exposed outer surfaces of the reef units). *Sagartiogeton undatus* preferred sheltered surfaces. *Metridium senile* formed local colonies of several dm^2 by means of cloning.

The outer surfaces of the reef blocks were 100% covered by epifauna during the whole sampling season. Only in October did this decrease to 95%, because dying barnacles left some open spaces that were not immediately colonized (winter approaching). This is a phenomenon that has been observed on other occasions in The Netherlands (personal observation). *Alcyonium digitatum* and *Psammechinus miliaris* were characteristic species on the undersides of reef blocks where these were not in contact with the seabed. Two cephalopod species (*Alloteuthis subulata* and (probably) *Loligo vulgaris*) deposited their egg capsules on the reef during the summer of 1993. In September scyphistomae (benthic stage) of the moon jelly (*Aurelia aurita*) were found.

After 1993 each sampling recorded uncolonized areas of the reef blocks, total cover being between 75% and 95%. Species composition stayed more or less the same. Each winter, low temperatures and storms halted biological succession, which started anew each spring. This led to a low-species-diversity epifaunal community (seven attached species in 1993, increasing to 17 in 1994 and decreasing to 14 in 1995 (Fig. 3)), similar to that found in the mouth of the Oosterschelde at 20 m depth (Aquasense, 1995a).

The number of fish species increased over time. Hundreds of individuals formed the schools of *Trisopterus luscus* which circled the reef units at a few metres distance, while *T. minutus* moved much closer or even in between the stones. The former species dominated until October each year, while the latter became a little more abundant. *Callionymus lyra* was found in high densities on the sandy seabed just beside the reef units. In 1992 four fish species were observed on the reef, increasing to 12 in 1993.

In addition to the fish, records of mobile animals increased from five species in 1992 to 21 species in 1993. Numbers of individuals also increased (e.g. *Cancer pagurus* up to about 30 per reef unit in 1993, maximum carapace width 30 cm). The total number of mobile species (fish and others) observed on the reef units was 25 in 1992 and 50 in 1993. Unfortunately, fish and other mobile animals were not recorded after 1993, because funds were not able to support this activity.

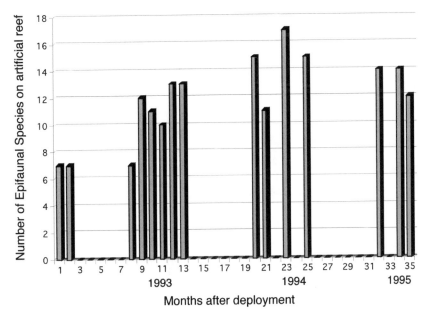

Figure 3. Number of sessile epifaunal species on the reef over time.

Two observations are particularly noteworthy. In May 1993 specimens and egg masses of the nudibranch *Polycera* sp. were found. This was the first record of this genus from the Dutch coast and interestingly, all collected specimens showed characteristics intermediate between two species: *P. quadrilineata* and *P. faeroensis*. Furthermore, on one of the reef units the 'Ross worm' (*Sabellaria spinulosa*) was found in considerable numbers. This species used to form banks in the German part of the Wadden Sea, but disappeared from there in the beginning of the 20th century, cause unknown. An artificial reef may provide an opportunity for reintroduction.

Biomass

Epifaunal samples from reef surfaces in June 1993 showed a mean (AFDW) biomass of 11 g m^{-2}. This increased to 28 g m^{-2} in July, and stabilized in September at about 40 g m^{-2}. This was 3–4 times the biomass found from the boxcore samples from the surrounding seabed. The epifaunal biomass values were similar to those found for a hydroid-dominated community on a shipwreck, one year after sinking (Van Moorsel *et al.*, 1991). This value may well be characteristic for such a pioneer community. In 1994 biomass values on the reef units, were higher, giving values of 184 g AFDW m^{-2} in May, 142 g m^{-2} in June and 125 g m^{-2} in August. In the same months in 1995 the values were 121 g m^{-2}, 145 g m^{-2} and 153 g m^{-2}, respectively (Fig. 4). Biomass in more mature shipwreck communities in the North Sea reaches values of 500 g m^{-2} and more. Further details, such as the biomass per species, are given in Van Moorsel (1994).

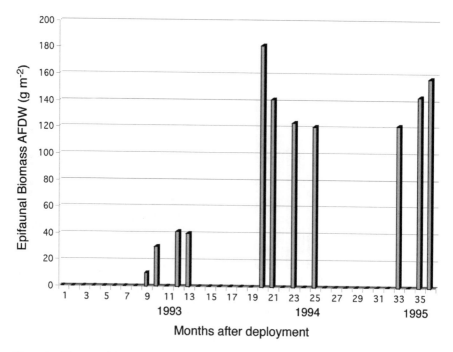

Figure 4.   Biomass of the sessile epifaunal species on the reef over time.

Trawl results

In 1993, 17 species of fish were found, an increase compared with 1992 (10) and the baseline assessment (14). In the next 2 years the fish species numbers were 16 and 15, respectively.

Pomatoschistus minutus was most abundant, while in August through October *Trisopterus luscus* and *T. minutus* were the most important species. They were found in particularly high numbers at 20 and 40 m north of the reef units, with a maximum at 20 m. In August, the number per haul here was 147, in October 574 during which time their mean size also increased by 4 cm. *Callionymus lyra, Limanda limanda* and *Pomatoschistus pictus* appeared to be common species in the area. *Trachurus trachurus,* another pelagic species, was also found in considerable numbers. About 22 other species were present in the trawl nets during 1993 while in 1994 and 1995 the numbers of these 'others' were 27 and 14, respectively, strongly dominated by *Crangon crangon.* Figure 5 shows the length distribution of *Trisopterus* spp. in August and October 1993.

Stomach contents

Forty-five specimens of *Trisopterus luscus* and 29 *T. minutus*, between 53 and 237 mm total length, were dissected and the stomach contents analysed. Shrimp appeared to be the numerically important food source. This was reflected in the

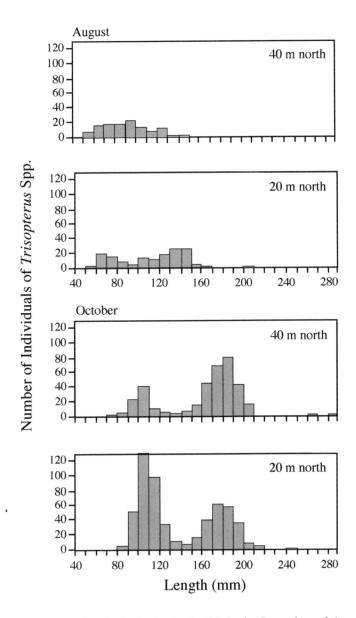

Figure 5. Trisopterus spp. length distribution in the fish hauls. Comparison of August and October 1993 results.

development of the numbers: when *Trisopterus* spp. numbers increased, shrimp numbers decreased. *T. minutus* showed a preference for small tube-dwelling crustaceans on the reef. *T. luscus* had a more mixed diet than *T. minutus*, except for the largest specimen, which appeared to have eaten mainly from the reef. This might indicate a change of diet with increasing age/size.

Seabed cores

No evidence was found of any reef-related changes in the faunal species composition at 20 m and further away from the reef. The numbers of species found in the
box cores in 1993 through 1995 were 46, 74 and 34, respectively. The numbers per
sampling cruise showed a decreasing trend (Fig. 6). Biomass, however, increased
(Fig. 7). The 'normal' value in the area was about 13.5 g AFDW m^{-2}, the biomass
values measured in 1994 and 1995 were much higher (maximum 60.7 g AFDW m^{-2}
in October 1994). This can be attributed to the recruitment and rather patchy
occurrence of the bivalve *Ensis arcuatus* in Dutch coastal waters. Van Moorsel and
Munts (1995) working in a nearby sea area, found a similar increase of biomass
(16 g AFDW m^{-2} in 1993 and 1994 increasing to 55 g m^{-2} in 1995), the 1995 value
being virtually all *Ensis arcuatus*. If *Ensis arcuatus* is left out of the samples, the
biomass values correspond well to the normal values found in the Dutch coastal
area.

Settlement bricks

The results of this experiment are not readily interpretable. Details can be found
in Van Moorsel (1994).

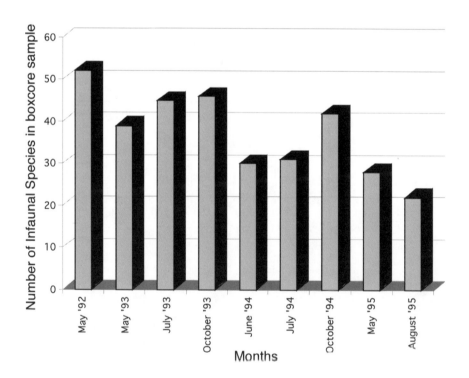

Figure 6.   Number of infaunal species in the box cores.

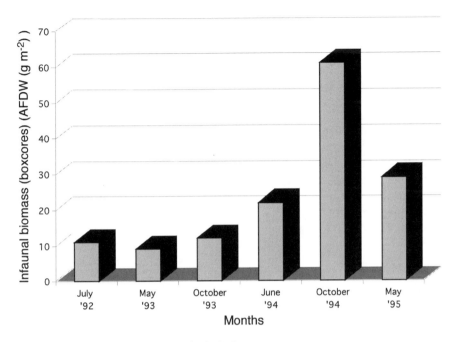

Figure 7. Biomass of the infaunal species in the box cores.

Reactions from Society

Both sport fishermen and professional fishermen showed a keen interest in the reef, especially during a workshop in October 1993. The sport fishermen were pleased with this opportunity of catching more fish on a readily recognizable location (the site was originally buoyed). The professionals considered it a major obstruction for their (trawl) fishing, and complained that they were not consulted about the project at an early stage.

Nature conservation groups also had problems with the reef concept. They argued that the Dutch foreshore consists uniquely of sand and mud, that this is the only area of the country still relatively unchanged by man, and that it therefore should not be acceptable to place 'unnatural' constructions there, even if such constructions would enhance species diversity and biomass. Only in combination with another function (like coastal protection) could it be accepted. It was argued that this was ultimately the aim but to understand the impacts of such actions an experiment was necessary, to record and assess what actually happens when constructions of a given shape and size in this particular habitat.

Reactions from the general public were published in newspaper interviews around the time of deployment. A minority of the public thought that it was interesting to set up an experiment like ours. They doubted, however, whether it would produce more fish and they had questions concerning the contribution to sustainable development.

Politicians tended, as often in such cases, to consider the experiment a waste of money, although lately (1997) there is a perceived change in attitude, now that several plans for large-scale constructions on and off the Dutch coast have been presented (landfills for building houses and creating room for recreation and terrestrial nature; airport on an offshore island).

Discussion and Conclusions

The four reefs still exist and they have survived several severe storms. Although the original plan was to continue monitoring until 1997, opposition against the reef appeared so strong that it was decided to shorten the experiment. Plans to extend the reef with more units, of another construction, have been abandoned. After publication of the final report (Leewis *et al.*, 1997) it was decided to leave the reef in place, so that it will be possible to study the developments since 1995 when the need is felt. This may be important, as research on shipwrecks shows that it takes 5 or 6 years to attain a stable community (Leewis *et al.*, chapter 25, this volume).

The reef gradually attracted more organisms until monitoring was concluded in 1995. However, a large part of the reef was depopulated by the rather rough circumstances in winter, which means that colonization has almost to start again the following spring. Community development will, therefore, be slow, and may stop at a low level of diversity (Warner, 1984).

Little influence on the benthos and fish in the area around the reef was found, except very close to the reef. The number of epifaunal species on the reef remained lower than that in the benthos of the surrounding area. As the species on the reef units were mainly species other than those found in the benthos from the surrounding area, total diversity increased. Biomass became much higher. In 1994 and 1995, the biomass of the sessile fauna on the reef, mainly consisting of primary consumers, was 140 g AFDW m^{-2}, with a seasonal variation between 120 and 180 g AFDW m^{-2}. This corresponds to 56 g C m^{-2} (C to AFDW ratio = 0.4 approximately). Assuming a ratio of production over biomass of approximately 1 on an annual basis (P/B = 1 y^{-1}), the annual secondary production on the reef can be estimated at 56 g C m^{-2} y^{-1}. This is 6–8-fold higher than the benthic biomass of the area surrounding the reef. It is felt that macrobenthic production on the reef was even higher, because the biomass per m^2 of reef surface was used in the calculation, while a better measure for comparison would be biomass per m^2 of original seabed, i.e. of the whole reef surface area over the m^2 of seabed occupied by the reef units. This difference is called the 'available surface increase (ASF) factor'. Van Moorsel *et al.* (1991) measured this factor on two 'representative' shipwrecks off the Dutch coast, and found it to be between 4 and 7. On the reef the ASF factor was not measured, but tentatively estimated at 2. Taking this into account means that the difference in biomass between the reef and the seabed around it amounted to a factor of 12–16 in the last 2 years of the study, or to about 20 compared to that of the normal soft-seabed fauna around the reef and in the southern North Sea in general. Assuming that a fish species inhabiting the reef actually feeds from the structure, and applying an ecological efficiency of 10% to these figures (justified on the basis of NSTF, 1993) and a P/B ratio of 1, the reef could support a fish production of 11.2 g C m^{-2} y^{-1}. This is more than five times the average fish production in the southern North Sea, which amounts to 2 g C m^{-2} y^{-1}. This is probably a conservative estimate, because the concentrated and easily accessible food source on the reef could well lead to a higher ecological efficiency, while, on top of that, the fish on the reef can also use the food in the immediate surroundings of the reef. Therefore, a 10-fold increase of fish production related to the presence of an artificial reef is considered possible.

A simple calculation with data from pouting caught near the reef shows that the biomass (and production) of this species actually amounts to about 20 g C m^{-2} y^{-1}.

One of the ideas at the outset of the project was to investigate whether (chemically) stabilized waste materials could be safely used for reef construction. Many studies all over the world confirm that this is a feasible option (see e.g. Collins *et al.*, 1993; Falace and Bressan, 1993; Stanton *et al.*, 1985; Van der Sloot *et al.*, 1989; Yasuda and Sawada, 1985). This could prove to be both environmentally and financially profitable, reducing the need for landfill sites and waste-processing costs while at the same time enhancing fisheries as well as biodiversity in the sea.

In the Netherlands there is resistance from several social groups to this natural stone reef. The resistance to the idea of using waste materials is even stronger. Dutch legislation forbids putting anything into the sea that can be considered 'chemical waste'. The definition of chemical waste is such, that it includes nearly everything that is not totally natural ('precautionary principle'). The argument that our plans involved stabilized waste materials, and that many experiments abroad had already shown that leaching from some of those is less in sea water than in fresh water (and therefore also in applications on land subjected to rainfall), was not sufficient. Up until now funds for repeating some of those experiments in our laboratory, let alone in the field, have not been made available.

The added functionality of artificial reefs will not, in any single way, have an influence on the North Sea as a whole. It is to be expected that building reefs with habitat management as its sole purpose will not be feasible in Dutch waters. It is, however, important that policy makers and coastal managers are conscious of the possibilities of artificial reefs, so that they can apply them whenever a situation gives reason to do so.

Acknowledgements

We thank Bureau Waardenburg for developing most of the methodology for this project, and for carrying out the monitoring and research until the end of 1993. Monitoring in 1994 and 1995 was carried out by Aquasense. Dr G. Van Moorsel critically read the manuscript and kindly discussed some aspects.

References

Aquasense. 1995a. Monitoring kunstriffen Noordzee, 1994. In opdracht van Rijkswaterstaat, Directie Noordzee. Rapport 950575: 1–51.

Aquasense. 1995b. Monitoring kunstriffen Noordzee, 1995. In opdracht van Rijkswaterstaat, Directie Noordzee. Rapport 95 0756: 1–36.

Braun-Blanquet, J. 1964. *Pflanzensoziologie – Grundzuege der Vegetationskunde*. 3rd edn. Springer Verlag, Vienna-New York.

Collins, K.J., A.C. Jensen, A.P.M. Lockwood and W.H. Turnpenny. 1993. Evaluation of stabilised coal-fired power station waste for artificial reef construction. *Bulletin of Marine Science*. **55**(2–3): 1265–1278.

De Kluyver, M.J. and R.J. Leewis. 1994. Changes in the sublittoral hard substrate communities in the Oosterschelde estuary (SW Netherlands), caused by changes in the environmental parameters. *Hydrobiologia*. **282/283**: 265–280.

De Looff, A.P. de 1993. Kunstrif achtergronddocument lokatiekeuze en ontwerp. (Working document) WWS-93. 112x, Rijkswaterstaat, Tidal waters division: pp. 1–14.

Falace, A. and G. Bressan. 1993. Récifs artificiels en Méditerranée. Liste bibliographique. *Bolletino di Oceanologia Teorica de Applicata*. **XI**: 247–256.

FUGRO 1991. Haalbaarheidsstudie kunstriffen in de Noordzee. Fase 1: Literatuuronderzoek naar de toepassingsmogelijkheden van restoffen. Report FUGRO B.V., Nieuwegein: pp. 1–31+ appendices, pp. 1–4 and 1–51 (Feasibility study, concentrating on technical and environmental properties of selected waste materials).

Groenewold, A. and Y. Van Scheppingen. 1988. De ruimtelijke verspreiding van het benthos in de zuidelijke Noordzee. Report No. 02 (88-14) project MILZON-benthos. North Sea Directorate, Tidal waters Division and SBNO, The Hague: pp. 1–19.

Groenwold, A. and Y. Van Scheppingen. 1989. De ruimtelijke verspreiding van het benthos in de zuidelijke Noordzee, voorjaar 1988. Report No. 90-01, project MILZON, North Sea Directorate/Tidal waters Division/SBNO: pp. 1–27.

Intron, B.V. 1986. Kunstmatige riffen van gestabiliseerde reststofblokken. Intron-rapport No. 86012: pp. 1–38.

Leewis, R.J. 1986. Kunstmatige riffen en alternatieve bouwmaterialen. Notitie GWAO-86.402, Rijkswaterstaat, Tidal Waters Division: pp. 1–7.

Leewis, R.J. and C. ter Kuile. 1985. Ecotoxicologische verkenningen m.b.t. ertsslakken in waterstaatswerken. *Vakbl. v. Biol.* **65**: 43–49.

Leewis, R.J. and H.W. Waardenburg. 1989. The flora and fauna of the sublittoral part of the artificial rocky shores in the south-west Netherlands. *Progress in Underwater Science*. **14**: 109–122.

Leewis, R.J. and H.W. Waardenburg. 1991. Environmental impact of shipwrecks in the North Sea; positive aspects: epifauna. *Water Science and Technology.* **24**: 297–298.

Leewis, R.J., R. Misdorp, J. Al and Tj. de Haan. 1984. Shore-protection – a tension field between two types of conservation. *Water Science and Technology.* **16**: 367–375.

Leewis, R.J., H.W. Waardenburg and A.J.M. Meijer. 1989. Active management of an artificial rocky coast. *Hydrobiology.* **23**: 91–99.

Leewis, R.J., H.W. Waardenburg and M.W.M.v.d. Tol. 1994. Biomass and Standing stock on sublittoral hard substrates in the Oosterschelde, south-west Netherlands. *Hydrobiologia.* **282/283**: 397–412.

Leewis, R.J., I. de Vries, H.C. Busschbach, M. de Kluyver and G.W.N.M. Van Moorsel. 1997. Kunstriffen in Nederland. Final report project Kunstrif. Rijkswaterstaat, North Sea Directorate, Den Haag: pp. 1–31.

MTZM (Marine Techniques, Sea Measurements and Instrumentation). 1993a. Projekt Kunstriffen 1992. Report North Sea Directorate.

MTZM. 1993b. Project Mustrif, Monitoring 1993. Report North Sea Directorate: pp. 1–30.

NSTF (North Sea Task Force). 1993. North Sea Subregion 4 Assessment Report. OSPARCOM, London. 194 pp.

Stanton, G., D. Wilber and A. Murray. 1985. Annotated bibliography of artificial reef research and management. Florida State Univ., Talahassee (USA). Sea Grant Coll. Program: 1–275.

Van der Sloot, H.A., G.J. de Groot and J. Wijkstra. 1989. Leaching characteristics of construction materials and stabilization products containing waste materials. In Coté, P.L. and T.M. Gilliam. (eds.): *Environmental Aspects of Stabilization and Solidification of Hazardous and Radioactive wastes.* ASTM STP. **1033**: 125–149.

Van Moorsel, G.W.N.M. 1991. Literatuurstudie betreffende biologische aspecten van kunstriffen in de Noordzee. Report No. 91.21, Bureau Waardenburg, Culemborg: pp. 1–72+ app. and Book of Abstracts.

Van Moorsel, G.W.N.M. 1992. Literatuurstudie betreffende biologische aspecten van kunstriffen in de Noordzee. Report No. 92.41, Bureau Waardenburg, Culemborg: pp. 1–29+ app.

Van Moorsel, G.W.N.M. 1993. Vooronderzoek kunstriffen Noordzee 1992. Report No. 93.02, Bureau Waardenburg, Culemborg: pp. 1–37+ app.

Van Moorsel, G.W.N.M. 1994. Monitoring kunstriffen Noordzee 1993. Report No. 94.05, Bureau Waardenburg, Culemborg: pp. 1–55 + app.

Van Moorsel, G.W.N.M. and R. Munts. 1995. Effecten van zandoverslag met de 'Punaise-II' op sediment en macrobenthos. Report Bureau Waardenburg, Culemborg.

Waardenburg, H.W., A.J.M. Meijer, R.J.L. Philippart and A.C. Van Beek. 1984. The hard bottom biocoenoses and the fish fauna of Lake Grevelingen, and their reactions to changes in the aquatic environment. *Water Science and Technology.* **16**: 677–686.

Warner, G.F. 1984. *Diving and Marine Biology. The Ecology of the Sublittoral.* Cambridge University Press, Cambridge, 210 pp.

Yasuda, M. and T. Sawada. 1985. Study of FGC concrete and its use for artificial fishing reefs. Proc. 7th Int. Ash Utilization Symposium. DDE/METC-85-16018, vol. 1. Orlando, Florida. pp. 763–775.

18. Environmental Effects of Artificial Reefs in the Southern Baltic (Pomeranian Bay)

JULIUSZ C. CHOJNACKI

Department of Marine Ecology and Environmental Protection, University of Agriculture of Szczecin, Faculty of Marine Fisheries and Food Technology, Ul.K.Królewicza 4/H-19, 71-550 Szczecin, Poland

Background

For many years the Baltic has received sewerage discharges which, in general, have been untreated or subject only to mechanical treatment. In consequence there has been a considerable increase in the concentrations of macro and micro-elements, including nutrients, toxic substances, pesticides, PAHs and other organic and inorganic contaminants (Chojnacki, 1986, 1989; Kubiak, 1980; Protasowicki, 1993; Tadajewski *et al.*, 1989). Water quality, as evaluated by microbiological parameters, has deteriorated and the geographical distribution of zooplankton, benthos and fish has altered (Bojanowski and Szymanowicz, 1989; Chojnacki, 1986, 1987, 1989, 1991).

Anthropogenic pressure is considered to be the major cause of this progressive environmental degradation of the Baltic Sea. Efforts on the part of the Baltic countries to ameliorate the situation have so far proved unsuccessful and are rendered more and more difficult by the widening gap between the scale of the degradation processes and the technological and economic limitations of the remedial action.

The catchment area of the Baltic is some 1 380 900 km², spread between nine countries. Pollution of the freshwater run-off of this area is the main factor responsible for the critical condition of the sea.

The total annual freshwater discharge from the area of Poland (311 900 km²) in the catchment basin was 39.8 km³ in 1991 (Glówny Urzad Statystyczny, 1992). The River Odra contributed 13.3 km³, of which 9.88 km³ was treated mechanically (55.6%), chemically (6.4%) or biologically (38%). The organic load of that discharge, expressed as BOD, was estimated at 69 200 tonnes. The N_{total}, cadmium, copper, zinc, and lead loading were 3 922 815; 90.4; 791; and 125 tonnes per annum, respectively. Additional sources of pollution originating from gas and dust emissions from distant areas in western Europe are difficult to quantify and originate. All of these allochthonous and autochthonous pollutants have contributed to the critical condition of the Baltic ecosystem (Chojnacki, 1993; Kubiak, 1980; Tadajewski *et al.*, 1989; Wirdheim and Chojnacki, 1992). Eutrophication of the Baltic is continuously increasing, as is the area of the seabed affected by anoxia. Oil spills and accidents involving tankers are frequent. The only positive outcome of the efforts made by the Baltic countries to reduce pollution is a decrease in DDT and PCB levels, although those substances are

A.C. Jensen et al. (eds.), Artificial Reefs in European Seas, 307–317
© *2000 Kluwer Academic Publishers. Printed in Great Britain.*

being replaced by other organic chemicals of unknown toxicity (Larsson *et al.*, 1989; Luther, 1990; Protasowicki, 1993; Wirdheim and Chojnacki, 1992). Nevertheless scientists and politicians alike still see a chance for rescuing the Baltic. The 'therapy' was prescribed by the Helsinki Convention of 1974 and the Ronneby declaration made by the Prime Ministers of the seven Baltic countries in 1990. However, administering the 'therapy' has proven more difficult in practice as it has to be enacted by countries differing widely in the level of their economic advancement.

One of the means investigated to ameliorate the situation has been the use of artificial structures composed of old nets or rigid materials. Szlauer (1979, 1980) showed that eutrophication of inland water bodies can be treated by means of flexible barriers made of discarded fishing nets, and his methods have been applied, with varying degrees of success, to other water bodies (Ciszewski and Kruk-Dowgiallo, 1991; Korolev *et al.*, 1991; Piesik, 1992; Szatybelko, 1994).

Rigid artificial reefs have been used in Japan, North America, Korea, China, Taiwan, Great Britain, France, Italy, Spain, Latvia, and Bulgaria, but the objectives of reef deployment differ. Their numerous functions include reefs being treated as an aid in mariculture; creation of fish attracting devices (FAD); generation of new food chains; modification of environmental parameters; reduction of mortality levels in eggs, larvae, juveniles and adult aquatic animals, particularly fish; prevention of illegal fishing; and setting up new opportunities for game fishing (Bojanowski and Szymanowicz, 1989; Chojnacki, 1991; Doumenge, 1990; Korolev *et al.*, 1990, 1991; Relini and Relini, 1990; Stone, 1985a, 1985b; Szlauer, 1979).

The degradation of the Baltic coastal ecosystem has spurred an experimental deployment of artificial reefs to enhance the natural self-purification processes. The deployment of artificial reefs in the southern Baltic has been planned to create hard substrata on the otherwise sandy seabed of Pomeranian Bay (this study), Puck Bay (Ciszewski and Kruk-Dowgiallo, 1991), Gulf of Riga (Korolev *et al.*, 1990) and Vistula Lagoon (Korolev *et al.*, 1990). Such hard surfaces were expected to provide a surface suitable for the settlement of autochthonous sessile organisms whose biology involves suspension feeding. Previous studies with reefs suggested that biological production would be enhanced and biofiltration by micro and macro-organisms would facilitate natural self-purification (Chojnacki, 1989, 1993; Relini and Cormagi, 1990; Luther, 1990; Piesik, 1992). Following this principle the Baltic reefs have been designed to serve as a tool to be used in active restoration and conservation efforts.

Practical Logistics

Initially, artificial reefs, consisting of 12 units providing about 5500 m² of substrata (later one reef was expanded to 20 000 m²) were deployed at two sites in Pomeranian Bay, within the River Odra plume. The reefs were deployed at different locations to facilitate study of settlement preferences and rates as a function of distance from shore, depth, and construction shape. The deployment sites were selected with consideration for seabed topography (post-glacial sand flats interspersed with stones and boulders) and depth (9–13 m). This type of seabed, unstable and mobile, particularly in the area located close to the shore, is not amenable to the settlement of sessile epibenthos. Solitary boulders or groups of stones form 'oases' on which different organisms compete for space. The artificial reefs considerably enlarged the extent of this stable substratum and immediately became biologically active.

Site 1, located 0.5 km off the island of Wolin cliff shore, is situated at 53° 58′32″N, 15° 30′E, while Site 2, 11 km from the mouth of the River Dziwna, has the following co-ordinates: 54° 05′38″N, 15° 33′12″E (Fig. 1). The typical salinity of the area is 6–8%.

On 26 June 1990, artificial reefs were constructed at both sites by divers who assembled concrete modules transported by a supply vessel. The reef elements were shaped as pyramids, stars, tunnels and car tyre bundles. The modules themselves were concrete pipes (0.8 × 1 m and 1.8 × 2.5 m in size; see Colour Plate 24) and used car tyres. Each pyramid-type reef consisted of three layers of modules, while the six-rayed stars were built from six modules alternately arranged around a circle, 1 m in diameter. The 10 m long tunnels were formed by 10 modules, placed one next to another.

Materials and Methods

The study focused on determining the settlement rate, taxonomic structure of fouling communities and assessment of the rate and sequence of succession on

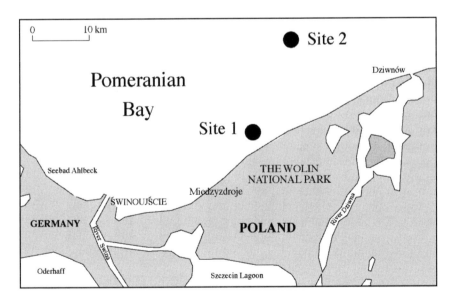

Figure 1. Inshore (Site 1) and offshore (Site 2) artificial reefs in Pomeranian Bay (southern Baltic).

different types of artificial reef. Soon after the deployment of concrete structures (after 145 days) studies on settlement began on the three reef types used. Samples were collected from outer and inner surfaces of module walls as well as from the inlet and outlet portions of a module.

For these ecological studies, fouling organisms were serially sampled by divers who, on each occasion, scraped epibiota from three 100 cm² sampling areas per module with a 10 cm long metal blade. The samples were placed in plastic bags and returned aboard the ship where they were preserved with buffered formalin. In the laboratory, the samples were placed on plastic trays and the macrobenthic animals were sorted, identified, and enumerated. Whenever the water visibility permitted, the biological succession on the reefs was documented by underwater photography, using a Nikonos V amphibious camera.

Results

The quantitative data obtained were extremely satisfactory: mean densities of the fouling communities at all reef types at Site 1 and Site 2 were about 110 000 individuals m⁻² and 180 000 individuals m⁻² of the substratum area, respectively. The dominant (*Mytilus edulis*) vs. subdominant (*Balanus improvisus*) species ratio was 8:1 (Fig. 2).

The following species were recorded at Site 1: *Mytilus edulis, Balanus improvisus, Gammarus zaddachi, Gammarus salinus, Gammarus oceanicus, Gammarus locusta, Gammarus duebeni, Gammarus inaequicauda, Chaetogammarus stoerensis, Corophium volutator, Melita palmata, Leptocheirus pilosus, Jaera* spp., *Pygospio*

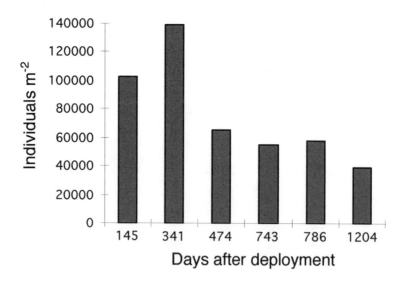

Figure 2.   Changes in fouling community density on artificial reefs 1204 days after deployment.

elegans, Nereis diversicolor, Turbellaria, Cordylophora lacustris as well as larval and adult fishes: cod, clupeids, lumpsucker, sculpin, perch, pikeperch, ruffe and sand eel.

The Site 2 reefs were inhabited by *Mytilus edulis, Balanus improvisus, G. zaddachi, G. salinus, G. locusta, Melita palmata, Jaera* spp., *Nereis diversicolor, Turbellaria, Cordylophora lacustris*, Bryozoa, *Halichondria panicea, Halisarca dujardini*, and larvae of freshwater and marine fish.

The colonization dynamics on all reef types at both sites, in terms of fouling community density changes, was followed over more than 1204 days of exposure. Initially, the most pronounced effects were visible on the pyramid-type reefs. Subsequently, however, the tunnel-type reefs became the preferred habitat maintaining the most stable densities of animals (Figs 3, 4). Reef colonization stabilized, both quantitatively and qualitatively, at about 40–50 000 individuals m^{-2} some 2 years after deployment. While showing similar dynamics, the densities at Site 2, about 11 km away from shore, were much higher than at Site 1 (Figs 2, 3).

As the dominant organisms grew in size and their biomass increased, larval settlement was observed to occur every spring. After 1204 days mean densities differed between Site 1 and Site 2 reef types and were as follows:

(1) Site 1, pyramid-type reef: 39 175 individuals m^{-2}, star-type reef: 40 750 individuals m^{-2}; tunnel-type reef: 185 266 individuals m^{-2}.
(2) Site 2, pyramid-type reef: 141 850 individuals m^{-2}; star-type reef: 87 110 individuals m^{-2}; tunnel-type reef: 158 900 individuals m^{-2}.

Colonization rate differed between sites: it was much lower at the inshore Site 1 than at the offshore Site 2. Comparison of mean densities after more than 1204 days of exposure showed the pyramid- and tunnel-type reefs at Site 1 and tunnel- and pyramid-type reefs at Site 2 to be most readily and efficiently colonized (Fig. 3).

Discussion

The experiment showed the rough surface of concrete constructions to be the best hard substratum that could be provided for animal settlement in the Baltic, which is in agreement with findings of Kung-Hsiung (1985), Relini and Cormagi (1990), and Collins *et al.* (1992). Smooth concrete pipes were tried as well, but settlement was greatly reduced, some surfaces remaining unsettled after several seasons of exposure.

Prior to deployment there was some concern that the mobile, sandy seabed of Pomeranian Bay would affect the stability of the reef constructions. It was felt that when deployed at a depth of 9–13 m, reef units could be easily buried by sand, and/or moved along the shore by currents and storm waves. To check the extent to which the reef units were burial-prone, several differently shaped modules were deployed and placed parallel to the shore to form contracted sections with respect to the flow. Additionally, concrete pipes were positioned vertically, sticking out of the seabed. To determine the spatial arrangement of modules which would be least

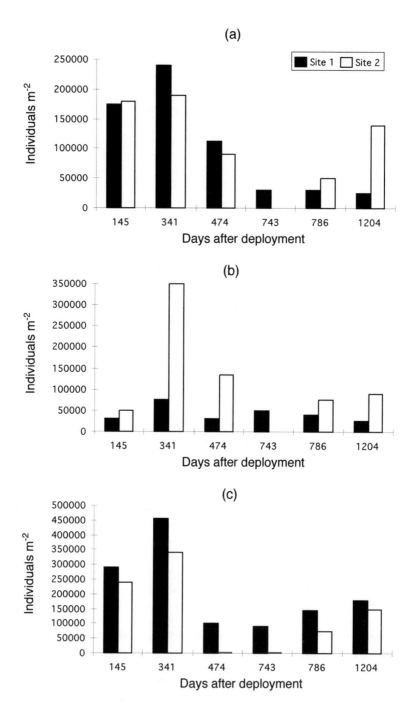

Figure 3. Changes in *Mytilus edulis* and *Balanus improvisus* densities with time on (a) pyramid, (b) star and (c) tunnel type reefs at Sites 1 and 2.

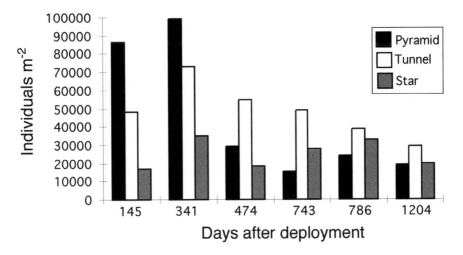

Figure 4. Densities of *Mytilus edulis* and *Balanus improvisus* on artificial reefs 1204 days after deployment.

susceptible to burial, a six-ray, star-shaped construction was deployed. During the initial months of observations, small amounts of sand accumulated on the bottom of all reef modules. After the first autumn/winter season, however, sand accumulation was 10 cm high around a reef module or the bottom part of a module was exposed. The tube-shaped reefs and those made from vertical concrete pipes were rapidly buried, e.g. the pipes accumulated sand to a depth of 80 cm; the remaining elements vanished, covered by the sand, so that their presence was given away only by the uneven surface of the seabed. The lower parts of the concrete modules showed a settlement-free belt extending up to 10–15 cm above the seabed, created by shearing and abrasive action of the sand. Detailed observation of the sand inside and next to a star-type reef failed to reveal restrictive sections in which the flow rate would increase; the sand would be usually removed from the inside of the pipes, its abrasive action precluding any settlement.

The initial colonization rate on the artificial reef structures was very rapid: densities of fouling communities after 145 days of exposure were as high as 400 000 individuals m^{-2} (mainly juvenile *M. edulis* and *B. improvisus*). A reduction in the densities of the sessile fauna was recorded after 474 days. Subsequently, an equilibrium of a kind was attained, the density level, however, being much lower than the initial one (Fig. 2). Modules deployed in the autumn were not colonized by *M. edulis* before late spring of the following year, the surfaces being meanwhile covered by sessile algae, sponges, and detritus (Chojnacki, 1993; Chojnacki and Ceronik, 1997; Ciszewski and Kruk-Dowgiallo, 1991; Piesik, 1992; Stone, 1985b).

Little succession has been observed on the artificial substrata located in the Southern Baltic as only one numerically dominant (*M. edulis*) and one subdominant species (*B. improvisus*) were present on the reefs from the beginning. The small changes in species composition involved an increase in biodiversity which was higher at the inshore site (Site 1) than at the offshore one, the reefs of Site 2 being inhabited by much denser communities.

Similar observations were reported by Korolev *et al.* (1990) from the Vistula Lagoon and Gulf of Riga, and by Ciszewski and Kruk-Dowgiallo (1991) and Szatybelko (1994) from Puck Bay. Extensive sand flats, as found in Pomeranian Bay, are also typical of the Gulf of Riga, while increased productivity is usually associated with a hard, stony seabed. Earlier in 1984, a project aimed at increasing biological productivity by artificial extension of a herring spawning area was started in the Gulf of Riga. Open-work constructions of various shapes (including pyramids), weighted and covered with used nets were deployed at various depths down to 10 m, similar to the design used in 1990 for reefs in Pomeranian Bay. The nets proved very efficient artificial spawning sites; moreover, artificial reefs were found to be of a great use when re-creating biotic elements of the seabed after tanker accidents. Artificial reefs were also found to double the filtration efficiency, compared with natural conditions.

Several stages of colonization were observed to take place in the Pomeranian Bay reefs:

(1)   Spontaneous mass settlement (*M. edulis, B. improvisus*).
(2)   Settlement of phytobenthos on very small surfaces (mainly associations of *Enteromorpha* spp. with *Cordylophora lacustris* and *Halisacra dujardini*).
(3)   Quantitative and qualitative equilibrium after 400 days; climax after 2 years.

Subsequently, simple food chains were established, involving primary producers, suspension feeders, interstitial invertebrates, nektobenthos and fish. The community included bryozoa, *Cordylophora* sp. and sponges. Divers observed a shoal of cod consisting of about 60 adults, close to the reef. This shoal may have been semi-resident. Demersal fish, typical of coastal Baltic waters, were abundant. In spring and early summer, very abundant schools of larvae and fry were photographed by divers, thus documenting the role of the reef as a refuge for juvenile and adult indigenous fish fauna. Such properties of artificial reefs have also been reported by Ciszewski *et al.* (1991) and Korolev *et al.* (1990, 1991) in the Gulf of Riga.

Apart from providing settlement sites for epibiota, the reefs have played another role, that of accumulating detritus, nutrients, and heavy metals. The reefs will not be removed in order to maintain the equilibrium already attained and to study any biological development beyond the present colonization. Because of the accumulation of heavy metals the reefs are evoking some negative comments to the effect that, due to their accumulative properties, they could become a potential environmental danger to Pomeranian Bay. Such comments seem to be a gross exaggeration because any human intervention into the natural environment can raise the same objections, e.g. trawling activities which destroy benthic communities for long periods of time, harbour construction, oil rig deployment and shipwrecks. We are far from having a full understanding, as yet, of the environmental responses to artificial substrata and, if for this reason alone, studies should continue to unravel the role of artificial reefs in coastal zone management.

Artificial reefs have served to revitalize benthic communities and combat a decrease in sea water quality, caused by man's activities (sewage run-off, dredge spoil dumping, sediment aggregate extraction and oil pollution related to drilling, transportation and accidents). Rapid and dynamic colonization of artificial substrata, regardless of the materials used to construct the reefs, their shapes and surfaces, demonstrates their high revitalizing potential (Chojnacki, 1993; Doumenge, 1990; Korolev *et al.* 1990, 1991). System-oriented actions to save the marine ecosystem can be possible when optimal methods are developed and the time needed to improve degraded biotic and abiotic ecosystem components is determined.

The function of artificial reefs as a means by which to increase biological production of the sea seems to be still underestimated, although there is ample evidence that the biomass of benthos and fish associated with artificial reefs is much higher than under natural conditions (Chojnacki and Ceronik, 1997; Doumenge, 1990; Kun-Hsiung, 1985; Piesik, 1992; Relini and Relini, 1990; Stone, 1985b; Szlauer, 1980). Fishing nets that surround Pomeranian Bay artificial reefs in the fishing season (to the extent of producing an effective obstacle to a research vessel transporting divers and sampling equipment) are proof that the fish yield has indeed increased. Local fishermen who at the beginning of the experiment were against it and tended to destroy buoys marking the reef deployment sites, today are the best allies of the scientists because the fish catches have increased considerably.

Conclusions

(1) Artificial reefs, deployed in Pomeranian Bay during 1990, were incorporated into the local ecosystem as evidenced by a high settlement rate and the history of natural biological processes.

(2) Turbulence and near-bottom currents, stronger at the inshore site, result in a higher biodiversity and density of the fouling communities at Site 1, compared with the offshore Site 2.

(3) The macrobenthic dominant and subdominant species at both sites were *Mytilus edulis* and *Balanus improvisus*.

(4) Artificial reefs in the Baltic seem to play multiple roles: they augment the seabed topography, enhance biodiversity of the benthos, and intensify the self-purification processes of the sea, which has been the major objective of attempts to revitalizing Pomeranian Bay.

Acknowledgements

I thank Mrs Dorota Aniol, M.Sc. and Mrs Beata Sobolewska, M.Sc. for sharing their data on changes after 1204 and 1170 days at Sites 1 and 2, contained in their M.Sc. thesis.

The pilot studies were financially supported by the Ministry of National Education, while the further support was granted to the J.C. Chojnacki by the Ministry of Environmental Protection, Natural Resources and Forestry, and by the National and District Funds for Water Management and Nature Conservation.

The Nikonos V amphibious camera was purchased from funds provided by grants Nos. 898/89 and 938/91.

References

Bojanowski, A. and A. Szymanowicz. 1989. Przybrzeżne rybołówstwo łodziowe w polskiej strefie rybackiej. Proceedings of Polish–Swedish seminar: Baltic water protection. 27–28 September 1989. AR Szczecin, pp. 40–47.

Ceronik, E. 1989. Możliwości pozytywnego oddziaływania na mikroflorę wód przybrzeżnych Bałtyku Południowego. Seminarium: Mikrobiologia wód Bałtyku i Zatoki Gdańskiej, Gdańsk, 5–6 June 1989, Politechnika Gdańska, pp. 59–66.

Chojnacki, J.C. 1986. Strukturalne zmiany zooplanktonu ze strefy przybrzeżnej Zatoki Pomorskiej. Materiały 13 Zjazdu PTH w Szczecinie 16–19 September 1986. pp. 31–32.

Chojnacki, J.C. 1987. Sukcesja sezonowa zoocenoz planktonowych Południowego Bałtyku. Szczecińskie Roczniki Naukowe. STN. **2**: 29–44.

Chojnacki, J.C. 1989. Hydrobionty Zatoki Pomorskiej i estuarium Odry. Proceedings of Polish–Swedish seminar: Baltic water protection. 27–28 September 1989. AR Szczecin, pp. 21–27.

Chojnacki, J.C. 1991. Variability of telson observed in population of *Neomysis integer* (Leach 1815) (Crustacea, Mysidacea) from inshore waters of southern and north-eastern Baltic Sea, *Zeszyty Naukowe AR. Szczecin, Ser Ryb Mar Techn Żywn*. **18**: 31–35.

Chojnacki, J.C. 1993. Man-made reefs. An alternative for recultivation and protection of the Pomeranian Bay. *Aura*. **6**: 11–13.

Chojnacki, J.C. and E.J. Ceronik. 1997. Artificial reefs in the Pomeranian Bay (Southern Baltic) as biofiltration sites. Proceedings of the 13th Baltic Marine Biologists Symposium 31st Aug–3rd Sept. 1993, Riga-Jurmala.

Ciszewski, P. and L. Kruk-Dowgiallo. 1991. Elements of the Lagoon of Puck degradation and some proposals for several biotechnical measures aiming at this sea area restoration. 12th Baltic Marine Biologists Symposium poster abstract, 28–29th Aug. Helsingor, pp. 7.

Collins, K.J., A.C. Jensen and A.P.M. Lockwood. 1992. Stability of coal waste artificial reef. *Chemistry and Ecology*. **6**: 79–93.

Doumenge, F. 1990. Problématique des récifs articles. FAO Fisheries Report 428. The first session of the working group on artificial reefs and mariculture, Ancona, 27–30 Nov. 1989. pp. 52–69.

Główny Urząd Statystyczny. 1992. Ochrona środowiska 1992. Materiały i opracowania statystyczne. Warszawa, pp. 385.

Korolev, A.P., V.B. Kadnikov, T.A. Kuzniecova, V.I. Muravskij, A.N. Nazarienko, A.A. Zake and V.G. Drozdietskij. 1990. Issledovania primenenia iskustvennykh soorhuzei v pribereznoi zone Baltiskogo moria. Iskustvennyie rify dla rybnogo khozaistva. WNIRO, Moskva. pp. 166–177.

Korolev, A.P., T.A. Kuzniecova and V.G. Drozdietskij. 1991. Kunstijehe riffe in der Kustenzone der Ostsee. *Fisculterei Farscliung.* **29**: 57–60.

Kubiak, J. 1980. Studia nad eutrofizacją Zatoki Pomorskiej w rejonie odpływu wód Świny i Dziwnej. Doctoral thesis, University of Agriculture Szczecin (in Polish) pp. 156.

Kun-Hsiung, C. 1985. Review of artificial reefs in Taiwan: emphasising site selection and effectiveness. *Bulletin of Marine Science.* **37**(1): 143–150.

Kun-Hsiung, C. and J. Rong-Quen. 1984. Artificial reef project in Taiwan. TML Conference Proceedings. **1**: 51–55.

Larsson, U., R. Elmgrem and F. Wulff. 1989. Eutrofication and the Baltic Sea: causes and consequences. *Ambio.* **14**(1): 9–14.

Luther, G. 1990. Erhalt von Fauna und Flora in küstennahen Ostseegewässern durch Verlangern der Übergangszeit. Ideen des Vereins Ostseesanierung. Nationale Konferenz – Schutz der Meeresumwelt – Ostsee 19.05.1990 Verein Ostsee Sanierung, Rostock, Xerocopy p. 5.

Piesik, 1992. Biologia i ekologiczna mia organizmów poroślowych (perifiton) zasiedlających sztuczne podłoża w różnych typach wód. *Uniw Szczecin, Studio i rozprawy.* **122**: 262.

Protasowicki, M. 1993. Preliminary studies on heavy metals content in biota and sediments from 'artificial reef' area. Studio i Mat. *Oceanology.* **64**(3): 77–83.

Relini, G. and P. Cormagi. 1990. Colonisation patterns of hard substrata in the Loano artificial reef (Western Ligurian Sea). FAO Fisheries Report of the first session of the working group on artificial reefs and mariculture, Ancona, 27–30 Nov. 1989. **428**: 108–113.

Relini, G. and L.O. Relini. 1990. Artificial reefs in the Ligurian Sea. FAO Fisheries Report of the first session of the working group on artificial reefs and mariculture, Ancona 27–30 Nov. 1989. **428**: 114–119.

Stone, R.B. 1985a. National artificial reef plan. NOAA Technical memorandum NMFS OF-6. USA, 98 pp.

Stone, R.B. 1985b. History of artificial reef use in the U.S. In D'Itri, F.M. (ed.) *Artificial Reefs: Marine and Freshwater Applications.* Lewis, Chelsea, pp. 3–9.

Szatybelko, M. 1994. Biotechniczne możliwości usuwania zanieczyszczeń z obszarów silnie degradowanego środowiska morskiego. Morsk. Instyt. Rybacki, sprawozdanie Gdynia. 40 pp.

Szlauer, L. 1979. Możliwości zastosowania barier do ochrony urządzeń hydrotechnicznych przed racicznicę oraz usuwania biogenow z wody. *Żesz Nauk A R Szczecin Ser Ryb Morsk Techn Żywn.* **75**: 29–38.

Szlauer, L. 1980. Oczyszczanie zbiorników wodnych przy pomocy sztucznych barier. *Gosp Wodna,* **8/9**: 255–256.

Tadajewski, A., M. Knasiak and T. Mutko. 1989. Warunki hydrochemiczne ekosystemu estuarium Odry. Proceedings of Polish–Swedish seminar: Baltic water protection. 27–28th September 1989. AR Szczecin, pp. 5–14.

Wirdheim, A. and J.C. Chojnacki, (eds.) 1992. Co się dzieje z Morzem Bałtyckim. CCB and Naturskydds foreningen, Stockholm. 85 pp.

Wulff, F., A. Stigebrandt and L. Rahm. 1990. Nutrient dynamics of the Baltic Sea. *Ambio.* **19**(3): 126–133.

19. Employment of Artificial Reefs for Environmental Maintenance in the Gulf of Finland

ALEXANDER ANTSULEVICH[1], PASI LAIHONEN[2] and ILPPO VUORINEN[3]

[1]Department of Hydrobiology and Ichthyology, St. Petersburg State University, 16 Linia, 29, St. Petersburg, 199178, Russia; [2]Southwest Finland Regional Environment Centre, PO Box 47, FIN-20801, Turku, Finland; [3]Archipelago Research Institute, University of Turku, FIN-20014, Turku, Finland

Introduction

The main goal of the Finnish Archipelago Sea artificial reef project was to study whether the growth capacity of fouling communities in the northern Baltic Sea was high enough to capture significant amounts of nutrients released by fish farms. Fish farming is an increasing industry in the area and nutrients released by over-feeding and fish faeces are becoming a serious problem, causing eutrophication of the Archipelago Sea. So far no effective technology is available to treat waste waters released by fish farms. The local fish-farming technology is especially problematic, since the fish are kept in open net cages, effectively creating a non-point source of pollution.

The Baltic Sea is, in part, a large brackish water basin and, given its size, a unique environment in Europe. Results from artificial reef research in a full marine environment cannot necessarily be extrapolated to the Baltic. Fluctuation of salinity in different parts of the Baltic affects the structure and biomass of animal and plant communities.

The need to use artificial reefs is largely dependent on the amount of filter-feeding animals and filamentous algae that already live on natural hard substrata in the ecosystem. If hard habitat is a limiting resource, then artificial reefs may serve to increase the filter-feeding and nutrient-capturing biomass of epibiotic organisms.

In the northern Baltic the first artificial reef experiments were started in Russian parts of the Gulf of Finland in 1992. In south-western Finland an artificial reef project was started in 1993 (Fig. 1). Co-operation between Russian and Finnish artificial reef research groups started in 1993. In a separate initiative, promising results from artificial reef experiments had been obtained by a Polish research group working since 1990 in Pomerian Bay, southern Baltic (Chojnacki, chapter 18, this volume; Chojnacki et al., 1993). In all projects, the main goal has been to study the role artificial reefs might play in water purification.

A.C. Jensen et al. (eds.), Artificial Reefs in European Seas, 319–329
© 2000 Kluwer Academic Publishers. Printed in Great Britain.

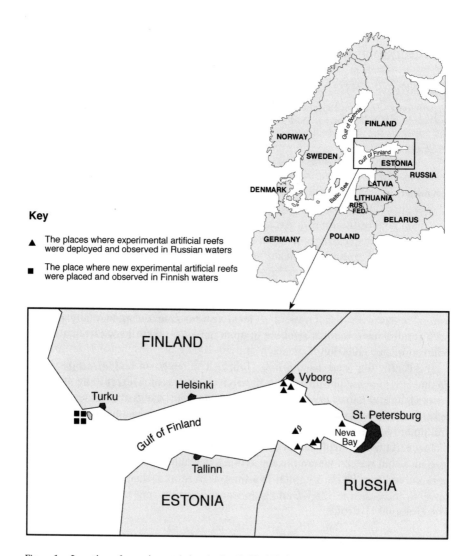

Key

▲   The places where experimental artificial reefs were deployed and observed in Russian waters

■   The place where new experimental artificial reefs were placed and observed in Finnish waters

Figure 1.   Location of experimental sites in the Gulf of Finland.

Artificial Reefs and the Baltic Sea

Salinity as a Limiting Factor in the Gulf of Finland

Salinity in the Gulf of Finland increases gradually from practically zero in the Neva Estuary up to 6 ppt in the western part of the area. Horizontal variation of salinity is high. Significant changes in the general character of the biota have been observed in the salinity interval between 3 and 5 ppt (Jarvekulg, 1979). Plant and animal communities are mainly represented by algae, colonial hydroids, barnacles and bivalves attached to hard substrata. In spite of the low number of species (usually only 1–2 species from each group) the abundance of individuals can be high. In the eastern part of the Gulf (Russian waters) only five fouling species are usually responsible for more than 90% of the total biomass. The species are *Cladophora glomerata* and *Enteromorpha intestinalis* (filamentous algae), *Cordylophora caspia* (colonial hydroid), *Dreissena polymorpha* (the zebra mussel) and *Balanus improvisus* (barnacle). In Finnish waters the zebra mussel is substituted by the blue mussel (*Mytilus edulis*) and other brackish water species appear (Laihonen and Vuorinen, 1981; Lietzen *et al.*, 1984). The algae are capable of utilizing dissolved nutrients from the water column whereas the invertebrate animals feed on bacteria and suspended organic matter.

The Organisms

The Filamentous Algae

The biomass of green filamentous algae on artificial reef substrata can reach 2 kg m^{-2} in the study area. The removal of 500 tonnes of *Enteromorpha intestinalis* biomass (approximately corresponding to an area of 250 000 m^2) is equivalent to the removal of 500–4000 kg of nitrogen and 50–100 kg of phosphorus from a water body (Parchevsky and Rabinovich, 1989). The algae can also remove toxic substances, such as heavy metals and polycyclic aromatic hydrocarbons from the water (Irha and Kirso, 1993; Saenko, 1989). The distribution of algae is limited by light. Since the transparency of water in the Gulf of Finland is relatively low, algae are limited to the uppermost water layer. According to our observations, artificial reef constructions for cultivation of algae cannot be 'multi-stored'. This is especially true in turbid waters, where the top reef layer would prevent the penetration of light to the lower levels.

The Colonial Hydroids

Colonial hydroids are abundant on artificial reef substrata in the Gulf of Finland especially in sites with high water turbidity. In Russian waters only *Cordylophora caspia* Pallas is common. *Gonothyraea loveni* (Allman) has recently been recorded for the first time in the Russian part of the Gulf (Gogland Island). The species has not been reported earlier in such a low salinity (3.7–4.3 ppt). In Finnish waters both *C. caspia* and *G. loveni* are present.

An exponential growth pattern is a feature of hydroid colonies when food and space are unlimited. Thus, the potential for colonial hydroid biomass production on artificial substrata is high. According to our experience the study area can reach a biomass of 0.3–0.4 kg m^{-2} in 2 months. In favourable conditions the biomass can be much higher. *C. caspia* can catch prey up to 1 mm in size. *Gonothyraea* prefers smaller prey, planktonic larvae, dinoflagellates and detritous particles (Marfenin and Homenko, 1987).

322 A. Antsulevich et al.

The Barnacle

Balanus improvisus Darwin is the only barnacle species in the northern Baltic. In Russian parts of the Gulf of Finland the species appears approximately 40 miles off the outfall of the River Neva. The abundance increases, with salinity, towards the west. From a salinity of 3.5–4.0 ppt onwards the barnacle is a common biofouler on artificial reef and natural hard substrata.

Buoys were used to support and provide settlement surfaces allowing the biomass and abundance of the species close to the Russian–Finnish border to be recorded. A maximum of 6000 individuals (1.3 kg m^{-2}) settled and grew between May and November. *B. improvisus* is euryphagous. The animal can trap food particles up to several millimetres in size. The distribution pattern of the barnacle on seabed and floating artificial reefs seems relatively indifferent to light or local turbulence. Most abundant aggregations of barnacles were observed in the inner parts of the artificial reefs.

The Zebra Mussel

The bivalve *Dreissena polymorpha*, or zebra mussel, was first reported in the Russian part of the Gulf of Finland recently (Antsulevich and Lebardin, 1990; Antsulevich and Chiviliov, 1992). The species is not abundant in the study area, being relatively intolerant of the lower temperatures (bij de Vaate, 1991; Mikheev, 1964).

First experiments with small artificial reef models demonstrated a good zebra mussel colonization density. On artificial reefs there was a density of 5100 individuals m^{-2} compared to 800 m^{-2} on nearby natural hard substrata. However, the growth rate of zebra mussels seems to be lower in the Gulf of Finland than in southern areas of Russia, Ukraine and Western Europe (bij de Vaate, 1991; Kachanova, 1962; Mikheev, 1964).

Zebra mussel is one of the most effective filtrators among fresh- and brackish-water animals. The mussel is capable of trapping small particles, down to 1–3 μm in size. The filtration rate according to many authors and various methods is approximately 40–75 ml mussel^{-1} hour^{-1}. Several equations in relation to environmental conditions, shell length, individual or total biomass have been proposed for the description of the filtration process (Karatayev and Burlakova, 1993; Lvova-Kachanova, 1971; Mikheev, 1967; Reeders *et al.*, 1993; Smit *et al.*, 1992). Zebra mussels can contribute to the control of eutrophication by filtering phytoplankton out of the water column. Mussel beds can also contribute to the sedimentation of suspended matter by producing faeces and pseudofaeces (bij de Vaate, 1991; Nalepa and Schlesser, 1993; Reeders and bij de Vaate, 1990; Reeders *et al.*, 1989; Smit *et al.*, 1992). The zebra mussel may even provide a 'new perspective for water quality management' by biomanipulation (see references above). The environmental implications of *Dreissena* colonization have been schematically described by Reeders and bij de Vaate (1990) as shown in Fig. 2.

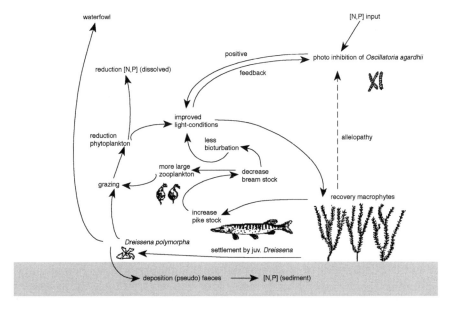

Figure 2.   Environmental implications of the presence of the zebra mussel *Dreissena polymorpha*.

The Blue Mussel

The blue mussel (*Mytilus edulis*) has a similar role in ecosystems to the zebra mussel (Zaitsev, 1989). Laine (1990) has reviewed the role of blue mussel in the northern Baltic Sea. In the Gulf of Finland the species has been studied only sporadically (Öst and Kilpi, 1997; Rosenius, 1964; Sunila, 1981). The blue mussel is a dominant species in the Baltic Sea; its biomass may form 90% of the total epibiotic biomass in water shallower than 25 m (Jansson and Kautsky, 1977). Below the bladder-wrack zone the hard substrata are dominated by a community of blue mussels and red algae (Kautsky and Wallentinus, 1980). Large number of blue mussels may also be found on sedimentary seabeds. Together with the Baltic tellin (*Macoma baltica*) bivalves may form 70% of the total sedimentary seabed biomass. In the northern Baltic Sea the distribution of blue mussels is controlled by low salinity (4–4.5 ppt) (Lassig, 1965; Segerstråle, 1942; Vuorinen *et al.*, 1986).

The Distribution of the Species

On the Finnish side of the Gulf the salinity is 1–3 ppt higher than on the Russian side. This causes remarkable changes in the species that form the fouling communities. The most important difference on hard substrata is the appearance of the blue mussel (*Mytilus edulis*) on the Finnish side of the border. The line of distribution between zebra and blue mussels almost coincides with the Russian–Finnish border on the northern part of the Gulf. The two species are not symptric in the region discussed. However, Jarvekulg (1979) has documented a sympatric distribution pattern near the southern coast of the Gulf of Finland.

Materials and Methods

The first artificial reefs were deployed by the Russian research group in 1992 and the experimental work started in 1993. Several small artificial reef modules were placed at four sites (Fig. 1). Each module consisted of a steel frame, $3 \times 1.5 \times 1$ m in size, furnished with 40–60 10×10 cm fouling panels. The panels were made of plain and enamelled steel, plastic (PVC), asbestos-concrete and ceramics. The underwater field work was done by SCUBA divers. The stands were used for preliminary investigations of fouling community succession. Additionally 15 hydrographical buoys (7 m high with 4 m underwater) were monitored in the Russian part of the Gulf.

The Finnish research group started in 1993. In the first phase a plan was made to assess the feasibility of using artificial reefs in the Archipelago Sea of south-western Finland. Experimental reefs were constructed from buoyed ropes attached to an aluminium frame. The purpose of the experiment was to test the structure of the model reef in practice. The reefs were deployed at three different sites. Relatively dense epifaunal animal communities developed on the reef surface. However, filamentous algae did not attach during the experiment.

In the second phase, experimental artificial reefs were placed in three separate sites in the Archipelago off the city of Turku. The purpose of the experiment was to test the capability of the model reefs to develop biomass in a moderately eutrophicated area where a fish farm was providing a nutrient source. Reef A was used as a control, situated in an area with no specific emission sources affecting water quality. Reef C was placed in the immediate vicinity of an experimental fish farm of the Finnish Game and Fisheries Research Institute, and site B in the intermediate zone between sites A and C (Fig. 1). The biomass of the experimental artificial reefs was removed and both wet and dry weight biomass was measured twice during the experiment. The results of the experiment were compared statistically using ANOVA. A comparative calculation was made between annual nutrient emission rate of an average fish farm and the amount of nutrient captured in the experiments.

Results and Discussion

The employment of artificial reefs for environmental management can be regarded as a new kind of biomanipulation technique (Antsulevich and Bugrov, 1991; Bombace, 1989; Lee-Shing Fang, 1989; Zaitsev, 1987). Artificial reefs are considered able to increase the biological productivity of ecosystems (Bohnsack and Sutherland, 1985; Bombace, 1989). The intensity of biofiltration depends on the taxonomic composition and abundance of the community, the physiological con-

Table 1. The average biomass and P-contents of the filamentous algae in the experimental sites of the second experimental phase in Finnish waters.

	biomass (g rope⁻¹)	s.d.	range	biomass (g m⁻²)	s.d.	range	P content (g m⁻²)	s.d.	range
21.8.1996									
Site A (reference)	1.61	0.35	1.13–1.96	10.06	2.19	7.06–12.25	0.007	0.001	0.005–0.008
Site B	8.40	3.51	4.94–13.29	52.50	21.94	30.88–83.06	0.035	0.015	0.021–0.056
Site C	3.50	1.16	2.64–5.12	21.88	7.25	16.50–32.00	0.05	0.005	0.011–0.021
20.9.1996									
Site A (reference)	1.12	0.92	0.26–3.21	7.00	5.75	1.63–20.06	0.005	0.004	0.001–0.013
Site B	8.98	4.54	4.33–18.99	56.13	28.38	27.06–118.69	0.038	0.019	0.018–0.080
Site C	4.10	1.97	1.49–7.68	25.63	12.31	9.31–48.00	0.017	0.008	0.006–0,032

Table 2. The average biomass and N-contents of the filamentous algae in the experimental sites of the second experimental phase in Finnish waters.

	biomass (g rope⁻¹)	s.d.	range	biomass (g m⁻²)	s.d.	range	N content (g m⁻²)	s.d.	range
21.8.1996									
Site A (reference)	1.61	0.35	1.13–1.96	10.06	2.19	7.06–12.25	0.024	0.05	0.17–0.30
Site B	8.40	3.51	4.94–3.29	52.50	21.94	30.88–83.06	1.27	0.53	0.75–2.01
Site C	3.50	1.16	2.64–5.12	21.88	7.25	16.50–32.00	0.53	0.18	0.40–0.77
20.9.1996									
Site A (reference)	1.12	0.92	0.26–3.21	7.00	5.75	1.63–20.06	0.17	0.14	0.04–0.48
Site B	8.98	4.54	4.33–8.99	56.13	28.38	27.06–18.69	1.36	0.69	0.65–2.87
Site C	4.10	1.97	1.49–7.68	25.63	12.31	9.31–48.00	0.62	0.30	0.23–1.16

Table 3. The comparison of biomasses of the first measurement (one-way ANOVA).

Source	Sum of squares	df	mean square	F-ratio	p value
1st Measurement	5.182	2	2.591	24.532	0.000
error	0.951	9	0.106		

Table 4. The comparison of biomasses of the second measurement (one-way ANOVA).

Source	Sum of squares	df	mean square	F-ratio	p value
2nd Measurement	20.671	2	10.335	29.981	0.000
error	7.239	21	0.345		

dition of the organisms, abiotic environmental factors and shape and material of the artificial reef (Antsulevich and Bugrova, 1989; Baynes and Szmant, 1989; Bohnsack and Sutherland, 1985; Damman, 1974; Hixon and Brostoff, 1985; Khailov *et al.*, 1987).

In the first experimental phase filamentous algae did not attach to the reefs (Laihonen *et al.*, 1996). The results of the second experimental phase when algae did colonize the reefs are shown in Tables 1 and 2. The results show a statistically significant difference between all experimental sites during both measurements (Tables 3 and 4). However, the maximum algal biomass values from the experiment were measured on site B. The result was unexpected, since nutrient emissions were highest at site C. Possible justifications for the result were varying hydrographic conditions between sites and influence of the antifouling treatment applied to the net cages of the experimental fish farm at site C.

A calculation on the basis of the results of Tables 1 and 2 was made to evaluate, theoretically, the surface area needed to capture nutrient emissions of an average fish farm. The annual emissions from an average Finnish fish farm producing 40 tonnes of rainbow trout are 3000 kg of nitrogen and 40 kg of phosphorus. About 345 000 m^2 of reef surface area (about 70 football pitches) would be needed to remove 12–25% of annual nutrient emissions.

These trials to assess artificial reefs as a tool in the management of non-point source nutrient releases have shown that, in the Finnish context, reefs fouled with algae do not remove sufficient nutrients to be of practical use.

Acknowledgements

Initial funding received from the Russian Ministry of Science and Technical Policy is greatly appreciated. We are also grateful to The John D. and Catherine T. MacArthur Foundation for substantial support of the project. Dr A. bij de Vaate kindly permitted to use the original drawing of Fig. 2. Professional discussions with Dr Nicolaj Maximovich (St. Petersburg University) were very useful.

References

Antsulevich, A.E. and L.A. Bugrova. 1989. Gidrobiologicheskie kriterii proektirovaniya iskusstvennykh rifov (Hydrobiological criterias for design for artificial reefs). SU Conference for Nauchno-teonicheskie problemy marikultury v strane (Scientific and Technical Problems of Mariculture in the Country), Vladivostok, pp. 220–221 (in Russian).

Antsulevich, A.E. and L.Y. Busgrov. 1991. Artificial reefs project for improvement of water condition in Neva Estuary close by Leningrad Dam. *5th Int. Conference for Aquatic Habitat Enhancement*. Long Beach, California, pp. 4–5.

Antsulevich, A.E. and S.M. Chiviliov. 1992. Modern state of Luga Inlet benthic fauna of the Gulf of Finland. *Vestn St.-Petersb Univ Biol Ser 3*. **3**(17): 37–110 (in Russian with English summary).

Antsulevich, A.E. and M.V. Lebardin. 1990. 'Wandering shell' *Dreissena polymorpha* (Pall.) is close to Leningrad. *Vestn Leningr Univ Biol Ser 3*. **4**(24): 109–110 (in Russian with English summary).

Baynes, T.W. and A.M. Szmant. 1989. Effect of current on the sessile benthic community structure of an artificial reef. *Bulletin of Marine Science*. **44**(2): 547–566.

bij de Vaate, A. 1991. Distribution and aspects of population dynamics of zebra mussel, *Dreissena polymorpha* (Pallas, 1771), in the lake Ijsselmeer area (The Netherlands). *Oecologia*. **86**: 40–50.

Bohnsack, J.A. and D.L. Sutherland. 1985. Artificial reef researches: a review with recommendations for future priorities. *Bulletin of Marine Science*. **37**(1): 11–39.

Bombace, G. 1989. Artificial reefs in the Mediterranean sea. *Bulletin of Marine Science*. **44**(2): 1023–1032.

Chojnacki, J.C., E. Ceronik and T. Perkowski. 1993. Artificial reefs – an environmental experiment. *WWF Baltic Bulletin*. 1/1993.

Damman, A.E. 1974. Some problems that may be faced in the construction of an artificial reef. In Colunga, L. and R. Stone. (eds.) *Proc Int Conf Art Reefs*. Houston, Texas, pp. 19–20.

Hixon, M.A. and W.N. Brostoff. 1985. Substrate characteristics fish grazing, and epibenthic reef assemblages off Hawaii. *Bulletin of Marine Science*. **37**(1): 200–213.

Irha, N.I. and U.A. Kirso. 1993. Rol vodoroslei v samoochischenii vodoemov ot kantserogennykh politsiklicheskikh aromatischeskikh uglevorodorodov (The role of algae in the self-purification of the basins from carcinogenic polycyclic aromatic hydrocarbons). *Russian Journal of Ecological Chemistry*. **1**: 27–31 (in Russian).

Jansson, A.M. and N. Kautsky. 1977. Quantitative survey of bottom communities in a Baltic Archipelago. In Keegan, B.F., P.O. Ceidigh and P.J.S. Boaden. (eds.) *Biology of Benthic Organisms*. 11th European Symp Mar Biol., Galway, Oct 1976. Pergamon press, Oxford, pp. 359–366.

Jarvekulg, A. 1979. Donnaya fauna vostochnoy chasti Baltiiskogo morja (Bottom fauna of the Baltic Sea eastern part), Valgus, Tallinn (in Russian).

Kachanova, A.A. 1962. The ecology of Dreissena polymorpha in the Uchinsk reservoir. *Voprosy Ecology Vysshaya Shkola Kiev*. **5**: 94–95 (in Russian with English title).

Karatayev, A.V. and L.E. Burlakova. 1993. Filtratsionnaya dejatelnost dreisseny i ejo vozdeistviye na troficheskuju strukturu soobshestv planktonnykh i donnykh bespozvonochnykh (Filtrational activity of dreissena and its impact on planktonic and benthic invertebrate communities trophicstructure), *Proc. 6th Meeting on Species and its productivity in the distribution area*, Gidrometeoizdat, St. Petersburg, pp. 211–212 (in Russian).

Kautsky, N. and I. Wallentinus. 1980. Nutrient release from a Baltic *Mytilus*-red algal community and its role in benthic and pelagic productivity. *Ophelia*. Suppl. **1**: 17–30.

Khailov, K.M., S.E. Zavalko and Y.G. Kamenir. 1987. Biologicheskiye i fizicheskiye parametry obarastaniya v more i konstruirovaniye iskustvennykh rifov (Biological and physical parameters of marine fouling and artificial reef design). Abstracts of SU Conf. Iskusstvennive rify dlja rybnogo khozjaistva (Artificial reefs for fishery),VNIRO, Moscow, pp. 35–37 (in Russian).

Laihonen, P. and I. Vuorinen. 1981. Fouling-habitat Suomessa. Esitutkimus ongelman laajuuden selvittämiseksi ja torjuntakeinojen löytämiseksi (A preliminary research of fouling problems on Baltic coast of Finland-Summary). *Turun yliop Biol laitoksen julk.* **2**: 1–42.

Laihonen, P., J. Hänninen, J. Chojnacki and I. Vuorinen. 1996. Some prospects of nutrient removal with artificial reefs. In Jensen, A.C. (ed.) *European Artificial Reef Research.* Proceedings of the 1st Conference of the European Artificial Reef Research Network, Ancona, Italy, 26–30 March 1996, pp. 85–96.

Laine, A.O. 1990. Prediumin venesataman vaikutuksista vesistöön sinisimpukkaa (*Mytilus edulis* L.) bioindikaattorina käyttäen. Master's thesis in hydrobiology. University of Helsinki, Finland. (in Finnish).

Lassig, J. 1965. The distribution of marine and brackish water lamellibranchs in the northern Baltic area. *Comm Biologicae.* **28**: 1–41.

Lee-Shang Fang. 1989. A theoretical approach to estimation of the productivity of an artificial reef. *Bulletin of Marine Science.* **44**(2): 1066.

Lietzen, E., P. Laihonen and I. Vuorinen. 1984. Fouling jatkotutkimus Suomessa (Summary: Fouling research in Finland: Occurrence of fouling animals and some control methods). *Turun yliop Biol laitoksen julk.* **8**: 1–56.

Lvova-Kachanova, A.A. 1971. On the role of *Dreissena polymorpa* Pallas in processes of self-purification of the water of Uchinskoye reservoir. In Complex Investigations of Water Reservoirs, Moscow University Press, **1**: 196–203 (in Russian with English title).

Marfenin, N.N. and T.N. Homenko. 1987. Spektr pitanija razlich nykh gidroidov na Belom more (The spectrum of nutrition of various hydroids on the White Sea), 3rd SU regional Conf. on the White Sea problems, Kandalaksha, pp. 199–201 (in Russian).

Mikheev, V.P. 1964. On linear growth of *Dreissena polymorpha* Pallas in some water reservoirs of the European part of the USSR, in Biologia dreisseny i bor'ba s ney Nauka Publisher Moscow-Leningrad, pp. 55–65 (in Russian with English title).

Mikheev, V.P. 1967. Filtration nutrition of the *Dreissena.* In Voprosy prudovogo rybovodstva, 15, Pischevaya Promyshien nost Publisher, Moscow, pp. 117–129 (in Russian with English summary).

Nalepa, T.F. and D.W. Schlesser. (eds.) 1993. *Zebra mussels, Biology, Impacts, and Control,* Lewis Publisher, Int.

Öst, M. and M. Kilpi. 1997. A recent change in size distribution of common mussels (*Mytilus edulis*) in the western part of the Gulf of Finland. *Ann Zool Fennici.* **34**: 31–36.

Parchevsky, V.P. and M.A. Rabinovich. 1989. The artificial cultivation of the green alga *Enteromorpha intestinalis* (L.) link in coastal waters near sewage outfalls. Abstract to Int. Symp for Modern Problems for Mariculture in Socialist Countries, VNIRO, Moscow, pp. 161–162 (in Russian with English title).

Reeders, H.H. and A. bij de Vaate. 1990. Zebra mussels (*Dreissena polymorpha*): a new perspective for water quality management. *Hydrobiologia.* **200/201**: 437–450.

Reeders, H.H., A. bij de Vaate and F.J. Slim. 1989. The filtration rate of *Dreissena polymorpha* (Bivalvia) in three Dutch lakes with reference to biological water quality management. *Freshwater Biol.* **22**: 133–141.

Reeders, H.H., A. bij de Vaate and R. Noordhius. 1993. Potential of the Zebra mussel (*Dreissena polymorpha*) for water quality management. In Nalepa, T.F. and Schloesser, D.W. (eds.) *Zebra Mussel Biology, Impacts, and Control.* Lewis Publisher, Int. pp. 439–451.

Rosenius, H. 1964. Ekologiska och morfologiska undersökningar över *Mytilus edulis* L., vid sydvästkusten av Finland. Master's thesis in Zoology. University of Helsinki (in Swedish).

Saenko, G.N. 1989. The role of mariculture in biosphere protection. Abstracts to Int. Symp. for Modern Problems of Mariculture in Socialist Countries, VNIRO, Moscow, pp. 17–18 (in Russian with English title).

Segerstråle, S. 1942. Ein Beitrag zur Kenntniss der Östlichen Verbreitung der Miesmuschel (*Mytilus edulis* L.) an der Sudküste Finnlands. *Memoranda Society Fauna Flora Fennica*. **19**: 5–7.

Smit, H., A. bij de Vaate and A. Fioole. 1992. Shell growth of the Zebra mussel (*Dreissena polymorpha*) (Pallas) in relation to selected physicochemical parameters in the Lower Rhine and some associated lakes. *Archives of Hydrobiology*. **124**(3): 257–280.

Sunila, I. 1981. Reproduction of *Mytilus edulis* L. (Bivalvia) in a brackish water area, the Gulf of Finland. *Ann Zool Fennici*. **18**: 121–128.

Vuorinen, I., P. Laihonen and E. Lietzen. 1986. Distribution and abundance of invertebrates causing fouling in power plants on the Finnish coast. *Memoranda Society Fauna Flora Fennica*. **62**: 123–125.

Zaitsev, Y.P. 1987. Iskusstvennye rify instrument pravlenija ekologicheskimi protsessami v pribrezhnoi zone morja (Artificial reefs is a tool for ecological processes regulation in marine coastal zones), in Abstracts of SU Conference Iskusstvenniye rify dlja rybnogo khozjaistva (Artificial reefs for fishery), Moscow, VNIRO, pp. 3–5 (in Russian).

Zaitsev, Y.P. 1989. Cultivation of hydrobionts in hypertrophic waters of the Black Sea. Abstracts to Int. Symp. for Modern Problems of Mariculture in Socialist Countries, VNIRO, Moscow, pp. 25–26 (in Russian with English title).

20. Rigs to Reefs in the North Sea

GORDON PICKEN[1], MARK BAINE[2], LOUISE HEAPS[3] and JONATHAN SIDE[2]

[1]Cordah, Kettock Lodge, Aberdeen Science and Technology Park, Bridge of Don, Aberdeen, AB22 8GU. [2]International Centre for Island Technology, Heriot-Watt University, The Old Academy, Back Road, Stromness, Orkney, KW16 3AW. [3]MRAG Ltd., 47 Princes Gate, London, SW7 2QA

Introduction

In the next two decades, the UK will have to decommission most of its offshore platforms, when they cease cost-effective production and become redundant. This has resulted in considerable discussion over the past 10 years about the engineering, legal, financial and environmental aspects of decommissioning (Read, 1984, 1985; Side, 1993; Side et al., 1993). A framework of regulations and standards for North Sea platform decommissioning is in place including guidelines drawn up by the International Maritime Organisation in 1989. In the wake of the Brent Spar incident, however, there has been considerable debate on the disposal at sea of offshore installations. This has resulted in the contracting parties to the 1992 OSPAR Convention for the Protection of the Marine Environment of the North-East Atlantic agreeing a Decision (98/3) on the Disposal of Disused Offshore Installations, to enter into force in February 1999. This decision prohibits the leaving wholly, or partly in place, of disused offshore installations within the maritime area, the only potential exceptions being the footings of steel installations weighing more than 10 000 tonnes in air; gravity based and floating concrete installations; concrete anchor bases; and any other installation suffering exceptional or unforeseen circumstances resulting from structural damage, deterioration or equivalent difficulties. This decision, however, does not cover those installations which serve another legitimate purpose in the maritime area authorized or regulated by the competent authority of the relevant contracting party. These installations are subject to Annex III (Article 8) of the 1992 OSPAR Convention and other relevant UK and international legislation and guidelines, the former requiring authorization from the contracting party in accordance with relevant applicable criteria, guidelines and procedures adopted by the Commission, with a view to preventing and eliminating pollution.

Several options for the re-use of platforms have been suggested, including alternative uses in situ such as marine research stations, search and rescue facilities, energy generation, fish farming or for waste disposal (Side and Johnston, 1985). Although some of these are attractive in theory, two major drawbacks would be the questions of maintenance and liability. Few activities would justify the cost of continued offshore inspection and maintenance, and it therefore seems likely that when platforms are no longer required for oil-related activities they will be decommissioned and dismantled to a greater or lesser extent.

A.C. Jensen et al. (eds.), Artificial Reefs in European Seas, 331–342
© 2000 Kluwer Academic Publishers. Printed in Great Britain.

The potential creation of artificial reefs from decommissioned platforms, however, remains one of the few viable uses for these structures.

The Concept of 'Rigs to Reefs'

By far the greatest concentration of oil- and gas-platform-related artificial reefs is found in the Gulf of Mexico. These reefs consist of steel latticework jackets or topsides which have been totally submerged *in situ* or at another, chosen location, for the specific purpose of creating an artificial habitat for marine life. The Gulf of Mexico has over 4000 oil and gas structures in water depths ranging from 30 m to 100 m, at distances of 2.5–143 km from the shore (Bleakley, 1982; Ditton and Falk, 1981; Driessen, 1985; Driessen, 1986; Harville, 1983; Reggio *et al.*, 1986). It has been estimated that these platforms provide 4500 acres of hard substrata, 28% of the known hard bottom habitat in the Gulf of Mexico (Reggio *et al.*, 1986; Driessen, 1985). Off the coast of Louisiana 3700 of these platforms provide an estimated 90% of coastal hard bottom habitat (Stanley and Wilson, 1990; Reggio, 1987). These structures have a significant direct effect on offshore recreational fishing, commercial hook-and-line fishing, and SCUBA diving activities in the USA (Reggio, 1987).

Direct evidence for the success of platforms as artificial reefs has come from the sport fishing industry, where anglers have caught consistently bigger and more desirable catches around these installations (Reggio, 1987) than elsewhere and, since the first platform was installed, a multi-million dollar recreational fishery has developed off the coast of Louisiana (Reggio *et al.*, 1986). Platform operators even provide mooring facilities for fishing boats. Commercial fish and shellfish landings have also shown a substantial increase, particularly the shrimp harvest, suggesting at the very least that the reefs have not had an adverse impact on the fishing industry (Driessen, 1989). These fishing activities have also had a profound beneficial impact on the local economy through the growth of shore-based ancillary industries.

By the year 2000 it is expected that 1625 platforms will have been removed from the continental shelf of Louisiana. The high cost of removal and the potential loss of this commercially important habitat have been the incentives for rigs-to-reefs initiatives, and under new federal policy guidelines many of these platforms will be used in artificial reef construction (Reggio *et al.*, 1986). In general it is thought that platforms make good artificial reefs because they provide an abundant food supply, reference points, structural openness, large surface areas, physical design complexity, a range of habitats and spawning and nursery areas.

Platforms are unique as artificial reefs because they extend throughout the water column, providing benthic, midwater and surface habitats. Gallaway and Lewbel (1982) have estimated that a typical platform standing in 50 m of water provides 1.2–1.6 ha of hard substrata.

Fish studies around Gulf of Mexico platforms have revealed that fish are present at all depths with the greatest variety in the range 30–70 m depth (Ditton and Falk, 1981). The most desirable species (for angling) rarely venture more than 20 m from the platform (Love and Westphal, 1990). Scarborough-Bull (1989) has classified platform-associated fish into transient or resident species. Resident demersal fish are thought to use the hard substrata communities as food or cover and may expand their vertical range to the midwater depths. Pelagic fish, however, are thought to be merely transient and may use the reefs as reference points and a potential source of food. It is also thought that pelagic species may be attracted to the hydrographical instabilities created when the current impinges on the reef (Sheehy and Vik, 1982).

Assessments on fish activity around working platforms in the USA have relied on visual observation, diver visual census, and monitoring the returns from commercial and sport fishing. It has been estimated that there are 20–50 times more fish at petroleum platforms in the Gulf of Mexico than at soft-bottom control areas of the same size, and 5 times more fish than on nearby natural reefs (Driessen, 1985). Several investigations have revealed that trophically independent transient fish are often dominant in terms of biomass, but habitat faithfulness has been identified for some pelagic and benthic species (Seaman *et al.*, 1989). One hundred and fifty species that were unknown in the Gulf of Mexico 30 years ago have now been identified at platforms (Driessen, 1985). Numerous species of fish and invertebrates have been observed at Gulf of Mexico platforms in hatching, juvenile and nesting stages (Driessen, 1989) and one platform appeared to act as a nursery ground for juvenile rockfish, which moved away after the second year (Love and Westphal, 1990).

Clearly, most of the work on the rigs-to-reefs concept to date has been carried out in the Gulf of Mexico and care must be taken when applying these findings directly to the North Sea. These qualitative and quantitative data may give great insight and set a precedent for oil-related reefs, but there is a need to investigate more specifically how effective high-profile, steel artificial reefs would be on the UK continental shelf. The North Sea is a very different environment from that of the Gulf of Mexico and so far there has been no specific rigs-to-reefs project undertaken in the North Sea. Several 'snapshot' studies have been carried out around platforms in the North Sea, however, to identify whether these structures are also acting as effective *de facto* artificial reefs and to examine the potential impact that the marine environment immediately around the platform may be having on fish that congregate there. It must be emphasized that this work, and that in the future, is focused on the steel lattice structures that make up the jacket or support structure of a platform and not concrete oil storage structures such as the Brent Spar.

Fish Studies Around North Sea Platforms

Abundance and Diversity of Platform-Associated Fish

A growing body of evidence confirms that oil and gas installations in the North Sea attract fish. During a study of fish densities at a complex of 13 structures in the Ekofisk Field in the Norwegian sector of the North Sea, timed fishing trials were conducted at varying distances from the structures using long-lines, gillnets and jiggers (Valdemarsen, 1979). The aggregation of cod around the platforms was particularly noticeable and it was found that fish densities decreased with distance from the platform. An average of 3.3 fish were caught over the 15-minute surveys 50 m distant from the complex. This catching rate was halved 50–100 m away and further reduced to 0.6 fish every 15 minutes 100–200 m away. It was observed that there was a dominance of krill (*Megacytiphanes norvegicus*) in the stomach contents of the cod. Krill aggregation in the vicinity of the platform was thought to be a result of artificial lighting and alterations in current patterns caused by the platforms.

A later study of fish activity around 15 platforms in the North Sea recorded substantial numbers of fish at all the structures, at all times of the day and night, and in historical videotape data spanning a period of six years (AUMS 1987a, 1987b). Of the 21 species identified (Table 1), saithe (*Pollachius virens*) was by far the most abundant in the central and northern North Sea. This gadoid is known to form large shoals and is often found around rocky outcrops, sandbanks and other seabed features. In the northern and central sectors shoals of up to 2000 individuals at densities of 3 m^{-3} were observed from a depth of 21 m down to the seabed. Individuals were generally in the size range 40–60 cm, indicating that they were probably 2–3 years old. Other fish seen at these depths included haddock (*Melanogrammus aeglefinus*) and whiting (*Merlangius merlangus*). Saithe shoals generally declined in density with increasing depth, but at the base of the jacket and up to 15 m from the seabed, further smaller shoals were found. Other species were also present in this lower zone, including cod (*Gadus morhua*), present in shoals of up to 100 individuals at densities of 0.2 fish m^{-3}, with individuals up to 100 cm long. Immediately above the seabed, and around any small pieces of debris lying within the confines of jackets, a variety of other species were identified including ling (*Molva molva*), wolf-fish (*Anarchichas lupus*), red fish (*Sebastes marinus*) and large shoals (5000 individuals) of Norway pout (*Trisopterus esmarkeii*) at densities of 5 m^{-3}.

Table 1.   Fish species identified around North Sea platforms.

Saithe (*Pollachius virens*)	Plaice (*Pleuronectes platessa*)
Cod (*Gadus morhua*)	Norway haddock (*Sebastes viviparous*)
Bib (*Trisopterus luscus*)	Haddock (*Melangrammus aeglefinus*)
Whiting (*Merlangius merlangus*)	Sole (*Solea solea*)
Ling (*Molva molva*)	Pollack (*Pollachius pollachius*)
Wolf-fish (*Anarchichas lupus*)	Torsk (*Brosme brosme*)
Dab (*Limanda limanda*)	Norway pout (*Trisopterus esmarkeii*)
Red fish (*Sebastes marinus*)	Poor cod (*Trisopterus minutus*)
Anglerfish (*Lophius piscatorius*)	Long rough dab (*Hippoglassoides platessoides*)
Hagfish (*Myxine glutinosa*)	Sea scorpion (*Taurulus bubalis*)
Dragonette (*Callionymus lyra*)	Sand eels (*Ammodytes* spp.)

In the southern North Sea, saithe and cod were less abundant, but the variety of species near the seabed in and around platforms increased, with species such as plaice (*Pleuronectes platessa*), dab (*Limanda limanda*) and bib (*Trisopterus luscus*).

The relationship between fish and North Sea platforms was further investigated (ICIT, 1991) using Simrad EK500 echosounder surveys at four selected offshore installations. Control sites over 40 km from the nearest platform, in water depths ranging from 120–194 m, were also examined.

Aggregations of fish around the offshore platforms were very closely associated with the structure. The echograms were visually assessed for the presence of pelagic fish shoals, individual pelagic fish, demersal fish and feeding layers (juvenile fish and zooplankton). Each of these was categorized according to the size/numbers and intensity of the fish and were converted to quantitative abundance estimates.

The pelagic shoals were mostly dominated by saithe, whereas the demersal shoals were dominated by cod. Distribution analyses showed an increase in the abundance of pelagic fish in the vicinity of three installations, with aggregations largely restricted to within 100 m of the structure. Demersal fish were less abundant but also exhibited aggregation. From the complete data set from all four platforms, it was estimated that about 70 000 pelagic fish and 9000 demersal fish aggregated within 100 m of each installation.

The numbers of fish found around working North Sea platforms, although locally high, are still very small in relation to the overall stocks of fish in the North Sea. Early data from studies at working platforms indicated that a large structure might hold 10 000 fish (AUMS 1987a, 1987b). Even if one assumes that these are all 'new fish', fish that would not be present but for the existence of the structure, then the 40 platforms in the central and northern North Sea would hold about 400 000 fish. The total North Sea stocks of cod, haddock and saithe are roughly 3.7×10^7 tonnes, 5×10^9 tonnes and 7.7×10^8 tonnes, respectively, so in this estimate the additional fish that may be provided by platforms represented an insignificant proportion (less than 0.001%) of the total North Sea stock (Davies *et al.*, 1987).

ICIT (1991) provided further data which gave a slightly higher figure. Fish abundance estimates indicated that the majority (50–80%) of installations in the northern and central North Sea could account for between 4000 and 18 000 tonnes of saithe and between 450 and 3600 tonnes of cod. The overall stocks of saithe and cod in the North Sea were each estimated at 600 000 tonnes, so these aggregations may account for 0.7% to 3% of total saithe stocks and 0.08% to 0.6% of total cod stocks. These quantitative estimates should, however, be treated with caution since they are indicative estimates and not accurate predictions.

Attraction Criteria

It has been suggested that working oil platforms attract fish because of the presence of factors such as noise, heat, light and/or additional sources of food, and that these attractive features might disappear once a platform was decommissioned. To consider this the wreck of Transocean III, lying in 108 m of water 7 km south-west of 'Beryl Alpha' (AUMS, 1987b) was surveyed. Video surveys revealed that the wreck and its immediate surroundings had a greater number of fish species and individuals than adjacent 'open sea' sites 1 km away. It was concluded that the physical presence of the structure itself was an attraction for these fish.

Migratory species of pelagic fish can detect artificial reefs along their migratory routes from great distances even when up-current and well out of visual range (Mottet, 1985). The physical presence of a natural or artificial structure and its interaction with waves and currents can generate current shadows, sounds, pressure variation, lee waves and internal waves, which can be detected by fish auditory and lateral line systems (Grove and Sonu, 1985; Nakamura, 1985; Sato, 1985). It appears, therefore, that neither visual nor chemical clues are required for fish to find and maintain their association with a seabed structure.

Several studies around the world have reported that artificial structures may attract large numbers of fish within days or hours of placement (Bohnsack and Sutherland, 1985). Clearly, the presence of fouling on the submerged structure is not essential in attracting or retaining these initial populations of fish, although the structure does need to be located in areas where appropriate food organisms occur (Mottet, 1985). Rough calculation with optimistic assumptions about fouling productivity and fish assimilation efficiency shows that the epibiota on the 40 steel platforms in the central and northern sector of the UK North Sea could at best support only about 100 tonnes of fish (Picken and McVicar, 1986). ICIT (1991) reported that the stomach contents of the fish caught around three platforms were dominated by sand eels (*Ammodytes* spp.), krill (*Meganyctiphanes norvegica*), salps and other jelly-like organisms. That the surfaces of offshore structures are unlikely to make a significant contribution to the food available to fish does not appear to detract from their aggregating effect.

If favourable changes in the hydrodynamic regime around platforms were important in maintaining fish at these installations, it would be expected that the fish would lie in a particular region in relation to the current movement. ICIT (1991) report no apparent environmental directionality in fish distribution, although further investigation is necessary. Typical swimming speeds for gadoids such as cod, haddock and whiting are in the range 1.6–2 ms^{-1} (Thurman and Webber, 1984) and maximum current speeds in the vicinity of the installations are around 0.3 to 0.5 m s^{-1}, so it would appear that these species have the capacity to take up station at any point around a structure in the North Sea.

No single factor appears to be dominant in determining fish distribution around platforms. Indeed, it is likely that a combination of factors gives rise to the observed distributions and that species-specific aspects of behaviour will be at least as important as installation-related features. It has been suggested that the only benefit that pelagic fish can derive from a fixed structure is a spatial reference around which they can orientate in an unstructured environment (Klima and Wickham, 1971). It is widely accepted, however, that most artificial reefs created for the attraction of demersal species are successful if they are sited correctly on migration routes, or provide appropriate habitats and conditions for certain phases of their lifecycle (Mottet, 1985).

Flavour and Tissue Hydrocarbon Content of Fish Caught at Platforms

In order to determine if the marine environment close to the platform, and in particular the presence of the cuttings pile, was having any measurable effect on fish quality, AUMS (1989) analysed the tissue hydrocarbon content and the flavour of saithe caught near platforms. 'Platform fish' were caught at a multi-well production platform in the northern North Sea where low-toxicity oil-based mud had been used during development drilling. 'Open sea' fish were also caught at a control site about 40 km SE of Sumburgh Head, Shetland Islands, where there was no industrial activity.

The results revealed that fish caught in the vicinity of both working platforms and at platforms where no drilling had taken place, had elevated levels of hydrocarbons in their flesh and liver (AUMS, 1989). Analysis indicated that the source of these hydrocarbons was general industrial activity, including diesel fuel, rather than the discharges of low-toxicity oil-based cuttings. All the platform fish were in good condition when landed and none was classified as having an oily taint when subjected to a taste panel assessment. The results of this study are consistent with those carried out by McGill *et al.* (1987), who found only a 'tendency towards tainting' in dabs (*Limanda limanda*) caught 500–800 m away from the Beatrice oil platform in the Moray Firth.

The Growth and Condition of Fish Found at Platforms

The subtle effects that the marine environment around platforms may be having on individual fish were investigated by comparing, using biochemical measures, the relative condition and growth rate of fish at oil platforms with those of fish from open-sea control sites. The results indicated that saithe caught in the vicinity of the Beryl platforms were in good condition and growing at least as well as fish in the open sea (Mathers *et al.*, 1992a, 1992b). There were no indications in any of the parameters measured of any adverse effect on the saithe from industrial activity associated with the offshore installation.

Conclusions on Fish Studies

Several commercially important fish species are found close to many working platforms in the UK North Sea. On the basis of qualitative and quantitative observations using historical videotape material, and ship-based sonar survey, it seems likely that these sightings are not merely chance observations of migratory shoals that happened to encounter offshore structures, but represent individuals that are spending a proportion of their lifetime in the vicinity of offshore structures. Such a finding would be in agreement with the results of research by Japanese scientists on fish behaviour with respect to artificial structures.

Although these individuals may derive additional food from the structure it is likely that the major cause of the aggregating effect is the physical presence of a large structure at an offshore location. All the evidence obtained so far indicates that fish living around structures are in at least as good a condition as those collected from the open sea.

The immediate platform environment appears to be one which fish find acceptable and to which they are attracted and that produces no adverse effects on the fish in terms of physiology, biochemistry or commercial value. The structures scattered throughout the North Sea provide local 'reef habitats' utilized by fish for a time, and the evidence suggests that the existing working platforms are having a small, beneficial effect on local fish populations.

The benefits of fish aggregations around offshore structures have not gone unnoticed by fishermen. The financial implications of the aggregating effect of structures were vividly demonstrated in June 1987, when a fisherman was fined £8000 for repeatedly fishing within the 500 m safety exclusion zone around platforms, during which time over £200 000 worth of fish were caught. Offshore structures are clearly fulfilling one of the criteria of successful artificial reefs, namely that of aggregating fish so that they may be caught more economically.

Building Artificial Reefs in the UK from Platforms

The 'materials of opportunity' furnished by the decommissioning and removal of UK North Sea platforms could provide ideal components for artificial reefs. Whole structures, or parts of structures could be deployed as artificial reefs either by toppling them *in situ*, or removing them to designated inshore reef sites. The creation of inshore reefs may provide an attractive habitat for the safe aggregation and growth of juveniles and adults of certain species of fish and shellfish, protect parts of inshore grounds from the effects of mobile fishing gear, and become the focus of a reef fishery which would provide easier access to fishing grounds at less expense. Any fishery would have to be properly managed to ensure that the boats frequenting it took a sustainable level of catch with careful attention to the size and age of individual fish and shellfish caught.

The growth of any reef fishery may necessitate complementary changes in fishing techniques and gear, but the industry itself is already reviewing such aspects and would no doubt readily adapt if the reef fishery were of long-term benefit. Designated inshore reef sites would have to be carefully selected and comprehensive planning would have to be undertaken involving all relevant parties. McIntyre (1987) maintains that before accepting or rejecting such proposals, pilot studies are required to provide information on how such reefs will develop in North Sea conditions. A recent initiative (Aabel et al., 1997) by the Offshore Decommissioning Communications Project (ODCP) has brought this research need to the fore and prepared outline proposals for scientific investigation into platform structures in both offshore and near shore environments.

If supported, these studies will establish the residence period of fish and correlate behaviour to the physical variables around a platform, establish size frequency distributions of fish around the structures, identify the diet of the fish species, estimate growth rates and investigate the time for recovery after a commercial fishing event. These data will facilitate modelling of possible artificial reef creation scenarios and contribute to the overall understanding of what role a reef can play in fisheries protection, management, and exploitation in both an inshore and an offshore area.

Conclusion

Artificial reef programmes in the Gulf of Mexico using materials of opportunity from decommissioned oil platforms have proved highly successful and cost-effective for the end user and the oil industry. Desirable fish species are associated with the structures, and there is even talk of the damage that could be done to fisheries by the total removal of platforms which are already acting as *de facto* reefs. Although some other materials of opportunity represent a compromise in design, oil platforms are apparently ideal reef components, similar to fish attraction reefs in Japan, and if carefully selected can satisfy many of the design criteria for artificial reefs constructed to attract finfish.

The North Sea fishing industry has made clear its preference for complete removal (Allan, 1986) of oil- and gas-production structures. Total removal of platforms would restore a small part of the sea to the operation of trawl fisheries but the magnitude of the engineering task and the great financial expenditure (both government and industry) required act against this. Toppling a structure *in situ* has been suggested as an alternative to total removal. Although controlled and precise toppling has been successfully completed in the Gulf of Mexico (Quigel and Thornton, 1989), the option of toppling distant offshore platforms in the North Sea *in situ* is not favoured by the fishing industry or by current UK government policy and is not provided for under the new OSPAR Decision (98/3) on the Disposal of Disused Offshore Installations. The creation of artificial reefs from platform components placed in an agreed location to achieve a declared aim may provide a positive fisheries management tool which may prove to be more acceptable than merely toppling the structure.

It is difficult at present to reach any conclusive decision as to whether total removal or artificial reef creation would be of greatest benefit to fishermen or fisheries. It has been estimated that complete disposal of thousands of tonnes of this high-grade steel, already fabricated into reef-type components and with a life expectancy of at least another 100 years, would cost about £4.4 billion (1988 prices; UKOOA, 1988).

Side (1993) underlines the importance and need for an integrated resource management approach to the rigs-to-reefs issue, highlighting the House of Commons' Environment Select Committee's report on Coastal Zone Protection and Planning. Final decisions on abandonment will be made within national and international restrictions on 'dumping at sea' and against a background of activity by environmental lobby NGOs (non-governmental organizations) which are opposed to the dumping of waste (both real and perceived) at sea. Fishermen's representatives will also have a crucial role in the debate on abandonment, and in particular will be central to any rigs-to-reefs initiative.

Acknowledgements

The authors would like to thank UKOOA Ltd and Mobil North Sea Ltd for financial support in carrying out some of the studies described in this chapter.

References

Aabel, J.P., S.J. Cripps, A.C. Jensen and G. Picken. 1997. Creating artificial reefs from decommissioned platforms in the North Sea: a review of knowledge and proposed programme of research. Report to the Offshore Decommissioning Communications Project (ODCP) of the E and P Forum from Dames and Moore Group, RF-Rogaland Research, University of Southampton and Cordah, 115 pp.

Allan, R. 1986. Abandonment – a fishing industry perspective. In *Proceedings of the Offshore Decommissioning Conference,* November 1986, London. Offshore Conferences and Exhibitions Ltd, pp. 16–19.

AUMS. 1987a. Fish activity around North Sea oil platforms. Unpublished Report by Aberdeen University Marine Studies Ltd, 46 pp.

AUMS. 1987b. Fish activity around North Sea oil platforms. Phase II: a survey of Transocean 3. Unpublished Report by Aberdeen University Marine Studies Ltd, 19 pp.

AUMS. 1989. Investigation into the flavour and tissue hydrocarbon content of fish caught at an oil production platform. Unpublished report by Aberdeen University Marine Studies Ltd for UKOOA, 78 pp.

Bleakley, N.B. 1982. Tenneco scores first with artificial reef. *Petroleum Engineering International.* **54**(14): 11–13.

Bohnsack, J.A. and D.L. Sutherland. 1985. Artificial reef research: a review with recommendations for future priorities. *Bulletin of Marine Science.* **37**(1): 1–39.

Davies, J.M., G.P. Arnold and G.B. Picken. 1987. The implications of partial platform and pipeline abandonment for fisheries. In *Proceedings of the Conference on Decommissioning and Removal of North Sea Structures.* IBC Technical Services Ltd, London.

Ditton, R.B. and J.M. Falk. 1981. Obsolete petroleum platforms as artificial reef material. In *Artificial Reefs:* Proceedings of a Conference. Report of the Florida Sea Grant Program, pp. 96–105.

Driessen, P.K. 1985. Oil platforms as reefs: oil and fish can mix. In *Coastal Zone Conference.* American Society of Civil Engineers. **2**: pp. 1417–1438.

Driessen, P.K. 1986. Offshore oil platforms; an invaluable ecological resource. In Oceans 1986 Conference Record, IEEE Publishing Service, New York, USA. pp. 516–521.

Driessen, P.K. 1989. Offshore oil platforms: Mini-ecosystems. In *Petroleum Structures as Artificial Reefs: A Compendium. Fourth International Conference on Artificial Habitats for Fisheries.* Rigs to Reefs Special Session, Miami, Florida, November 4 1987. pp. 3–5.

Gallaway, B.J. and G.S. Lewbel. 1982. The ecology of petroleum platforms in the northwestern Gulf of Mexico: A community profile. U.S. Fish and Wildlife Service, Office of Biological Sciences, Washington DC FWS/OBS-82/27. Bureau of Land Management, Gulf of Mexico OCS Regional Office, Open-file Report 82-03, 92 pp.

Grove, R.S. and C.J. Sonu. 1985. Fishing reef planning in Japan. In D'Itri, F.M. (ed.) *Artificial Reefs: Marine and Freshwater Applications.* F.M. Lewis Publishers Inc., Chelsea, Michigan, pp. 185–251.

Harville, J.P. 1983. Obsolete petroleum platforms as artificial reefs. *Fisheries.* **8**: 4–6.

ICIT. 1991. UKOOA survey of fish distribution and contamination. Report by the International Centre for Island Technology in association with Aberdeen University Research and Industrial Services Ltd for UKOOA (90/1212), December 1991, 94 pp.

Klima, E.F. and D.A. Wickham. 1971. Attraction of coastal pelagic fishes with artificial structures. *Transactions of American Fisheries Society.* **100**: 86–99.

Love, M.S. and W. Westphal, 1990. Comparison of fish taken by a sport fishing party vessel around oil platforms and adjacent natural reefs near Santa Barbara, California. *Fisheries Bulletin.* **88**: 599–605.

Mathers, E.M., D.F. Houlihan and M.J. Cunningham. 1992a. Nucleic acid concentrations and enzyme activities as correlates of growth rate of the saithe *Pollachius virens*: growth-rate estimates of open-sea fish. *Marine Biology.* **112**: 363–369.

Mathers, E.M., D.F. Houlihan and M.J. Cunningham. 1992b. Estimation of saithe *Pollachius virens* growth rates around the Beryl oil platforms in the North Sea: a comparison of methods. *Marine Ecology Progress Series.* **86**: 31–40.

McGill, A., P.R. Mackie, P. Howgate and J.G. McHenery. 1987. The flavour and chemical assessment of dabs (*Limander limander*) caught in the vicinity of the Beatrice Oil Platform. *Marine Pollution Bulletin.* **18**(4): 186–189.

McIntyre, A.D. 1987. Rigs and reefs. *Marine Pollution Bulletin.* **18**(5): 197–198.

Mottet, M.G. 1985. Enhancement of the marine environment for fisheries and aquaculture in Japan. In D'Itri, F.M. (ed.) *Artificial Reefs: Marine and Freshwater Applications.* Lewis Publishers Inc, Chelsea, Michigan. pp. 13–112.

Nakamura, M. 1985. Evaluation of artificial fishing reef concepts in Japan. *Bulletin of Marine Science.* **37**(1): 271–278.

Picken, G.B. and E.R. McVicar. 1986. The biological implications of abandonment options; the need for information. In *Proceedings of the Offshore Decommissioning Conference*, November 1986, London. Offshore Conferences and Exhibitions Ltd. pp. 22–27.

Quigel, J.C. and W.L. Thornton. 1989. Rigs to reefs – a case history. *Bulletin of Marine Science.* **44**(2): 799–886.

Read, A.D. 1984. The decommissioning of offshore installations – a world-wide survey of timing, technology and anticipated costs. Report No. 10.5/108. Oil Industry International Exploration and Production Forum, London.

Read, A.D. 1985. Platform decommissioning requirements – the way forward. Report No. 10.7/119. Oil Industry International Exploration and Production Forum, London.

Reggio, V. 1987. Rigs-to-reefs: The use of obsolete petroleum structures as artificial reefs. OCS Rep. MMS87-0015, US Dept. Imt., Minerals Manage. Serv., Gulf of Mexico OCS Region, New Orleans, 17 pp.

Reggio, V., V. Van Sickle and C. Wilson. 1986. Rigs to Reefs. *Louisiana Conservationist.* **38**(1): 4–7.

Sato, O. 1985. Scientific rationales for fishing reef design. *Bulletin of Marine Science.* **37**(1): 329–335.

Scarborough-Bull, A. 1989. Fish assemblages at oil and gas platforms compared to natural hard/live bottom areas in the Gulf of Mexico. *Coastal Zone '89.* pp. 979–987.

Seaman, W. Jr., W.J. Lindberg, R.G. Carter and T.K. Frazer. 1989. Fish habitat provided by obsolete petroleum platforms off southern Florida. *Bulletin of Marine Science.* **44**(2): 1014–1022.

Sheehy, D.J. and S.F. Vik. 1982. Artificial reefs – a second life for offshore platforms? *Petroleum Engineer International.* May: 40–52.

Side, J.C. 1992. Rigs to reefs: Obstacles, responses and opportunities. In Baine, M.S.P. (ed.) *Artificial Reefs and Restocking.* 1st British Conference on artificial reefs and restocking. Stromness, Orkney Islands, Scotland, 12 September 1992. pp. 53–60.

Side, J.C. and C.S. Johnston. 1985. Alternative uses of offshore installations. Institute of Offshore Engineering, Heriott-Watt University.

Side, J.C., M. Baine and K. Hayes. 1993. Current controls for abandonment and disposal of offshore installations at sea. *Marine Policy.* **17**(5): 354–362.

Stanley, D.R. and C.A. Wilson 1990. A fishery dependent based study of fish species composition and associated catch rates around oil and gas structures off Louisiana. *Fisheries Bulletin.* **88**: 719–730.

Thurman, H.B. and H.H. Webber. 1984. *Marine Biology.* Charles E. Merril, Colombus, Ohio, USA.

342   *G. Picken et al.*

UKOOA. 1988. The abandonment of offshore installations and pipelines. Factsheet on oil and gas activities. United Kingdom Offshore Operators Association, London.
Valdemarsen, J.W. 1979. Behaviour aspects of fish in relation to oil platforms in the North Sea. ICES Fishing Technology Committee, CM 1979/B:27, 6 pp. (mimeo).

21. Coal Ash for Artificial Habitats in Italy

GIULIO RELINI

Laboratori di Biologia Marina ed Ecologia Animale, Università degli Studi di Genova, DIP.TE.RIS, Via Balbi, 5, 16126 Genova, Italy

Introduction

In Italy artificial reefs are generally multipurpose structures (Relini and Orsi-Relini, 1989) placed in near-shore areas to prevent illegal trawling in water shallower than 50 m, so protecting the natural environment and special biocoenosis (such as *Posidonia oceanica* beds). Such reefs provide attachment surfaces, protection and shelter for eggs and juveniles of commercial species and enhance local fish populations for capture by sport and professional fishermen. Local fishing communities are increasingly requesting deployment of artificial reefs in 'their' areas of operation. This increase in demand implies a need for large quantities of materials for reef construction if requests are to be satisfied.

In common with other maritime works, concrete is normally used to build blocks for artificial reef construction. In the last few years interest in the utilization of cement-stabilized recycled waste material has increased. A major source of material suitable for cement stabilization is ash from coal-fired power stations. There are two ecological advantages to the use of cement-stabilized coal ash in reefs: the need to dispose of ash in land disposal sites or by dumping loose ash at sea is reduced and the demand for quarry extraction of limestone and subsequent cement production is lessened: cement-stabilized ash blocks use less cement than concrete blocks.

If cement-stabilized coal ash is to become widely used in the production of blocks for artificial reefs it must fulfil two criteria: it must be physically stable over time, maintaining its mechanical properties during decades of immersion, and it must be environmentally benign. Environmental concerns focus on the heavy-metal content of the ash (derived from the heavy-metal content of coal) and whether cement stabilization is effective in binding these metals into the block matrix, so preventing leaching.

Scientific investigations have been undertaken to verify these two criteria. These are also linked with studies to ensure compliance with environmental protection legislation and rational location of a reef, to meet the requirements of not compromising the quality of the marine ecosystem and promoting the development and growth of fish by using aquatic habitat enhancement techniques.

Several non-Mediterranean countries have successfully tested cement-stabilized coal ash blocks in artificial habitats (Shao *et al.*, 1994 (China); Chen, 1987 (Taiwan); Woodhead *et al.*, 1982, 1985 (USA)), prior to 1987 when the Italian Electricity Board (ENEL) initiated the CENMARE project. (CENMARE is the name of the project and means 'ash at sea').The UK initiated a programme in 1988 with reef

A.C. Jensen et al. (eds.), Artificial Reefs in European Seas, 343–364
© 2000 Kluwer Academic Publishers. Printed in Great Britain.

deployment in 1989 (Collins *et al.*, 1990, 1991, 1994a, 1994b; Jensen *et al.*, 1994, chapter 16, this volume). In Japan (Suzuki 1985, 1995) and in Taiwan (Kuo *et al.*, 1995) large artificial reefs utilizing coal ash concrete have been deployed in the sea for fishery enhancement.

The CENMARE programme, linking ISMES (Istituto Sperimentale Modelli e Strutture), CISE (Centro Informazioni Studi e Esperienze) and the Institute of Zoology, Genoa University, the ENEL Research and Development Department, Ash Research Centre in Brindisi (at present Centro Ricerca, Valorizzazione e Trattamento Residui), was developed to investigate the possibility of using coal ash derived from ENEL power stations, (which produce about 1.5×10^6 tonnes of ash per year) as a component of blocks for artificial habitats being deployed along the Italian coast.

The main aim of these studies was to ascertain, within the limits of existing legislation, the environmental compatibility of the blocks when placed in the sea. Priority was given to establishing whether the quality of the coastal water was maintained, describing and quantifying the pattern of biological colonization on and around the reefs and in ascertaining whether any bioaccumulation of dangerous chemicals occurred within the artificial reef communities.

Prior to full-scale reef deployment a pilot experiment was considered appropriate to provide initial data that could be used to support applications, and allay environmental concerns relating to deployment of cement-stabilized coal ash reefs in Italian coastal waters.

The CENMARE Project

The research programme was implemented in three stages.

Stage 1 (1987–88)

Laboratory tests of a physical, mechanical and chemical nature were conducted to decide which materials were to be used in the construction of artificial reefs and to assess their compatibility with the marine environment. This preliminary stage allowed the identification of a coal ash base mixture suitable for use and ascertained that there was no significant chemical leaching from blocks made of these mixtures which might be of environmental concern.

Physical properties

During stage 1, six ash mixtures were studied (Relini *et al.*, 1995b; Sampaolo and Relini 1994). Each mix used a high percentage of fly ash (fine dust from the power station chimney) and bottom ash (from the furnace) with a small percentage addition of a hydraulic binder (Portland cement and/or lime) (Table 1). Each mix was chosen in the belief that it would show good mechanical properties suitable for the intended use and have a low production cost, comparable to that of producing artificial reef blocks from conventional materials.

Table 1. Composition of the six mixtures tested. Each component is expressed as a percentage of the total mix.

	1	2	3	4	5	6
Fly ash (C.L.)	61.0	49.4	69.2	69.1	73.1	71.1
Bottom ash (C.P.)	9.7	24.7	–	–	–	–
Portland cement 325	1.9	–	6.3	3.1	–	2.1
Hydrated lime	4.8	4.9	3.1	6.3	5.3	5.3
Water	22.6	21.0	21.4	21.5	21.6	21.5

For a mixture to be used in block construction the following characteristics were considered desirable:

(1) compressive strength > 5 Mpa;
(2) tensile strength > 0.5 Mpa, necessary to give the mass good adhesion (good mechanical properties) and reduced effective porosity (low permeability and good resistance to environmental stress).

Mixture number 2 (Table 1) was judged suitable for economic reasons, and was chosen to make the blocks used in the experimental test rig because both bottom and fly ash could be incorporated in the mix. The uniaxial compressive strength (ASTM C39.86 American standard) value reached 10 Mpa; optimum dry density was 1415 g cm^{-3} (wet density = 1730 g cm^{-3}). The compressive strength measured on samples of concrete used as reference after 28 days of curing exceeded 50 Mpa (Sampaolo and Relini, 1993, 1994).

Chemical properties

After preliminary investigations to ascertain the leaching range for various chemical elements over time and establish specific analytical methods, leaching tests were carried out on three experimental blocks immersed in seawater.

Each block, a 20 cm sided cube, was immersed in a volume of seawater equal to 20 l (solid/liquid ratio 1:2.5) held in a polyethylene container. The water was continuously agitated over a 90-day period: previous leaching values ascertained by laboratory tests (in the order of ppb) indicated that there would not be any saturation of the solution. The water was sampled and analysed periodically (0, 1, 2, 5, 15, 30, 45, 60 and 90 days) for Si, Al, As, Cr, Se, Tl, Na, K, Cu, Zn, Fe, Cd, Pb, Hg, Be, Sb and Te concentrations. The seawater was then replaced with clean water for a second 90-day cycle of analysis.

The total quantity of material leached into solution during the two contact cycles (180 days) was about 140 mg Si and 7 mg Al, which was a negligible fraction of the quantity present in the solid phase (about 3 kg of Si and 2 kg of Al in each individual block; average weight 14 kg). Of the trace elements, As, Cr, Se and Tl were measured in solution. The maximum concentrations in solution were, however, in the order of tens of mg l^{-1} for As, Cr, and Se, and less for Tl. The release rate proved to be fairly constant over time. The remaining elements tested (Na, K, Cu, Zn, Fe, Cd, Pb, Hg, Be, Sb, Te) were not found to have leached into the seawater.

No appreciable changes in chemical composition were seen in the block; an increase in the concentration of Mg and a corresponding decrease in Ca was seen, but only in surface sections.

Stage 2 (1988–92)

An experimental through-flow test-rig (Fig. 1) was constructed at the ENEL 'Torrevaldaliga South' coastal power station (mid-Tyrrhenian sea) 50 km NW of Rome (Fig. 2). Small (20 × 20 × 20 cm) ash-based and concrete blocks (as controls) were used to build model reefs (Figs 2, 3). Carrying out tests in an experimental rig, prior to a study in the natural open-sea ecosystem presents undoubted advantages in terms of understanding and modelling of the leaching of pollutants from the experimental artificial blocks. It was possible to measure chemical leaching without the overlap of uncontrollable natural processes, such as the hydrodynamics of the water masses, which could lead to difficulties in interpretation of such data.

The test rig was designed around two fibre-glass-lined trial tanks (2 × 10 × 1.5 m) (Fig. 1), which received a constant through-flow of coastal seawater. Water was initially pumped from the sea through two $100\,l\,s^{-1}$ submersible pumps into an epoxy-lined sedimentation tank (surface area 45 m^2, 3 m deep). From this tank six $33\,l\,s^{-1}$ pumps sent the water via two pipes into the trial tanks 400 m away. To ensure uniformity of water flow (3–5 cm s^{-1}) along the transverse section, a hydraulic damper was installed in the water inlet of each tank using a series of

Figure 1.   Two through-flow test rigs at Torrevaldaliga power station.

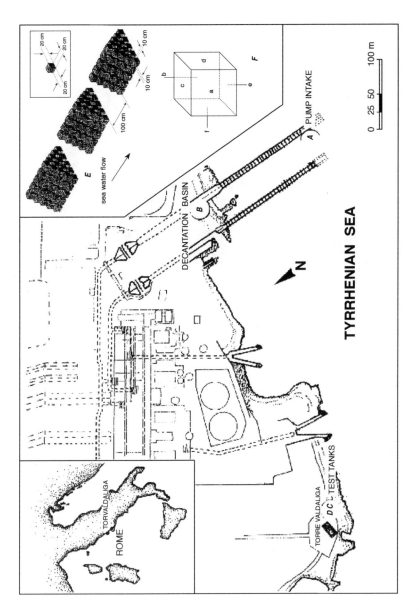

Figure 2. Map of the Torrevaldaliga south ENEL power stations. Location of experimental plant (C,D) and stations for control of settlement (A,B). Arrangement of three pyramids of each tank (E). Key to the labelling of the faces of the cubes (F).

Figure 3.   Pyramids of blocks in one of the two test tanks at Torrevaldaliga.

fibre-glass grates. Water left the trial tanks over a rectangular weir, which stretched the entire width of the tank. Two grates were placed in front of the weir.

Models of underwater reefs, made up of a total of 225 blocks (20 × 20 × 20 cm) were set up in each trial tank using a pyramid configuration (Figs 2, 3) with a solid to liquid ratio of about 1:20. Coal ash blocks were immersed in one trial tank while concrete blocks made of pozzolanic cement, sand and gravel, were placed in the other tank, which was used as a control. All the blocks were cured in a climatic cell for 28 days. The blocks were prepared in a rigid chamber of known volume by pressing a known quantity of mixture whose optimal values of density and humidity had been established by the Proctor test.

The main aims were to obtain data about the behaviour of such materials and the interaction between the blocks and marine organisms, in particular the macro-benthos, during a 2-year experimental programme (April 1990–March 1992). Every 3 months a series of blocks of both materials were recovered for physical, chemical and biological investigations and replaced with new ones (Relini *et al.*, 1995a, 1995b; Relini and Patrignani 1992, 1993). In addition seawater flowing through the rig was sampled and analysed periodically.

Physical monitoring

During the 2 years of experimentation in the test-rig, 44 ash-based blocks and 44 concrete blocks were subjected to non-destructive tests (geometry, swelling, density, sonic velocity). There were no variations in shape, no swelling and no appreciable variations of density in either the ash or the concrete samples. No external signs of weathering, such as cracking or bits of block breaking off were seen.

The values recorded for sonic velocity in the ash-based blocks showed a consistent increase with time (from 2200 m s^{-1} to 3300 m s^{-1} after 6 months) which implied curing had continued during immersion resulting in a more compact internal structure. Results for the concrete control blocks showed a similar pattern with a practically constant VL value of around 5000 m s^{-1} resulting after 6 months.

Block compressive strength values were obtained, giving values of 10, 25.8, 29.4, 31.0 and 31.4 Mpa (average of three samples) respectively at 0, 3, 6, 9 and 12 month immersion periods for ash-based blocks. Concrete blocks showed values of 51.7, 71.1 and 72.0 Mpa respectively after 0, 6 and 12 months in seawater.

The ash-based samples immersed in water showed a constant permeability value (K) of 10^{-7}–10^{-8} cm s^{-1}.

Chemical monitoring

Interest focused on monitoring elements of concern listed in Italian environmental protection regulations: Cu, Mn, Zn, Fe, Cd, Pb, As, Tl, Cr, Se, Sb and Be. Experiments relating to the leaching of these elements from blocks into seawater (Be was only investigated in the suspended solid sampled with seawater) were undertaken.

In general, ash blocks showed a higher level of trace metals than did concrete. Looking at the overall chemical analysis, a few generalities can be clearly demonstrated.

(1) There was a clear increase in the concentration of Na within the blocks, dependent on the period submerged, the effect of continual soaking by seawater.
(2) An increase, even though moderate, in the Mg level was noted, as was expected following the process of dolomitization, previously seen in laboratory tests.
(3) An increase in the Mn level was noted, dependent on the period of time submerged; there is no obvious explanation for this phenomenon.
(4) No differences in chemical composition were observed between block sides.

In all the chemical element studies, no variations in concentration were measured that could not be explained by the heterogeneity of the samples and the analytical techniques used.

These data relating to the water phase do not show any appreciable leaching (in the order of ppb) compared to the intrinsic variability of the analytical methods used for the elements As, Cd, Cr, Fe, Mn, Sb, Se and Tl (Relini *et al.*, 1995b; Sampaolo and Relini 1993, 1994).

Biological monitoring

Biological studies focused on the colonization patterns of benthic organisms (in particular the aim was to highlight whether coal ash was selected preferentially by the settling larvae of sessile organisms) and the subsequent biological community development (Relini and Patrignani 1992, 1993; Zamboni and Relini, 1992). Investigations were also carried out to assess the possibility that block heavy metals might be bioaccumulated by the colonizing benthic macrofauna.

Colonization of block surfaces

Biological settlement was assessed on six sides of each cube according to the following parameters:

(1)   species richness (number of species per surface unit);
(2)   covering index (see below);
(3)   dry and wet weight of sessile macrofauna;
(4)   density (number of individuals per dm^2).

The six sides of each block (Fig. 2F) were photographed, and then the macro-benthos described and recorded using a microscope. Each systematic group was assigned a value of covering index. Covering (C) means the approximate percentage of the substratum surface which was covered as a projection of the total number of individuals of a species, and was assessed for each species.

The covering index (C.I.) is expressed using Braun-Blanquet's scale of the coefficient of abundance–dominance: + (negligible presence), 1 (<5% surface covered), 2 (5–25%), 3 (25–50%), 4 (50–75%), 5 (>75%).

Wet weight, dry weight (100°C) and ash-free dry weight (500°C) were measured. In order to study the biological communities which formed on the ash and cement blocks the work schedule involved quarterly checks on the blocks immersed for 3, 6, 9, 12, 15, 18, 21 and 24 months.

The main groups identified were algae, protozoans, sponges, hydroids, serpulids, spirorbids, bivalves, gastropods, cirripedes, amphipods, encrusting bryozoans, non-encrusting bryozoans and ascidians. The main and heavy settlement occurred during summer, with maximum values in August and high values in July and September.

Over the 2-year period biological settlement on the ash blocks proved to be greater in quantity (biomass) and better in quality (number of species) than that on the concrete blocks. On ash blocks 63 taxa were found, compared to 53 on concrete (Table 2). Algae, together with serpulids were the dominant components on the ash blocks during the whole cycle of observations; on the other hand, Corallinaceae and spirorbids peaked in 9-month-old communities, decreasing in the 12-month community analysis. Sponges, gastropods and bivalves were always represented in low numbers, while ascidians, which showed a reasonable settle-ment on substrata immersed for 3 months (3 M June '90), decreased in number as the experiment continued. Colonies of encrusting bryozoans were always present, with a less marked seasonal fluctuation than seen in non-encrusting bryozoans.

Communities on the concrete blocks had fewer species than the ash blocks. Algae, Corallinaceae, serpulids and spirorbids were the dominant components: sponges were represented in lower numbers than on the ash blocks. Gastropods and bivalves settled in small quantities, while the settlement of encrusting

Table 2. List of taxa found on the blocks of ash and concrete exposed at Torvaldaliga trial plant.
*indicates presence.

	Ash	Concrete
Macroalgae	*	*
Corallinaceae	*	*
Protozoa Foraminifera	*	*
Protozoa Folliculinidae	*	*
Sycon spp.	*	*
Leucosolenia spp.	*	*
Laomedea calceolifera (Hincks, 1871)	*	*
Platyhelmintha	*	
Serpula concharum (Langerhans, 1880)	*	*
Serpula vermicularis (Linnaeus, 1767)	*	*
Hydroides elegans (Haswell, 1833)	*	*
Hydroides spp.	*	*
Vermiliopsis striaticeps (Grube, 1862)	*	*
Pomatoceros triqueter (Linnaeus, 1767)	*	*
Pomatoceros lamarckii (Quatrefages, 1865)	*	*
Spirobranchus polytrema (Philippi, 1844)	*	*
Protula spp.	*	*
Josephella marenzelleri Caullery-Mesnil, 1896	*	*
Filograna spp.	*	*
Pileolaria militaris (Claparéde, 18709)	*	*
Pileolaria pseudomilitaris (Thiriot-Quiévreux, 1965)	*	*
Janua pseudocorrugata (Bush, 1904)	*	*
Polychaeta errantia	*	*
Vermetus triquetrus (Ant. Bivona, 1832)	*	*
Osilinus turbinatus (Monodonta) (Von Born, 1778)	*	*
Bittium reticulatum (Da Costa, 1778)	*	*
Fissurella nubecula (Linnaeus, 1758)	*	*
Patella ulysipponensis Gmelin, 1791	*	*
Mytilus galloprovincialis Lamarck, 1819	*	*
Ostrea edulis Linnaeus, 1758	*	*
Anomia ephippium Linnaeus, 1758	*	*
*Modiolarca subpicta (*Cantraine, 1835)	*	
Hiatella spp.		
Chama gryphoides Linnaeus, 1758	*	*
Clamys varia Linnaeus, 1758	*	
Balanus perforatus Bruguiére, 1789	*	*
Balanus trigonus Darwin, 1854	*	
Pantopoda	*	
Amphipoda	*	
Isopoda	*	
Walkeria uva (Linnaeus, 1758)		*
Bowerbankia gracilis (Leidy, 1855)	*	*
Aetea spp.	*	*
Cabera boryi (Auduoin, 1826)	*	*
Bugula neritina (Linnaeus, 1758)		*
Bugula stolonifera (Ryland, 1960)	*	
Watersipora subovoidea (D'Orbigny, 1852)	*	*
Watersipora complanata (Norman, 1864)	*	*
Schizobrachiella sanguinea (Norman, 1868)	*	*
Schizoporella errata (Waters, 1878)	*	*
Schizoporella longirostris Hincks, 1886	*	*
Schizoporella unicornis (Johnston in Wood, 1844)	*	

Continued on next page

Table 2.   Continued

	Ash	Concrete
Schizoporella mutabilis Calvet, 1927	*	*
Schizoporella spp.	*	
Microporella marsupiata (Busk, 1860)		*
Chorizopora brogniartii (Audouin and Savigny, 1826)	*	*
Celleporina spp.	*	*
Cellepora spp.	*	*
Turbicellepora magnicostata (Barroso, 1919)	*	
Turbicella crenulata (Hayward, 1978)	*	*
Crisia spp.	*	*
Echinodermata		*
Didemnum granulosum (Von Drasche, 1833)	*	
Ciona intestinalis (Linnaeus, 1767)	*	*
Ascidiella aspersa (Müller, 1776)	*	
Botryllus schlosseri (Pallas, 1776)	*	*
other Ascidiacea (Piuridae)	*	
Total	63	53

bryozoans was greater on the control blocks especially in 6-month-old communities. Colonies of non-encrusting bryozoans were particularly evident on blocks submerged for 3 months, but ascidians were missing although abundant on the corresponding ash substratum.

The increase in biological community over time was more evident and gradual for the ash blocks than for concrete, at least up to 15 months' immersion (Tables 3, 4; Fig. 4).

Biomass measurements confirmed the qualitative and quantitative differences expressed in the cover indices between the communities which settled on the ash blocks and those on the concrete cubes. During this stage the accumulation of fouling (total wet weight but also confirmed by the dry weight and the ash-free dry weight) was greater on the ash-based substrata for all periods of immersion except on the blocks exposed for 21 months (Fig. 4), where biomass was greatest on concrete blocks. On substrata immersed for 24 months (Fig. 5) there was a definite recovery in the ash block biomass values; these values are similar to those obtained after the first year of study.

There were concerns that the test rig itself might influence the settlement of macrobiota on the test blocks as the long pipe runs gave ample opportunity for larval settlement. In order to ascertain the possible selectivity of the plant (pumps, long tubes) with regard to settlement in the trial tanks, test panel cages used as references were positioned at four different locations (Fig. 2.A,B,C,D).

The settlement showed a clear qualitative and quantitative reduction from station A to B and subsequently from C (tank with concrete blocks) to D (tank with ash blocks). This was evident from the decrease in the values of covering indices for the corresponding periods of exposure as well as values for species richness and biomass (Relini *et al.*, 1995b; Sampaolo and Relini, 1994).

Table 3. Coverage indices of epibiota on different sides of ash blocks after 3, 6, 9, 12, 15, 18, 21 and 24 months immersion. (Key: 1 = <5%; 2 = 5–25%, 3 = 25–50%; 4 = 50–75%, 5 > 75%)

	3M JUNE 1990					6M SEPT. 1990					9M DEC. 1990					12M MAR. 1991					15M JUNE 1991					18M SEPT. 1991					21M DEC. 1991					24M MAR. 1992				
	a	b	d	e	f	a	b	d	e	f	a	b	d	e	f	a	b	d	e	f	a	b	d	e	f	a	b	d	e	f	a	b	d	e	f	a	b	d	e	f
Microalgae	4	4	4		2	4	3	1		5	1	4	3			4	2	+		3	2	3	+		3	3	4	5		1	5	5	+		5	5	5	5		
Macroalgae	+	1				+	+				5	4	5		1	3	3	1		5	4	3	+		5	4	3	5		1	3	4	+		5	4	4	5		1
Corallinaceae											4	4	5			5	2	2		3	2	4	2		2	3	5	5		2	4	5	+		5	4	5	4		2
Protozoa																				+					1					+					+	+	+	+	+	+
Porifera	+				+	1	1	1	2	+																+					+									1
Hydroidea																																				+	+			
Serpulidae	2	2	1	4	4	3	2	3	2	2	3	3	1	4	4	4	3	5	4	1	3	2	3	3	3	1	1	+	4	4	2	1	4	4	1	2	1	+	4	3
Spirorbidae	2	1	2	+	2	3	3	3	3	3	2	2	1	2	3	3	2	2	3	2	1	1	2	1	1	2	1	+	3	3	1	1	4	2	+	3	2	+	2	2
Polychaeta errantia																										+												+		
Bivalvia	+	+				1	+	1			1		+	1		+	1	3			2	+				2	+	+	1		1	2				+		+	1	2
Gasteropoda						1	2	1	+	2	+		2	+		+	+	+	+	2	1	+	+	+	2	1	+	2	1		1	+				1	1		+	
Cirripedia	+	1	+	+	+	+			+	+					+					+					+															+
Amphipoda																									1	+									+	+	+		+	+
Bryozoa (encrusting)	+	+	1	3	1	2	+	2	1	2	2	+		2	1	1	+	2			2	1	2	3	1	1			+	2	1	3			2					
Bryozoa (non-encrusting)	+	+	2	1		4	4	4	5	4	+	+	+	+	+	+																								
Ascidiacea	1	1	1	+	2																				1			+	+											2

Table 4. Coverage indices of epibiota on different sides of concrete blocks after 3, 6, 9, 12, 15, 18, 21 and 24 months immersion. (Key: 1 = <5%; 2 = 5–25%, 3 = 25–50%; 4 = 50–75%, 5 > 75%)

	3M JUNE 1990					6M SEPT. 1990					9M DEC. 1990					12M MAR. 1991					15M JUNE 1991					18M SEPT. 1991					21M DEC. 1991					24M MAR. 1992				
	a	b	d	e	f	a	b	d	e	f	a	b	d	e	f	a	b	d	e	f	a	b	d	e	f	a	b	d	e	f	a	b	d	e	f	a	b	d	e	f
Microalgae	3	4	4		2	4	3	1		4	1					2	3	1	+	3	2	2	2		1	5	5	4		3	5	3	3		+	5	5	5		1
Macroalgae	1	1	1		1	2	2	1		3	3	4	3		1	2	3	1		4	4	3	+	+	5	4	4	5		2	4	+	+			4	2	5		
Corallinaceae											2	4	4		2	3	3	2		2	3	3	2		4	4	4	5		2	4	3	2		1	4	4	4	+	+
Protozoa																							+	+	+				+	+					+				+	+
Porifera	+				+	1					+	+	+											+	+	+	1	+	1	+	+	2	2	+	2			+	+	1
Hydroidea																									+															
Serpulidae	1	3	2	2	3	1	2	3	3	1	3	3	2	3	3	2	1	2	4	1	2	1	3	4	2	+	1	+	3	1	2	3	3	3	3	2	1	1	4	4
Spirorbidae	+	+	1	+	2	1	2	3	1	2	2	2	1	1	3	2	1	1	1	1	2	2	2	2	2	1	2	1	2	3	4	3	3	4	4	2	1	1	4	4
Polychaeta errantia																				+					+	1														
Bivalvia	+		+		+	1	1				1			+	+					+	+			+	+	2				+										
Gasteropoda						1				+	1	1			+	+		+	+	1	1	+	+	+	1	1		2			+	1	1			+	1	1		+
Cirripedia	+		+			+			+		+													+				+			+	+	+		2	1	1	+		
Amphipoda																					+	+	+	+										+					+	
Bryozoa (encrusting)	1	+	1	2	1	2	2	2	4	+	3	1	2	4	3	2	+	2		3	1	3	2	3	1	3	+	+	1	1	2	3	4		2	2				4
Bryozoa (non-encrusting)	2	3	2			1						+							+					+	2				+											
Ascidiacea	+		+		1									+						1	1		+	+	1				+	+				+						1

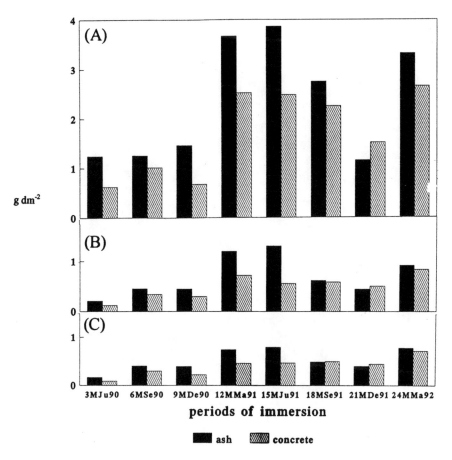

Figure 4. Average of wet (A), dry (B) and ash (C) weights (g dm^{-2}) of the fouling settled on the blocks of two materials.

Bioaccumulation

These studies utilized the naturally settled species *Ostrea edulis* (bivalve), the gastropod *Bittium reticulatum* and some macroalgae. In addition, mussels (*Mytilus galloprovincialis*) were placed on the experimental blocks and examined after 3 and 6 months' exposure. The molluscs were left for 18 h in filtered water to eliminate ingested and non-ingested particles. Samples were stored at –20°C. After lyophilizing the tissue 2.5 g were taken, and incinerated at a low temperature in oxygen plasma supported by radio frequency with the aim of destroying the organic matrix. The samples were taken in solution by means of a treatment in a microwave oven with a mixture of acids using closed Teflon containers. The elements considered were: Al, As, Cd, Cu, Cr, Fe, Pb, Se and Zn; they were analysed at CISE by atomic absorption spectrometry.

Figure 5. Top: Ash blocks submerged for 12 (left) and 24 (right) months. Bottom: Concrete blocks submerged for 12 (left) and 24 (right) months.

No significant differences were found in the tissue content of elements from animals taken from concrete and ash blocks (Table 5).

Stage 3 (1992–)

Following the 3-year programme of experimental research, coal-ash reef modules were placed in the sea to confirm or refute the positive results from the Stage 2 experiment and to assess (at full scale) the acceptability of using coal ash for the construction of artificial reefs in Italian coastal waters.

After an initial 12-month delay, field investigations into the effectiveness and acceptability of coal-ash artificial reef modules are progressing at sites in the Adriatic Sea and in the Ligurian sea.

Table 5. Metal concentrations (μ g^{-1} dry weight) in algae, mussels, gastropods (*Bittium reticulatum*) and oysters after different periods of exposure.

			Al	As	Cd	Cr	Cu	Fe	Pb	Se	Zn
Algae	15 months	Concrete	11600	30	0.3	28.3	22	11300	3.9	0.4	251
		Ash	12100	26	0.3	24.1	15	9700	3.3	0.5	356
	18 months	Concrete	2400	21	0.2	10.4	18	3900	18	0.9	312
		Ash	2500	31	0.2	8.5	17	3400	15	0.9	300
	21 months	Concrete	24730	43	0.2	40	19	14660	30	<1.0	275
		Ash	23750	59	0.2	30	24	14530	21	<0.6	232
	24 months	Concrete	12750	26	0.2	20	25	14550	38	0.7	110
		Ash	9050	18	0.2	14	22	11450	26	0.9	170
Mussels	3 months	Concrete	21	33	0.9	20.1	58	325	3.1	3.8	178
		Ash	33	36	0.9	19.1	56	314	3.6	5	220
	6 months	Concrete	65	13	0.4	4.9	18	206	1.3	2.4	309
		Ash	75	16	0.4	5.4	21	226	2.2	3.7	363
	6 months	Concrete	89	17	0.9	5	20	171	0.6	3	246
		Ash	42	19	0.8	3	16	116	0.2	3	216
Bittium	15 months	Concrete	110	4.5	0.03	1.8	16	131	1.2	0.9	137
		Ash	180	3.7	0.03	2.5	17	155	1.5	0.8	138
Oysters	15 months	Concrete	283	30	2.1	0.8	292	463	5.3	2.9	3500
		Ash	398	26	3.1	1.3	315	457	3.5	4.2	3000

Tests in Loano artificial reef
For testing coal ash blocks directly in the natural environment different modules
were immersed at 12 m depth in the Loano artificial reef.

Ash (70.8% fly ash, 6.8% hydrated lime and 22.4% water) and concrete blocks
of different sizes (20 × 20 × 20 cm, 20 × 20 × 40 cm, 100 × 100 × 100 cm) were used
both separately and assembled in various shapes.

In order to study the settlement and development of the macrobenthos, small
cubes (20 × 20 × 20 cm) of ash and cement were checked at quarterly intervals
after 3, 6, 9 and 12 months' immersion from November 1993 to November 1994
and then after 6, 12, 24 months up to November 1995. After retrieval the six sides
of each block were photographed, the macrobenthos was examined using a
microscope and then it was scraped away for weighing. The methods for
description and evaluation of the settlement are as described above.

Colonization of block surfaces
The main taxa found on small blocks (20 cm side) made with two materials are
listed in Table 6. Most are identified at species level. In total 99 taxa were recorded,
75 on ash and 91 on concrete blocks.

Population evolution on substrata submerged for 3, 6, 9, 12 and 24 months is
also described by means of covering indices (Tables 7 and 8) and the biomass
values (Table 9) (Relini et al., 1995a).

Some differences, either in quantitative or in qualitative terms, were found in
the settlement on the cubes made of cement or of ash (see Tables 7, 8, 9); probably
caused mainly by the friability of ash blocks.

The cubes collected after 12 months' (beginning of December) immersion
supported a good algal settlement made up of green and brown algae (attached to
the substratum, especially on side 'f' (Fig. 2F). Animal settlement proved, in
general, to be less than previously observed, particularly in the dominant groups.
Serpulids were reduced to a few individuals on each block side (P. triqueter, J.
marenzelleri, Filograna sp., H. pseudouncinata, S. polytrema, Hydroides spp.) and
there were a few colonies of small-sized encrusting bryozoans (T. magnicostata, S.
sanguinea, W. complanata, S. errata, U. ovicellata, S. longirostris), except on surface
'e', which was almost completely covered by them. On all block sides there were
hydroids (C. hemisphaerica, B. ramosa) and few non-encrusting bryozoans (Aetea
spp.). Also present were bivalves (A. ephippium, H. arctica) and on side 'd' three
large individuals of O. edulis. The presence of small-sized cirripedes (B. trigonus)
on two sides of the block were noted. There was a notable difference in the
settlement of serpulids between the ash and the cement cubes; the poor settlement on the ash contrasts with a population rich in species and individuals on the
control cube.

The 24-month blocks were covered by a good layer of algae, in particular face
'd' was settled by brown algae. Serpulids (P. triqueter, S. polytrema, S. concharum
and some species of Hydroides) and encrusting bryozoans (T. magnicostata) occurred
on all the sides of the cubes, hydroids in lesser numbers. There were also some
barnacles (B. perforatus and B. trigonus) while ascidians were lacking. There were

Table 6. List of taxa found on concrete (C) and ash (A) blocks immersed in the Loano Artificial reef for two years. *indicates presence.

	Ash	Concrete
Macroalgae	*	*
Algal film	*	*
Corallinacea	*	
Protozoa Foraminifera	*	*
Protozoa Folliculinidae	*	*
Sycon sp.	*	*
Other Porifera	*	*
Bougainvillia sp.	*	*
Bougainvillia ramosa (Van Beneden, 1844)	*	*
Tubularia crocea Agassiz 1862		*
Eudendrium sp.	*	*
Eudendrium armatum Tichomiroff, 1887	*	
Eudendrium capillare Alder, 1856		*
Obelia sp.	*	*
Obelia dichotoma (Linnaeus, 1758)	*	*
Obelia bidentata Clarke, 1875	*	
Clytia hemisphaerica (Linnaeus, 1767)	*	*
Clytia linearis (Billard, 1904)		*
Sertularella gaudichaudi (Lamouroux, 1824)	*	*
Sertularella polyzonias (Linneus, 1758)		*
Plumularia setacea (Linnaeus, 1758)		*
Other Hydrozoa (Campanulinidae)	*	*
Madreporaria		*
Pomatoceros triqueter (Linnaeus, 1767)	*	*
Pomatoceros lamarckii (Quatrefages, 1865)	*	*
Serpula concharum (Langerhans, 1880)	*	*
Serpula vermicularis (Linnaeus, 1767)	*	*
Spirobranchus polytrema (Philippi, 1844)	*	*
Semivermilia cribrata (O.G. Costa, 1861)	*	
Vermiliopsis striaticeps (Grube, 1862)		*
Hydroides sp.	*	*
Hydroides elegans (Haswell 1883)	*	*
Hydroides stoichadon Zibrowius, 1971	*	*
Hydroides pseudouncinata Zibrowius, 1971	*	*
Hydroides helmata Iraso, 1921	*	*
Josephella marenzelleri Caullery-Mesnil, 1896	*	*
Filograna sp.	*	*
Janua sp.	*	*
Lepidonotus clava (Montagu, 1808)		*
Eunice harassii (Audouin and Milne-Edwards, 1834)		*
Lysidice ninetta (Audouin and Milne-Edwards, 1833)		*
Nereis sp.	*	*
Other Polychaeta	*	*
Anomia ephippium Linnaeus, 1758	*	*
Mytilus galloprovincialis Lamarck, 1819	*	*
Mytilaster minimus (Poli, 1795)	*	*
Hiatella sp.	*	*
Hiatella rugosa (Linnaeus, 1767)		*
Hiatella arctica (Linnaeus, 1767)	*	*
Modiolarca subpicta (Cantraine, 1835)	*	*
Ostrea edulis Linneaus, 1758	*	*
Cardium sp.	*	*

Continued on next page

Table 6.   Continued

	Ash	Concrete
Chama gryphoides Linnaeus, 1758	*	*
Chlamys sp.		*
Arca noae Linnaeus, 1758	*	*
Gasteropoda	*	*
Nudibranchia	*	
Balanus sp.	*	*
Balanus perforatus Bruguiére, 1789	*	*
Balanus trigonus Darwin, 1854	*	*
Balanus eburneus Gould, 1841		*
Amphipoda	*	*
Amphipoda Caprellidae	*	*
Isopoda	*	*
Pycnogonida	*	*
Macrura		*
Brachyura	*	*
Cryptosula pallasiana (Moll, 1803)	*	*
Schizobrachiella sanguinea (Norman, 1868)	*	*
Schizoporella errata (Waters, 1848)	*	*
Schizoporella longirostris (Hincks, 1886)	*	*
Turbicellepora magnicostata (Borroso, 1919)	*	*
Umbonula ovicellata Hastings, 1944	*	*
Conopeum reticulum (Linnaeus, 1767)	*	*
Callopora lineata (Linnaeus, 1767)		*
Pentapora ottomulleriana (Moll, 1803)	*	*
Pentapora fascialis (Pallas, 1766)	*	*
Watersipora complanata (Norman, 1864)	*	*
Microporella ciliata (Pallas, 1766)	*	*
Parasmittina sp.		*
Lichenopora radiata (Audouin, 1826)		*
Bugula sp.		*
Bugula stolonifera Ryland, 1960		*
Aetea spp.	*	*
Nolella gigantea Busk, 1856	*	*
Scruparia chelata (Linnaeus, 1758)	*	
Scruparia ambigua (D'Orbigny, 1841)	*	*
Scrupocellaria scruposa Linnaeus, 1758		*
Savygniella lafontii (Audouin, 1826)	*	
Caberea boryi (Audouin, 1826)	*	*
Sertella septentrionalis (Harmer, 1933)		*
Crisia sp.	*	*
Filicrisia geniculata (Milne-Edwards, 1838)		*
Other Bryozoa (Ctenostomata)		*
Diplosoma listerianum (Milne-Edwards H., 1841)	*	*
Didemnum maculosum Milne-Edwards, 1842	*	*
Other Didemnidae	*	*
Other Ascidiacea	*	*
	75	91

Table 7. Covering indices of main taxa settlement on concrete cubes exposed for 3, 6, 9, 12 and 24 months at Loano.

	February 3M 1994						May 6M 1994						August 9M 1994						November 12M 1994						November 24M 1995					
side	a	b	c	d	e	f	a	b	c	d	e	f	a	b	c	d	e	f	a	b	c	d	e	f	a	b	c	d	e	f
Algae							1	+	+	1	1		+	1	1	+	+	+	1	+	3	2			+	2	2			+
Algal film	3	2	2	2			3	3	2	2	1	3	4	3	4	3	2	3	4	3	2	3	4	3	1	2	+	2	1	
Protozoans							+	+	+				+	+		+			+	+	+	+								
Sponges																														
Hydroids	1	1	1	1	2	1	+	+		+			+	+			+		+			1	1	+	+	+	+	+	1	+
Serpulids	2	2	1	2	4	2	5	5	3	5	5	5	4	4	2	4	3	4	4	4	2	3	3	4	4	4	2	3	4	3
Bivalves				+			1	+	+	1	1	+	+	1			+	1	+	+			1	+	+	1	2	+	1	1
Barnacles	+	+		1			1	+	+		+		1	1	1	1	1		+			+	1	+	1	+	+	1	+	
Bryozoans (encrusting)	1	+			1	1	4	3	1	4	3	3	3	3	2	3	3	3	2	3	1	3	4	4	3	3	2	3	4	4
Bryozoans (non-encrusting)		+					+	+	+	1	1	+	1	1	1	1	1	+	1	+	1	1	+	+	+					
Ascidians				+									1		3								+		+	+	1			+

Table 8. Covering indices of main taxa settlement on ash cubes exposed for 3, 6, 9, 12 and 24 months at Loano.

	February 3M 1994						May 6M 1994						August 9M 1994						November 12M 1994						November 24M 1995					
side	a	b	c	d	e	f	a	b	c	d	e	f	a	b	c	d	e	f	a	b	c	d	e	f	a	b	c	d	e	f
Algae		+					1	1	1	1		+	2	1	3	2		2	3	2	1	3		4	1	1	2	4		+
Algal film	2	2	2	2			2	3	2	3	3	3	3	3	4	3	2	1	2	+	2	1	1	1	+	+	1	1	2	1
Protozoans							+	+		+	1	+	+	+					+	+	+	1			+	+	+			+
Sponges										+						+													+	+
Hydroids	1	+	1		+	+	1				+	1	1	+					+	1	+	1	1	1	+	+	+	+	+	+
Serpulids	1	+	1	+	2	1	1	+	2	1	3	1	4	3	2	3	3	3	1	1	1	1	1	1	2	1	1	1	1	2
Bivalves				+			+	+	1	+	+	+	+	+		+	1	+	+	1	2	1		1	1	+	2	+		
Barnacles			+				+	+					1	1	1	+	1	+	+					1	1	+		1	1	1
Bryozoans (encrusting)		+					2	1	1	1	3	2	3	4	2	3	3	4	1	2	2	1	4	1	3	2	2	1	2	2
Bryozoans (non-encrusting)							+	+	1	+	1	+	1	1	1	+	+	+	+	+	+	+	+	+	+	+	+	+	+	
Ascidians													1						1								2			

no substantial differences between the two materials in species richness but concrete cubes were much more heavily covered by serpulids and bryozoans than those of ash.

There was no substantial difference in the settlement on other reef modules but the overall shape of the reef units was important in attracting fish.

Biomass development on small cubes was higher on the concrete than on the ash blocks (Table 9).

Table 9. Total wet weight, dry weight and ash weight in g (average between two blocks) of the biomass settled on blocks immersed at Loano for 3, 6, 9, 12 and 24 months..

Duration and month	Wet weight	Dry weight	Ash weight
3m Feb. 1994			
ash	7.03	2.48	2.27
concrete	31.34	13.58	12.10
6m May 1994			
ash	96.52	47.01	44.34
concrete	313.19	165.44	152.45
9m Aug. 1994			
ash	293.72	198.47	142.37
concrete	323.55	164.17	151.18
12m Nov. 1994			
ash	133.35	64.10	49.65
concrete	340.4	200.07	160.58
24m Nov. 1995			
ash	245.75	109.15	100.05
concrete	664.20	369.96	340.59

Conclusions from the Loano tests

No evidence of bioaccumulation of 10 elements of environmental interest was found from samples of mussels, sea urchins and fish (personal communication) after 6 and 12 months' exposure.

The ash blocks used in Loano seem to be composed of a more friable material than that used in Torrevaldaliga experiments, where ash blocks showed better settlement in terms of quality and quantity than did concrete blocks (Relini and Patrignani, 1992, 1993; Sampaolo and Relini 1994). In the Loano experiments the opposite results were obtained and it is believed that friability of the surface strongly influenced the settlement. Nevertheless it can be confirmed that coal-ash stabilized material may be used advantageously for artificial fish habitats without any environmental problem; the friability of the surface is a technical problem which is not difficult to solve.

Conclusions

Results show that coal ash may be used as a material for artificial reef blocks without giving rise to any specific technical or environmental problems. The physical and mechanical results demonstrate that neither distortions in form, nor swelling or even deterioration, were noted. Importantly, a progressive increase in compressive strength values and sonic velocity was seen during immersion in the Torrevaldaliga tests.

Periodic sampling for some chemical elements observed did not show any appreciable leaching given the intrinsic variability of the analytical method used.

The diffractometric tests performed on core sections drawn from blocks immersed in seawater at varying exposure intervals did not produce diffractograms that differed significantly from those obtained prior to immersion. No bioaccumulation of metals was found in the animals monitored in both the Torrevaldaliga and Loano experiments.

The lower level of biological colonization on ash blocks tested in Loano was caused by the friability of the material. In Stage 2 tests (Torrevaldaliga) species diversity on the ash-based blocks was higher than on those made of concrete, and biomass expressed in wet, dry and ash-free dry weight was always higher on the ash blocks than on concrete blocks. The ash-based material is therefore considered to be more suitable than concrete for the settlement of benthic organisms, when the problem of friability is solved.

References

Chen, G. 1987. Feasibility of using coal ash for artificial reef application. Taiwan Power and Light Company, Technical Report No. 088-2.

Collins, K.J., A.C. Jensen and A.P.M. Lockwood. 1990. Fishery enhancement reef building exercise. *Chemistry and Ecology*. **4**: 179–187.

Collins, K.J., A.C. Jensen and A.P.M. Lockwood. 1991. Artificial reefs: using coal-fired power station wastes constructively for fishery enhancement. *Oceanologica Acta*. **11**: 225–229.

Collins, K.J., A.C. Jensen and A.P.M. Lockwood. 1994a. Coastal structures, waste materials and fishery enhancement. *Bulletin of Marine Science*. **55**(2–3): 1240–1250.

Collins, K.J., A.C. Jensen and A.P.M. Lockwood. 1994b. Evaluation of stabilised coal-fired power station waste for artificial reef construction. *Bulletin of Marine Science*. **55**(2–3): 1251–1262.

Jensen, A.C., K.J. Collins and A.P.M. Lockwood. 1994. Colonization and fishery potential of a coal-ash artificial reef, Poole Bay, United Kingdom. *Bulletin of Marine Science*. **55**(2–3): 1263–1276.

Kuo, S.T., T.C. Hsu and K.T. Shao. 1995. Experiences of coal ash artificial reefs in Taiwan. *Chemistry and Ecology*. **10**: 233–247.

Relini, G. and L. Orsi Relini. 1989. The artificial reefs in the Ligurian Sea (N-W Mediterranean): Aims and Results. *Bulletin of Marine Science*. **44**(2): 743–751.

Relini, G. and A. Patrignani. 1992. Coal ash for artificial habitats in Italy. *Rapp. Comm. int. Mer. Médit.* **33**: 378.

Relini, G. and A. Patrignani. 1993. Accumulo della biomassa su blocchi a base di cenere di carbone. Atti 230 Congresso SIBM, Ravenna, 1992. *Biologia Marina*, suppl. Notiziario SIBM. **1**: 189–193.

Relini, G., M. Relini, G. Torchia, F. Tixi and C. Nigri. 1995a. Coal ash tests in the Loano artificial reef. *Proceedings of ECOSET '95*. Japan International Marine Science and Technology Federation. **1**: 107–118.

Relini, G., A. Sampaolo and G. Dinelli. 1995b. Stabilized Coal Ash Studies in Italy. *Chemistry and Ecology*. **10**: 217–231.

Sampaolo, A. and G. Relini. 1993. Environmental implications of coal use for artificial reef construction. *Proceedings of the Tenth International Ash Use Symposium*. EPRI TR 101774 **1**(28): 1–11.

Sampaolo, A. and G. Relini. 1994. Coal ash for artificial habitats in Italy. *Bulletin of Marine Science*. **55**(2): 1279–1296.

Shao, K.T., H.I. Peng, S.T. Kuo and L.S. Chen. 1994. Evaluation of the effectiveness of the coal ash reefs in Wan-li, Northern Taiwan, Republic of China. Fifth International Conference on Aquatic Habitat Enhancement, Long Beach, California 1991. Summary in *Bulletin of Marine Science*. **55**(2–3): 1352.

Suzuki, T. 1985. A concept of large artificial ridges using a new hardened product made from coal ash. In Kato, W. (ed.) *Ocean Space Utilization '85*. Springer Verlag, Tokyo, pp. 611–618.

Suzuki, T. 1995. Application of high-volume fly ash concrete to marine structures. *Chemistry and Ecology*. **10**: 249–258.

Woodhead, P.M.J., J.H. Parker and I.W. Duedall. 1982. The Coal-Waste Artificial Reef Program (C-WARP): A new resource potential for fishing reef construction. *Marine Fisheries Review*. **44**: 16–23.

Woodhead, P.M.J., J.H. Parker and I.W. Duedall. 1985. The use of by-products from coal combustion for artificial reef construction. In D'Itri, F.M. (ed.) *Artificial Reefs: Marine and Freshwater Applications*. Lewis Publishers, Michigan, pp. 265–292.

Zamboni, N. and G. Relini. 1992. Macrofouling su manufatti a base di ceneri di carbone. *Oebalia*. **17** (suppl.): 433–434.

22. Effects of Artificial Reef Design on Associated Fish Assemblages in the Côte Bleue Marine Park (Mediterranean Sea, France)

ERIC CHARBONNEL[1], PATRICE FRANCOUR[2], JEAN-GEORGES HARMELIN[3], DENIS ODY[4] and FRÉDÉRIC BACHET[5]

[1]Gis Posidonie, Faculté des Sciences de Luminy, 13288 Marseille, Cedex 9 France; [2]Gis Posidonie, Laboratoire Environnement Marin Littoral, Faculté des Sciences, Université Nice-Sophia Antipolis, Parc Valrose, 06108 Nice Cedex 2, France; [3]Centre d'Océanologie de Marseille, Station Marine d'Endoume, CNRS UMR Dimar, 13007 Marseille, France; [4]890, Carraire de Salins, 13090 Aix-en-Provence, France; [5]Parc Régional Marin de la Côte Bleue, Maison de la mer, BP 37, 13960 Sausset, France.

Introduction

The Côte Bleue marine park, located east of Marseille (Fig. 1) was created in 1983 with the objectives of promotion (increasing public awareness) of the marine system, protection of the natural environment and the preservation of traditional fisheries. An artificial reef programme was initiated in 1983, with the deployment of 225 m³ of experimental artificial reefs, made from bricks and breeze blocks. Since 1985 the programme has continued using industrially made concrete reefs. Two types of small cubic modules (Fig. 2) have been used: 1.7 m³ (390 units) and 2 m³ (90 units). These CARU modules (cubic artificial reef units) have been arranged in piles of 20–120 m³ (Fig. 3). Eight 158 m³ volume modules (LARU; large artificial reef unit, Figs. 2 and 3) were individually placed near the CARU reefs. The whole programme comprised 2332 m³ of artificial reef deployed in three sites of the marine park. Most of these reefs are open to fishing. This programme was financed by several organizations (UE-FEOGA, France, Région Provence-Alpes Côte d'Azur, Département des Bouches-du-Rhône and towns associated with the marine park management). Investments made between 1985 and 1989 reached around FF 1 700 000 (average reef cost: FF 510 m⁻³).

The objectives of these reefs were:

(1) to increase the biological production in impoverished marine areas (sandy seabeds);
(2) maintenance and promotion of artisanal coastal fishing.

This programme was complemented by the deployment of heavy reefs designed to protect against illegal trawling within the three nautical mile offshore limit (83 rock blocks each weighing 12 tonnes deployed in 1986, 100 sea rock modules (each 1.6 m³) placed in 1990 and 91 modules made with concrete telegraph poles deployed in 1997 (each module weighing 10 tonnes and having a volume of 13 m³)).

These anti-trawling reefs are located along an 11 km stretch off the Côte Bleue. Since they have been installed, a significant decrease in illegal trawling has been noted (Bachet, 1992).

A.C. Jensen et al. (eds.), Artificial Reefs in European Seas, 365–377
© 2000 Kluwer Academic Publishers. Printed in Great Britain.

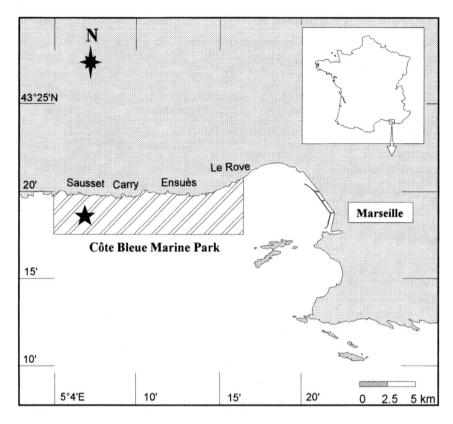

Figure 1.   Location of the Côte Bleue Marine Park (Mediterranean Sea, Bouches-du-Rhône, France) and of the artificial reefs studied (indicated by a star).

The aims of this paper are to compare the fish assemblages associated with both CARU and LARU reef types 4 years after their deployment (Charbonnel and Francour, 1994), and to describe the trends in the development and maturity of the fish assemblage through comparisons with data recorded in 1986 (Ody, 1987) on the same reef types.

Materials and Methods

Reefs Studied

The study site is located off Sausset; it involved 1321 m³ of LARU and CARU artificial reefs deployed at a depth of 27–30 m on sandy areas strewn with rocky boulders, near the lower limit of *Posidonia oceanica* seagrass beds. Both LARU and CARU reef types were surveyed in 1986 (November 1985–January 1987) and 1993 (July–December 1993):

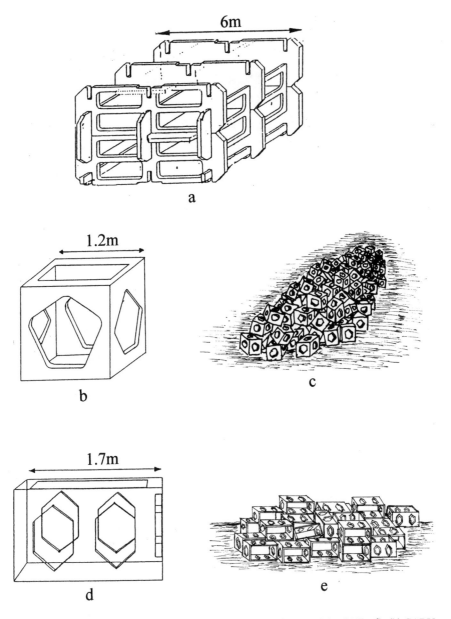

Figure 2. Reef modules studied using scuba: (a) LARU reef (one module of 158 m³); (b) CARU reef (modules of 1.7 m³) and (c) CARU reef (chaotic heap of 70 modules) studied in 1993; (d) CARU reef studied in 1986 (2 m³) and (e) the 30 module CARU reef (from Charbonnel and Francour, 1994).

Figure 3. Scientific diver working on the two types of reef, (a) CARU modules and (b) the LARU module (photography: E. Charbonnel).

- 1986: 1 LARU reef (1 module of 158 m³), 1 CARU reef (30 modules each 2 m³).
- 1993: 1 LARU reef (1 module of 158 m³), 1 CARU reef (70 modules each 1.7 m³).

Data Collection

Data were collected using SCUBA diving and visual census (Francour, 1990; Harmelin-Vivien and Harmelin, 1975; Harmelin-Vivien *et al.*, 1985) adapted to the specificity of artificial reefs (Charbonnel *et al.*, 1995, 1997). For each type of reef, 17 censuses were undertaken in 1993 during summer (July–August) and late autumn (November–December). The fish assemblages were characterized according to:

(1) Species composition, species richness (total number of species found on each reef type), mean species richness (mean number of species per census).
(2) Frequency of species occurrence. Four classes were considered: class I (75–100% presence: permanent species), class II (50–75%: frequent species), class III (25–50%: scarce species), class IV (<25%: rare species). The relative importance of each class in the fish assemblage is an estimate of its temporal variability and is a good indication of assemblage stability.
(3) Trophic structure of the assemblages. Five trophic categories were considered according to the classification proposed by Bell and Harmelin-Vivien (1983): Ma = Macrocarnivore; Mi = Microcarnivore; M1 = Mesocarnivore 1; M2 = Mesocarnivore 2; He = Herbivore.
(4) Abundance: the number of individuals in groups larger than 10 individuals was estimated according to six abundance classes: 11–30; 31–50; 51–100; 101–200; 201–500; >500 (Harmelin-Vivien and Harmelin, 1975).
(5) Size of individuals: the total length of individuals was referred to in 3 size categories: small (S), medium (M) and large (L). The size ranges of these categories were adapted to each species; they corresponded respectively to <1/3; 1/3–2/3; >2/3 of the maximal size commonly accepted for each species (Whitehead *et al.*, 1995).
(6) Biomass: calculated from abundance estimates of each size category using a species-specific size–weight relation (Francour, 1990; Ody, 1987).

As artificial reefs have a three-dimensional structure, abundance and biomass data are referred to the reef volume rather than to its surface area, as usual in natural biotopes (Charbonnel *et al.*, 1995, 1997). Density (number of individuals) and biomass (wet weight, in g) data are given per 10 m³ and per m³, respectively.

Data Processing

Comparisons of means were conducted with a non-parametric ANOVA (Kruskal-Wallis test) in order to avoid data transformations in case of heteroscedasticity. If the null hypothesis (equality of means) is rejected, the non-parametric test of Newman-Keuls-Student enables significantly different averages to be identified (Zar, 1984). Before carrying out a factor-by-factor (date or season) analysis, the independence of both factors was tested with a two-factor ANOVA. This last result is not detailed because in all cases, both factors were independent ($P > 0.75$).

Results

Spatial Distribution of Fish Assemblages

The different fish species were not randomly distributed on the artificial reefs. Position has therefore been classified with respect to four main strata related to home range categories of species (Fig. 4, modified from Charbonnel, 1990):

(1) Water column area: located above and around the reef. This mid-water zone was occupied by schools of microcarnivores such as Centracanthidae (*Spicara* spp.), Pomacentridae (*Chromis chromis*) and two Sparidae (*Boops boops, Oblada melanura*). This area was sporadically frequented by transient pelagic species (Carangidae, Scombridae), which may have been attracted by certain physical modifications brought about by the reef structure or by the occurrence of prey aggregating around the reef.

(2) Transit area: the general horizontal area around the reef frequented by high-priced necto-benthic species (mainly sparids), which moved from one reef to another.

(3) Reef area: narrow peripheral zone inhabited by Labridae, Serranidae, and some Sparidae (*Diplodus annularis, D. sargus* and *D. vulgaris*). In most cases, these species were resident and constituted the main component of the permanent fish assemblage of the reef.

(4) Contact area: habitat of benthic species living in close association with the modules (Scorpaenidae, Blenniidae, Gobiidae, Tripterygiidae) and of noctur-nal species which occupied cryptic shelters during daytime (e.g. *Conger conger, Phycis phycis*).

Composition and Species Richness

The whole assemblage recorded on both artificial reef types in 1993 comprised 36 species and 13 families (Table 1). Only three families were represented by more

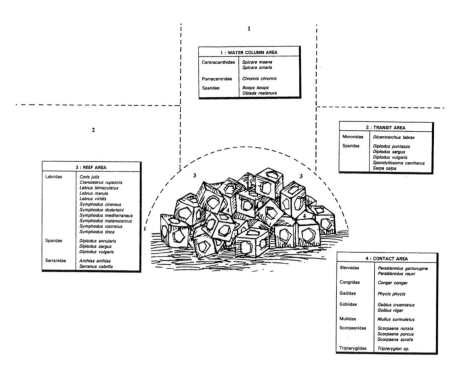

Figure 4. Spatial distribution of fish assemblages during daytime on CARU artificial reef related to home range categories of species (modified from Charbonnel, 1990).

than two species: Labridae (11 spp., >30% of the species pool), Sparidae (8 spp., >22%), and Scorpaenidae (3 spp.).

The global species richness was clearly higher on the multimodular reef (CARU: 35 spp., i.e. 97.2% of the whole species pool versus LARU: 24 spp., 66.7%; Table 1). This result also indicates that the LARU reef had no specific habitat resources and that the assemblage recorded on it was only an impoverished replicate of that sheltered by the CARU reef. The mean number of species per census was also greater on the CARU reef (13.6 ± 0.4 spp. versus 8.0 ± 0.6 spp.).

Frequency of Species

Eleven species on the CARU reef and seven on the LARU reef had a frequency of occurrence greater than 75% (class I). These permanent species represented 33% (LARU) or 38% (CARU) of the total species number (Table 2). The between-reef difference in assemblage stability was shown when the permanent species category was restricted to a 90–100% frequency of occurrence (CARU: 10 spp., LARU: 3 spp.). Frequent and scarce species (Classes II and III) were poorly represented on both reefs (3.5–14%; Table 2), while rare species (Class IV) represented the most important part of both reef assemblages (CARU: 45%, LARU: 52%). This high proportion of rare species was mainly due to two home-range categories:

(1) cryptic and/or inconspicuous species, which can be missed by visual census (e.g. *Conger conger, Phycis phycis, Scorpaena* spp., *Parablennius* spp.);
(2) species with an extended home-range, whose presence on the reef was sporadic, such as some sparids (e.g. *Diplodus puntazzo, Oblada melanura, Sarpa salpa, Boops boops*) or the sea-bass (*Dicentrarchus labrax*).

Table 1. Species composition of ichthyofauna and frequency of occurrence (%) of species sampled in 1993 on CARU and LARU reefs (Charbonnel and Francour, 1994).

| | | Reef type | CARU | LARU |
| | | Volume: unit/reef | 1.7/119 m³ | 158/158 m³ |
Families	Species	Feeding habit		
Blenniidae	*Parablennius gattorugine*	M2	+	+
	Parablennius rouxi	M2	+	+
Centracanthidae	*Spicara maena*	Mi	71%	29%
	Spicara smaris	Mi	71%	29%
Congridae	*Conger conger*	Ma	6%	6%
Gadidae	*Phycis phycis*	Ma	6%	0%
Gobiidae	*Gobius cruentatus*	M2	+	0%
	Gobius niger	M2	+	0%
Labridae	*Coris julis*	M1	100%	100%
	Ctenolabrus rupestris	M1	94%	88%
	Labrus bimaculatus	M1	88%	76%
	Labrus merula	M1	59%	18%
	Labrus viridis	M1	24%	0%
	Symphodus cinereus	M1	6%	0%
	Symphodus doderleini	M1	71%	24%
	Symphodus mediterraneus	M1	94%	53%
	Symphodus melanocercus	M1	100%	100%
	Symphodus rostratus	M1	18%	18%
	Symphodus tinca	M1	100%	88%
Moronidae	*Dicentrarchus labrax*	Ma	0%	18%
Mullidae	*Mullus surmuletus*	M2	24%	24%
Pomacentridae	*Chromis chromis*	Mi	100%	100%
Scorpaenidae	*Scorpaena notata*	Ma	24%	6%
	Scorpaena porcus	Ma	18%	0%
	Scorpaena scrofa	Ma	12%	0%
Serranidae	*Anthias anthias*	Mi	94%	0%
	Serranus cabrilla	Ma	100%	88%
Sparidae	*Boops boops*	Mi	6%	12%
	Diplodus annularis	M2	41%	0%
	Diplodus puntazzo	M2	6%	0%
	Diplodus sargus	M2	100%	18%
	Diplodus vulgaris	M2	100%	18%
	Oblada melanura	M2	6%	0%
	Sarpa salpa	He	6%	6%
	Spondyliosoma cantharus	M2	65%	29%
Triptergiidae	*Tripterygion* sp.	M2	+	+
Total number of species			35	24

+ = sporadically censused species, not considered for the frequency or occurrence.
Feeding habit (from Bell and Harmelin-Vivien, 1983): Ma = Macrocarnivore; Mi = Microcarnivore; M1 = Mesocarnivore 1; M2 = Mesocarnivore 2; He = Herbivore.

Density and Biomass

The density of the total fish assemblage (Table 3) was significantly higher on the CARU reef than on the LARU reef (summer: 27.7 versus 11.8 individuals 10 m^{-3}; Kruskal-Wallis test, $P < 0.05$; late autumn: 58.4 versus 19.7 individuals 10 m^3; Kruskal-Wallis test, $P < 0.05$). Data on biomass (Table 3) indicated the same between-reef difference (Kruskal-Wallis test, $P < 0.05$).

The dominant families in terms of numerical density were Sparidae (CARU reef: late autumn) and Pomacentridae (CARU reef: summer; LARU reef: summer + late autumn, Table 3). Sparidae were dominant in biomass terms (80%) on the CARU reef whereas this family represents only 3–9% of the total biomass on the LARU reef. On this large module, the families which dominated the biomass were Pomacentridae (summer) and Centracanthidae (late autumn, Table 3).

Table 2. Proportion of the 4 classes of occurrence frequency on LARU and CARU reefs sampled in 1993. Missing species (0% presence) in Table 1 were not used in the calculation.

Reef type	LARU	CARU
Volume: unit/reef	158/158 m^3	1.7/119 m^3
Class 1 (75–100% presence)	33.3%	37.9%
Class 2 (50–75% presence)	4.8%	13.8%
Class 3 (25–50% presence)	9.5%	3.5%
Class 1 (<25% presence)	52.4%	44.8%

Table 3. Mean density (number of individuals 10 m^{-3}) and mean biomass (g m^{-3}, wet weight) of the main families censused on artificial reefs of Côte Bleue Marine Park, in summer and late autumn 1993.

	Density				Biomass			
	Summer		Late autumn		Summer		Late autumn	
Families	LARU	CARU	LARU	CARU	LARU	CARU	LARU	CARU
Coris julis	1.02	6.85	1.75	14.40	3.10	14.70	5.50	26.30
Other Labridae	0.83	1.79	1.04	2.17	5.60	16.30	10.70	17.60
Sparidae	0.02	7.68	0.86	22.63	0.90	237.70	8.90	367.20
Serranidae	0.09	0.53	0.20	0.69	1.30	3.30	2.70	3.80
Centracanthidae	0.00	0.06	5.45	2.20	0.00	0.30	38.10	7.30
Pomacentridae	9.78	10.31	10.34	16.11	12.70	17.20	15.80	20.70
Scorpanidae	0.01	0.04	0.00	0.07	0.10	1.30	0.00	1.50
Other species	0.02	0.01	0.09	0.09	3.20	1.00	18.20	3.40
Total	11.77	27.70	19.73	58.36	26.90	291.80	99.90	447.60

Trophic Structure

The two artificial reef types clearly differed in the trophic structure of their fish assemblages. On the LARU reef, the midwater microcarnivores (planctivores) were the dominant trophic category in both sampling periods, on a numerical basis (71% and 91% of the total density) and on a ponderal basis (53% and 57% of the total biomass, Table 4). On the CARU reef, the mesocarnivores 2 (mainly nectobenthic Sparidae) dominated (>85% of total biomass in both seasons; 52% of total density in late autumn; Table 4). However, the microcarnivores did not differ in numerical density and biomass on either of the reef types (Kruskal-Wallis test, $P > 0.15$), contrary to the mesocarnivores 2, which were more abundant on the CARU reef (Kruskal-Wallis test, $P < 0.03$). The three other trophic categories contributed very little to the density or biomass.

Assemblage Changes Between 1986 and 1993

On the LARU reef, the same global species richness was recorded in 1986 and 1993 (24 species), and the assemblage composition was similar (21 species in common). Very few changes were observed between 1986 and 1993 in mean species richness (1986: 7.3 spp. per census; 1993: 8.0 spp. per census) and in density (1986: 0.12 individuals m^{-3}; 1993: 0.14 individuals m^{-3}), but the mean species richness was markedly less variable in 1993 (Table 5). On the other hand, the biomass was noticeably higher in 1993 than in 1986 (21.5 g m^{-3} versus 5 g m^{-3}). As the dominant species were the same in both sampling periods, this increase in biomass can be attributed to an increase of the mean weight of individuals associated with the reef.

On the CARU reef, all assemblage parameters showed higher values in 1993 than in 1986 (Table 5): global species richness (+20%), mean number of species per census (+30%), density (+60%) and biomass (+160%).

Discussion

The fish assemblages associated with the artificial reefs located along the French Mediterranean coasts are similar in species composition, density and biomass to those occurring on natural rocky areas (Ody, 1987). On both natural and artificial habitats, the assemblage is qualitatively dominated by sparid and labrid species (>50% of the species richness) (Charbonnel, 1990; Ody, 1987; Ody and Harmelin,

Table 4. Mean density (number of individuals/10 m^{-3}) and mean biomass (g m^{-3}, wet weight) of the different trophic categories. Mediterranean wrasses (*Coris julis*) are excluded from mesocarnivores 1.

	Density				Biomass			
	Summer		Late autumn		Summer		Late autumn	
Feeding habit	LARU	CARU	LARU	CARU	LARU	CARU	LARU	CARU
Macrocarnivores	0.11	0.24	0.24	0.35	3.82	5.15	19.54	6.00
Mesocarnivores 1	0.83	1.79	1.03	2.17	5.57	16.26	10.67	17.59
Mesocarnivores 2	0.03	7.68	0.91	22.72	1.72	237.27	10.28	369.25
Microcarnivores	9.78	10.71	15.79	18.72	12.74	17.98	53.83	28.64
Herbivores	0.00	0.01	0.00	0.00	0.00	0.42	0.00	0.00

Table 5. Evolution of the ichthyofauna assemblages between 1986 and 1993 on LARU reef (158 m³ module) and CARU reefs (1.7 and 2 m³ modules). Data collected during summer and late autumn 1993 were pooled. In order to compare 1986 and 1993 data of density and biomass, *Coris julis, Boops boops, Spicara* spp. and *Chromis chromis* were not taken into account. Numbers in square brackets [ ] show standard deviation.

	LARU reef		CARU reef	
	1986	1993	1986 (2 m³)	1993 (1.7 m³)
Total number of species	24	24	28	35
Mean species richness	7.3 [1.8]	8.0 [0.55]	10.8 [3.4]	13.6 [0.4]
Density (individuals m⁻³)	0.12	0.14	0.93	1.52
Biomass (g m⁻³)	5.00	21.50	116.00	306.00

1994) but the proportion of resident species is much higher in natural rocky areas (Harmelin, 1987).

In addition to these basic characteristics, the structure of the fish assemblages clearly differed according to the reef type. The comparison undertaken in 1993 between a multimodular CARU reef (70 × 1.7 m³ modules) and a single-chambered LARU reef (158 m³ module) deployed in the same environment confirms that, among the habitat-specific factors, the structural complexity (rugosity sensu Luckhurst and Luckhurst, 1978) of artificial reefs has the most influence on richness and abundance of the fish assemblages (Bohnsack, 1991; Helvey and Smith, 1985; Hixon and Beets, 1989). CARU reefs, which offer numerous, relatively small, interconnected chambers and an extended substratum area, are obviously preferred by demersal fish to the more simply designed reefs with the same gross volume, as seen in LARU reefs. CARU reefs present a greater species richness and much increased density and biomass (×10) than the LARU reefs. The attractiveness of multimodular reefs seems to be enhanced when the modules are arranged chaotically in heaps instead of being regularly stacked and when the internal space is divided according to particular species-specific habitat requirements. Thus, Ody and Harmelin (1994) noted that the density of the two-banded sea bream *Diplodus vulgaris* on a small CARU reef substantially increased (×3–6) after the introduction of an *in situ* increase in complexity of the basal chambers that tended to imitate the specific shelters of this sparid.

Therefore, the reef architecture and module design determine not only the global performance of the reef (species richness, density and biomass) but also the identity of species that are particularly fitted for exploiting this new resource.

The artificial reefs of the Côte Bleue marine park sheltered a higher proportion of residential species than other artificial reefs located in the same region (Charbonnel, 1990: mean = 27%; Ody and Harmelin, 1994: 18% to 25%). However, the proportion of residential species was much higher in natural rocky areas (70% to 84%; Harmelin, 1987).

An increase of the mean individual weight was evident on both reef types (CARU and LARU) when comparing the 1986 and 1993 fish assemblages. This evolution suggests that reef maturation induces an increase in the stock of large adults. This phenomenon may have particular consequences on hermaphrodite species with sexual inversion (Sparidae and Labridae), where the absence of older individuals may correspond to absence of a gender. The increased number of large potential spawners in old artificial reefs is a phenomenon similar to that observed in marine reserves (e.g. Côte Bleue reserve: Harmelin *et al.,* 1995). The same supposed beneficial effect to fish numbers seen in the peripheral areas can be attributed to artificial reefs, the spawning productivity of larger individuals being much higher than that of small ones (Bohnsack, 1993; Buxton, 1993; Francour, 1994; Roberts and Polunin, 1991). Furthermore, it is interesting to note that artificial reefs, by virtue of their physical presence, may develop a refuge function without imposing the need for any particular protective management (e.g. fishing restriction).

The clear differences in species richness, density and biomass observed on the CARU reefs between 1986 and 1993 can be explained either by the age of the reefs (1986 reef: 2 years versus 4 years for 1993 reef) or by the more chaotic spatial disposition of the modules in the reef studied in 1993. The temporal evolution of the LARU reefs was less noticeable and only affected the size of individuals present. Ody (1987) showed that colonization of small reefs by fishes appeared to stabilize after 1–2 years' immersion. The present results suggest that the duration of the assemblage maturation depends on the size and complexity of the reef: colonization of LARU reefs appeared to stabilize rapidly because of their too simple structure (only a very large single chamber with a wide mesh aperture). In more complex reefs (e.g. CARU type), the greater diversity in habitat resources allowed colonization to proceed over a longer period of time, resulting in greater species richness, density and biomass.

Conclusions

Artificial reefs should be designed according to the habitat requirements of fish living in rocky areas from the bathymetric zone concerned by the reef programme (Harmelin and Bellan-Santini, 1987). Large artificial reefs which comprise vast undivided empty chambers have no natural equivalent and are ineffective; evolution of their fish assemblages seems to stop rapidly. Smaller modules arranged in chaotic heaps appear to be ideal for their biological 'efficiency' (species diversity, abundance and size of individuals) as well as for practical reasons (relatively easy to manufacture and to deploy). The present study confirms that design of reefs is a crucial factor in their effectiveness.

Marine reserves and artificial reefs may function as efficient refuges for large adults of over-fished species, and thus can be used as management tools for exploited stocks of some benthic or necto-benthic species. However, artificial reefs should be only one facet of the overall management of coastal resources, which must take into account all phases of the life history of the over-fished species and more especially their spawning areas and nurseries.

References

Bachet, F. 1992. Evaluation des retombées économiques du Parc Régional Marin de la Côte Bleue. In Olivier, J., N. Geradin and A.J. de Grissac. (eds). *Impact Économique des Espaces Côtiers Protégés de Méditerranée*, Ajaccio. MEDPAN Secrétariat publ., pp. 43–46.

Bell, J.D. and M.L. Harmelin-Vivien. 1983. Fish fauna of French Mediterranean *Posidonia oceanica* seagrass meadows. 2. Feeding habits. *Tethys*. **11**(1): 1–14.

Bohnsack, J.A. 1991. Habitat structure and the design of artificial reefs. In Bell, S.S., E.D. McCoy and H.R. Mushinsky. (eds.), *Habitat Structure: The Physical Arrangement of Objects in Space*. Chapman and Hall, London, pp. 412–426.

Bohnsack, J.A. 1993. Marine reserves: they enhance fisheries, reduce conflicts, and protect resources. *Oceanus*. **36**: 63–71.

Buxton, C.D. 1993. Life-history changes in exploited reef fishes on the East coast of South Africa. *Environmental Biology Fishes*. **36**: 47–63.

Charbonnel, E. 1990. Les peuplements ichtyologiques des récifs artificiels dans le département des Alpes-Maritimes (France). *Bulletin Société Zoologique de France*. **115**(1): 123–136.

Charbonnel, E. and P. Francour. 1994. Etude de l'ichtyofaune des récifs artificiels du Parc Régional Marin de la Côte Bleue en 1993. Report. GIS Posidonie publ., Marseille, France, 66 pp.

Charbonnel, E., P. Francour, J.G. Harmelin and D. Ody. 1995. Les problèmes d'échantillonnage et de recensement du peuplement ichtyologique dans les récifs artificiels. *Biologia Marina Mediterranea*. **2**(1): 85–90.

Charbonnel, E., P. Francour and J.G. Harmelin. 1997. Finfish population assessment techniques on artificial reefs in the European Union. In Jensen, A.C. (ed.) *European Artificial Reef Research*. Proceedings of the first EARRN conference, Ancona, Italy, pp. 261–275.

Francour, P. 1990. Dynamique de l'écosystème à *Posidonia oceanica* dans le parc national de Port-Cros. Analyse des compartiments matte, litière, faune vagile, échinodermes et poissons. Doct Univ. Univ. P.M. Curie, Paris. 373 pp.

Francour, P. 1994. Pluriannual analysis of the reserve effect on ichthyofauna in the Scandola natural reserve (Corsica, Northwestern Mediterranean). *Oceanologica Acta*. **17**(3): 309–317.

Harmelin, J.G. 1987. Structure et variabilité de l'ichtyofaune d'une zone rocheuse protégée en Méditerranée (Parc national de Port-Cros, France). *PSZNI Marine Ecology*. **8**(3): 263–284.

Harmelin, J.G. and D. Bellan-Santini. 1987. Modèles naturels pour les récifs artificiels en Méditerranée. Coll. France-Japon. Océanogr. Marseille 16–21 Sept. 1985, **6**: 85–92.

Harmelin, J.G., F. Bachet and F. Garcia. 1995. Mediterranean marine reserves: fish indices as tests of protection efficiency. *PSZNI Marine Ecology*. **16**(3): 233–250.

Harmelin-Vivien, M.L. and J.G. Harmelin. 1975. Présentation d'une méthode d'évaluation *in situ* de la faune ichtyologique. *Trav sci. Parc nation. Port-Cros*. **1**: 47–52.

Harmelin-Vivien, M.L., J.G. Harmelin, C. Chauvet, C. Duval, R. Galzin, P. Lejeune, G. Barnabé, F. Blanc, R. Chevalier, J. Duclerc and G. Lassere. 1985. Evaluation visuelle des peuplements et populations des poissons: problèmes et méthodes. *Revue Ecologie (Terre et Vie)*. **40**: 467–539.

Helvey, M. and R.W. Smith. 1985. Influence of habitat structure on the fish assemblages associated with two cooling-water intake structures in southern California. *Bulletin of Marine Science*. **37**(1): 189–199.

Hixon, M.A. and J.P. Beets. 1989. Shelter characteristics and Caribbean fish assemblages: experiments with artificial reefs. *Bulletin of Marine Science*. **44**(2): 666–680.

Luckhurst, B.E. and K. Luckhurst. 1978. Analysis of the influence of the substrate variables on coral reef fish communities. *Marine Biology*. **49**: 317–323.

Ody, D. 1987. Les peuplements ichtyologiques des récifs artificiels de Provence (France, Méditerranée Nord-Occidentale). Thèse 3ème cycle. Univ Aix-Marseille II. 183 pp.
Ody, D. and J.G. Harmelin. 1994. Influence de l'architecture et de la localisation de récifs artificiels sur leurs peuplements de poissons en Méditerranée. *Cybium.* **18**: 57–70.
Roberts, C.M. and N.V.C. Polunin. 1991. Are marine reserves effective in management of reef fisheries? *Review of Fish Biology and Fisheries.* **1**: 65–91.
Whitehead, P.J.P., M.L. Bauchot, J.C. Hureau, J. Nielsen and E. Tortonese. (eds.) 1986. *Fishes of the North-eastern Atlantic and the Mediterranean.* UNESCO, Paris. Vols. I, II and III. 1173 pp.
Zar, J.H. 1984. *Biostatistical Analysis.* 2nd edn. Prentice-Hall International. 718 pp.

23. The Potential Use of Artificial Reefs to Enhance Lobster Habitat

ANTONY JENSEN[1], JOHN WICKINS[2] and COLIN BANNISTER[3]

[1]School of Ocean and Earth Sciences, University of Southampton, Southampton Oceanography Centre, European Way, Southampton SO14 3ZH, UK; [2]CEFAS Conwy Laboratory, Benarth Road, Conwy, N. Wales, LL32 8UB, UK; [3]CEFAS Lowestoft Laboratory, Pakefield Road, Lowestoft, Suffolk, NR33 0HT, UK

Introduction

Artificial reefs provide habitat which is exploited in a variety of ways by the marine life associated with them. UK researchers interested in lobster (*Homarus gammarus* (L.)) stock enhancement, artificial reefs, and design criteria for artificial reefs, are collaborating to address the possibility of designing, building and stocking artificial reefs to enhance lobster populations. As lobster habitat requirements are determined so the assessment of natural habitat to hold lobsters is facilitated. The knowledge gained will allow fishery managers to evaluate the worth of possible restocking programmes to supplement or extend the range of local lobster populations.

UK lobster fisheries are locally important, both socially and economically. Lobster fishing occurs in most UK coastal waters where the seabed provides the shelter necessary for the juvenile and adult stages. Adults and juveniles, however, are not necessarily found in the same place since habitat requirements vary as lobsters grow: those under 35 mm carapace length (CL) are considered to be burrow dwellers, those above, crevice dwellers. In general, the fishery is undertaken by small boats under 10 m length, working daily from small harbours or even launched from the beach. The fishery is concentrated inshore, 20 km from the coast at the most, and the boats are generally crewed by two or three fishermen. In the last 10 years, a number of modern, fast (25 kn plus) boats have entered the fishery working further offshore, particularly off the English east coast and in the English Channel, where, with the aid of GPS navigation, it has been possible to catch lobsters from previously unfished grounds and wreck sites (Grey, 1995). At some east coast ports, larger fishing boats have recently begun to convert from fin-fishing to lobstering, and there is concern about the potential effect of this new increase in fishing effort on stocks. Any removal of the present 6 mile limit, which prohibits non-UK-registered fishing vessels from working within 6 miles of the shore, could result in increased fishing effort.

It is recognized that the present management of crustacean fisheries by conventional legislation (in the UK, for example, a national minimum landing size (MLS) of 85 mm CL, with some regional variations) is more likely to sustain, rather than increase, production from this fishery. Nevertheless, opportunities to increase lobster stocks, rather than to concentrate existing ones, could arise if new

A.C. Jensen et al. (eds.), Artificial Reefs in European Seas, 379–401
© 2000 Kluwer Academic Publishers. Printed in Great Britain.

populations can be established where there is little or no natural settlement or where existing fishing grounds can be extended. This represents new opportunities for commercial activity (hatchery production of juveniles, new fishing prospects) and the possible amelioration of some impacts on the coastal zone.

Lobster Habitat

H. gammarus require shelter at all benthic stages of their life history. Eggs hatch in the summer and develop through three planktonic instars (often called 'stages'). Metamorphosis to the fourth instar is followed by active settlement into protective crevices between rocks and cobbles and is usually followed by creation or modification of a burrow in the seabed. The natural density of *H. gammarus* appears to be quite low, and as a result, UK fishery landings only average about 1200 tonnes per annum. Unlike the American lobster (*Hommarus americanus*), adult European lobsters do not migrate offshore and inshore with the seasons because UK coastal water temperatures only vary between 4 and 20°C.

Adult lobsters occupy shelters in a wide variety of habitats. Preferred lairs include the crevices in natural rock, boulder and scree formations but lobsters also readily occupy suitable holes in man-made structures. Divers report the occurrence of lobsters in wrecks, harbour walls and under pipelines, for example, and frequently note that occupation can originate within a very short period of time following the deployment of an artificial structure (e.g. 3 weeks in the case of the Poole Bay artificial reef (Jensen *et al.*, chapter 16, this volume). The implication is that in at least some areas, a proportion of the lobster population is always on the move, possibly foraging within a local area and frequently returning to lairs occupied on previous occasions (Ennis, 1984; Turner and Warman, 1991). Work in the Poole Bay fishery area (north of the artificial reef – Jensen *et al.*, 1994a; Jensen *et al.*, chapter 16, this volume) and in Bridlington Bay (Bannister *et al.*, 1994a) shows that lobsters rarely move more than 4 km over several years and the majority move less than 2 km.

Research on the burrowing habits and within-burrow behaviour of juvenile lobsters, and on the critical features of adult lobster habitat, is increasing our understanding of their environmental needs and preferences (Bertran *et al.*, 1986; Howard and Bennett, 1979; Wickins and Barry, 1996; Wickins *et al.*, 1996). Young *H. gammarus* are cryptic and spend their initial years of life almost totally underground in burrows feeding on infauna and possibly bacteria. Food quantity, availability and nutritional variety within the burrow are major factors influencing when the juvenile lobster emerges to forage, so becoming subject to competition and more vulnerable to predation (Heasman, 1991; Parsonage, 1993; Roberts, 1990).

Although small lobsters are well adapted to a burrowing existence, there comes a time when the food reserves within and close to the burrow become increasingly incapable of providing complete sustenance (Lawton, 1987; Wahle, 1992a). The lobster then starts to forage further afield seeking shelter whenever necessary to avoid strong currents and predators (Howard and Nunney, 1983; Wahle and Steneck, 1991, 1992).

Crevices in most natural and many artificial reefs are fractal in character, i.e. there are far more small crevices than large ones in natural or random rock reefs, and an obligate crevice-dwelling animal like the lobster may experience a decline in the number of suitable-sized shelters as it grows (Caddy, 1986). If this happens, members of the population will be obliged to migrate to another locality, suffer higher mortality due to competition for a declining number of shelters, or reduce growth rate by delaying moulting until a larger shelter becomes available. Circumstantial evidence suggests that these factors can contribute to the distribution of natural lobster populations and also provide mechanisms whereby the size of a lobster does not necessarily indicate its chronological age (Sheehy *et al.*, 1996; Wickins and Sheehy, 1993).

Lobster Stock Enhancement

History

Addison and Bannister (1994) review the many attempts to enhance lobster stocks in Europe and North America, dating back to at least the 1850s. These included large-scale releases of larvae from North American hatcheries, numerous releases of larvae or translocation of adults in Britain and the creation of breeding sanctuaries in France. The efficacy of these exercises could not be ascertained, primarily because the released stock was not distinguishable within the wild population and because natural fluctuations in abundance were greater than could credibly be attributed to the releases.

Experimental Methodology Used in the UK

Lobster stock enhancement experiments took place between 1981 and 1995 (Anon, 1995; Bannister and Addison, 1998; Bannister and Howard, 1991; Bannister *et al.*, 1994a, 1994b; Wickins *et al.*, 1995). The results supply evidence that hatchery-reared lobsters released into the wild can survive to maturity and beyond, and therefore provide a basis for the concept of stocking artificial reefs.

Beard *et al.* (1985), had previously shown that it was technically possible to rear lobsters to adult size year-round in controlled environment systems but this proved to be too expensive to be economically viable. Based on this expertise, an experiment was set up to evaluate the prospects for augmenting UK lobster stocks through the release of tagged, hatchery-reared juveniles onto carefully selected seabed areas (Bannister and Howard, 1991; Beard and Wickins, 1992; Howard, 1988). The key steps in the experiment were as follows:

(1) Eggs from wild broodstock were hatched and reared through the larval and juvenile instars in warm water at 18°C for 3 months by which time most had reached instar XII.
(2) A small microwire tag (adapted from Jefferts *et al.* (1963)) bearing a batch code was implanted in each lobster at the base of the fifth walking leg where

it would not be shed at moulting, or be eaten later if the lobster was captured. Experiments showed that the retention of tags inserted by careful operators varied between 85 and 100% (Wickins *et al.*, 1986).

(3) After careful transport to the coast, juveniles were released onto selected seabed areas where the cobble-boulder habitat was known to be suitable for wild lobsters. Most lobsters were released underwater from trays carried to the seabed by diver. Where large, contiguous areas of suitable habitat existed, lobsters were also released using a delivery pipe running from the boat to the seabed (Cook, 1995). Every effort was made to release lobsters when sea temperatures were above 11°C, in order to ensure that juveniles would be sufficiently active to burrow or enter crevices quickly, and to choose release times when presence of fish predators such as cod and wrasse would be at a minimum. Lobsters were seeded at a density of about 1–2 m^{-2}.

(4) A recapture programme was subsequently established to test for the presence of microtagged lobsters in landings and in lobsters caught at sea during field sampling in the release area (Bannister *et al.*, 1994a, 1994b). Lobsters tested at the landing place were legal-sized animals only, and their precise position of capture was not individually known. Lobsters from field sampling came from known positions, and included both legal-sized and sub-legal-sized animals. Lobsters with a microtag were measured, labelled and frozen for return to the laboratory, where they were X-rayed to confirm the location of the microtag which was removed. This was then decoded to identify the year of release.

This pattern of work was adopted by two other research groups in Britain, working in three other areas (Fig. 1). Broadly similar rearing and release programmes were carried out from 1983 to 1988 on the west coast of Wales, (Cook, 1995), at Ardtoe in west Scotland, and at Orkney in north Scotland (Burton, 1993; Burton *et al.*, 1994). From 1988 to 1995, recapture programmes were undertaken using field sampling at sea (offshore at Aberystwyth, and at Ardtoe and Orkney) and by monitoring fishery landings (Aberystwyth and Orkney).

It is important to emphasize that across the four experimental locations, lobsters were not only distributed under different hydrological conditions but also over a number of different grounds. At Bridlington, nearly 50 000 lobsters were released in batches ranging from 200 to 2000 at 80 individual release spots throughout a 100 km^2 area. At Orkney, Ardtoe and Aberystwyth the number of local sites and animals may have been more restricted.

Principal Results

The combined number of lobsters released at all sites was approximately 91 000 and over 1500 animals were recaptured. Table 1 shows the number of lobsters released and recaptured in each release and recapture year, and the overall recovery rates (number recaptured per 1000 released) (Anon, 1995). Recaptures began 3–5 years after release, and continued up to 9 years after release, after

Figure 1. Map of locations at which microtagged lobsters were released and recaptures monitored.

which experimental sampling ceased. The annual recovery rate (total number recaptured as a proportion of the number released) varied from 1 to 5.5% depending on year and site. Among other things this reflects the fact that only a proportion of the landings from an area was tested. For the Aberystwyth data an attempt was made to correct for this sampling effect, and also for the fact that some animals were probably landed before sampling began. This increased the average recovery rate in that experiment from 2.4% to 4.5% (Cook, 1995).

Lobsters grew to the legal size of 85 mm carapace (CL) 4–6 years after release, and exhibited the growth pattern illustrated in Fig. 2. The considerable differences in size at age show that the growth rate of juvenile lobsters can vary considerably.

Table 1. The number of hatchery-reared lobsters (*Homarus gammarus*) released and recaptured at four experimental study sites in the United Kingdom from 1983 to 1994.

Release year	Number released	1985	1986	1987	1988	1989	1990	1991	1992	1993	1994	Total	No. per 1000 released
Bridlington, E. England[1]													
1983	2390				14	19	6					39	16
1984	8616				12	67	97	25	14			215	25
1985	7979					24	112	21	5			162	20
1986	11562						1	28	39	5		73	6
1987	12629						2	41	72	20		135	11
1988	5952								22	7		29	5
Aberystwyth, Wales[2]													
1984	1250						33	6	3			42	34
1985	3760				5		67	74	31	11		188	50
1986	2438					3	5	28	26	2		64	26
1987	5079							2	36	23		61	12
1988	6706							3	40	44	11	98	15
Ardtoe, W. Scotland[3]													
1984	451	2	2	3	9	8	1					25	55
1985	1268				2	6	5	1				14	11
1986	513												
1987	553				1	4	8	4	1			18	33
1990	259									1		1	4
Scapa Flow, N. Scotland[3]													
1984	4469	3		18	3	16	84	68	23	1		216	48
1985	3800					1	10	3	12	1		27	7
1986	2356								5			5	2
1987	3610								37	5		42	12
1988	2260								8	4		13	6
1989	3025									1		1	0

[1]Released by Ministry of Agriculture, Fisheries and Food. (Bannister et al., 1994a)
[2]Released by North Western and North Wales Sea Fisheries Committee (Cook, 1985)
[3]Released by Sea Fish Industry Authority (Burton et al., 1994)

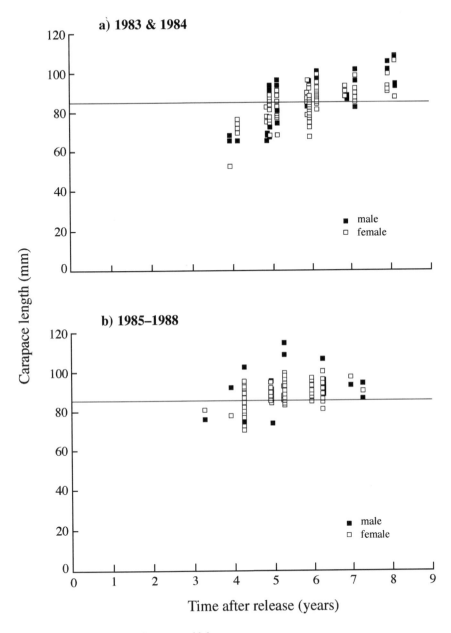

Figure 2.   Growth pattern of recaptured lobsters.

In the MAFF east coast experiment, which recaptured over 620 lobsters from 49 000 lobsters released, 18 female recaptures were carrying eggs. These animals had matured, mated and spawned, successfully demonstrating their potential to contribute to natural stocks.

Recaptures from known positions during the field sampling programme showed that most recaptures were clustered quite close to known release sites, and Fig. 3 illustrates this. At Aberystwyth some juveniles released within 5 km of the coast were later recaptured some 20 km offshore, but very few of the lobsters released offshore were recaptured inshore (Cook, 1995). At both Aberystwyth and Ardtoe one or two individuals moved up to 20 km (Burton, Cook, personal communication). Tagging experiments on wild lobsters also show that most lobsters stay in the vicinity of their release position and although a small proportion may move a considerable distance, *H. gammarus* does not appear to make extensive migrations. These results are important, showing that hatchery stock released onto good habitat will mostly stay where they are released, supporting previous research on adults (Anonymous, 1995; Bannister, *et al.*, 1994a, 1994b; Jensen *et al.*, 1994b), and could form a worthwhile component of the local catch (Anonymous, 1995).

Interpreting the Recapture Results

The gross recovery rates, ranging from 1 to 5%, are estimates obtained under experimental conditions. Because not all fishery landings were sampled, and because the intensity of field sampling was limited by finance and manpower, they are minimal estimates. They are however broadly similar at the four different geographical and fishery sites. At Bridlington, field sampling data provided estimates of the catch per unit effort of hatchery lobsters compared to wild lobsters at individual release sites. Hatchery lobsters were mostly caught singly, but on some occasions 4–8 individuals were caught in the same 'string' of 25 traps. The best field sampling result was obtained in 1989, when hatchery lobsters formed 30% of the lobsters caught experimentally in the 75–79 mm length group. In Norway, where the natural stock density of lobster in the eastern fjords is particularly low, experimental results indicate that hatchery-reared lobsters have contributed a similar, and increasing, proportion of the lobster catch landed by fjord fishermen (G. van der Meeren, personal communication).

The Bridlington field sampling results have been used to make two estimates of how well hatchery lobsters have survived. The first method used the catch rate of hatchery recaptures at individual release sites. This was converted to an estimate of the number of hatchery lobsters on the ground at the recapture site, using the results of another tagging experiment to estimate the capture efficiency of traps. The resulting point estimate of hatchery stock at that site was then compared with the known number released there in order to estimate survival. The method involved a number of untested assumptions but suggested that survival to legal size could be as high as 50–80% (Bannister *et al.*, 1994a).

An alternative approach starts with a range of assumptions about survival rate, fishing rate and recovery rate, and uses a statistical model to identify which combination best matches the observed pattern of recaptures in Table 1. With this approach, the best estimate of cumulative survival was about 37% over the five years from release to legal size, and the best recovery rate was about 2% per annum, a result which was similar for all four experiments (M. Bell, personal

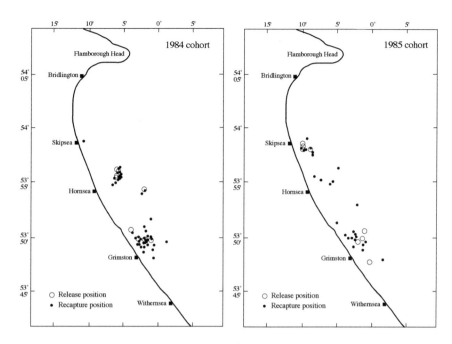

Figure 3. An example of recapture positions in relation to known release sites for a release on the English east coast.

communication). Biological survival still appears to be reasonably good, but lower than the estimate described in the previous paragraph, whilst the experimental recovery rate was confirmed to be low.

The results are the first to show that significant numbers of hatchery lobsters survived to adulthood at their release site, were subsequently detected in fishery landings and demonstrated their potential contribution to recruitment by carrying eggs. In the natural environment the validity of these results depends on whether such lobsters have actually added to the stock, as opposed to displacing the natural stock in the area. Displacement would be expected if habitat is limiting, but this would not become a factor until the lobsters were 35 mm CL or 2+ years old and began to forage outside burrows, unless the seabed into which they burrowed was nutritionally or physically inadequate. The importance of choosing fertile, productive substrata for releases of hatchery-reared lobsters is clear.

Concern whether juvenile releases enhance and/or displace wild lobster stocks raises the question whether additional competition was caused by the hatchery animals, leading to an increased death rate or emigration among the natural population. This was discussed at some length in Bannister and Addison (1998). Two North American studies provide additional information collected for *Homarus americanus*. Research in Northumberland Strait, Canada, suggested that when the number of naturally produced juveniles increased, the number of legal-sized lobsters did not increase, proposing that there was increased juvenile competition and hence an increased natural death rate. This effect was very strong (Fogarty and Idoine, 1986). Studies have also been carried out on the abundance and ecology of juvenile *Homarus americanus* in the state of Maine. When artificial cobble plots were deployed, the density of newly settled juveniles was high, but seeding experiments did not show evidence of increased mortality on plots seeded with artificially high densities of juveniles (Wahle, 1992). Further, after years of low landings, US lobster fishery landings increased substantially during the late 1980s, and whilst much of this was attributable to increased fishing effort, ground-fish surveys show that recruitment also increased, suggesting that the carrying capacity of lobster habitat had not been limiting (Wahle and Incze, 1997).

The scientific programmes have shown that hatchery-reared lobsters released at sea can enter the fishery and contribute to stocks. The question as to who is likely to benefit is of comparable importance to the magnitude of benefit. In addition to any direct financial revenue, the benefits might include sustaining jobs in rural communities, and mitigating measures to compensate for other planned uses of the seabed. Where natural recruitment is low, as in Norway, the potential impact of enhancement may be substantial, but elsewhere the commercial potential of enhancement remains unresolved (Bannister and Addison, 1998). In the future, economic evaluation of the ranching-for-profit question requires further experi-mentation. In particular, optimal stocking density can only be determined by comparing the survival and recapture rate of hatchery lobsters stocked in a range of densities and sizes with different background densities (including zero) of natural stock. Studies also need to consider season of year for stocking, frequency of restocking the same area, number of releases per year, and advantages of 'fallowing' (Wickins, 1997).

Artificial Reefs and Lobsters: Research to Date

At least four countries, Canada, USA, UK (clawed lobsters) and Israel (slipper lobster), have focused attention on artificial reefs as a specific lobster habitat, while nature conservation reefs off Monaco have been shown to be good spiny lobster habitat. In some warm-water regions of the world, artificial structures (e.g. those known as *Casas Cubanos* in the Caribbean) are specifically prepared to provide habitat for spiny or rock lobsters (Miller, 1982; Eggleston *et al.*, 1990a). This concentrates existing populations, thereby increasing fishing efficiency (and fishing mortality within the population) but may also provide new habitat for juveniles collected from elsewhere in smaller artificial habitats, (e.g. *Casitas Cubanos*: Eggleston, *et al.*, 1990b).

Canadian scientists built the first artificial reef specifically for lobster research in 1965. They used quarry rock placed 400 m away from minor lobster habitat, 2–2.5 km from major concentrations of lobsters (Scarratt, 1968, 1973). Over the following eight years the lobster population on this artificial reef was monitored by diving scientists. The reef was initially colonized by large lobsters (*H. americanus*; >41 mm CL). These were thought to have outgrown their burrows, so being forced to roam and seek new shelter. By 1973 the size frequency distribution of the artificial reef population was similar to that on natural reefs in the area. Scarratt (1973) concluded that the standing crop on the reef might be increased by using a different configuration of rocks, but that a cheaper source of reef material, or a multiple use reef, would be needed before an artificial reef could be considered an economically viable proposition.

In 1996 an experiment with reef ball modules (spherical, alveolar modules with a variety of diameters) began off the coast of Newfoundland (Ramea and Port au Port) (Fisher, personal communication). Results are still awaited on whether these structures will prove to be effective as lobster habitat, especially in areas with habitat damaged by bottom fishing techniques.

Artificial shelters have been investigated as lobster habitat in the USA (Sheehy, 1976). The number of lobsters inhabiting single and three-chambered shelter units exceeded those found in natural reefs, and densities were similar to, or greater than, the artificial reef populations described by Scarratt (1973). Abundance per unit area was a function of shelter spacing and number of compartments per shelter. The importance of inter-shelter spacing in determining lobster abundance suggests that the nearest-neighbour distances between juvenile lobsters may be an important factor in maximizing the carrying capacity of an artificial habitat.

In 1997, the World Prodigy artificial reef was placed off Rhode Island, USA, to mitigate the effects of oil spilled from a stranded tanker. The reef consists of six 10×20 m reef units, half of each unit constructed from cobble (10–20 cm diameter), the other from 20–40 cm diameter stone. This experimental reef will investigate whether artificial reef habitats will be effective for the clawed lobster *H. americanus*. A 5-year study will consider recruitment rate, speed of colonization, residence time, impact on natural colonization of releasing hatchery-reared juveniles (microwire tagged) on the reef and the overall implication of release on the lobster population of the area (Cobb and Castro, personal communication).

Since 1989 a small (30×10 m) artificial reef placed in Poole Bay on the central south coast of England (see Jensen *et al.*, chapter 16, this volume) has been the focus of lobster habitat and behaviour studies. Deployed on a flat, sandy seabed, one aim of the study was to assess the potential of reefs for fisheries enhancement. Within 3 weeks of reef deployment, clawed lobsters (*H. gammarus*) were present on the reef (Collins *et al.*, 1991a, 1991b) (Fig. 4). Tagging studies were initiated in 1990 and recapture data to September 1997 show that lobsters have found the artificial reef a suitable long-term habitat; the longest period of residence stands at 1050 days (Collins *et al.*, in press; Jensen *et al.*, 1994b).

Conventional tagging of sub-legal size (<85 mm CL) lobsters in the nearby Poole Bay fishery revealed that lobsters in this area did not undertake any seasonal migration, and that most movements averaged over time were less than 4 km in magnitude (Jensen *et al.*, 1994b). On the reef itself, a novel electromagnetic telemetry system has been used to track the movements of individual lobsters, revealing various patterns of local behaviour and movement (Smith *et al.*, 1998). Nocturnal movement dominated (peaking at dusk and dawn), but there were frequent changes of daytime refuge, and multiple occupancy of the conical

Figure 4.   Lobster (*Homarus gammarus*) in the Poole Bay artificial reef.

reef units (1 m high and 4 m base diameter). Some animals also left the reef site for up to 3 weeks then returned (unpublished observations).

Diver observations and evidence from pot-caught lobsters suggest that the Poole Bay artificial reef can support all phases of the benthic life cycle: berried females utilized the shelters, and have reproduced more than once on the reef whilst lobster instar IV larvae have been taken from the waters above the artificial reef. A 27 mm CL individual was caught in a 'prawn pot' on the reef (it is likely that this lobster settled on the reef as an instar IV larva), and a wide size range of juvenile and adult animals have been captured and/or observed by diving scientists. The size frequency distribution of the lobster population in the Poole Bay fishery is significantly different from that on the artificial reef, because of a greater proportion of 85 mm CL and above animals on the artificial reef. Whilst fishing mortality on the reef is lower than in the fishery (possibly zero), it is also felt that this difference is in some part due to the greater proportion of large niches observed on the artificial reef in comparison to the natural reefs in Poole Bay.

In Israel, efforts have focused on the slipper lobster, *Scyllarides latus*, an important commercial species found off the Mediterranean coast (Spanier, 1991). These unclawed lobsters were found to inhabit artificial reefs made from tyres. Research showed that slipper lobsters preferred horizontal shelters with two narrow entrances on the lower portion of the reef. Shelter response is believed to be a major defence mechanism for these animals (Spanier *et al.,* 1988) and the presence of the artificial reef provided new and suitable habitat for colonization. Slipper lobsters migrate into deeper water as the sea temperature rises in summer but tagged individuals were seen to return to the tyre reef over a 3-year period (Spanier *et al.,* 1988). Spanier (1991) suggests that, in the long term, populations of these heavily exploited animals could be safeguarded by building appropriately designed artificial reefs for slipper lobsters in protected areas such as underwater parks and reserves.

Can Habitat Creation and Juvenile Lobster Release be Usefully Combined?

Because of the high labour cost of rearing lobster to market size in conventional, tank-based culture systems (e.g. Lee and Wickins, 1992), lobster 'farming' is unlikely to become financially attractive in the near future. The release of juveniles into the wild at instar IV, however, is potentially an attractive way of increasing stock size, provided habitat is not limiting, and the displacement and/or enhancement question can be resolved. This type of release is practised in the USA where large numbers of instar IV juveniles of *H. americanus* have been released, without any attempt to monitor recruitment into the fishery (see review of enhancement of Addison and Bannister, 1994). This operation is essentially an act of faith which assumes that some action to replenish stocks must be better than no action at all.

If natural habitat lacks suitable features for lobster habitation, it could in theory be supplemented by targeted deployment of artificial reefs. The deployment of an artificial reef specially designed to provide habitat for wild lobsters where none had existed before could lead to the creation of a totally new fishery or the extension of an existing one. As far as is known, this has only been reported through the use of experimental reefs (Jensen *et al.,* 1994c; Scarrett, 1968).

The combination of hatchery and artificial reef technologies might also result in a ranching enterprise where specifically designed habitat would be deployed and seeded with juvenile lobsters. Success of such a venture will depend on an effective design providing habitat for both burrow- and crevice-dwelling phases, minimizing off-reef movement in search of food and shelter, low costs of reef deployment and juvenile production, the market price for harvested lobsters and the likely recovery rate. If the reef is not deployed for the general 'public good', the investment in reef materials and released lobsters needs to be protected by establishing ownership of the lobsters introduced, or the sole right to harvesting.

Research to date shows that artificial reefs can provide effective lobster habitat, and that hatchery-reared juvenile *H. gammarus* released into the wild can be harvested later in the fishery. In the UK recent revisions to the Sea Fisheries

(Shellfish) Act (1967) provide a means of establishing sole harvesting rights for lobsters and other crustacea by the creation of a 'Several Fishery' giving the licence holder exclusive rights to deposit, propagate and harvest crustaceans. There is therefore renewed interest in developing artificial reefs for lobsters, whether to improve or extend existing habitat and/or to create a public fishery or ranching enterprise where lobster habitat has not previously existed. In order to optimize this approach, however, research is still needed on lobster habitat requirements, reef design and materials and combined with a thorough cost–benefit analysis in order to facilitate rational decisions about investing in hatcheries and reef construction and deployment (Wickins, 1997).

Scientific Research

Reef design should provide a range of crevice sizes in order to maximize provision of lobster habitat in artificial reefs and minimize the need for recently moulted lobsters to leave a reef in search of new shelter. Hypothetically, a 'lobster reef' should contain approximately equal numbers of crevices throughout the size range, and their size and design should meet the habitat preferences of lobsters as they grow, taking into account such factors as entrance position and shape, local food availability and nearest neighbour preferences.

Work begun on the Poole Bay artificial reef (Jensen *et al.,* chapter 16, this volume) is being extended to study lobster shelter preferences in a large (12 × 6 m) indoor seawater tank (Smith, personal communication). Single lobsters are offered a choice of two simple shelters, differing in one attribute at a time (e.g. height, width, number of exits). The location of the lobster is monitored by an electromagnetic telemetry system with a grid of aerials laid on the floor of the tank, and shelter choice is assessed on the basis of the proportion of time spent in each shelter. The shelter dimensions being offered to lobsters relative to their body size and they bracket the preferred dimensions reported for *H. americanus* (Barry and Wickins, 1992; Cobb, 1971).

Data from a natural, unfished population of *H. gammarus* sheltering in boulder scree crevices in Lough Hyne, Ireland have revealed significant positive correlations between lobster size (carapace length) and crevice entrance area (Pearson's $r = 0.626$; $p < 0.01$), the shelter internal volume ($r = 0.686$; $p < 0.01$) and, to a lesser extent, entrance width (Robinson *et al.,* in preparation). Seventy percent of crevices had a low-profile entrance with a height-to-width ratio of less than 1.0. Figure 5 shows the relationship between the area of crevice entrance and the size of occupying lobster in Lough Hyne. For comparison, the entrance areas calculated from a model used by Barry and Wickins (1992) are included. This work will determine a relationship between body size and preferred shelter dimensions, with a measure of the degree of selectivity.

In addition to knowing shelter size requirements it is also necessary to know the likely frequency of shelter occupancy, and to consider that lobster density will also be affected by foraging distances, crevice entrance position and shape, local food supplies, and behavioural characteristics such as nearest neighbour preferences.

In the natural environment, it is assumed that for much of the time behavioural mechanisms will reduce the risk of harmful agonistic encounters. For the European lobster, estimates of adult (approximately 60–190 mm CL) density vary considerably, ranging from around 0.03 (R.J. Hanford cited in Howard, 1988) to 0.27 lobsters m⁻² (Jensen *et al.*, 1994b). In Lough Hyne observed nearest neighbour distances ranged from 0.75 m to 35 m, corresponding to a local density of 0.03 lobsters m⁻² (Robinson *et al.*, in preparation) and in a natural, horizontal rock crack in Scottish waters, from 0.4 to 5.0 m⁻¹ (mean 1.5 m⁻¹) (Comely and Ansell, personal communication).

Laboratory experiments using matched pairs of juvenile lobsters (15–35 mm CL) to investigate the occupancy of standardized 'reefs' (circular monolayers of cobbles) indicate that dominance was usually established at the first meeting and that once one lobster takes up residence in the reef, the second wanders around the tank rather than attempts to enter. Increasing the diameter of the reef in relation to the size of the lobster, however, increased the amount of time that a second lobster spends inside the reef (Lim, 1994). The density in these trials was high at 0.4 lobsters m⁻². In other experiments, increasing the complexity of the physical environment surrounding artificial shelters inhabited by lobsters decreased the probability of a dominant lobster encountering a submissive lobster when on a foray. Increasing the number of shelters available per lobster, however, increased the number of forays they made, as if to control the more valuable resources. Further analysis of foray frequency provided some evidence that the perceived value of an artificial shelter increased with increase in cohesiveness (mud > gravel) of the underlying substratum and with increasing complexity of

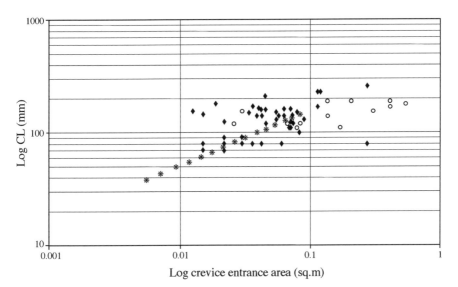

Figure 5. The relationship between crevice entrance area and the size of lobsters occupying the crevice (Robinson *et al.*, in preparation). *Calculated (Barry and Wickins, 1992); ◆ Lough Hyne 1995 (N.B. CL measurement includes rostrum) and ○ Lough Hyne 1996.

the surrounding habitat (Knock, 1996).

Telemetry data from the Poole Bay artificial reef is providing an insight into movement frequency, activity levels and multiple occupancy of reef units by adult lobsters in natural environmental conditions in relation to prevailing light levels, current speed and direction, and temperature in the sea (Collins *et al.*, 1997, in press; Smith *et al.*, 1998).

To utilize data on habitat requirements, mathematical models are being developed to estimate the number and sizes of crevices visible at the surface of submerged, artificial rock reefs intended to provide shelter for lobsters and crayfish (Barry and Wickins, 1992; Wickins and Barker, 1997). The initial model of Barry and Wickins (1992) assumed that rocks were discs, so it represented a more dense packing of rocks than might occur naturally with angular material. It therefore provides only a very crude approximation to the number of crevices per unit area, and no information on the internal galleries that would be generated in a real reef. Such models are specifically related to structures made of, or covered with, rocks or other block-like materials (rip-rap). Work is therefore in progress to achieve greater realism and, potentially, greater design flexibility, using computer-simulated rock assemblages (Wickins and Barker, 1997). The new model also has the potential to provide images and dimensions relating to the form and extent of the internal gallery systems within a reef, which seem fundamental to continued occupation by lobsters (Fig. 6). Ultimately these theoretical models will need to relate to the distribution of lobsters in natural habitat if they are to have practical value.

Economic Appraisal

Whitmarsh (1997) developed a cost–benefit analysis for a hypothetical artificial reef designed for lobster ranching or fishing (see also Whitmarsh and Pickering, chapter 27, this volume). The example used a reef 100 times the size of the Poole Bay artificial reef, a 100-year time horizon, and space for an annual harvest of 5000 lobsters derived from annual restocking of hatchery-reared juveniles. The conclusion was that such a reef, built from locally available quarry rock, could satisfy UK government criteria for project worth as a programme with social as well as economic benefit, i.e. for the public good. The possibility of additional earnings from tourism (from the hatchery and recreational angling on the reef) and commercial finfishing (close to the reef) were not factored into the analysis.

The economic appraisal was particularly sensitive to changes in the first sale price of lobsters, the level of sustainable production, and the costs of reef deployment and juvenile production in a hatchery (the latter being greatly influenced by whether the juveniles could be released at instar V rather than XII). As yet, there is no reasonable evidence to indicate that releasing juveniles into UK coastal waters soon after metamorphosis will be worthwhile in terms of either survival or the time taken to reach legal size.

On the important question of ownership of a lobster ranching site, the 1997 revision of the UK Sea Fisheries (Shellfish) Act 1967 appears in principle to offer

a

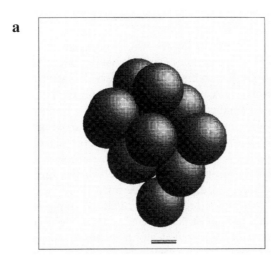

b

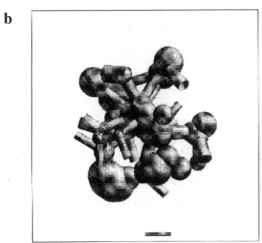

Figure 6.   Model reef formed from rocks of similar sizes (a). The internal chambers are shown as combinations of spheres and the connections between then as cylinders (b).

an investor in lobster ranching the means to protect any investment and retain the economic rewards from ranching. This legal revision has prompted at least one UK company to consider the possibility of lobster ranching using an artificial reef made from 'materials of opportunity', such as used tyres. This choice of material is interesting as used tyres are difficult to dispose of and it may be possible to charge for their disposal, thus reducing reef construction costs. It is expected, however, that the licensing authority would need to be convinced that such a structure is environmentally acceptable (Collins *et al.,* 1995) and that it provides for an increase in population levels of lobsters rather than merely concentrating existing lobsters.

Discussion

The role of artificial reefs in lobster stock enhancement is one of providing habitat. This can be the creation of lobster habitat where none existed before (perhaps in mitigation for anthropic environmental changes elsewhere), the reinstatement of damaged reefs or the modification of natural habitat to, say, provide an increased number of suitable shelters for lobsters of a specified range of sizes. It is likely that in the future it will be feasible to design a reef and provide biologically appropriate shelters in sufficient numbers and size to minimize excursions by large lobsters searching for a suitable size of shelter after moulting. Design features should not just consider lobsters; a structure on the seabed will attract fish to the area and different species will be preferentially attracted by different types of reef profile (e.g. Spanier, 1995). Complementary uses, such as tourist angling, could also boost the value of a reef to a coastal community.

Artificial reefs have been shown to support at least four species of commercially important lobster (two homarid species, spiny lobster and slipper lobster (e.g Spanier, 1991; Spanier et al., 1988). Questions have been raised about dilution of the natural lobster population if additional habitat placed as an artificial reef in a fishery area aggregates lobsters from nearby. Such an effect would occur initially, before all niches were occupied, and could be minimized by careful siting. An offshore artificial reef could be supplied with hatchery-reared juveniles that would have a good prospect of survival, as seen in experiments that released juveniles in the UK (Bannister et al. 1994a). At present the maximum lobster density that can be achieved has not been established, but data presented for *H. americanus* suggest that the Canadian quarry rock reef supported 1 lobster every 6 m^2 (Scarratt, 1973) while the Poole Bay reef is estimated to hold an individual *H. gammarus* for every 2 m^2 of reef surface area. Neither structure was designed to maximize lobster habitat and no attempt was made to provide additional food resources such as mussels (Lee and Wickins, 1992).

The recent revision to UK legislation has removed a major disincentive to the development of lobster ranching or stock enhancement programmes utilizing artificial reefs, but the economics of artificial reef construction are still being debated. The use of high-technology concrete and steel structures with large-scale construction techniques does not seem to be feasible in a UK context at present, where it is unlikely that the lobster fishing 'industry' (a collection of small 'one-person' businesses) would pay fully for such structures. Grant aid from the European Commission (EC) may be possible; the EC supported 50% of Italian and Spanish and 89% of French reef construction costs in the past (Bombace et al., 1993) but such funding has yet to be explored for reefs and lobsters in a UK context. From a fisherman's point of view, a more realistic approach could be the re-use of environmentally acceptable materials such as quarry rock and tyres, or low-cost, stabilized powders such as cement-stabilized pulverised fuel ash (PFA) (Collins and Jensen, 1995; Collins et al., 1994a, 1994b, 1995; Jensen and Collins, 1995; Jensen et al., 1994a, 1994b). Such materials could be deployed over a period of time by a combined effort from fishermen (plus industrial partner, if appropriate) in order to create properly planned, multiple-function fishing reefs at a low cost. An experiment is planned to start in Scotland, where cement-stabilized quarry slurry and PFA blocks will be deployed as a reef in the year 2000 in order to increase lobster habitat (Sayer and Wilding, personal communication).

While it has been shown that artificial reefs can support lobster populations over significant periods of time, many questions remain relating to the way lobsters utilize this habitat. In order to maximize stocking densities and minimize emigration, lobster behaviour still needs further detailed study. In the UK this may include deployment of artificial reefs designed to test hypotheses relating to shelter size and stock densities, as well as the continuation of telemetry studies to detail localized behaviour. Elsewhere, both spiny and slipper lobster (Spanier, 1991; Spanier *et al.*, 1988) are important catches in European wild fisheries. Both animals appear able to exploit artificial habitats, and research effort needs to investigate what ranching or stock enhancement opportunities exist for these species.

Although existing economic appraisals are somewhat marginal, it has to be stressed that most of the work done so far has been scientifically based on a modest scale: it is conceivable that this perspective could be redressed by large-scale commercially inspired ventures combining an entrepreneurial approach, economies of scale, and a long time horizon. Given the popularity of seafood in Europe and elsewhere, lobster ranching or stock enhancement using artificial reefs still seems to be a target worth pursuing in the quest to bring improved social and economic benefits to coastal communities.

Acknowledgements

National Power and MAFF supported the research activity on the Poole Bay artificial reef. MAFF, The Worshipful Company of Fishmongers and the Shellfish Association of Great Britain supported lobster hatchery development and juvenile deployment experiments at the CEFAS Conwy and Lowestoft Laboratories. We are grateful for the discussions with colleagues such as Ken Collins and Philip Smith (SOES, University of Southampton), Julian Addison (CEFAS), Craig Burton (SFIA), Bill Cook (NWNWSFC), Gianna Fabi (CNR), David Whitmarsh (CEMARE) and Ehud Spanier (University of Haifa) to name but a few in the development of these ideas.

References

Addison, J.T. and R.C.A. Bannister. 1994. Re-stocking and enhancement of clawed lobster stocks: a review. *Crustaceana.* **67**(2): 131–155.

Anonymous. 1995. Lobster stocking: progress and potential. Significant results from the UK restocking studies 1982 to 1995. MAFF Direct. Fish Res Lowestoft. 12 pp.

Bannister, R.C.A. and J.T. Addison. 1998. Enhancing lobster stocks: a review of recent European methods, results, and future prospects. *Bulletin of Marine Science.* **62**: 369–387.

Bannister, R.C.A. and A.E. Howard. 1991. A large-scale experiment to enhance a stock of lobster (*Homarus gammarus* L.) on the English east coast. *ICES Marine Science Symposium.* **192**: 99–107.

Bannister, R.C.A., J.T. Addison and S.R.J. Lovewell. 1994a. Growth, movement, recapture rate and survival of hatchery-reared lobsters (*Homarus gammarus*, Linnaeus, 1758) released into the wild on the English east coast. *Crustaceana.* **67**(2): 156–172.

Bannister, R.C.A., J.T. Addison and S.R.J. Lovewell. 1994b. Lobster stock enhancement experiments III: Estimating survival and assessing future fishery applications. In ICES Workshop to Evaluate the Potential of Stock Enhancement as an Approach to Fisheries Management. Charlottenlund, Denmark, 19–24 May 1994. ICES CM 1994/F:9: 41–53.

Barry, J. and J.F. Wickins. 1992. A model for the number and sizes of crevices that can be seen on the exposed surface of submerged rock reefs. *Environmetrics.* **3**(1): 55–69.

Beard, T.W. and J.F. Wickins. 1992. Techniques for the production of juvenile lobsters (*Homarus gammarus* (L.)). Fish Res Tech Rep MAFF Direct Fish Res, Lowestoft. (92), 22 pp.

Beard. T.W., R.R. Richards and J.F. Wickins. 1985. The techniques and practicability of year-round production of lobsters, *Homarus gammarus* (L.), in laboratory recirculation systems. Fish Res Tech Rep MAFF Direct Fish Res, Lowestoft. (79): 22 pp.

Bertran, R., J.Y. Gautier and J. Lorec. 1986. Caractéristiques de l'abri du jeune homard Européen (*Homarus gammarus*). ICES C.M. 1986/K13, 17 pp.

Bombace, G., G. Fabi and L. Fiorentini. 1993. Census results on artificial reefs in the Mediterranean Sea. *Bollettino di Oceanologia Teorica ed Applicata.* **11**(3–4): 257–263.

Burton, C.A. 1993. The United Kingdom lobster stock enhancement experiments. First British Conference on Artificial Reefs and Restocking, Stromness, Orkney. 12 Sept. 1992, ICIT Heriot-Watt University, Edinburgh. pp. 22–35.

Burton, C.A., R.C.A. Bannister, J.T. Addison and W. Cook. 1994. Lobster stock enhancement experiments II: Current experiments in the British Isles. In ICES Workshop to Evaluate the Potential of Stock Enhancement as an Approach to Fisheries Management. Charlottenlund, Denmark, 19–24 May 1994. ICES CM 1994/F:9: 37–40.

Caddy, J.F. 1986. Modelling stock-recruitment processes in Crustacea: some practical and theoretical perspectives. *Canadian Journal of Fisheries Aquatic and Science.* **43**: 2330–2344.

Cobb, J.S. 1971. The shelter related behavior of the lobster, *Homarus americanus*. *Ecology.* **52**: 108–115.

Collins, K.J. and A.C. Jensen. 1995. Stabilised coal ash reef studies. *Chemistry and Ecology.* **10**: 193–203.

Collins, K.J., A.C. Jensen and A.P.M. Lockwood. 1991a. Artificial Reefs: using coal fired power station wastes constructively for fishery enhancement. *Oceanologica Acta.* **11**: 225–229.

Collins, K.J., A.C. Jensen and A.W.H. Turnpenny. 1991b. The Artificial Reef Project, Poole Bay: A fishery enhancement experiment. In de Pauw, N. and J. Joyce. (eds.) *Aquaculture and the Environment.* European Aquaculture Society special publication No. 14, pp. 74–75.

Collins, K.J., A.C. Jensen, A.P.M. Lockwood and S.J. Lockwood. 1994a. Coastal structures, waste materials and fishery enhancement. *Bulletin of Marine Science.* **55**(2–3): 1253–1264.

Collins, K.J., A.C. Jensen, A.P.M. Lockwood and W.H. Turnpenny. 1994b. Evaluation of stabilised coal-fired power station waste for artificial reef construction. *Bulletin of Marine Science.* **55**(2–3): 1242–1252.

Collins, K.J., A.C. Jensen and S. Albert. 1995. A review of waste tyre utilisation in the marine environment. *Chemistry and Ecology.* **10**: 205–216.

Collins, K.J., A.C. Jensen and I.P. Smith. 1997. Tagging, tracking and telemetry in artificial reef research. In Jensen, A.C. (ed.) *European Artificial Reef Research.* Proceedings of the first EARRN conference, March 1996 Ancona, Italy, Southampton Oceanography Centre, pp. 293–304.

Collins, K.J., I.P. Smith and A.C. Jensen. (in press). Lobster (*Homarus gammarus*) behaviour studies using electromagnetic telemetry. Fifth European Conference on Wildlife Telemetry, Strasbourg, France, 25–30 August 1996.

Cook, W. 1995. A lobster stock enhancement experiment in Cardigan Bay. Final Report. North Western and North Wales Sea Fisheries Committee, Lancaster, UK. 33 pp.

Eggleston, D.B., R.N. Lipscius and D.L. Miller. 1990a. Stock enhancement of Caribbean spiny lobster. *The Lobster Newsletter.* **3**: 10–11.

Eggleston, D.B., R.N. Lipscius, D.L. Miller and L. Coba-Letina. 1990b. Shelter scaling regulates survival of juvenile Caribbean spiny lobster *Panulirus argue. Marine Ecology Progress Series.* **62**: 79–88.

Ennis, G.P. 1984. Territorial behavior of the American lobster *Homarus americanus. Transactions of the American Fisheries Society.* **113**: 330–335.

Fogarty, M.J. and J.S. Idoine. 1986. Recruitment dynamics in an American lobster (*Homarus americanus*) population. *Canadian Journal of Fisheries and Aquatic Science.* **43**: 2368–2376.

Grey, M.J. 1995. The coastal fisheries of England and Wales, Part III: a review of their status 1992–1994. Fish Res Tech Rep, MAFF Direct Fish Res, Lowestoft. (100): 99 pp.

Heasman, M.S. 1991. In-burrow feeding behaviour and the effects of current and starvation on the foraging of juvenile European lobsters *Homarus gammarus* (L.). M.Sc. Thesis, University of Wales, Bangor, 78 + 12 pp.

Howard, A.E. 1988. Lobster behaviour, population structure and enhancement. *Symposium of the Zoological Society of London.* **59**: 355–364.

Howard, A.E. and D.B. Bennett. 1979. The substrate preference and burrowing behaviour of juvenile lobsters (*Homarus gammarus* (L.)). *Journal of Natural History.* **13**: 433–438.

Howard, A.E. and R.S. Nunney. 1983. Effects of near-bed current speeds on the distribution and behaviour of the lobster *Homarus gammarus. Journal of Experimental Marine Biology and Ecology.* **71**: 27–42.

Knock, A.G. 1996. The effect of habitat complexity and other environmental factors on the behaviour of juvenile European lobsters (*Homarus gammarus* (L.)). M.Sc. Thesis, University of Southampton, 78 pp.

Jefferts, K.B., P.K. Bergman and H.F. Fiscus. 1963. A coded wire identification system for macro-organisms. *Nature, London.* **198**: 460–462.

Jensen, A.C. and K.J. Collins. 1995. The Poole Bay artificial reef project 1989 to 1994. *Biologia Marina Méditerranea.* **2**(1): 111–122.

Jensen, A.C., K.J. Collins, E. Free and J.J. Mallinson. 1994a. Poole Bay lobster and crab tagging programme, February 1992–January 1994. Report to Ministry of Agriculture Fisheries and Food, 17 Smith Square, London. SUDO/TEC/94/7C.

Jensen, A.C., K.J. Collins, E.K. Free and R.C.A. Bannister. 1994b. Lobster (*Hamarus gammarus*) movement on an artificial reef; the potential use of artificial reef for stock enhancement. *Crustaceana.* **67**(2): 198–211.

Jensen, A.C., K.J. Collins, A.P.M. Lockwood, J.J. Mallinson and A.H. Turnpenny. 1994c. Colonisation and fishery potential of a coal waste artificial reef in the United Kingdom. *Bulletin of Marine Science.* **55**(2–3): 1265–1278.

Lawton, P. 1987. Diel activity and foraging behaviour of juvenile American lobsters, *Homarus americanus. Canadian Journal of Fisheries and Aquatic Science.* **44**: 1195–1205.

Lee, D.O'C. and J.F. Wickins. 1992. *Crustacean Farming.* Blackwell Scientific Publications, Oxford, UK. 392 pp.

Lim Poh Yeong. 1994. Factors affecting the occupation of artificial indoor reefs by juvenile European lobsters (*Homarus gammarus* (Linnaeus)). M.Sc. Thesis, University of Stirling, 118 pp.

Miller, D.L. 1982. Construction of shallow water habitat to increase lobster production in Mexico. *Proceedings of the Gulf Caribbean Fisheries Institute.* **34**: 168–179.

Parsonage, J.R. 1993. Social interaction and resource sharing of juvenile European lobsters, *Homarus gammarus* (L.). M.Sc. Thesis, University of Southampton, 47 + 67 pp.

Roberts, J.C. 1990. Aspects of the within-burrow behaviour of juvenile European lobsters *Homarus gammarus* (L.). M.Sc. Thesis, University of Wales, Bangor, 47 pp.

Robinson, M., J.R. Turner and J.F. Wickins. (in preparation). Shelter occupancy and between-site movements by the European lobster, *Homarus gammarus* (L.) in Lough Hyne, Ireland.

Scarratt, D.J. 1968. An artificial reef for lobsters (*Homarus americanus*). *Journal of the Fisheries Research Board of Canada.* **25**(12): 2683–2690.

Scarratt, D.J. 1973. Lobster populations on a man made rocky reef. ICES CM 1973/K: **47**.

Sheehy, D.J. 1976. Utilization of artificial shelters by the American lobster (*Homarus americanus*). *Journal of the Fisheries Research Board of Canada.* **33**: 1615–1622.

Sheehy, M.R.J., P.M.J. Shelton, J.F. Wickins, M. Belchier and E. Gaten. 1996. Ageing the European lobster, *Homarus gammarus,* by the lipofuscin in its eyestalk ganglia. *Marine Ecology Progress Series.* **143**: 99–111.

Smith, I.P., K.J. Collins and A.C. Jensen. 1998. Movement and activity patterns of the European lobster *Homarus gammarus* (L.), revealed by electromagnetic telemetry. *Marine Biology.* **132**: 611–623.

Spanier, E. 1991. Artificial reefs to insure protection of the adult Mediterranean slipper lobster, *Scyllarides latus* (Latreille, 1803). In Boudouresque, C.F., M. Avon and V. Gravez. (eds.) *Les Espèces Marines à Protéger en Méditerranée.* Pub. GIS Posidonie, France, pp. 179–185.

Spanier, E. 1995. Do we need special artificial habitat for lobsters? *ECOSET'95 International Conference on Ecological System Enhancement Technology for Aquatic Environments* Tokyo. **2**: 548–543.

Spanier, E., M. Tom, S. Pisanty and G. Almog. 1988. Seasonality and shelter selection by the slipper lobster *Scyllarides latus* in the south-eastern Mediterranean. *Marine Biology Progress Series.* **42**: 247–255.

Turner, J.R. and C.G. Warman. 1991. The mobile fauna of sublittoral cliffs. In Myers and Little (eds.) *The Ecology of Lough Hyne,* Royal Irish Academy, Dublin, pp. 127–138.

Wahle, R.A. 1992a. Body size dependent anti-predator mechanisms of the American lobster. *Oikos.* **65**: 52–60.

Wahle, R.A. 1992b. Substratum constraints on body size and the behavioural scope of shelter use in the American lobster. *Journal of Experimental Marine Biology and Ecology.* **159**: 59–75.

Wahle, R.A. and L.S. Incze. 1997. Pre and post-settlement processes in recruitment of the American lobster. *Journal of Experimental Marine Biology and Ecology.* **217**: 179–207.

Wahle, R.A. and R.S. Steneck. 1991. Recruitment habitats and nursery grounds of the American lobster (*Homarus americanus* Milne Edwards): a demographic bottleneck? *Marine Ecology Progress Series.* **69**: 231–243.

Wahle, R.A. and R.S. Steneck. 1992. Habitat restrictions in early benthic life: experiments on habitat selection and *in situ* predation with the American lobster. *Journal of Experimental Marine Biology and Ecology.* **157**: 91–114.

Whitmarsh, D. 1997. Cost Benefit Analysis of artificial reefs. In Jensen, A.C. (ed.) *European Artificial Reef Research.* Proceedings of the 1st EARRN conference, Ancona, Italy, March 1996, Southampton Oceanography Centre, pp. 175–194.

Wickins, J.F. 1997. Strategies for lobster cultivation. *Shellfish News*, Centre for Environment, Fisheries and Aquaculture Science, CEFAS Conwy Laboratory, **4**: 6–10.

Wickins, J.F. and G. Barker. 1997. Quantifying complexity in rock reefs. In Jensen, A.C. (ed.) *European Artificial Reef Research.* Proceedings of the first EARRN conference, March 1996, Ancona, Italy, Pub Southampton Oceanography Centre, pp. 423–430.

Wickins, J.F. and J. Barry. 1996. The effect of previous experience on the motivation to burrow in early benthic phase lobsters (*Homarus gammarus* (L.)). *Marine and Freshwater Behaviour and Physiology.* **28**(4): 211–228.

Wickins, J.F. and M.R. Sheehy. 1993. Age determination in Crustacea. *The Lobster Newsletter.* **6**: 2–4.

Wickins, J.F., T.W. Beard and E. Jones. 1986. Microtagging cultured lobsters, *Homarus gammarus* (L.), for stock enhancement trials. *Aquaculture and Fisheries Management.* **17**: 259–265.

Wickins, J.F., R.C.A. Bannister, T.W. Beard and A. Howard. 1995. Lobster Stock Enhancement Investigations 1983–1993. Fish Res Video Rept, MAFF Direct Fish Res, Lowestoft. (1): 42 min.

Wickins, J.F., J.C. Roberts and M.S. Heasman. 1996. Within-burrow behaviour of juvenile European lobsters *Homarus gammarus* (L.). *Marine and Freshwater Behaviour and Physiology.* **28**: 229–253.

24. Physical Protection of the Seabed and Coast by Artificial Reefs

BEN HAMER[1], JOHN GARDNER[2] and ROBERT RUNCIE[3]

[1]*Halcrow Maritime, Burderop Park, Swindon, SN4 0QD, UK;* [2]*Mouchel International Consultants, West Hall, Parvis Road, West Byfleet, KT14 6EZ, UK;* [3]*Environment Agency: Anglian Region, Kingfisher House, Goldhay Way, Orton Goldhay, Peterborough, PE2 5ZR, UK*

Introduction

A detailed study was carried out in 1991 to develop a sea defence strategy for the 14 km of eroding coastline between the Norfolk villages of Happisburgh and Winterton, on the English east coast. The strategy was developed to provide protection to this coastline for a period of 50 years. Of the available options for sea defence, an offshore reef system was considered to form the best method to achieve the schemes objectives. The first stage of the strategy was to construct four offshore reefs, or barriers, to the south of Happisburgh (Fig. 1). The scheme will protect several villages and some 6000 ha of low-lying land, including access to the Norfolk Broads, from flooding by the sea. Following completion of the first stage, a review of the strategy has been carried out to compare the predicted performance of the barriers with observations made at the site. A revised strategy has been developed, with further construction phases over a period of 20 years. A series of 10 further reefs, exposed only at low water levels, will be constructed, along with a long-term commitment to beach nourishment. Stage Two of the strategy was completed in July 1997, and comprised five more reefs and beach nourishment with 10^6 m^3 of sand.

The Coastal Environment

Flood events are not new phenomena for this exposed and vulnerable coastline. Records show that in 1287 a large inland area was flooded extending to Hickling, some 5 km from the coast (Harland and Harland, 1980). Water levels rose to almost 0.5 m above the Priory's high altar at Hickling and 180 people were drowned. Three centuries later, massive seas destroyed part of the coastal village of Eccles and drowned many inhabitants.

Major breaching of the natural sand-dune defences occurred again in 1938, when an area of 3000 ha was deeply flooded. On the night of 31 January 1953, a surge tide broke through the dunes at Sea Palling, drowning seven people (Waverley Committee, 1954).

The original sea defence comprised a timber and steel groyne field, in varying states of repair, in front of a concrete seawall constructed after the 1953 floods.

A.C. Jensen et al. (eds.), Artificial Reefs in European Seas, 403–417

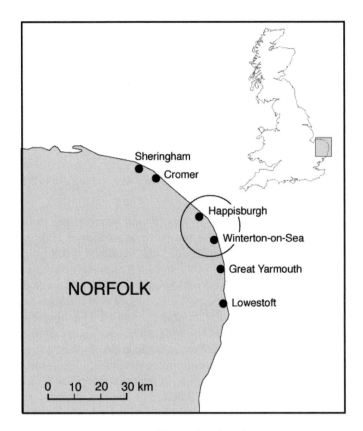

Figure 1. Happisburgh to Winterton sea defences: location plan.

The wall has a steel sheet-piled toe, penetrating the underlying clay, and an upper concrete surface of steps surmounted by a large recurve wall. A dune system backs on to the recurve wall and, in some locations, the dunes have covered the wall and steps.

The beach comprises a highly mobile sand (median particle size 0.5 mm) or shingle veneer, overlaying clay (Soil Mechanics, 1993). A high beach, to absorb wave action, is an important part of the natural sea defences. In places, the veneer of sand and shingle is 2–3 m thick, but during storms rapid short-term variations can result in beach level drops of 2 m or more as the sand is transported offshore and the clay is exposed.

In recent years, the Environment Agency has become increasingly concerned about the stability of the sea defences between Happisburgh and Winterton; particularly those between Eccles and Sea Palling. The 8.3 km of reinforced concrete seawall, built following the 1953 flood, is in excellent condition, and anchors the toe of the dune system against which it was built. However, the steel sheet-piled toe of the seawall became progressively more exposed and the clay, which underlies the beach sand veneer, was exposed for significant lengths of the frontage.

Historically, this coastline has been in retreat for many hundreds of years, and the sand dunes have progressed inland under the force of the sea whilst several small settlements (such as Dunwich and Eccles) have been lost. By fixing the line of defence with a concrete seawall, the coastal retreat has been restricted to a narrowing and steepening of the beaches which front the seawall. This is demonstrated by a study of Ordnance Survey maps available for over 100 years.

Studies of sediment input from the north Norfolk cliffs (Clayton *et al.*, 1983) show that these cliffs supply sediment to the beaches for up to 60 km downdrift, effectively to Great Yarmouth. The construction of coast protection measures to safeguard the cliffs, which started towards the end of the 19th century, reduced the sediment supply to about 75% of its natural level. Fixing the defence line whilst coastal retreat has continued, means the reduced sediment input has not been sufficient to maintain healthy beaches between Happisburgh and Winterton. The beach levels are now lower than at any time during the previous 5000 years. The ruins of Eccles Church are exposed frequently at low tide, and many bones have been discovered.

By virtue of its geographical location, the eroding beach frontage is exposed to waves from an unusually wide range of wave directions, with inshore wave heights dominated by wave-breaking effects during storms.

Coastal Management Policy

The management policy for this beach frontage is to maintain the position of the natural dune system, which provides flood defence to the hinterland. Additional aims are to preserve the environmental and amenity values of the beach and dune system, and to develop the ecology of the reefs to support a greater diversity of flora and fauna.

Sea Defence Options

Rock Toe Protection

Lowering of the beach, combined with down-cutting erosion of the clay foreshore, left the steel sheet pile toe of the seawall unprotected and vulnerable to wave attack (Fig. 2). Rock was placed to protect the seawall toe, to reduce wave reflection and provide toe weight to the seawall.

The rock toe protection was successful in preventing undermining of the seawall, and in some locations appeared to contribute to a modest improvement in beach levels. However, the wave-induced beach volatility had not been controlled, and sudden losses of up to 2 m of sand were still experienced in front of some lengths of rock protection, causing it to slump. While it was originally considered

Figure 2.   Exposure of steel sheet-piled toe of the seawall and of clay underlying the beach sand veneer.

that the effective life of the rock protection would be 10 years, additional works were necessary in practice after 2–5 years. This confirmed the conclusion that the continual upgrading of the toe protection would not be a viable long-term solution, dealing only with the consequence and not the cause of the problem.

Fishtail (Y-shaped) Groynes

Initial studies of the beach frontage recommended 13 fishtail groynes at 1 km spacing, and extending 280 m offshore from the seawall (Barber, 1989). Their purpose was to divert the longshore currents further offshore and, by redirecting incident waves, to induce the formation of bays between the groynes. The principal problem with such an approach at this site is that fishtails would interrupt the longshore drift, causing severe downdrift erosion. This problem can be alleviated by beach bypassing (transporting sediment artificially from one side of the groyne to the other), but the commitment would be large and permanent. Furthermore, the groynes would not prevent the substantial onshore/offshore movements.

Further studies (Halcrow, 1991) showed that 16 fishtail groynes would be required to provide the same degree of protection as 16 reefs, and the overall costs of such a scheme would be approximately 25% higher, including beach management requirements.

Beach Nourishment

Beach nourishment on its own cannot be successful on this beach frontage, as it does not deal with the underlying conditions which cause beach volatility. It is apparent that at locations where a wide beach was previously recorded there are now depleted beaches. Therefore, increasing the volume of beach will not prevent that volume being lost again very rapidly in the future. The wave and current climate has the potential to cause losses of the order of 10^6 m^3 of sand in a single year. Therefore, reliance on beach nourishment without control structures would be an extremely high-risk option.

Through mathematical modelling, it was found that large-scale nourishment consisting of coarser material than the present beach would tend to cause downdrift erosion, as it would block the natural sediment drift along the coast. Nourishment consisting of finer material, on the other hand, would suffer very large losses to offshore.

Offshore Reefs or Barriers

The aim of maintaining the beach and dune system, with minimum environmental impact, necessitates reduction of beach volatility and long-term beach losses, such that the current rate of sediment input can sustain the beaches with the minimum of artificial beach management. A linear system of reefs or barriers would reduce wave energy reaching the beach, without halting longshore drift. A reduction in beach volatility requires, principally, a reduction in the inshore wave climate and/or the provision of a much wider beach.

The normal constraint of 'no adverse effects on adjacent stretches of coastline' (Halcrow, 1991) requires that the effects of the sea defence scheme on longshore transport are minimized. Any shore-normal structures for beach control are, there-fore, not acceptable unless accompanied by beach management measures. It was decided that a reef system (i.e. a shore-parallel system) should be developed, with the aims of preventing beach losses offshore during storms, whilst having the minimum impact on longshore transport.

Design of Offshore Reefs

Design Philosophy

Offshore reefs reduce the wave energy reaching the shore, and dampen down beach volatility whilst not completely preventing longshore transport behind the reefs. The beach plan shape will adjust to form a series of cuspate bays with salients, or spit features, behind each reef and embayments in the gaps between each reef pair. In order to minimize the impact on longshore transport, the development of tombolos (when the salient becomes attached to the reef) should be prevented. Reefs should be porous, to minimize wave reflections into the busy shipping channel, and to reduce scour potential. The reefs will shelter the coastline, by dissipating wave energy on the structures themselves and within the multiple bay shapes of the beach. The beaches behind the reefs will, eventually, be free of groynes, which will allow free access for the public, maintenance plant and emergency vehicles.

Design Methods

Three principal factors control the beach response to offshore reefs or barriers:

(1)   ratio of reef length to offshore distance;
(2)   ratio of reef length to gap length;
(3)   reef crest elevation (which controls wave energy transmission coefficient, K_T).

K_T is defined as the ratio of incident significant wave height on the leeward side of the reef and the significant wave height offshore. Thus, $K_T = 1.0$ suggests that the reef does not modify wave heights, whilst $K_T = 0.0$ suggests that no wave energy passes through or over the reef.

The aim of the scheme was to build up salients, rather than allow tombolos to form. A desk study and a literature search of detached breakwater schemes, predominantly in the USA and Japan, resulted in the following parameters being selected for the preliminary design of the first four reefs:

- reef length/offshore distance   = 1.20
- reef length/gap length   = 0.86
- effective crest elevation   = +1.0 m Ordnance Datum (OD)

A mathematical modelling study was carried out to check the suitability of these values with regard to beach stability, and physical model studies were carried out to assess the wave transmission and structural stability.

The long-term change in beach plan shape was predicted using the Combined Onshore–Offshore and Alongshore Sediment Transport (COAST) model, for a range of reef layouts. This model was developed by Halcrow, based on research by Perlin and Dean in 1983. The main factors examined were:

(1)   the effect on the beach of the progressive development of a reef system;
(2)   the ability of the reef system to retain beach recharge;
(3)   the extent of material build-up on the updrift side of the reef system, and the corresponding downdrift erosion;
(4)   the effect of varying the wave transmission coefficient on each of the above.

Principal Conclusions of Modelling Work

The ability of the reef system to retain material is sensitive to changes in wave transmission (i.e. crest level). For high crest levels the formation of tombolos is highly likely, with the associated problems of longshore transport blockage. Conversely, for low crest levels, the protection afforded by the reefs to the beach is inadequate to prevent substantial losses of beach material during storms. It is clear, therefore, that a balance must be achieved between the prevention of large offshore losses during storms and minimizing the blockage to longshore transport during more typical conditions. For the first four reefs, it was decided to adopt a cautious approach, and minimize beach losses during storms as a priority.

When reef construction commences, the beach inshore of the reefs is not stabilized until at least four reefs have been constructed. When a smaller number of reefs have been completed, the ability to retain recharge material is markedly reduced. When four or more reefs have been completed, sediment transport within the reef system is reduced, and the protected shoreline quickly forms into a series of alternate bays and salients. The addition of recharge material becomes successful at this stage, and it is strongly recommended to prevent erosion of the beaches between reefs as material is naturally redistributed.

The updrift accretion and downdrift erosion volumes associated with the reef system were predicted to be about $1.5 \times 10^5 \, \text{m}^3$ per year due to longshore drift effects. Downdrift erosion may be much higher if the seawall becomes exposed and wave reflections increase volatility.

Physical model tests were commissioned to examine the structural stability and wave transmission of various reef cross-sections. This work was carried out in the wide flume at the Danish Hydraulic Institute (1991).

At the outset of modelling, a crest elevation of $+2.0$ m OD was examined, and showed that the wave transmission was higher than considered appropriate for storm conditions. By raising the crest level to $+3.0$ m OD, a wave transmission coefficient of about 40% was achieved during severe storms, which was considered adequate in terms of beach protection. The original coast protection strategy was developed to the form shown in Fig. 3.

Alternative Materials

Preliminary studies indicated that natural rock of sufficient size for stability as a cover layer might be difficult to obtain and handle. Rock up to 20 tonnes was considered a practical maximum and the physical model tests evaluated the performance of armour rock up to this limit. Based on the model tests, armour rock in the size range 8–16 tonnes was selected.

The alternative to rock armour is concrete units, and these were also tested in the physical model study. Accropodes, Akmons and cubes were studied, and of these alternatives, Accropodes of 20 tonne mass were found to be most suitable.

Financial Justification for Original Strategy

In order to attract grant-aid from the Ministry of Agriculture, Fisheries and Food (responsible for approving and funding coastal defence works in England and Wales) a financial justification was provided, on the basis of the direct benefits associated with the sea defence. As the scheme was justified by those direct benefits associated with flood prevention to property and agriculture alone, no assessment

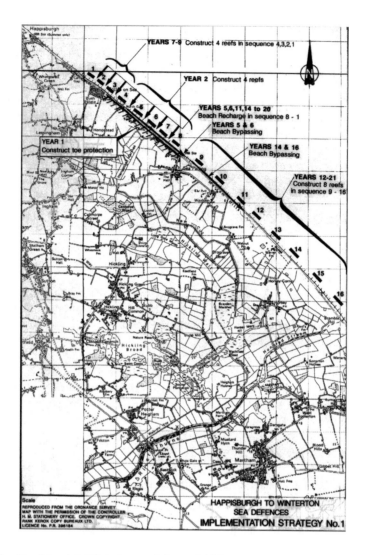

Figure 3.   Happisburgh to Winterton sea defences: initial strategy.

of environmental and recreational (indirect) benefits was necessary (Penning-Rowsell *et al.*, 1992).

Construction of the First Four Reefs

The first four reefs were constructed between 200 m and 300 m from the shore and were aligned to be shore-parallel. Reef construction comprised a geotextile mattress, laid on the seabed, on which was placed a 100–500 mm bedstone layer. The core of the reef was formed of 1–3 tonne rock, which was then covered with 8–16 tonne armour rocks.

Initially, this construction method involved placing a transhipment barge on the seabed. Rock was imported from Sweden and transferred from the delivery barge to the transhipment barge prior to placement on the reefs using a floating crane barge. The method was subsequently revised, such that the imported rock was unloaded and stockpiled in Great Yarmouth, prior to transhipment to the site by sea. Great Yarmouth is the nearest port to the site, located approximately 30 km to the south. At the site, the rock was unloaded and placed on the reef using a crane barge. Movement and final placement of the rock was carried out by two hydraulic excavators working on the reefs. The first four reefs were completed after 2 years, in 1995.

Strategy Review

Performance of First Four Reefs

Implementation of the first four reefs commenced one year later than envisaged. Toe protection works were carried out to provide short-term protection through this period of delay. The reef construction was sufficiently advanced by Autumn 1994 that a significant influence on the shoreline could be observed.

During the one year of delay, beach levels continued to fall so that they were lower than expected once the reefs began to influence the shoreline. The beach in the lee of the reefs took the predicted shape very quickly, with corresponding erosion in the gaps and on downdrift beaches. The general denudation of beach levels during the extended implementation period meant that the proposed bypassing could not be carried out, because of potential detriment to updrift beaches. Downdrift erosion was such that further rock revetment was required (Halcrow, 1995a, 1995d).

Beaches updrift of the reefs were steadily improving, as a consequence of the reduction in longshore transport which has been created. This suggested that the level of blockage to longshore drift was higher than necessary, given the adequate level of protection in the direct lee of the reefs (Halcrow, 1995c).

Erosion occurred in the gaps between the first four reefs. This was exacerbated by the lack of beach recharge behind the reefs, but is of a degree which indicates that a closer reef spacing is required (Fig. 4).

Since reef completion, several major storm events have been encountered, with return periods of up to about 10 years. It has been clear that the level of protection behind and updrift of the reefs is much greater than 3 years ago (Fig. 5).

Design Review

A review of technical papers and reports, which were produced since the original reef design, has been undertaken. This highlighted a significant discrepancy between

Figure 4.   Aerial view of first four reefs, showing erosion between the gaps.

various authors in the recommendations for the key design parameters (Halcrow, 1996a).

A reduction in the length of the gaps between the first four reefs is, however, consistent with much of the available guidance. A modified reef geometry was also determined, to fall closer within the best guidance for sites with a macro-tidal range. The revised parameters are:

- reef length/offshore distance = 0.80;
- reef length/gap length   = 1.00;
- effective crest elevation   = +1.2 m OD (1.0 m above mean sea level)

Modelling Studies

A detailed study of the revised reef design was carried out, incorporating cross-shore and longshore (beach plan shape) modelling (Halcrow, 1996a). The cross-shore development of typical beach profiles was examined using COSMOS-2D, whilst longshore beach development was studied using the Beach Plan Shape model (BPSM).

COSMOS-2D was developed jointly by Halcrow, HR Wallingford and Imperial College London (Wallace, 1994). Cross-shore modelling was used to determine the optimum crest elevation for the reefs. This was taken to be the minimum level at which the beach was protected against substantial losses during storms. A crest level of +1.2 m OD was selected as the optimum to prevent sand losses during extreme events, although a lower elevation was acceptable for most conditions.

Figure 5. Aerial view of the northernmost reef.

The BPSM was developed by Halcrow, and is described in Halcrow, 1996a. Longshore modelling was used to determine the optimum reef layout, given the crest elevation determined from the cross-shore model studies. Model predictions demonstrated that the 'do-nothing' option would result in excessive levels of beach erosion at localized areas, due to longshore transport alone. The amount of erosion also indicated that beach nourishment alone does not present a viable option.

Of three plan layouts for the reefs, a preferred option was developed with 160 m long reefs, 160 m apart, and located the same distance offshore as the first four reefs. Model tests also showed the significant beneficial effects of the first four reefs, whereby updrift beaches were improved for a substantial distance, could be reproduced further south, even with reefs at a lower crest level. A gap of almost 3.8 km may be left in the central portion of the strategy length, if a rock revetment is constructed to reduce wave reflections from the seawall. The BPSM results after one year are shown in Fig. 6 for this layout.

Revised Strategy

The revised strategy is shown in Fig. 7. Proposed amendments to the first four reefs comprise beach management together with small, shore-parallel, retaining structures in the gaps. Prospective modifications may include low-level extensions to the first four reefs. Future reefs will have a revised geometry (Halcrow, 1996a), as determined by these most recent studies.

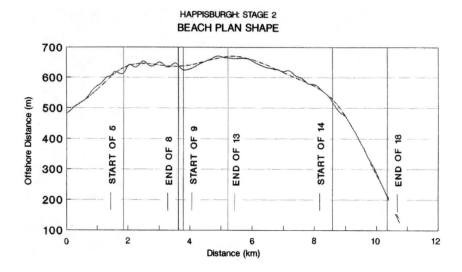

Figure 6. Happisburgh stage 2: Beach plan shape for proposed strategy. See Fig. 7 for reef locations.

Despite the improvements made to the reef system, a long-term commitment to beach recharge and bypassing will be necessary. This is demonstrated in the programme for the revised strategy (Table 1).

Monitoring

A monitoring strategy has been developed to run in parallel with the revised strategy for sea defence (Halcrow, 1995b). This monitoring will include twice-yearly measurement of beach and bathymetry profiles at 50 m spacing, to provide feedback to the model predictions made during the design stage.

Offshore wave conditions and water levels will be obtained as a 3-hourly time-series from the Meteorological Office, for use in any further model studies which may be required.

The biodiversity of the reefs will also be examined, possibly on a quarterly basis, by a study of recorded catches and beach/reef surveys. It is expected that the reefs will support crustaceans, such as crabs and lobsters, which will add to the local fishing catch (Halcrow, 1996b).

A colony of little terns is situated on the beach in this area, and numbers have reduced historically as beach erosion has taken place. Hence a study of bird diversity

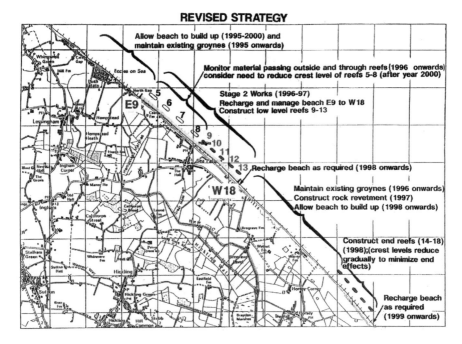

REVISED STRATEGY

Allow beach to build up (1995-2000) and
maintain existing groynes (1995 onwards)

Monitor material passing outside and through reefs (1996 onwards)
consider need to reduce crest level of reefs 5-8 (after year 2000)

Stage 2 Works (1996-97)
Recharge and manage beach E9 to W18
Construct low level reefs 9-13

13 Recharge beach as required (1998 onwards)

Maintain existing groynes (1996 onwards)
Construct rock revetment (1997)
Allow beach to build up (1998 onwards)

Construct end reefs (14-18)
(1998);(crest levels reduce
gradually to minimize end
effects)

Recharge beach
as required
(1999 onwards)

Figure 7. Happisburgh to Winterton sea defences: revised strategy.

Table 1. Revised strategy programme.

i	Immediate	Identify offshore source for future recharge material;
ii	1996–2000	North of reef 5, utilize partial blockage caused by reefs 5–8 to allow beaches to return to satisfactory levels. Maintain existing timber groynes in this area to assist with beach retention;
iii	1996/1997	Construct rock bunds in the gaps opposite reefs 5–8 to retain future recharge;
iv	1996/1997	Recharge beach (from offshore source) opposite gaps in reefs 5–8 and from reef 8 to groyne W18;
v	1996/1997	Construct reefs 9–13;
vi	1996/1997	Monitor scheme performance biannually;
vii	1997	Construct revetment along face of seawall between reef 13 and site of reef 14;
viii	1997	Recharge south of reef 13;
ix	1998	Monitor performance of reefs 9–13;
x	1998	Recharge south of reef 13;
xi	1998	Construct reefs 14–18 and extend reefs 5–8, if necessary;
xii	1999	Monitor performance of reefs 9–18;
xiii	1999	Recharge south of reef 13 and south of reef 18;
xiv	2000	Recharge south of reef 13 and south of reef 18;
xv	2000	Monitor performance of reefs 9–18;
xvi	2000	Consider the need to reduce the crest level of reefs 5–8;
xvii	2001–2015	Annual recharge south of reef 18 only. Monitor for natural bypassing of reefs 5–13, otherwise bypass artificially;
xviii	2016 onwards	Monitor performance of strategy biannually.

and numbers of breeding pairs will be undertaken, to determine the influence of the reefs on this important species.

Conclusions

The first four reefs have shown that the performance of reefs in a complex marine environment can be predicted, and monitoring results to date indicate that offshore reefs are the correct solution to protect this beach frontage. The consequences of failing to implement strategic works in a timely fashion have been evident.

A range of improvements have been highlighted by monitoring of the reefs, and these have been incorporated into a revised strategy. Stage Two comprised the construction of a further five reefs with revised geometry and major beach nourishment. These works commenced in September 1996 and were completed in July 1997. An ongoing monitoring strategy will continue to provide feedback on the reef and beach nourishment performance, to plan implementation of future recharge commitments and to monitor the environmental performance of the scheme. A second strategy review will take place in 1999 prior to stage three of the implementation.

References

Ahrens, J.P. and J. Cox. 1990. Design and Performance of Reef Breakwaters. *Journal of Coastal Research.* 61–75.

Barber, P. 1989. Anglian Water NRA Unit. Happisburgh to Winterton Sea Defences. Technical Strategy Report.

Brampton, A.H. and J.V. Smallman. 1985. Shore Protection by Offshore Breakwaters. HR Wallingford Report SR 8.

CIRIA. 1996. Beach management manual. Report 153.

Clayton, K.M., I.N. McCave and C.E. Vincent. 1983. The establishment of a sand budget for the East Anglian coast and its implications for coastal stability. In: *Shoreline Protection.* Thomas Telford Ltd., London, pp. 91–96.

Dally, W.R. and J. Pope. 1986. Detached Breakwaters for Shore Protection. Technical Report CERC-86-1. US Army Corps of Engineers.

Danish Hydraulic Institute. 1991. Happisburgh to Winterton. Sea Defences. Hydraulic Investigations, May 1991.

Halcrow. 1991. Sea Defences – Norfolk. Happisburgh to Winterton. Final Report on Proposed Sea Defences. Volumes 1 and 2.

Halcrow. 1995a. Sea Defences – Norfolk. Happisburgh to Winterton. Report on Urgent Works in Stage Two. Volume 1.

Halcrow. 1995b. Sea Defences – Norfolk. Happisburgh to Winterton. Strategic Monitoring. October, 1995.

Halcrow. 1995c. Sea Defences – Norfolk. Happisburgh to Winterton. Bathymetric and Beach Level Changes, November 1990 to January 1995.

Halcrow. 1995d. Sea Defences – Norfolk. Happisburgh to Winterton. Report on Urgent Works in Stage Two. Volume 2.

Halcrow. 1996a. Happisburgh to Winterton Sea Defences. Strategy Review.

Halcrow. 1996b. Happisburgh to Winterton Sea Defences: Stage Two. Environmental Statement.

Harland, M.G. and H.J. Harland. 1980. *The Flooding of Eastern England*. Minimax Books Ltd, Peterborough.

Ministry of Agriculture, Fisheries and Food. 1993. Flood and Coastal Defence: Project Appraisal Guidance Notes.

National Rivers Authority. 1989. Sea Defences, Norfolk. Happisburgh to Winterton. Engineer's Report. (As revised December 1989.)

National Rivers Authority. 1991. Update to Engineer's Report of December 1989.

Ozasa, H. and A.H. Brampton. 1980. Mathematical Modelling of Beaches Backed by Seawalls. *Coastal Engineering*. **4**:

Penning-Rowsell, E.C., C.H. Green, P.M. Thompson, A.M. Coker, S.M. Tunstall, C. Richards and D.J. Parker. 1992. *The Economics of Coastal Management: A Manual of Benefits Assessment Techniques*. Belhaven Press, London.

Perlin, M. and R.G. Dean. 1983. A numerical Model to simulate sediment transport in the vicinity of coastal structures. US Army Corps of Engineers, Misc Report No. 83-10.

Pope, J. and J.L. Dean. 1986. Development of design criteria for segmented breakwaters. *Proceedings, 20th International Conference on Coastal Engineering*. Taipei, Taiwan. American Society of Civil Engineers, 2144–2158.

Seiji, W.N., T. Uda and S. Tanaka. 1987. Statistical study on the effect and stability of detached breakwaters. *Coastal Engineering in Japan*. **30**: 131–141.

Soil mechanics. 1993. Geotechnical Investigation, Scheme 41306. Report 7797.

Suh, K. and R.A. Dalrymple. 1987. Offshore Breakwaters in Laboratory and Field. *Journal, Waterway, Port, Coastal and Ocean Engineering*. **113**: 105–121.

US Army Corps of Engineers. 1993. Engineering Design Guidance for Detached Break-waters as Shoreline Stabilization Structures. Technical Report CERC-93-19.

Wallace, H.M. 1994. COSMOS-2D (Version 2.3): Nearshore sediment transport model. Description of model structure and input. HR Wallingford Report IT388 (Revision B).

Wallingford, H.R. 1994. Effectiveness of Control Structures on Shingle Beaches. Physical Model Studies. Report. SR387.

Waverley Committee. 1954. Recommendations of the Waverley Committee. Report of the Departmental Committee on Coastal Flooding. Cmd 9165. HMSO.

25. Shipwrecks on the Dutch Continental Shelf as Artificial Reefs

ROB LEEWIS[1], GODFRIED VAN MOORSEL[2] and HANS WAARDENBURG[2]

[1]National Institute of Public Health and Environment, P.O. Box 1, 3720 BA Bilthoven, The Netherlands. [2]Bureau Waardenburg BV, consultants for environment and ecology, P.O. Box 365, 4100 AJ Culemborg, The Netherlands

Introduction

Shipwrecks are known to be places where sport and commercial fishing can yield a good catch; a practical demonstration of the fact that objects deposited in the sea attract a variety of organisms which make use of hard substrata as dwelling places or attachment sites. For example, Hiscock (1980) described hard-substratum communities containing 222 species on a 19 m deep wrecked coaster in the Bristol Channel, UK.

Hard substratum in the southern North Sea is mainly limited to man-made structures like wrecks, drilling platforms, pipelines and, along the shore, dyke-slopes and other coastal protection structures (e.g. see Hamer *et al.*, chapter 24, this volume).

Wrecks

More than 10 000 wrecks are known to lie in the Dutch sector of the North Sea. Interest in the fauna of the wrecks was stimulated in 1970 when Van 't Hof stressed the importance of such sites to ecological, environmental and mariculture studies. Later in the decade, mussels (*Mytilus edulis*) were collected from some Dutch wrecks by the biological working group of the Dutch Underwater Sports League for tissue analysis to test for contamination by polychlorinated biphenyls (PCBs) and heavy metals, although the results do not appear to have been published.

More detailed ecological research was undertaken in the 1980s and 1990s. Dekker (1985) mentioned several species found on unspecified wrecks; then from 1986 through to 1991 Bureau Waardenburg carried out research for the Tidal Waters Division of the Ministry of Transport and Public Works (DGW) (now the National Institute for Coastal and Marine Management (RIKZ)) on a number of wrecks in the North Sea. This study produced the first inventory of the distribution and abundance of hard-substratum fauna in the Dutch North Sea, later to be incorporated into the Dutch biological monitoring programme. Some results of the study have been reported (Leewis, 1986; Leewis and Waardenburg, 1991; Van Moorsel and Waardenburg, 1990, 1991; Van Moorsel *et al.*, 1989, 1991; Waardenburg, 1987a, 1987b, 1989).

A.C. Jensen et al. (eds.), Artificial Reefs in European Seas, 419–434
© 2000 Kluwer Academic Publishers. Printed in Great Britain.

Oil and Gas Production Platforms

There is more information about the communities on and around oil and gas production platforms (rigs) than wrecks; data describing epibiotic fouling on the legs of the supportive steel 'jackets' is important to an operating company because such growth influences the stability of the whole platform in rough weather. There may also be some public relations (PR) value in providing information about the ecological value of such a structure. A platform was included in our study (Van Moorsel *et al.*, 1991; Waardenburg, 1987a).

Dikes and Sea Walls

The hard-substrata flora and fauna of dikes and sea walls have been studied extensively. Early research on algal communities was carried out by Den Hartog (1959) and Nienhuis (1969, 1980). Several research projects have been carried out in the Delta waters of the south-west Netherlands by the Tidal Waters Division and Bureau Waardenburg (De Kluyver and Leewis, 1994; Leewis and ter Kuile, 1985; Leewis and Waardenburg, 1989, 1990; Leewis *et al.*, 1994; Meijer and Van Beek, 1988; ter Kuile and Waardenburg, 1986; Waardenburg, 1988; Waardenburg *et al.*, 1984).

Wrecks and Steel Jackets Studied

Research was undertaken during the summer months (weather conditions during the rest of the year being too rough to be able to work properly) by diving biologists. A total of 21 wrecks and a steel jacket were visited. A group of eight wrecks, lying 10 km offshore and parallel to the coast between Ostend and Rottum were dubbed 'group 1'. A second group of five wrecks were situated more or less at right angles to the coast off the Hook of Holland up to the Brown Bank (Fig. 1). Six wrecks lay within a few km of an offshore platform, bisecting wreck groups 1 and 2. Finally, a few wrecks were chosen 100 and 150 km offshore (wrecks U and Z), to obtain additional information about wreck communities distanced from coastal waters.

Most of the wrecks were surveyed at least once, some more frequently. Wreck C was visited 12 times in 1986. Several working methods were tested on the wreck in order to establish how information could be gained in the most efficient way.

Wreck M, which sank in 1985, provided an excellent opportunity to study community development and succession. An older wreck near M, designated M', provided comparative data on a 'mature' wreck community in the locality.

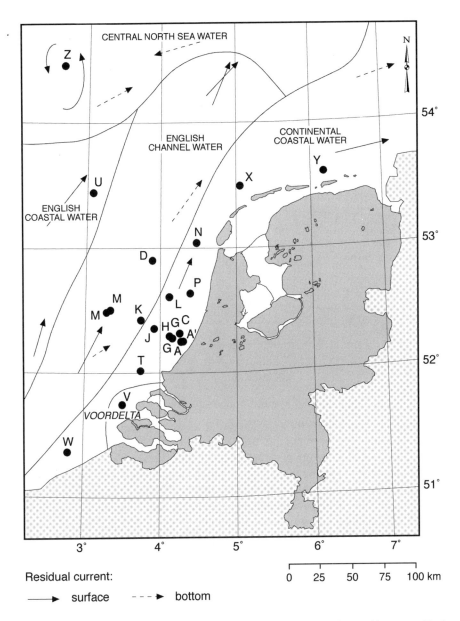

Figure 1. Chart of locations visited. The main water masses, and the surface and bottom residual current direction are indicated.

Table 1.   Locations and details of the sites visited.

Code	Name	Location		Depth (m)	When sunk
A		52 16 41 N	04 17 24 E	14–20	
A'	Noordwijk	52 16 43	04 17 25	18	
C		52 18 42	04 15 29	17–22	WW II
D	Submarine	52 55 32	03 54 13	28	
G	Delft (?)	52 17 12	04 09 33	23	27/10/41?
G'		52 17 086	04 09 22	20–24	WW II
H		52 17 54	04 08 54	20–21	
J		52 23 37	03 56 53	27	
K		52 26 146	03 44 03	22–27	
L	Leliegracht	52 36 061	04 07 39	21–28.5	28/09/73
M	Twin (Linda M)	52 30 314	03 19 18	26–32	23/11/85
M'		52 29 456	03 17 00	30–34	
N	Kiphissia (?)	53 03 40	04 30 46	22–26	17/05/43?
P	Cathar. Duyvis	52 39 534	04 24 179	20–24	30/01/53
T	Internos	52 02 107	03 44 13	20–28	4/80
U	Alpha H	53 27 22	03 03 49	30–34	26/06/83
V		51 48 585	03 29 319	20–30	
W	LCT 457?	51 24 455	02 43 483	20–23	5/44?
X		53 27 596	05 03 441	25–29	
Y	Donau	53 37 391	06 13 079	20–25	07/10/69
Z	oil/gas rig?	54 28 18	02 37 51	23–29	
▲	Platform	52 16 26	04 17 46	1–18	

Methods

Sampling was undertaken by scientific divers working in buddy pairs. One diver recorded all the species found and estimated their percentage cover on different parts of the wreck. The other diver collected small or cryptic organisms which might have been missed by the first diver and took samples for biomass analysis. These were taken by scraping a surface area of 625 cm^2 with a 'stopping-knife'. An attempt was made to scrape surfaces considered to provide a representative sample for the species distribution over the complete wreck surface. In the laboratory samples were dried at 80°C until constant weight was obtained and burned at 575°C to obtain ash-free dry weights (AFDW).

To obtain more accurate percentage cover estimates of the various fauna groups and communities, photographs were taken along transects laid across representative wreck areas. Various structures, such as vertical and horizontal plates of steel, wood, steel debris, insulation material, and nets lost by fishermen, were included. Photographs (colour slides) were taken sequentially along the transects to obtain overlapping records of the surface. Each image recorded approximately 1040 cm^2 surface area. After projection of the slides onto a screen, species cover was measured at an accuracy of 5–10%. An adaptation of the Braun-Blanquet scale (1964), was used to describe cover and frequency (Table 2).

Total cover could be higher than 100% when the cover of each species was summed with the others. This was an effect caused by epizoic animals using sessile animals as settlement surfaces. When describing mobile animals (such as fish) or organisms which were difficult to observe, i.e. small organisms, rather than percentage cover the scale was modified (0, ++ and +++) (Table 2) to provide presence and relative abundance data.

In total 17 photo transects were laid out with a total length of 54.8 m on the wrecks studied. From the records taken during the research, provisional faunal groupings and their characteristic or diagnostic species were identified. TWINSPAN two way indicator species analysis (Hill, 1979) was used to obtain insight into the spatial distribution of species over the wrecks. To obtain more data and to determine faunal communities, wreck C, situated relatively close to a harbour, was studied extensively.

Results

Species Composition

A total of 127 species or higher taxa (Table 3) were found on and associated with, the wrecks. Some hydroids, nudibranchs and small brachyurans were not identified to species level.

Table 2. Adapted scale of Braun-Blanquet used to describe species cover and abundance.

0	=	present, no covner determined
r	=	very few individuals, the species is rare
+	=	the species is present, cover less than 5%
1	=	individuals abundant, cover up to 5%
2	=	individuals random, cover 5–25%
3	=	individuals random, cover 26–50%
4	=	individuals random, cover 51–75%
5	=	individuals random, cover 76–100%
++	=	species is relatively abundantly present, no cover determined
+++	=	species is abundantly present, no cover determined

Table 3. Number of species per phylum on the wrecks.

Porifera	9
Coelenterata	18
Polychaeta	10
Mollusca	23 (including 11 nudibranchia)
Crustacea	25
Bryozoa	10
Echinodermata	4
Tunicata	5
Pisces	19
Other phyla	4

Species composition on the wrecks differed considerably from the infauna of the sedimentary seabed in the same area (see Bergman *et al.*, 1991) (Fig. 2).

Three things were obvious to the researchers: polychaetes were more abundant in the soft seabed infauna communities; the biological diversity on wrecks was much larger than in the surrounding sediments and the contribution of crustaceans to the population was about the same in both habitats.

The epifauna of the sedimentary seabed showed some resemblance to the wreck fauna, although only a few species occurred in both habitats. For convenience, species were divided into three categories: sessile fauna like hydroids, barnacles and tunicates, constituting 46% of all species; mobile fauna, like crabs, nudibranchs and sea stars, 39% of the species list; and fish, 15% of all species.

A few species were nearly always present on the wrecks: *Metridium senile, Asterias rubens,* cirripedia, tube dwelling Gammaridea, and *Cancer pagurus.* A little less important were: *Halichondria panicea, Caprellidae, Diadumene cincta, Hydractinea echinata,* and *Tubularia* spp.

Percentage Cover

Based on the photo-transects from the wrecks between 1986 and 1989 the mean percentage cover for all species were established. Data from wreck M were not included in this calculation because the structure of the communities on this wreck was still developing. The other wrecks were all more than 10 years old, except for T (sunk 1980) and V (sunk 1983).

Total cover was estimated at 103% with *Metridium senile* (37%), *Halichondria panicea* (11%) and an amphipod/turf community (32%) providing 80% of the total cover. Another 10% was dominated by *Hydractinea echinata, Diadumene cincta, Tubularia larynx* and *T. indivisa* and a final 21 species made up nearly 5%, the remaining substrata (8%) being uncolonized.

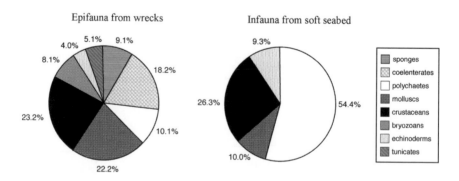

Figure 2. Percentage species number of the most important taxonomic groups. Left: epifauna from wrecks; right: infauna from soft seabed (after Duineveld *et al.*, 1990).

Faunal Communities

In general *Metridium senile, Jassa* spp. and *Halichondria panicea* appeared to be the dominant organisms. Information gained from wreck C was used as a starting point from which to compare communities found on other wrecks.

The following faunal communities were distinguished.

The *Metridium senile* community

This community, in which *Metridium senile* is dominant, excluding nearly all other species, was particularly obvious on horizontal surfaces. However, vertical and irregular structures could also be dominated by *Metridium senile*.

The Tube Dwelling Amphipod Community

This community was mostly dominated by *Jassa* spp. but *Corophium sextosa* and *C. acherusicum* were also involved. The community could be found anywhere on a wreck. Sometimes tube mats were less than 1 mm thick, but layers more than 10 cm thick were found. These layers consisted of small tubes built of silt and detritus and sometimes trapped a lot of sediment. On horizontal surfaces *Sagartia troglodytes* was often found associated with the sediment trapped by the amphipod tubes.

The *Halichondria panicea* community

This was predominantly found on vertical structures. Other conspicuous species were *Diadumene cincta* and *Sagartia troglodytes*. Caprellidae were often found holding on to the surface of the sponge while feeding. *Ophiothrix fragilis* was often found on this sponge.

The *Hydractinia* spp.–*Cuthona nana* community

This community, mostly found on surface edges and protruding structures, like masts and railings, was usually only several millimetres thick. On wreck M, it had developed into branched layers several centimetres thick. It also covered flat surfaces of several dm² up to several m², and attracted attention because it excluded all other species, except for the nudibranch *Cuthona nana*, which feeds upon *Hydractinia* spp. and was sometimes present in considerable numbers.

The Campanularidae–*Tubularia larynx* community

This was found as an early colonization stage on the recently sunk wreck (M). It was also found on lost fishing nets, edges of substratum and new substratum that had become available when layers of rust peeled off the wreck. It was considered to be a pioneer community.

The *Psammechinus miliaris* community

This species was not dominant, but was characteristic (on wreck M) as a stage in the succession following the Campanularidae–*Tubularia larynx* community.

The *Mytilus edulis* community

This community, in particular, was found on the upper part of the platform steel jacket. It was characteristically intertidal and covered the jacket in a very thick layer, about 65 cm deep (it appeared that a thicker layer would fall off under its own weight).

Attempts were made to establish whether the type of substratum affected the type of cover. Four types of substratum were compared: metal, rust, paint and wood. Differences in growth between the first three were small, while there was not enough wood available on the studied wrecks to obtain an adequate sample from this substratum.

Fish Species

Benthic (i.e. near hard substrata) species such as gobies (*Pomatoschistus* spp.) and two types of sea scorpion (*Taurulus lilljeborgi*, *Taurulus bubalis*) and the bullrout (*Myoxocephalus scorpius*) were found, though not on all the wrecks. A variety of fish species were found in and around the wrecks; numbers varied considerably. Members of the cod family were common; pouting (*Trisopterus luscus*), whiting (*Merlangius merlangius*), cod (*Gadus morhua*), and pollack (*Pollachius pollachius*). Near some wrecks scad (*Trachurus trachurus*), bass (*Dicentrarchus labrax*) and wrasse (Labridae) were found. No mackerel (*Scomber scombrus*) were seen by divers, but they were present in great numbers close to several wrecks and all through the water column, as boat-based catches showed.

The absence of cod (*Gadus morhua*) around wrecks in the southern part of the study area may have been due to the higher than normal (+3°C above average) water temperatures recorded in that part of the North Sea while the survey was carried out. Monitoring of the fish fauna in the Eastern Sheldt has shown that during the summer months cod move out of this basin (Meijer and Philippart, 1982).

Examination of the stomach contents of a number of fish showed that they were apparently partly dependent on the wrecks for food. The extent of this dependence is not yet clear (see Van Moorsel *et al.*, 1991).

Spatial Variation

All Braun-Blanquet observations were analysed using TWINSPAN, allowing wrecks to be grouped according to the abundance and percentage cover of the species recorded at each site. Ten different partial analyses were done, all reported in Van Moorsel *et al.* (1991).

The offshore platform separated from all the wrecks, the steel jacket having a very different species composition. The main features were the high mussel cover and the presence of three algal species.

The second partition distinguished a group of wrecks in coastal water (see Fig. 1): P, N, X, Y, V, W, A, A', G, G', H, D, L, J, and T (characterizing species *Halichondria panicea* and *Hyas araneas*) and an offshore group: K, M, M', U and Z (characterizing species *Sagartia elegans, Urticina felina, Alcyonium digitatum, Psammechinus miliaris* and *Pomatoceros triqueter*).

Within the offshore group, wreck Z separated from the rest because a number of species were unique to this wreck. Wreck M showed some features characteristic of incomplete species/assemblage succession.

Within the coastal water group, there was a separation between the southern wrecks V and W and the northern wrecks P, N, X, and Y. Typically in the southern group *Sagartiogeton undatum, Gadus morhua* and *Hyas araneas* were absent, and the communities on wrecks V and W grouped them weakly to A, C, G and H. On northern wrecks *Ophiothrix fragilis, Haliclona oculata* and *Hyas coarctatus* were conspicuously absent and diversity of fish was relatively high. These northern wrecks were weakly connected to A', G', T, J, L, and D.

Temporal Variation

Some of the TWINSPAN analyses showed evidence of a temporal variation in species abundance and cover. The winters of 1985–1987 were cold, while those of 1988–1990 were quite mild. Post-1987 observations reported the presence of *Sagartia elegans, Pomatoceros triqueter, Pisidia longicornis, Galathea squamifera, Diplosoma listerianum* and *Aplidium glabrum* which had been rare or absent previously. These species also appeared inshore. On wreck M the anemone *Actinothoe sphyrodeta* appeared after 1987, while *Liocarcinus puber* and *Ophiothrix fragilis* were observed more frequently. The latter species cannot cope with very low water temperatures (below 3°C). Research on hard substratum in the Eastern Scheldt has shown that after a series of very cold winters *Ophiothrix fragilis* (brittle star) disappeared, whilst it returned, and often in great numbers, after mild winters (Leewis and Waardenburg, 1990; van der Hurk and Waardenburg, 1987). This apparent temperature-related change in species composition happened in parallel to the variation on the wrecks. A temperature limitation may also explain why this species of brittle star, and some others, are absent from the northern wrecks.

Biomass

A minimum of two and a maximum of six biomass samples were collected from each wreck until 1990. These were insufficient to determine clear differences between the wrecks. In 1990, however, 20 samples were taken on each of the three wrecks visited, and in 1991 six samples were taken at wreck C and four at wreck M (Fig. 3). The 139 samples that were taken during the whole study did give an

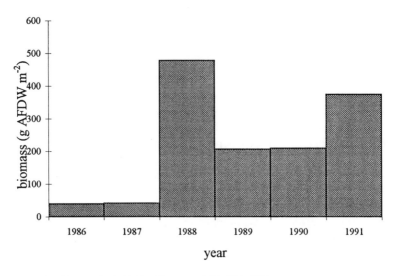

Figure 3.   Biomass (g AFDW m⁻²) on wreck M 1986–1991.

indication of relatively high biomass values on the hard substrata. *Metridium senile* constituted the bulk of the biomass. Only the jacket of the offshore platform (called 'Meetpost Noordwijk') showed biomass values of *Mytilus edulis* in approximately the same range as those of *Metridium senile* on the various wrecks.

The average biomass of samples taken in 1986 and 1987 was 1072 g AFDW m⁻². In the years after that, sample biomass appeared to be much smaller. It was assumed that this apparent decrease was partly caused by the increased number of samples (reducing error), and particularly by the fact that in the first period most samples were taken on plumose anemone or sponge-dominated surfaces. In later years the divers tried to select sampling areas with an overall representative cover. Changes in species composition may also have played a role in this decrease, these were paralleled by the hard-substrate faunal communities on estuarine dikes in the south-west of The Netherlands (Leewis and Waardenburg, 1990).

Wreck Biomass Compared to Sedimentary Seabed Biomass

Available surface area per wreck was determined by measuring all inner and outer surfaces available for growth on two wrecks, W (from 1944 and collapsed) and T (from 1980 and intact), of average dimensions (ca. 40 m long and 8 m beam). These were supplemented by calculations made using theoretical data provided by a shipyard. In this way the available surface of the two wrecks was estimated to be 3730 and 1911 m² being, respectively, 33% and 51% of their potential available surface. The mean value from wrecks T and W was 2821 m². Taking (from Table 3) the mean biomass m⁻² on the wrecks (642 g m⁻²), a total biomass per wreck was estimated at about 1.8 tonnes. Literature and archive research suggests that some 10 000–15 000 wrecks may exist in Dutch coastal waters (see e.g. Leewis and

Table 4. Biomass on the wrecks (g AFDW m^{-2}). Numbers in brackets are the number of samples taken.

Wreck	1986/87	1988	1989	1990	1991
A'			395 (2)		
C	1071 (4)		286 (6)	518 (20)	532 (6)
G'	333 (2)		645 (4)		
J	1033 (2)				
K	452 (2)		408 (5)	428 (18)	
L			358 (5)		
M	41 (2)	478 (1)	207 (2)	210 (20)	375 (4)
M'			774 (2)		
W	1025 (2)				
V	1236 (1)				
T	1954 (3)		627 (3)		
P	1458 (3)				
N	1377 (2)				
X	802 (2)				
Y	630 (2)				
D	1549 (2)				
U	542 (2)				
Z	260 (2)				
Mean*	1027 (27)		477 (33)	437 (38)	545 (10)

*Mean 1986–87 excludes data from M.

Waardenburg, 1991; Van Moorsel *et al.*, 1991). Total wreck biomass could be in the region of 18 000–27 000 tonnes. Groenewold and Van Scheppingen (1988) estimated average biomass of the seabed (sand and silt) in Dutch coastal waters of 12.3 g AFDW m^{-2}. As the total surface of the Dutch continental shelf is 56 000 km^2, an estimate of the soft seabed biomass amounts to 688 800 tonnes. Thus, the biomass on the wrecks would be some 2.6–3.9% of total benthic biomass. The distribution of soft seabed biomass is patchy; generally it decreases with distance from the coast (see e.g. Rachor, 1982), as does the number of wrecks.

Another way of expressing the impact of a wreck is by comparing the potential biomass of the seabed area on which it lies (the projection surface area) with the estimated biomass for the total surface area of that wreck. For the two measured wrecks W and T the total surface area was between 4 and 7 times the projection surface area, respectively. Biomass per m^2 on the wrecks was 52 times that on soft seabed. This means that the biomass per wreck was 209–364 times that of the projection surface of soft seabed.

Wrecks can also be quantified by the quantity of available hard substrata that their presence adds to the marine environment. Area estimates suggest that Dutch wrecks could provide 28.2–42.3 km^2 of hard substrata.

Succession

Wreck M (the M.V. Twin) sank in November 1985. The wreck was visited annually. The species composition found on the wreck differed from other wrecks surveyed. In 1986 Campanulariidae and *Tubularia larynx* dominated the fauna. During 1987 a big change in the composition of species and cover took place. A large number of, apparently full-grown, sea urchins (*Psammechinus miliaris*) were noted as were mussels, plumose anemones (*Metridium senile*) and dahlia anemones (*Taelia felina*). Fully-grown specimens of *Metridium senile* sometimes detach themselves and are transported by currents to new locations. This would explain the sudden appearance of full-grown specimens, something that also occurred on the Noordwijk artificial reef (Leewis and Hallie, chapter 17, this volume). Juvenile *Asterias rubens*, found in great numbers in 1986 on wreck M, were not seen in 1987 or thereafter, whereas full-grown specimens were recorded in 1987 and 1988. Campanularidae and *Tubularia larynx* were not observed after 1986. These developments indicate that a process of succession was occurring.

The number of species in the Braun-Blanquet samples from wreck M increased gradually from 6 in 1986 to 35 in 1991 (Fig. 4). If the species found in the biomass samples and on the photographs were included, the increase was from 6 to 44 species. The number of new species recorded each year decreased, only three additions were found in 1991. It was therefore assumed that population succession was approaching stability.

TWINSPAN analyses show that from 1990 onwards wreck M was in the same group as other wrecks in its (relative) vicinity (M', K, and U). A separate TWINSPAN comparison of the wreck M and M' datasets showed three stages in wreck M faunal development: the observations from 1986 were clearly unique to M; in the period 1987–1989 the observations on M were still separated from those on M', but they show similar trends; the 1990 observations on M grouped with the 1989 and 1990 data from M', indicating that, after 5 years, species composition on M was approaching that of the older wreck. However, in 1990 M still had a lower percentage cover of sponges and *Pomatoceros triqueter* than wreck M', and *Metridium senile*, dominant on M', only started to become important in 1990. In 1991 the sponge species *Halichondria panicea* and *Mycale macilenta* were found, suggesting that the sponge fauna was developing towards the type of community found on other wrecks studied.

The communities on the other wrecks leave the impression that succession is a continual process because the wrecks themselves are continuously changing as the structure rusts, layers peel off and superstructure collapses. These effects, modified by rough weather conditions, create new settlement sites allowing new colonizers to recruit to the wreck community.

Discussion and Conclusions

This work has created a basic wreck fauna data set which has potential as a tool for monitoring environmental change; the whole network of thousands of wrecks is available for the study of anthropogenic effects. TWINSPAN analyses showed that the largest differences in species composition between the wrecks were found relative to different water masses in the Southern Bight of the North Sea.

Further work on the bioaccumulation abilities of various species such as *Metridium senile* is important to establish if wreck organisms can be used as suitable indicator organisms for ecological monitoring.

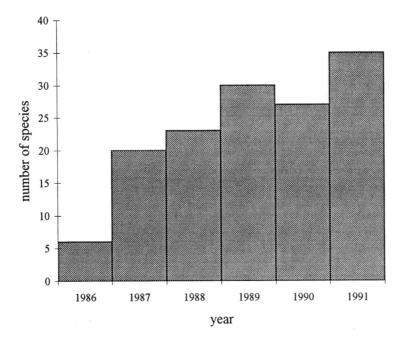

Figure 4.   Increase in the number of species found on wreck M.

The study of the species composition and biomass on one recently sunk wreck has revealed a pattern of succession, leading to a more or less stable situation after 5 or 6 years. Although the population may be slightly different from that in other water masses or depths (this wreck was in English Channel water at 35 m depth), observations of the fauna on other wrecks confirm the overall pattern of biological succession.

The species composition of the typical hard-substratum fauna on wrecks differs greatly from that of sedimentary sea floor. The greater biodiversity was an obvious feature of the wreck communities when compared to the surrounding seabed fauna. Biomass on the wrecks was also much larger than on/in the same area of soft sea floor.

As a consequence of the chosen research methods, aiming at an overall impression of the fouling communities on as large a number of wrecks as possible in different water masses, the range in biomass values found per wreck was large (179–754 g m^{-2}). In 1990, a more accurate picture was obtained on three wrecks, by taking 20 samples from each site. It is noteworthy that the overall mean biomass for 1990 was nearly the same as that for 1989, in which year 10 wrecks were sampled, the maximum number of samples per wreck being only six.

Allowing for the large range in biomass values the relative importance of wrecks on the Dutch continental shelf in terms of biomass has been shown to be in the order of 2.6–3.9% of the total benthic biomass. Wrecks are not major contributors to Dutch coastal zone biomass. However, it is considered that the local impact of wrecks is significant, particularly when the influence on fish populations is included. The question of whether wrecks attract and concentrate fish or provide circumstances which promote growth (such as food, shelter from predators and currents) is one of the most important to be settled in the context of the study of wrecks and artificial reefs. A study of productivity and the rate that produced biomass enters into and passes through the food chain may provide an insight into the importance of a wreck to its inhabitants.

Acknowledgements

The authors wish to thank the crew of the M.S. Volans and M.S. Octans, and J.H. Kamphuis en C. v. d. Hil from the Ministry of Transport, Public Works and Water Management, North Sea Directorate, for their technical assistance and co-ordination. Diving biologists were: P. van der Hurk, E.J. Kruijswijk, G.W.N.M. v. Moorsel, H.J.J. Sips, and H.W. Waardenburg.

References

Bergman, M.J.N., H.J. Lindeboom, G. Peet, P.H.M. Nelissen, H. Nijkamp and M.F. Leopold. 1991. Beschermde Gebieden in de Noordzee – noodzaak en mogelijkheden. Report Neth. Inst. of Sea Research (NIOZ) and Ministry for Agriculture, Nature Management and Fisheries, dept. NMF, Texel and The Hague. NIOZ-report nr. 1991-3.
Braun-Blanquet, J. 1964. *Pflanzensociologie, Grundzüge der Vegetationskunde*. Springer, Wien/New York.
Den Hartog, C. 1959. The epilithic algal communities occurring along the coast of the Netherlands. *Wentia*. **I**: 1–241.
Dekker, R. 1985. Over de fauna op en rond scheepswrakken. *Het Zeepaard*, mei 1985: 104–105.
De Kluyver, M.J. and Leewis, R.J. 1994. Changes in the sublittoral hard substrate communities in the Oosterschelde estuary (SW Netherlands), caused by changes in the environmental parameters. *Hydrobiologia*. **282–283**: 265–280.
Duineveld, G.C.A., P.A.W.J. de Wilde and A. Kok. 1990. A synopsis of the macrobenthic assemblages and benthic-ETS activity in the Dutch sector of the North Sea. *Netherlands Journal of Sea Research*. **26**: 125–138.
Groenewold, A. and Y. Van Scheppingen. 1988. De ruimtelijke verspreiding van het benthos in de zuidelijke Noordzee. Report no. 02 (88-14) project MILZON-benthos, North Sea Directorate, Tidal waters Division and SBNO, The Hague: pp. 1–19.
Groenewold, A. and Y. Van Scheppingen. 1989. De ruimtelijke verspreiding van het benthos in de zuidelijke Noordzee, voorjaar 1988. Report nr 90-01, project MILZON, North Sea Directorate/Tidal waters Division/SBNO: pp. 1–27.
Hill, M.O. 1979. A Fortran Program for Arranging Multivariate Data in an Ordered Two-way Table by Classification of the Individuals and Attributes. Ecology and Systematics, Cornell University, Ithaca, New York.
Hiscock, K. 1980. Marine life on the wreck of the M.V. 'Robert'. *Reports of Lundy Field Society* 32: 40–44.

Leewis, R.J. 1986. Epifauna of North Sea Shipwrecks and Offshore Platforms. Contribution no. 14 Quality Status Report North Sea, subregion 4: 2 pp.

Leewis, R.J. and C. ter Kuile. 1985. Ecotoxicologische verkenningen m.b.t. ertsslakken in waterstaatswerken. *Vakbl. v. Biol.* **65**: 43–49.

Leewis, R.J. and H.W. Waardenburg. 1989. The flora and fauna of the sublittoral part of the artificial rocky shores in the south-west Netherlands. *Progress in Underwater Science*. **14**: 109–122.

Leewis, R.J. and H.W. Waardenburg. 1990. Flora and fauna of the sublittoral hard substrata in the Oosterschelde (The Netherlands) – interactions with the North Sea and the influence of a storm surge barrier. *Hydrobiologia*. **195**: 189–200.

Leewis, R.J. and H.W. Waardenburg. 1991. Environmental impact of shipwrecks in the North Sea; positive aspects: epifauna. *Water Science and Technology*. **24**: 297–298.

Leewis, R.J., H.W. Waardenburg and M.W.M. van der Tol. 1994. Biomass and standing stock on sublittoral hard substrates in the Oosterschelde, south-west Netherlands. *Hydrobiologia*. **282–283**: 397–412.

Meijer, A.J.M. and R.J.L. Philippart. 1982. Monitoring-onderzoek aan de visfauna van de Oosterschelde, resultaten 1979–1982. Bureau Waardenburg BV, Culemborg, unpublished report.

Meijer, A.J.M. and A.C. van Beek. 1988. De levensgemeenschappen op harde substaten in de getijdezone van de Oosterschelde; typologie, kartering, relaties met substraat, oppervlakte-berekeningen, gevolgen van dijkaanpassingen. Bureau Waardenburg BV, Culemborg, unpublished report.

Nienhuis, P.H. 1969. The significance of the substratum for intertidal algal growth on the artificial rocky shore of The Netherlands. *Intern. Rev. ges. Hydrobiologia*. **54**: 207–215.

Nienhuis, P.H. 1980. The epilithic algal vegetation of the SW Netherlands. *Nova Hedwigia*. **33**: 1–94.

Rachor, E. 1982. Biomass distribution and production estimates of macro-endofauna in the North Sea. ICES, C.M. 1982/L:2.

Ter Kuile, C. and H.W. Waardenburg. 1986. Kolonisatie van natuurlijke en kunstmatige harde substraten in de Oosterschelde. Bureau Waardenburg BV, Culemborg, unpublished report.

Van 't Hof. 1970. Biologisch onderzoek bij scheepswrakken in de Noordzee. In: Yearly report 1970 Foundation Diving Research. Also in: J.R. Ferwerde (ed.): *Sportduiker Complet*: pp. 27–32.

Van der Hurk, P. and H.W. Waardenburg. 1988. De Brokkelster in de Oosterschelde. Bureau Waardenburg BV, Culemborg, unpublished report.

Van Moorsel, G.W.N.M. and H.W. Waardenburg. 1990. De fauna op en rond een aantal wrakken in de Noordzee in 1989. Bureau Waardenburg, Culemborg.

Van Moorel, G.W.N.M. and H.W. Waardenburg. 1991. De fauna op wrakken in de Noordzee in 1990. Report Bureau Waardenburg, Culemborg.

Van Moorsel, G.W.N.M. and H.W. Waardenburg. 1992. De fauna op wrakken in de Noordzee in 1991. Report Bureau Waardenburg, Culemborg.

Van Moorsel, G.W.N.M., H.J.J. Sips and H.W. Waardenburg. 1989. De fauna op en rond wrakken in de Noordzee in 1988. Bureau Waardenburg, Culemborg.

Van Moorsel, G.W.N.M., H.W. Waardenburg and J. van der Horst. 1991. Het leven op en rond scheepswrakken en andere harde substraten in de Noordzee (1986 tot en met 1990) – een synthese. Bureau Waardenburg, Culemborg.

Waardenburg, H.W. 1987a. De fauna op een aantal scheepswrakken in de Noordzee 1986. Bureau Waardenburg BV, Culemborg, unpublished report.

Waardenburg, H.W. 1987b. De fauna op een aantal scheepswrakken in de Noordzee 1987. Bureau Waardenburg BV, Culemborg, unpublished report.

Waarden, H.W. 1988. Autoecologische gegevens van diverse aan harde substraten gebonden mariene organismen. Bureau Waardenburg BV, Culemborg, unpublished report.

Waardenburg, H.W. 1989. De fauna op een aantal scheepswrakken in de Noordzee 1988. Bureau Waardenburg BV, Culemborg, unpublished report.

Waardenburg, H.W., A.J.M. Meijer, R.J.L. Philippart and A.C. van Beek. 1984. The hard bottom biocoenoses and the fish fauna of Lake Grevelingen, and their reactions to changes in the aquatic environment. *Water Science and Technology*. **16**: 677–686.

26. 'Periphyton' Colonization: Principles, Criteria and Study Methods

ANNALISA FALACE and GUIDO BRESSAN

Department of Biology, University of Trieste, via L. Giorgieri, 10, 34100 – Trieste, Italy

Introduction

When submerged in sea water, all objects, whether artificial or natural, are soon covered by fouling. Some workers consider sediment and silt deposited on substrata to be 'fouling' and use the term 'biofouling' for animal and plant aggregates. In terms of a biocenosis biofouling cannot be defined as a distinctive biological entity since it varies in accordance with a multitude of environmental factors (Relini, 1977).

Defining 'Periphyton'

The term 'periphyton' is often used to describe vegetal biofouling but, as noted by Round (1981), has been variously applied to epiphytic, epizoic and epilithic algal communities. Further generalization of the term has occurred since its application by Vismara (1989) to aggregates of both animals and plants attached to rooted aquatic plants and outcrop from the seabed; and by Marchetti (1993), noting its use in relation to the colonization of various substrata by bacteria, fungi, moulds, algae, protozoa, sponges, nematodes and oligochaetes. The precise meaning of 'periphyton' is thus no longer unequivocal without further definition.

In the context of this chapter the term is only used to describe the algal community of artificial substrata, itself a relatively neglected aspect of artificial reef studies.

Processes of Colonization

Colonization of artificial reefs can be described as the development of a fouling community which is conditioned by a series of environmental factors (biotic and non-biotic). This conditioning may influence the presence/absence of species (e.g. their capacity to adhere to the substratum), their luxuriance and their reproduction. The combined influence of these factors produces biotic diversity, which takes place over a series of temporal phases.

It is therefore possible to distinguish events in terms of space and time.

A.C. Jensen et al. (eds.), Artificial Reefs in European Seas, 435–449
© 2000 Kluwer Academic Publishers. Printed in Great Britain.

Processes of Colonization in Space

A variety of environmental factors condition, more or less directly, the settlement of a fouling community and regulate its development (Pérès and Devèze, 1963).

Edaphic Factors

The relationships between organisms and substratum are greatly influenced by the nature of the substratum itself; thus various materials have been used in experimental studies of biofouling, e.g. cement, glass, stone, wood, metal and plastic panels.

The main physical characteristics of a substratum which influence species settlement are (Relini, 1974): surface texture, slope, outline (surface shape), electrical charge, colour and light reflection.

Taking these influences into account, various types of material have been used to build artificial reefs (Grove and Sonu, 1991). The most commonly used reef module is made of prefabricated concrete, in the shape of a cube or cylinder with triangular or round sections. Tyres and recycled materials are also used (especially in the USA), as is rock (Philippines). More recently, materials derived from electro-deposition of natural elements found in the sea and materials based on pulverized fuel ash (PFA) have been used (Bohnsack and Sutherland, 1985; Jensen *et al.*, chapter 16, this volume; Relini, chapter 21, this volume; Relini and Partrignani, 1992; Relini and Zamboni, 1991; Sampaolo and Relini, 1994).

In Italy the most widely used material for artificial reefs is concrete. Groups of 8 m^3, concrete cubes ($2 \times 2 \times 2$ m) are considered particularly effective (Various authors, 1992; Bombace *et al.*, chapter 3, this volume). Pipes are less stable than these blocks, and may be moved by currents, damaging the population living on them (Fitzhardinge and Bailey-Brock, 1989). The design and layout of reef units on the seabed must be in accordance with the objectives set, the ecology of the target species and the physical and chemical parameters of the location. If the reef is built to promote repopulation of an area by species requiring a hard substratum, as many microhabitats as possible must be created and the surface area of the structure maximized to promote benthic colonization.

To facilitate the algal component of this colonization it is particularly important to:

(1)   provide optimum surface texture in order to maximize spore adhesion;
(2)   ensure a suitably continuous boundary layer so that algae may develop quickly;
(3)   prevent deposition of suspended particles as this impedes spores from adhering to the surface and/or covers germinating algal discs;
(4)   prevent sea urchins from grazing by reducing the surface area available for these animals.

To a certain extent a reef's productivity is a function of the amount of surface area available for the settlement and growth of benthic organisms (Riggio, 1988). Thus, structures with overlapping components, such as the 4–6 m high pyramids used in

Italy, are effective as their elevation from the seabed means that varying degrees of light, temperature and other chemico-physical parameters facilitate a wide range of species (Bombace, 1977, 1981).

Critical zones for benthic settlement are surface breaks (e.g. corners and edges), giving rise to the famous 'border effect' of colonization. Initial colonization occurs in these border areas and then proceeds towards the centre of the reef unit surface. Increasing the frequency and length of corners and edges respectively, up to a certain point (which must be found experimentally for each case), will lead to a significant, short-term increase in settlement (Riggio and Provenzano, 1982).

Reef module shape is also considered to be important. Reefs with oblique or almost vertical sides are considered effective for algae. Sloping surfaces and nets may enhance algal colonization by reducing sediment accumulation on the settlement surface, which is a significant limiting factor for spore adhesion in sites with a high regimen of sedimentation (Falace and Bressan, 1990, 1994).

As far as the orientation of the structure is concerned the axis of an artificial reef should be perpendicular to the prevailing current in order to maximize the interception of nutrients, organic material and planktonic stages of colonizing biota.

Environmental Factors

The main chemical and physical factors which are important in the development of biofouling (Pérès and Picard, 1964) are:

- water temperature (which relates to depth, latitude and season);
- light (related to water transparency, slope, exposure, latitude, season and depth);
- pH, salinity, nutrients;
- currents (source, direction and strength).

Currents are one of the most important parameters in ensuring that a reef develops a biological community since they transport organic material, nutrients, spores and planktonic larvae (Baynes and Szmant, 1989). Turbulence, caused when seabed currents meet submerged bodies, may also favour the settlement of benthic organisms and the growth of macroalgae, thus increasing biodiversity (Takeuchi, 1991).

Biogeographical location

At any given site, the biological community, and particularly the algae, is affected by pre-existing ecological conditions which may be said to predestine the artificial reef towards a certain type of community development. The final result will be a population characterized by organisms typical of a given biogeographical area (Huvé, 1977 in Bressan, 1988).

When deploying an artificial reef it is important to choose the location carefully with due regard to the objectives of the placement, since the same modules, located in different environments, may develop distinct communities (Various authors, 1992; D'Anna *et al.*, chapter 6, this volume). Moreover, it is not desirable to build artificial reefs everywhere because of interaction between existing species and species typical of artificial substrata. In some sites it may even be detrimental where, for example, biological communities have already reached climax equilibrium or where there are sensitive ecosystems (e.g. coralligenous and pre-coralligenous ecosystems) (Relini, 1990).

Fouling communities vary greatly from place to place with regard to organism type and the maturity and intensity of development. On the basis of temperature, currents, salinity, degree of pollution and exposure, it is possible to distinguish (Montanari, 1991):

(1)   fouling typical of ports, bays and estuaries, characterized by populations with a reduced number of species which are relatively homogeneous (Pérès, 1961). Typical of these are cosmopolitan species (Relini and Montanari, 1973) which are more tolerant of changes in the environment (nitrogen loving, euryhaline and eurythermic species);

(2)   fouling of offshore areas where percentage cover and biomass values rarely reach the levels observed in coastal areas.

Colonization

Colonization of a hard substratum takes place in two fundamental stages which vary in temporal periods and forms according to the seasons (Persoone, 1971): these are primary slime and secondary cover.

Primary Slime

The first organisms to appear (in less than an hour) are bacteria, attracted by the high concentration of nutrients on the substratum. At first their settlement takes a reversible form (Berk *et al.*, 1981) which later becomes irreversible through the secretion of extracellular polymeric substances (Marshall, 1976).

Within a few days Cyanophyceae appear, to be later replaced by Bacillario-phycineae (diatoms). Ciliate protozoans are the last to appear, after the bacterial film is well established.

Primary slime is a very effective form of microscopic life as it is capable of adapting to extreme environmental conditions. It plays an important role in nature and its composition depends on environmental conditions (Flemming, 1990).

Secondary Cover

Secondary cover is formed after the first month of immersion and is characterized by macrofouling made up of multicellular algae and invertebrates. The first to colonize the artificial reef are rapidly growing pioneer organisms which are then replaced by more slowly growing organisms (Montanari, 1991).

Ecological Succession

According to Huvé (1971) biotic succession is a transformation that is produced through an interdependency between two successive stages. By creating new conditions, animals and plants enable a climax community to be formed. In the opinion of Huvé (1971) if there is no close relationship or interdependency between the colonizers then that is an example of community evolution or, as Sheer (1945) defined it, a 'seasonal progression' in which the order of organism appearance is the result of a progression which depends on seasonal rotation, biological types, abiotic and biotic factors.

The concept of biotic succession is controversial: the fact that primary slime is indispensable for successive development of a fouling community has yet to be demonstrated. According to some authors, primary slime appears to be a determining factor in modifying the substratum for successive colonization since it helps to trap larvae, is a source of food, modifies smooth surfaces which deter settlement, influences the chemical and electrical potential of the surface, protects organisms from any toxic substances the substratum may emit, and increases alkalinity, thus encouraging the settlement of organisms with a calcareous shell (ZoBell, 1939).

Whedon (1939–1941) states that primary slime is an essential prerequisite for successive settlement of benthic microorganisms; on the other hand, Miller and Cupp (1942) maintain that primary slime may only accelerate or facilitate the adhesion of larger organisms. However, for Huvé (1977, in Bressan 1988) the initial stage is characterized by a fortuitous meeting of colonizing species which are not interconnected by reciprocal dependency. The initial stage is, therefore, a transitory period in the genesis of a community, completely based on chance and on the absence of effective biotic interactions which selectively move it towards its final result. According to Huvé (1970) a fouling community's final form would depend on environmental, and especially chemical, characteristics. Relini (1974) recognizes that the different stages of primary slime are strictly dependent, one on the other, rather than being a simple chronological sequence.

Many studies have analysed the settlement and dynamics of benthic colonization of artificial structures although few studies have considered the algal component compared with those describing the fauna.

Algae usually colonize new reefs rapidly, even if it may take several years to reach an equilibrium (Fager, 1971). The same species may be recorded each season or a successional pattern may appear when dominant species continuously change. Studies of macrofouling on artificial reefs have highlighted how abundance and final composition of the algal community depend on the composition of the substratum, on the season when the reef is deployed (availability of spores), on environmental variables (temperature, currents) and on the vicinity of natural substrata which favour migration of species.

According to Dayton (1971), algae and sessile invertebrates that directly occupy the substratum function as important 'habitat formers' and provide secondary biotic space in the form of 'lower-storey' habitats. These lower-storey habitats provide additional space and shelter on natural and artificial reefs for fish prey items including crustaceans, polychaetes and molluscs.

It has been observed that the foraging species may contribute to the persistence of the early stages of what would otherwise be successional communities on a reef. One example comes from the artificial reefs in the Gulf of Trieste (Falace and Bressan, 1995) where, 8 years after immersion, the algal community on the reef had yet to reach a climax stage because of sea urchin grazing and the presence of phytophagic fish (Labridae) which used the algae (particularly Rhodophyceae) to build their nests. The algal population had also been affected both qualitatively (presence–absence of species) and quantitatively (coverage and algal biomass) by the inhibitory effect of excessive sedimentation (which is a feature of the Gulf of Trieste), by seasonal environmental changes and, more recently, by sea urchins grazing on the reef. An experiment was conducted to reduce the surface available for sea urchins by using nets (Falace and Bressan, 1997).

The evolution of an algal population may be modified by inoculating reef surfaces with algal spores, in the example of algal cultivation (Carter *et al.*, 1985), or by transplanting adults, during reproduction, onto purpose-built structures. Grazing species may be removed (e.g. sea urchins) or nutrients added in order to foster algal growth. Reef design features, such as irregular surfaces, will help to limit the access of grazers such as urchins to reef communities.

Methods of Study of the Algal Component

A census conducted in Italy (Falace and Bressan, 1993a) and in the Mediterranean (Falace and Bressan, 1993b) showed that studies of artificial reefs were mainly concerned with animal populations and few studies had been carried out into periphyton of artificial structures. This is still the case.

Environmental Characteristics

Benthic algae on reefs can provide useful environmental data because, being attached to a reef surface and unable to escape the effects of ecological variables, they provide an ecological indicator at the population level. In addition, the substrata of artificial reefs often present very large homogeneous areas (4 m^2 in the case of Italian pyramids), which means they are particularly suited to studies where basic experimental conditions must be the same.

Experimental Conditions

Experiments involving artificial reefs need to satisfy the following criteria for success:

(1)   Experiments need to be representative. To have a representative sampling programme and in order to minimize experimental error, a minimum sampling area has to be determined.

(2) Data must allow tests of significance. The minimum number of replicates for meaningful statistical analysis has to be determined.
(3) Methodology needs to be comparable. Techniques must ensure that biotic and abiotic samples are taken from a definite site at the same time and that compatibility with other studies is maintained, both in methodology and in data format.

Fulfilling all of these conditions means that, at the end of the study, there is the best possible picture of the environment, allowing effective testing of a working hypothesis.

Experimental Procedures

The study of macrophytobenthos on an artificial substratum can be carried out by describing the structural aspects of the algae or by assessing their ecological role in the marine ecosystem.

This kind of research can be carried out by looking at time-series data, the heterogeneity of the environment or by combining these approaches. Choosing the right experimental procedures will enable a better understanding of inter-actions between the various environmental factors: biotic and abiotic, climatic and edaphic.

Characterization of Floristic/Vegetational Components

The classic approach to studying benthic populations requires, as a first step, the creation of an inventory of all species present in the study area. After this first phase of study, it is possible to determine the synergic relationships between species within a population (in some cases a true association) by using phyto-sociological methods.

Species Inventory

To carry out a study of the flora, a herbarium should be set up in order to develop a check-list of all species present. The study ought to be carried out with a phenological approach in all seasons and over several years. Such data will provide an appreciation of: seasonal change in algal presence; changes in luxuriance of individual species; and the fertility and reproductive cycle of a species.

The phenological study (be it morphological or reproductive) requires a frequent (at least monthly) sampling programme in order to detect those species only present for short periods, to find fertile talli of those species with short reproductive cycles and to discover those species with abnormal reproductive cycles (Cormaci, 1995).

There are experimental limitations to studies of the flora. Objectively, it is difficult to collect entire algal specimens (including the attachment point of the tallus), and especially those of the 'hard' species (for example encrusting calcareous red algae, endolythic cyanophyceae). From a subjective point of view it is important to have an even representation of all species: there is a danger of over-representing well-known species and neglecting those that are less familiar. It is also vital to be up-to-date with classification nomenclature, since the criteria for identifying species change continuously as taxonomic research progresses.

Phytosociological study

A phytosociological study aims to determine the synergetic relationship between the different species of an algal community. The spatial arrangement of the vegetation is rather complex, as algal cover stratifies into layers depending on their need for light. Photophylic ('light-loving') components form the upper layers of the canopy and sciaphylic or 'shade-loving' components of the community form a sublayer, composed essentially of species (soft or calcareous) that adhere to the substratum.

This algal community structure becomes even more complex as it is, at the same time, colonized by photophylic epiphytes (on the fronds of the species of the upper layer), by sciaphylic epiphytes (in the sublayer), and also by intermediate species in a complicated ecological pattern (Cormaci, 1995).

In order to obtain significant and defendable data, it is very important to standardize sampling methods, especially as far as the minimum qualitative area and the frequency of sampling are concerned.

Sampling technique (minimum area of sampling)

Braun-Blanquet (1964) showed that the minimum area has more than just practical importance as a sampling surface, since it represents the smallest possible surface on which a community can develop and reach an equilibrium in a particular location under given environmental variables.

The technique most correctly used to study algal communities is to sample the substratum by scraping all the algae from a given, generally square, surface area. The size of the sample area must be neither too small (sampling not representative) nor too large (considerable use of time for a negligible gain of information).

As far as the minimum qualitative area is concerned, various methods have been used (Ballesteros, 1992; Bourdouesque, 1974; Bourdouesque and Belsher, 1979; Cormaci, 1995), which differ essentially in the sequence and size of the quadrants used. Several authors have also tried to define a precise point on the 'species number–surface area' sampled curve which positively defines the minimum area (Boudouresque and Belsher, 1979; Cain, 1938, 1943; Calleja, 1962; Gounot, 1969; Gounot and Calleja, 1962; Tuxen, 1970; Vestal, 1949), because the visual determination of such a point on the graph is too empirical and subjective. These attempts have not been totally satisfactory because none of the methods allow minimum area to be determined beyond any doubt. It is, however, possible to determine the limits between which it is probably situated.

With all methods it is necessary to test the homogeneity of the sampling on the basis of two fundamental criteria: the nature of the substratum (which can influence the recruitment of juveniles and maintenance of adult individuals) and the spatial distribution of quadrants. It is of fundamental importance to determine the minimum qualitative area, especially when similar populations are sampled in disparate geographical areas, and different minimum areas will have been estimated (Ballesteros, 1988; Cinelli et al., 1977). This situation also occurs in the case of artificial reefs in different biogeographical regions. According to Dhondt and Coppeljans (1977) the differences in minimum area estimations depend mainly on the method used, but also on the edaphic characteristics of the population studied (Cormaci, 1995).

The quantitative minimum area is the smallest representative surface as far as the abundance of a given community is concerned (percentage cover, biomass, etc.). It can be determined experimentally, as it was during the study of the colonization of the artificial reef in the Gulf of Trieste (Falace and Bressan, 1994). In this study the quantitative minimum area was established by using a large number of samples (44 replicates on a surface of 2.4 × 2.4 m); this choice of method, only applicable on artificial reefs (extensive homogeneous surfaces) has provided representative and significant results, thereby reducing the margin of error caused by the subjectiveness of the phytosociological method.

The minimum qualitative area must be larger than, or equal to, the quantitative minimum area in order to avoid experimental error. If the quantitative minimum area was bigger than the minimum qualitative area, the sampled surface area would, in qualitative terms, not be sufficiently representative for all the species present. Niell (1974) suggested the use of a surface–index of diversity curve in order to define the quantitative minimum area.

Once the minimum experimental area has been determined, the phytosociological analytical parameters can be recorded: percentage cover of substratum occupied by the different species, the degree of associability (density of the individuals of the respective species) and the frequency of occurrence of the different algal species.

Role of Algae in the Marine Ecosystem

To optimize sampling methods for the flora present on artificial reefs the fundamental role these organisms play in the marine ecosystem must be considered, be it as primary producers, or as principal components of environmental heterogeneity.

Primary productivity

To determine the biological production of an artificial reef, the study of primary productivity is the fundamental starting point, because algae are the first step in the formation of a food chain. Since the study of energy fluxes in an ecosystem is based on parameters that cannot be easily measured directly, the common practice is to measure parameters that are obtainable (e.g. dry weight, organic carbon, oxygen production and respiration, CO_2) and then transform these results into calories with the use of appropriate coefficients (Marchetti, 1993).

Regrettably, throughout the literature, the terms 'production' and 'productivity' are often used as synonyms (Pignatti, 1995). In this chapter the terminology of Wood (1967) is used: production is the amount of organic matter produced per unit time and productivity is the capability to produce organic matter per unit time. Production is, therefore, an absolute parameter while productivity is a relative parameter.

In an aqueous environment, primary production by phytoplankton is generally determined by measuring parameters such as O_2, ^{14}C and CO_2 (using dark and light bottles). These methods are not applicable in the study of macrophyto-benthos as it is more useful to measure the changes in biomass over time.

The estimation of organic matter at any point in time on a certain surface is defined as standing crop. The plant biomass can be expressed as wet weight (blotted dry) or dry weight. In order to measure plant biomass, it is preferable to refer to dry weight, since the wet weight is subject to differences in the state of hydration of the alga and does not always give data that can be reproduced. Dry weight is obtained by drying the plant in an oven (60°C until constant weight or 105°C for 24 h). The ash weight is obtained by placing the dried sample in a muffle furnace at 500°C for 24 h.

Another parameter that may give an indirect estimate of the primary productivity is the chlorophyll concentration per unit of surface area, which gives a good evaluation of the potential photosynthesis in an ecosystem.

The amount of plant biomass available becomes especially significant in the study of the relationship between plant and animal populations. The growth of grazers, in effect, is often limited by the available biomass of the plant species that they target. Furthermore, if the amount of biomass is compared with data obtained in a study of phenological morphology of the key species (such as *Sargassum* sp. or *Cystoreira* sp.) results can describe the developmental stage and/or vigour of a population compared to that of another location (Cormaci, 1995).

An experimental protocol to evaluate artificial reef primary production using biomass and chlorophyll concentration has been proposed (Falace *et al.*, 1998) which will ensure a reproducible and standardized procedure, from the collection of algae to the carrying out of the diverse measurements.

The methods used up to now are destructive and dubious because they result in a change in the ecosystem: the measurement itself results in a modification of the system studied. In order to avoid this problem, non-destructive methods are being tested, for example a biomass estimate done 'by eye' whilst submerged (Boudouresque, 1971), although this method is quite difficult to carry out because it is very subjective.

Environmental heterogeneity; studying biodiversity

Over time, isolation, speciation and mutations have produced very important changes in organisms which have adapted to life in specific environmental conditions. The mutual dependency between organisms and their environment has given a fundamental impetus towards a multitude of life strategies, determining a wide variety of biological life forms as a response to the physiological needs of the individual.

This variety of form represents the richness of an environment, or better still of the genome itself, and can be measured by applying the appropriate indices which synthesize the information into a single 'number' which summarizes the mean conditions of the community.

Despite adverse views aired over the decades, in general a diversity index expresses the degree of structural evolution, the maturity and the stability of an ecosystem by means of the distribution of individuals within taxonomic units.

Some authors apply diversity indices as a measure of community equilibrium, some have shown that pollution or environmental stress leads to a reduction in diversity index, both for phytobenthos and zoobenthos. In reality, the relationship between stress and diversity index is not straightforward: at low stress levels, the diversity index may increase rather than decrease. Furthermore, indices calculated with diverse algorithms and/or applied to different classes of organisms can give different results (Marchetti, 1993).

The number of species and their respective abundance are the components of the specific diversity: each species is therefore taken into account only for their presence and abundance. The number of species that are present in a community is referred to by the term 'richness', while the relative abundance (i.e.: number of individuals, cover, biomass) is referred to as 'evenness'. It is on the basis of these two variables that communities can be distinguished, because the diversity increases when the number of species and their evenness increase.

There are several techniques for measuring the specific diversity in order to quantify the complexity of an ecosystem:

- indices that measure the uniformity or non-uniformity of distribution of the relative abundances;
- indices that supply an evaluation of the disorder (entropy of a system);
- graphic techniques that are used for comparing similar habitats.

For simplicity, only the three principal categories and respective authors are mentioned below:

(1) Richness indices, which essentially measure the number of species within a precise sampling unit (Hurlbert, 1971; Margaleff, 1957, 1972; Menhinick, 1964).

(2) Evenness indices, which measure the equidistribution of abundance values (Hill, 1973; Hurlbert, 1971; Magurran, 1988; Pielou, 1969);

(3) Indices that combine the two components and therefore derive, in one way or another, from the aforementioned categories. These are the indices that should really be called 'diversity indices'. In the literature, it is this diversity which is frequently referred to as heterogeneity. In these indices the following aspects are prevalent: diversity (Brillouin, 1962; Gini, 1939; Hill, 1973; Hurlbert, 1971; Levins, 1968; McIntosh, 1967; Shannon and Weaver, 1949) and dominance (Berger and Parker, 1970; Simpson, 1949).

It is necessary to remember that the results of an analysis depend on the initial data and that the nature of sampling can introduce analytical artefacts.

Since each index 'filters' information, it is difficult to recommend the use of one or several particular indices when various situations are to be studied. Consequently it is necessary to critically apply several analytical methods to the whole of the data set, provided that these data were correctly collected.

References

Ballesteros, E. 1988. Estructura y dinàmica de la comunidad de *Cystoseira mediterranea* Sauvageau en el Méditerràneo nordoccidental. *Inv. Pesq.* **53**(39): 313–334.

Ballesteros, E. 1992. Els vegetals i la zonaciò litoral: espècies, comunitats i factors que influeixen en la seva distribuciò. Institut d'estudis Catalans. Barcelona, pp. 616.

Baynes, T.W. and A.M. Szmant. 1989. Effect of cement on the sessile benthic community structure of an artificial reef. *Bulletin of Marine Science*. **44**(2): 545–566.

Berger, W.H. and F.L. Parker. 1970. Diversity of planktonic Foraminifera in deep sea sediment. *Science*. **168**: 1345–1347.

Berk, S.G., R. Mitchell, R.J. Bobbie, J.S. Nickels and D.C. White. 1981. Microfouling on metal surfaces exposed to sea water. *Intem. Biodet. Bull.* **17**(2): 23–27.

Bohnsack, J.A. and D.L. Sutherland. 1985. Artificial reef research: a review with recommendations for future priorities. *Bulletin of Marine Science*. **37**(1): 11–39.

Bombace, G. 1977. Aspetti teorici e sperimentali concernenti le barriere artificiali. *Atti IX Cong. SIBM*. Ischia, 29–42.

Bombace, G. 1981. Note on experiments in artificial reefs in Italy. *CGPM, Etudes et Revues*. **58**: 309–324.

Boudouresque, C.F. 1971. Méthodes d'étude qualitative et quantitative du benthos (en particulier du phytobenthos). *Thethys*. **3**(1): 79–104.

Boudouresque, C.F. 1974. Aire minimale et peuplements algaux marins. *Société Phycologie de France*. **19**: 141–155.

Boudouresque, C.F. and T. Belsher. 1979. Le peuplement algal du Port-Vendres: recherches sur l'aire minimale qualitative. *Cahiers de Biologie Marine*. **20**: 259–269.

Braun-Blanquet, J. 1964. *Pflanzensoziologie Grundzüge der Vegetationskunde*. Springer. Vienna and New York. 864 pp.

Bressan, G. 1988. Appunti sulla fattibilità di una barriera artificiale sommersa nel Golfo di Trieste. Processi di colonizzazione e fitocenosi guida. *Hydrores*. **5**(6): 47–56.

Brillouin, L. 1962. *Science and Information Theory*, 2nd edn. Academic Press, New York.

Cain, S.A. 1938. The species–area curve. *Amer. midl. Natur. USA*. **19**: 573–581.

Cain, S.A. 1943. Sample-plot technique applied to alpine vegetation in Wyoming. *American Journal of Botany*. **30**: 240–247.

Calleja, M. 1962. Etude statistique d'une pelouse à *Brachypodium ramosum*. VI. Etude de la courbe aire-espèces et de l'aire minimale. *Bull. Serv. Carte phytogéogr.* **2**(B): 161–179.

Carter, J.W., W.N. Jessee, M.S. Foster and A.L. Carpenter. 1985. Management of artificial reefs designed to support natural communities. *Bulletin of Marine Science*. **37**(1): 114–128.

Cinelli, F., E. Fresi, E. Idato and I. Mazzella. 1977. L'aire minimale du phytobenthos dans un peuplement à *Cystoseira mediterranea* de l'île d'Ischia (Golfe de Naples). *Rapp. Comm. int. Mer. Médit*. **24**(4): 113–115.

Cormaci, M. 1995. Struttura e periodismo dei popolamenti a Cystoseira (Fucophyceae) nel Mediterraneo. *Giorn. Bot. It.* **129**(1): 350–366.

Dayton, P.K. 1971. Competition, disturbance and community organization: the provision and subsequent utilisation of space in a rocky intertidal community. *Ecology Monographs*. **41**: 351–389.

Dhondt, F. and E. Coppejans. 1977. Résultats d'une étude d'aire minimale de peuplements algaux photophiles sur substrat rocheaux à Port-Cros et à Banyuls (France). *Rapp. Comm. int. Mer. Médit.* **24**(4): 141–142.

Fager, E.W. 1971. Pattern in the development of a marine community. *Limnology and Oceanography*. **16**: 241–253.

Falace, A. and G. Bressan. 1990. Dinamica della colonizzazione algale di una barriera artificiale sommersa nel Golfo di Trieste: macrofouling. *Hydrores*. **7**(8): 5–27.

Falace, A. and G. Bressan. 1993a. Strutture artificiali dei mari italiani – Biblografia analitica (a tutto il IX/1992). *Boll. Soc. Adr. Sci.* **74**: 41–72.

Falace, A. and G. Bressan. 1993b. Récifs artificiels en Méditerranée. Liste bibliographique. *Bollettino Oceanografia Teorica ed Applicata*. **XI**: 247–255.

Falace, A. and G. Bressan. 1994. Some observations on periphyton colonization of artificial substrata in the Gulf of Trieste (N Adriatic Sea). *Bulletin of Marine Science*. **55**(2–3): 924–931.

Falace, A. and Bressan, G. 1995. Esperienze di strutture artificiali nel Golfo di Trieste (Nord Adriatico). *Biologia Marina Mediterranea*. **II**(I): 213–128.

Falace, A. and G. Bressan. 1997. Adapting an artificial reef to biological requirements. In Hawkins, L.E., S. Hutchinson with A.C. Jensen, M. Sheader and J.A. Williams. (eds.) *The responses of marine organisms to their environments*. Proceedings of the 30th EMBS, Southampton, September 1995, pp. 307–311.

Falace, A., G. Maranzana, G. Bressan and L. Talarico. 1998. Approach to a quantitative evaluation of benthic algal communities. In Jensen, A.C. (ed.) Report of the results of EARRN workshop 4: Reef design and materials. Report to the European Commission, contract number AIR-CT94-2144. 70pp.

Fitzhardinge, R.C. and J.H. Bailey-Brock. 1989. Colonization of artificial reef materials by corals and other sessile organisms. *Bulletin of Marine Science*. **44**(2): 567–579.

Flemming, H.C. 1990. Biofouling in water treatment. *Proc. of Intern. Workshop on Indr. Biofouling and Biocorr.*, Stuttgard: pp. 47–80.

Gini, C. 1939. *Variabilità e Concentrazione. Memorie di Metodologia Statistica. Vol. 1.* Milano.

Gounot, M. 1969. *Méthodes d'etude quantitative de la végétation*. Masson et Cie éd., Paris, 314 pp.

Gounot, M. and M. Calleja. 1962. Coefficient de communauté, homogénéité et aire minimale. *Bull. Serv. Carte phytogéogr.* **7**(2B): 181–200.

Grove, R.S. and C.J. Sonu. 1991. Artificial habitat technology in the world: Today and Tomorrow. Japan-U.S. Symp. on Artif. Habitat, Tokyo, pp. 3–10.

Hill, M.O. 1973. Diversity and evenness: a unifying notation and its consequences. *Ecology*. **54**: 427–432.

Hurlbert, S.H. 1971. The nonconcept of species diversity: a critique and alternative parameters. *Ecology*. **52**: 577–586.

Huvé, P. 1970. Récherches sur la génèse de quelques peuplements algaux marins de la roche littorale dans la région de Marseille. Thèse Doctorat d'Etat Sciences-Naturelles, 1–470.

Huvé, P. 1971. Sur le concept de succession en écologie littorale marine. *Thalassia Jugoslavica*. **7**(1): 123–129.

Levins, R. 1968. *Evolution in changing environments: some theoretical explorations*. Princeton University Press, Princeton.

Magurran, A.E. 1988. *Ecological diversity and its measurement*. Croom Helm, London.

Marchetti, R. 1993. Ecologia applicata. *Soc. Ital. Ecol.* Ed. Città Studi.

Margaleff, R. 1957. La teoria de la información en ecologia. *Mem. Real. Acad. Cienc. Artes Barcelona*. **32**: 373–449.

Margaleff, R. 1972. Homage to Evelyn Hutchinson, or why is there an upper limit to diversity. *Trans. Connect. Acad. Arts. Sci.* **44**: 211–235.

Marshall, K.C. 1976. *Interfaces in Microbial Ecology.* University Press, Cambridge, Mass.

McIntosh, R.P. 1967. An index of diversity and the relation of certain concepts to diversity. *Ecology.* **48**: 392–404.

Menhinick, E.F. 1964. A comparison of some species – individuals diversity indices applied to samples of field insects. *Ecology.* **45**: 859–861.

Miller, M.A. and E.E. Cupp. 1942. The development of biological accelerated tests (Progress report). Ann. Report San Diego Naval Biol. Lab. to Bureau of Ships, Navy Department.

Montanari, M. 1991. Biologia ed ecologia del fouling. A.I.T.I.V.A.–I.C.M.M., Atti Giornata di Studio XXXI Salone Nautico, Genova, 1–22.

Niell, X. 1974. Les applications de l'index de Shannon a l'étude de la végétation interdale. *Société Phycologie de France.* **19**: 238–254.

Pérès, J.M. 1961. *Océanographie Biologique et Biologie Marine.* Presse Université de France, Paris.

Pérès, J.M. and L. Devèze. 1963. *Océanographie Biologique et Biologie Marine. II. La Vie Pélagique.* Press Université de France, Paris.

Pérès, J.M. and J. Picard. 1964. Nouveau manuel de Bionomie bentique de la Mer Méditerranée. *Riv. Extr. Rec. Trav. Stat. Mar. Endoume, France.* **31**(47): 5–138.

Persoone, G. 1971. Ecology of fouling on submerged surfaces in a polluted harbour. *Vie et Milieu.* **22** (suppl. 2): 613–636.

Pielou, E.C. 1969. *An Introduction to Mathematical Ecology.* Wiley, New York.

Pignatti, S. (ed.) 1995. *Ecologia Vegetale.* UTET.

Relini, G. 1974. La colonizzazione dei substrati duri in mare. *Mem. Biol. Mar. e Oceanogr., N.S.,* **4**(4,5,6): 201–261.

Relini, G. 1977. Le metodologie per lo studio del fouling nell'indagine di alcuni ecosistemi marini. Atti XLIV Conv. U.Z.I., *Boll. Zool.* **44**: 97–112.

Relini, G. 1990. Una gestione razionale dell'ambiente marino. Le Pietre ed il Mare, *Riv. Prov. Liguri.* **3**(3): 27–36.

Relini, G. and M. Montanari. 1973. Introduzione di specie marine attraverso le navi. *Atti III Simp. Naz. Conserv. Natura.* 263–280.

Relini, G. and A. Patrignani. 1992. Coal ash for artificial habitats in Italy. *Rapp. Comm. Int. Mer Médit.* **33**: 378.

Relini, G. and N. Zamboni. 1991. Macrofouling su manufatti a base di cenere di carbone. Atti XXII Congr. SIBM, Cagliari.

Riggio, S. 1988. I ripopolamenti in mare. Atti IV Conv. Sicil. Ecologia, Porto Palo di Capo Passero. 223–250.

Riggio, S. and G. Provenzano. 1982. Le prime barriere artificiali in Sicilia: ricerche e progettazioni. *Naturalista Siciliano.* **4,6** (suppl.)3: 627–659.

Round, F.E. 1981. *The Ecology of Algae.* Cambridge University Press, Cambridge.

Sampaolo, A. and G. Relini. 1994. Coal ash for artificial habitats in Italy. *Bulletin of Marine Science.* **55**(2–3): 1279–1296.

Shannon, C.E. and W. Weaver. 1949. *The Mathematical Theory of Communication.* Urbana IL. Univ. Illinois Press. 117 pp.

Sheer, B.T. 1945. The development of marine fouling communities. *Biology Bulletin.* **89**(1): 103–121.

Simpson, E.H. 1949. Measurements of diversity. *Nature.* 163–688.

Takeuchi, T. 1991. Design of artificial reefs in consideration of environmental characteristics. *Japan-U.S. Symp. on Artif. Habitat,* Tokyo, pp. 181–184.

Tuxen, R. 1970. Bibliographie zum Problem des Minimalareales und der Art-Areal Kurve. *Excerpta Botanica.* **10**(4b): 291–314.

Various authors. 1992. Rapporto sul workshop barriere artificiali tenutosi ad Ancona il

14/2/92. Not. SIBM. **22**: 36–56.

Vestal, A.G. 1949. Minimum areas for different vegetations. Their determination from species-area curves. *Illinois Biology Monography*. **20**(3): 1–129.

Vismara, R. 1989. *Ecologia Applicata*. Ed. Hoepli.

Whedon, W.F. 1937–1941. Investigations pertaining to the fouling of ships' bottoms. Reports from Scripps Institute of Oceanography. La Jolla, California to U.S. Navy, Bureau of Ships.

Wood, E.J.F. 1967. *Microbiology of Oceans and Estuaries*. Elsevier, Amsterdam, 319 pp.

ZoBell, C.E. 1939. Occurrence and activity of bacteria in marine sediments. In Trask (ed.) *Recent Marine Sediments*. pp. 416–427.

27. Investing in Artificial Reefs

DAVID WHITMARSH and HELEN PICKERING

University of Portsmouth, CEMARE, Milton Campus, Locksway Road, Southsea, Portsmouth PO4 8JF, UK.

Introduction

Economic appraisal of an artificial reef proposal is an essential step in determining the role such a structure may play in, say, fishery enhancement. Reef projects may be appraised from the perspective of either a private commercial firm or society as a whole, though the emphasis of the present chapter will primarily be on the latter. The need for economic appraisal is likely to assume greater importance in the future, given that artificial reef deployment is increasingly undertaken in many countries on often loosely defined 'public interest' grounds. Without a clear notion of what costs and benefits are likely to arise, however, there is a danger that artificial reefs will be constructed in circumstances which do not justify them. Artificial reefs may often be quite expensive and difficult to construct, and given that they have the potential to cause over-exploitation (Grossman *et al.*, 1997; Pickering and Whitmarsh, 1997) there is clearly no guarantee that their economic effects will be positive. What, then, is required in order to establish whether an artificial reef proposal constitutes a 'good investment' from a social perspective?

Principles of Investment Decision Making

Appraising Investment Projects

The main characteristic of investment is that the costs incurred and the benefits generated arise over a period of time. With artificial reef projects this feature is of central importance, given that the biological processes underlying fisheries yield do not occur instantaneously and also that the physical life of a reef may extend over several decades. In order to appraise an investment project, and to help decide whether or not it should be undertaken, some procedure is needed to enumerate and evaluate the costs and benefits associated with the project over its expected life (Milon *et al.*, 2000: in press).

A point which needs to be emphasized at the outset is that any investment appraisal always involves a comparison of some kind. This is because even if there is only a single version of a project, for example, constructing a reef in a given location using one type of material, the implicit alternative is the 'do nothing' option. In this instance, in order to assess the incremental benefits of the project, the relevant comparison is between the 'with project' and 'without project' (= do

A.C. Jensen et al. (eds.), Artificial Reefs in European Seas, 451–467
© *2000 Kluwer Academic Publishers. Printed in Great Britain.*

nothing) situations. Another type of comparison might be between investment scenarios which involve alternative assumptions about the regulation of fishing activity on the reef. Specifically, it may be necessary to find out how the economic benefits of the project might change if (say) a reef were to be regulated using a different set of control measures whose effect would be to alter the harvesting pressure on the fishery resource.

Enumerating Costs and Benefits

The first step is to decide which costs and benefits are to be included, and this in turn will depend upon the project planner's objectives. For a private-sector organization aiming to maximize the return on capital, a project would normally be evaluated on the basis of its effects on the firm's balance sheet (Perkins, 1994). Assessment would define the relevant costs and benefits which, unless they directly impinged upon the financial cash flow, would not be included in the analysis. In the terminology of economics, a firm concerned simply with profit maximization would logically disregard any externalities generated by a project. For example, 'intangible' recreational benefits arising from the creation of an artificial reef would not be included in a financial analysis unless they could be appropriated by the reef owner in the form of revenue (i.e. by charging anglers). Likewise any environmental pollution costs which might be created by the reef (e.g. as a result of the leaching of heavy metals from the substratum) would not figure in a financial appraisal if the firm believed that it could avoid such costs with impunity. Financial appraisal of a reef project would similarly disregard any of the inter-active bioeconomic effects of artificial habitat creation, even though these might be quite significant. Such effects could take the form of external benefits (e.g. stock recovery resulting from the diversion of effort away from an over-exploited fishery) or external costs (e.g. stock depletion resulting from intensive harvesting pressure near a reef established in an already heavily exploited area). Conversely, taxes and grants would need to be considered by a private commercial firm since these would impact on the flow of net financial benefits associated with the project.

A public-sector organization is likely to have a wider remit than a private firm as to the costs and benefits to be included, and might be required to carry out not a financial appraisal but an economic appraisal. Here the question being considered is whether society as a whole is expected to be better or worse off if the project were to be undertaken and, accordingly, the costs and benefits of an artificial reef project need to be assessed at a national (or possibly international) level. This means that externalities cannot be ignored as they can in a private-sector financial appraisal. An economic appraisal of a public artificial reef project would need to include any recreational benefits since, even though these may not be appropriated ('internalized') by the reef operator in situations where access to the reef is free, they would nevertheless have an economic value to anglers. In the same way, external costs, such as those associated with pollution or over-exploitation, would have to be incorporated into the analysis. Taxes and grants, on the other hand, should not be included in an economic appraisal since for society as a whole they do not represent a net cost or benefit; they are merely a transfer payment from one group to another.

Valuation of Costs and Benefits

The next step is to value the costs and benefits that have been selected as being relevant. If a project is conceived of as an activity for converting inputs into outputs, then at its simplest this involves measuring the inputs and outputs in physical terms and subsequently converting them to monetary values through the use of a suitable numeraire or measuring rod.

There are a number of practical problems here. The physical effects of a project may be difficult to establish, especially where these involve complex long-term environmental changes that impact on the productivity of other sectors. For example, artificial reefs may provide important functional benefits whose effects are largely indirect (e.g. habitat restoration, coastal protection, nutrient capture) and so are not easily quantified. A further problem regarding the expected physical productivity of reef projects concerns the assumption made about the intensity of harvesting. If access to a reef is uncontrolled then stock externalities will cause catch rates to fall below their potential maximum. As Milon (1989a) has pointed out, artificial habitat plans that fail to recognize the need for effort limitation are likely to overstate the expected economic performance of a project.

Another difficulty is that not all inputs and outputs have a market price attached to them which can be used as the basis of monetization. Where a project produces 'non-traded' outputs, for example, recreational opportunities for which no charge is made, then some alternative means of valuation needs to be found. A variety of techniques are available for this purpose, most of which involve an assessment of people's 'willingness to pay' for the unpriced benefit in question (Milon *et al.*, 2000 (in press)). One which has a long history of use in recreational benefit assessment, and which has been applied to measure the recreational value of marine artificial habitats in the USA, is the travel cost method (TCM). The logic behind TCM is that the travel costs incurred by sea anglers and divers will reflect their willingness to pay (WTP) for recreational enjoyment, and hence the value of the reef or FAD for that purpose (Bockstael *et al.*, 1986; Milon, 1988). Unpriced benefits such as recreation may also be valued using what is termed the contingent valuation method (CVM), which is based on direct questioning of respondents as to the monetary value that they as individuals derive from the use (or in some cases the existence) of an environmental amenity or resource. A number of CVM studies of artificial habitats have now been undertaken (Bockstael *et al.*, 1986; Milon, 1989b; Roberts and Thompson, 1983; Samples, 1986) and these confirm that sport fishermen are willing to pay for access to artificial habitats. Rather encouragingly, it appears that CVM and TCM approaches are able to produce similar estimates of WTP for reefs or FADs in a given location (Milon, 1991).

Even where market prices are available, however, they may not be appropriate for the valuation of outputs and inputs in the economic appraisal of public-sector projects. This is because the market system itself may suffer from distortions that result in prices failing to reflect the real scarcity value to society of the product or input in question. For example, the existence of high unemployment in certain sectors means that the real cost of labour to society as a whole is likely to be somewhat less than that shown by current rates of pay. The reason for this is that unemployed labour is (by definition) unproductive, and where an investment project makes use of such labour it imposes no real cost on society since no other output has been sacrificed. Economic appraisal would need to correct for this distortion by revaluing labour so as to reflect its true opportunity cost to society, a procedure known as shadow pricing. The need for shadow pricing is especially important for project appraisal in less developed countries where market prices often give an especially poor reflection of social value.

The final problem associated with the valuation of outputs and inputs is that relative prices may change throughout the life of a project. This may arise as a consequence of the project itself, since if the scale of investment is large in relation to the existing output of the industry then the additional supplies may cause market prices to fall. This is something which needs to be anticipated at the project planning stage and requires a knowledge of the price-quantity relationship for the market concerned, what is termed in economics the elasticity of demand. Even if the project itself is expected to have no market impact, potential changes in price still need to be considered since a change in relative prices may occur for a number of other reasons. If a rise in the price of output relative to other commodities is forecast, this would affect (favourably) our assessment of the project's worth. A project that is assessed as being marginally unprofitable, based on the assumption of constant relative prices, might prove to be worth undertaking if relative prices are expected to rise. Failure to anticipate price rises thus implies that some projects will not be undertaken when in reality it might have paid for them to go ahead. Conversely, a failure to foresee a fall in the relative price of output means that there will be occasions when projects are undertaken which turn out to be losers. Firms concerned with minimizing risk therefore need to anticipate if and when other investors may be planning to enter the same market, because this is likely to cause the relative price of the product to fall. The farming of Atlantic salmon in Europe and Scandinavia provides a salutary illustration of this. The fall in the relative price of salmon that has occurred throughout the 1980s and 1990s has no doubt given cause for thought to those who had established salmon farms at a time when market conditions were more buoyant.

Discounting

Once the enumeration and valuation steps are complete the analyst should be in a position to draw up a profile of the costs and benefits associated with each year of the life of the project. Typically this might show that in the early years the net benefits are strongly negative (because of heavy capital expenditure unmatched by economic output), and only later is the project expected to yield a positive net return. The task is to find some way of reducing this irregular stream to a single figure, and having done so to decide whether the project constitutes a 'good investment'.

The standard procedure for evaluating the stream of net benefits associated with an investment project is to undertake what is termed discounting. The rationale behind discounting is that money has a time value, inasmuch as the resources allocated to a given project could alternatively be used in some other (possibly more worthwhile) activity. The longer it takes for a project to yield a return, the more these returns need to be devalued (i.e. discounted) in order to reflect the benefits which could have been earned elsewhere but have now been forgone. In practice this requires the selection of a suitable percentage figure (termed the discount rate) which can then be applied to the net benefits arising in each and every year of the project's life. The simplest interpretation of the discount rate is the cost of capital used to finance the project, measured either as the lowest interest rate at which funds can be obtained from the capital market (in situations where the project implementor is a net borrower), or as the highest rate of return which could be obtained if the funds were invested elsewhere (in situations where the project implementor is a net lender). For a private-sector project, where the objective is purely commercial, the discount rate would be based on the financial circumstances of the individual firm and the level of real interest rates prevailing in the capital markets. It is recognized, however, that in practice capital markets operate imperfectly and hence do not provide a reliable guide to choosing an appropriate discount rate for public-sector projects. This is because market interest rates suffer from various distortions which mean that they do not accurately measure the scarcity of capital to the community as a whole. For this reason they usually need to be adjusted in order to derive the correct 'social' discount rate for use in public-sector cost-benefit analysis.

How the discount rate is chosen is of great significance, but for the moment it is more important to consider what the implications of discounting will be. Computationally it is straightforward, and requires the calculation of net discounted benefit (otherwise termed net present value, or NPV) which is based on the summation of the series:

$$\text{NPV} = (B_0 - C_0) + \frac{(B_1 - C_1)}{(1 + r)^1} + \frac{(B_2 - C_2)}{(1 + r)^2} + \frac{(B_3 - C_3)}{(1 + r)^3} + \dots + \frac{(B_n - C_n)}{(1 + r)^n}$$

where: $B_0 \dots B_n$ = revenue expected in each year 0 to n, $C_0 \dots C_n$ = costs incurred in each year 0 to n, r = discount rate.

It should be apparent that for any given value of the discount rate, the further away in time net benefits are received, the more heavily they will be discounted. At a discount rate of 3%, £100 earned in 25 years time is the equivalent of £47.80 ($= 100/1.03^{25}$) earned today; at the same discount rate, £100 earned in 50 years time is only worth £22.80 ($= 100/1.03^{50}$) in present value terms. The effect of discounting will obviously be greater at higher discount rates, which will be clear from our example if the discount rate were to be raised to 5%; the present value of £100 earned in year 25 and year 50 now becomes £29.50 and £8.70 respectively. This result is more than just an artefact of the calculation but an inherent part of the economics, and needs to be clearly understood. The more rapidly a project starts to generate net benefits, the sooner they can be re-invested in some alternative productive use; conversely, the greater the delay in generating a return, the more heavily should those returns be penalized (i.e. discounted) in order to reflect the forgone opportunities for re-investment.

The conclusion of this, therefore, is that even if an investment project is expected to produce a stream of net benefits over an extended period of time it does not guarantee that it will be a worthwhile use of funds. It may be tempting to think that an artificial reef built to last 100 years must eventually pay for itself if it generates an annual flow of income over such a long period, but this is not necessarily the case once the effects of discounting are taken into account. In artificial reef projects (as with other uses of scarce resources), time is money.

Appraising a Reef Project: An Example

To illustrate these principles, and to show how the discounting procedure can be used as the basis for project selection, a hypothetical example is considered of a reef project with a time horizon of 50 years built for the purposes of fisheries enhancement. It is supposed that construction of the reef incurs an initial capital outlay of £100 000 (in year 0), while incremental benefits (attributable to higher vessel catch rates) are estimated to be £12 000 per annum over the 50-year life of the project (starting in year 1). For simplicity it will be assumed that these benefits are fully appropriated by the reef owners (a fishermen's organization) and that there are no externalities. If the NPV of this project were to be calculated over a range of discount rates (say, 0% to 20%), the graphical relationship would appear as in Fig. 1. At a discount rate of zero the NPV is at its maximum value of £500 000, which is derived quite simply from the annual return (£12 000) times the productive life of the reef (50 years) minus the initial capital outlay (£100 000). As the discount rate increases from zero, reflecting the increasing importance

NPV Curve for Hypothetical Reef Project

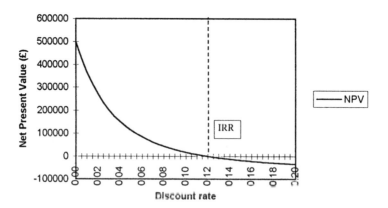

Figure 1.   NPV curve for hypothetical reef project.

attached to receiving economic benefits sooner rather than later, so NPV declines, and in this example switches from positive to negative once the discount rate rises above 12%.

There are three basic decision rules that can be used in order to evaluate a project. The first of these, the NPV criterion, states that an investment project is acceptable if the NPV is greater than zero. In this example, therefore, the reef is worth investing in provided that the discount rate is below 12%. If, for example, a fishermen's organization were able to obtain funds at an interest rate of (say) 6%, then it would pay to borrow the necessary capital and undertake the reef project. For a financial appraisal of this kind, 6% would then be the appropriate discount rate. Using this figure it can be shown that the NPV is £89 142, which gives a measure of the absolute worth of the project in present value terms. A variant of this approach is the benefit–cost ratio (BCR), which is calculated by taking the present value of future benefits and dividing by the initial investment. If the BCR is greater than 1.0 it implies that the total discounted stream of future benefits is greater than the original capital outlay, and hence the project should be accepted. In the present example, at a discount rate of 6% the BCR would be 1.89, confirming that the reef project should go ahead. The third criterion involves calculating the project's internal rate of return (IRR) and then comparing this figure with the cost of capital. IRR is defined as the discount rate which makes the NPV equal to zero, and in our example this is shown graphically in Fig. 1 as the point where the NPV curve cuts the horizontal axis, in this case, at approximately 12%. If the fishermens' organization is being charged 6% interest it clearly makes sense to go ahead with the reef project since the financial yield is expected to exceed the cost of capital. In most situations of this kind where the decision is simply one of 'accept or reject', the three decision criteria can be relied upon to produce the same answer. In other situations, for example where a choice has to be made between mutually exclusive projects or where the decision-maker faces a capital constraint, the criteria may not always produce identical results. To be precise, the ranking of projects may be different depending upon which criterion is used. It is beyond the scope of this chapter to explore this issue, but the advice given by economists is that as a general rule the NPV criterion should be adopted.

Risk and Uncertainty

The need to forecast future costs and benefits means that investment appraisal is always an exercise in decision-making under uncertainty. Various techniques are available for dealing with this, one of the most widely used being sensitivity analysis. Sensitivity analysis is appropriate in situations where it is not possible to ascribe numerical probabilities to the variables which make up NPV (Perkins, 1994), the most that the analyst can reasonably do being to try to establish how changes in the underlying assumptions may affect a project's viability. Examples showing the application of sensitivity analysis to fisheries stock enhancement include work by Ungson *et al.* (1995) on red sea bream ranching in Kagoshima Bay (Japan) and that by Sproul and Tominaga (1992) on the Japanese flounder ranching programme in Ishikari Bay (Japan). However, while sensitivity analysis is useful for answering 'what-if?' questions, it can tell us nothing about a project's chances of succeeding since it makes no assumptions regarding the likelihood of the parameters deviating from their expected values. Though investors in a proposed reef project may find it useful to know that a fall in product price of (say) 10% will cause the NPV to switch from positive to negative, what they also need to know is whether such a price fall is likely. Sensitivity analysis in its simplest form does not address this, and ideally needs to be supplemented by other techniques which truly account for risk.

In recent years the advent of suitable computer software has made it possible to undertake risk analysis, which involves the estimation of a probability distribution of all possible project outcomes on the basis of information about the component benefit and cost elements. The latter may be derived from various sources. Where published historical data are available then the appropriate means and standard deviations can be calculated, but for other variables a subjective judgement regarding the range of values (lowest probable, highest probable) may be the only information that can be used. A particular difficulty with risk analysis is the need to identify inter-relationships among the project variables, and failure to do so may distort the results and lead to misleading conclusions (Savvides, 1994). For example, where a project is large enough to have a significant impact on the market it is supplying, price and output may be negatively correlated, lower yield being associated with higher prices, and vice versa. In these circumstances treating the two variables as if they were independent is likely to result in risk being overestimated, since in the event of production being lower than expected the effects on NPV would be partially offset by above-average prices.

If no significant correlations exist between any of the selected risk variables it is possible to use the information on their range values to simulate a probability distribution of project outcomes. A traditional technique for this purpose is Monte Carlo simulation, which draws sets of values at random from each of the distributions of risk variables which combine to determine NPV. Several hundred iterations may be necessary if the 'true' probability distribution is to be approxi- mated. While the information required to undertake a full risk analysis may be difficult and expensive to obtain, the results of the exercise may often be justified. This is likely to be the case in situations where the expected NPV of a project is above the threshold of acceptance, but there is a potentially wide range of outcomes either side of this figure; in other words, where there is a significant risk of failure. The case study which follows illustrates precisely this problem.

Artificial Reefs for Lobster Production: A Case Study

Background

This case study, modified from Whitmarsh and Pickering (1995) and Whitmarsh (1997), builds on recent initiatives within the UK to explore the potential of artificial reefs as a medium for the enhancement of lobster stocks. In the UK, the European lobster (*Homarus gammarus*) is a high-unit-value species with a strong local market. It grows well in UK waters and has many of the physical and economic characteristics required for a successful enhancement exercise. Significant research has been undertaken into the viability of stock enhancement for this species, particular attention having been paid to the production, transportation and release of juveniles (Wickins, 1994). Enhancement trials have identified that hatchery-reared juvenile lobsters released into the marine environment grow normally, reaching the minimum landing size at 4–5 years old. They have also shown that hatchery-reared juveniles are recruited into commercial lobster fisheries (Addison and Bannister, 1994). Though recapture rates are generally low, the clustering of recaptures near release sites suggests that most of the adult lobsters are likely to remain in the vicinity provided the substrata are suitable (Burton, 1992). Running in parallel with these studies, there have been a number of artificial reef trials in the UK which, while not primarily for the purpose, have demonstrated that lobsters seem to be attracted to such structures. For example, on the Poole Bay reef, lobsters were found to start the process of colonization within 3 weeks of the reef's deposition (Collins *et al.,* 1992; Jensen and Collins, 1996; Jensen *et al.*, chapter 16, this volume). Unfortunately, it would appear from the Poole Bay reef and the Torness reef in Scotland (Kinnear, personal communication) that the extent of this natural colonization is insufficient by itself to support a commercial lobster fishery or to make the construction of a reef an economically viable investment. A suitable direction for future research, therefore, might be to combine the experience in artificial reef construction with that of lobster stock enhancement.

The key question is, under what circumstances might the construction of an artificial reef for lobster production be regarded as a worthwhile social investment? To answer this an economic appraisal has been undertaken of a proposed reef project, following the principles outlined earlier. Because the project is treated as a UK public-sector investment we have used the discount rate recommended by H.M. Treasury, which is currently set at 6% (H.M. Treasury, 1991).

The data sets used in the appraisal have been obtained from unpublished and published sources, the latter including MAFF Sea Fisheries Statistical Tables and CSO Economic Trends. These data have been used to generate expected values for the parameters of the model, though in all cases a range of values has been estimated. Assumptions have also had been made about the probability distribution underlying these estimates. It has been assumed that for lobster prices the average annual unit values are normally distributed, while for the other parameters the range estimates have been based on a modified triangular distribution with the centre point corresponding to the 'most likely' value. These probability distributions form the cornerstone of the risk analysis that is undertaken later. NPV and IRR are used as the criteria for project selection.

Benefits

Benefits are assumed to take the form of revenue received from the first sale of lobsters. Following the methodology outlined earlier, this requires a forecast to be made of the production of lobsters from the reef and their unit value (i.e. price).

The assumption is made that the reef project is sufficiently small in relation to the size of the market that the additional output has no significant impact on quayside prices. It is recognized that the creation of a reef might generate non-traded benefits to certain groups and individuals, particularly to sport fishermen, but because of data problems (and the limited scope of the project) no attempt has been made to quantify these.

Production

The reef is assumed to be a scaled-up version of the Poole Bay artificial reef, currently monitored by Southampton University School of Ocean and Earth Science (see Jensen *et al.,* chapter 16, this volume). The production system envisaged here is based on a 'pump priming' strategy in which hatchery-reared juvenile lobsters are released onto the reef and then targeted for recapture once they have reached the minimum landing size. While natural colonization of the reef by wild lobsters would be likely to occur, hatchery-reared lobsters were expected to account for the majority of catches after the first few years. The numbers of lobsters that could be accommodated at any one time would be determined by the size of the reef, but as numbers are removed by fishing so crevice space would become available. The implication of this is that the rate of input (i.e. the numbers released) can be tailored to the rate of output (i.e. harvesting). While it is recognized that, given the assumed number of juveniles released each year, the reef may initially become overstocked, the aim is to ensure a high level of recruits for subsequent harvest.

The project assumptions are given below, most of the figures being derived from information supplied by MAFF (Wickins, personal communication) and Southampton University (Collins and Jensen, personal communication).

(1) Property rights: Ownership of the reef and exclusive rights of harvest are assumed to be clearly defined.
(2) Reef size: 5000 tonnes of quarry rock arranged in piles over an area of 30 000 m^2, constructed in year 0. Niche space for 5000 market-size animals.
(3) Release of hatchery juveniles: 5000 in year 1, and annually thereafter.
(4) Natural colonization: Taking place in increasing numbers until year 3, but diminishing thereafter as crevice space becomes occupied by the growing numbers and sizes of hatchery lobsters. From year 7 onwards it is assumed that natural colonization will occur at the rate of 100 lobsters per year.
(5) Entry to the fishery: Hatchery releases enter the fishery at 85 mm CL (carapace length) in year 5 and 90 mm CL in year 6. Lobsters that have colonized the reef naturally are all assumed to be over 85 mm CL (from year 1).
(6) Fishing: Taking place from year 1, though catches are expected to be low since they are limited by both the rate of natural colonization and catch controls. From year 5 onwards, however, it is assumed that these controls are relaxed so as to coincide with the first batch of hatchery-reared lobsters that will by then be available for capture. It is assumed that all lobsters caught are between 85 mm and 100 mm CL.

(7) Project time horizon: 100 years.
(8) Size distribution and mean weights: It is assumed that numbers are uniformly distributed within each of the two size categories, the mean weights being 1.415 lb per lobster for animals in the range 85–90 mm CL, and 1.83 lb per lobster for animals in the range 90–100 mm CL.
(9) Variability: Because of fluctuations in the natural environment and in the intensity of fishing effort it is assumed that catches may vary by as much as ± 50%.
(10) Catches: Expected catch levels are shown in Table 1.

Prices

The average annual quayside price of lobsters landed in the UK over the 3 years 1992–4 has been taken as the basis of valuation of harvests from the reef; adjusted for inflation, this gives a figure of £4.14 per lb at 1994 equivalent prices. Whether or not the trend in real prices is likely to be upwards or downwards in the future is difficult to assess, since the movement of prices over the last 20 years has been somewhat erratic. Forecasting future prices is made more difficult by the UK's increasing dependence on international trade, which is likely to cause the domestic UK market for lobsters to become more strongly influenced by export and import prices. For this reason there is no alternative but to take the most recent price data and roll these forward in the projection of future values. The variability of price around the expected value has been calculated from the standard deviation of annual prices over the period 1974–1994, giving a coefficient of variation (s.d./mean) of 0.2.

Costs

The costs of the project are associated with constructing the reef, stocking it with juvenile lobsters and harvesting the stock once the lobsters have reached the minimum landing size. External costs are assumed to be zero, since the most likely adverse effect of the reef, damage to trawling gear, can in practice be minimized or avoided by the simple expedient of siting the reef away from trawling areas.

Table 1. Expected catch levels from reef project for lobster production.

Year	Size range 85–90 mm	Size range 90–100 mm	Total number	Total weight (lb)
1	120	80	200	316
2	120	80	200	316
3	300	200	500	791
4	300	200	500	791
5	1350	200	1550	2276
6	1350	900	2250	3557
100	1350	900	2250	3557

Construction

To illustrate the basic methodology, the case of a reef constructed of quarry rock is considered. Limestone rock could be delivered for a price in the range of £4.25 and £7.00 tonne^{-1}, ex-quarry, giving an expected value of £5.63 tonne^{-1}. In addition to that there is the cost of sea transport and placing the rock in position; to estimate that it was assumed that the material would be transported by barge or similar vessel to relatively deep water approximately 30 km from the jetty point. An expected cost of £5.10 tonne^{-1} was indicated, though here again the range was wide (£2.2–8.00 tonne^{-1}). Assuming the source of supply to be reasonably close to the jetty, the total cost of constructing a 5000 tonne reef from quarry rock was thus estimated to be £53 625.

Harvesting

The costs of harvesting the reef and making the lobsters available for consumption are an essential element in the economic analysis, but arguably the most difficult to assess with any certainty. On the assumption that the reef would be fished by vessels equipped with pots, the major operating expenses associated with this method of harvesting would be fuel, bait, repairs and maintenance. Data for a sample of English potters operating in 1993 indicated that fuel accounted for 4.5% of total earnings and vessel repairs 3.1%. Of the latter figure, roughly half could be regarded as a variable cost, and so the proportion of repairs that can be directly attributed to lobster fishing is taken as 1.55%. Expenditure on bait was not separately identifiable, but evidence from other lobster fisheries suggests that this item typically accounts for some 3% of total receipts. On the basis of the estimated catches from the sample of potters, these proportions suggest that the direct operating costs per lobster are £0.59.

As well as operating expenses, however, some allowance needs to be made for the costs of labour. Since crew payments are likely to overstate the real opportunity cost of labour employed in fishing, it is necessary to impute what is termed a shadow wage to reflect the alternative employment opportunities. Given that the occupational mobility of much of the labour employed in the UK fishing industry is extremely low, and that for many crewmen the alternative to lobster fishing would be unemployment, one might be justified in imputing a shadow wage of zero. This indeed is the approach taken by Radford *et al.* (1991) in their economic evaluation of the salmon fisheries of England and Wales. This may be regarded as too extreme, however, and arguably a more realistic state of affairs is one where perhaps 20% of those employed in lobster fishing might otherwise be gainfully employed in alternative occupations. Taking this proportion as the relevant shadow wage gives a labour cost of £0.49 per lobster. Putting this together with the estimate of direct operating expenses (£0.59) provides a figure of £1.08 per lobster, (estimated range: ± 20%) which can be taken to represent the cost to society of harvesting lobsters from the reef.

Cost of Juvenile Lobsters

Previous work on the economics of lobster stock enhancement (Whitmarsh, 1994) used an implied figure for the cost of juveniles released from a hatchery which was in excess of £1.30 per lobster, but more recent research indicates that this estimate is probably too high. All project appraisals are *ex ante* projections, and it is therefore necessary to assess what the cost of juveniles is likely to be in future. Evidence suggests that new hatchery production methods and economies of scale should enable the cost to be brought down to about £1.00 per juvenile (estimated range: ± 10%).

Results

Measures of Project Worth

The key results are summarized in Fig. 2, which shows the NPV curves for a 5000 tonne reef constructed of quarry rock. NPV declines as discount rate increases, and the point where the curve cuts the zero axis defines the IRR: in this case, 8.0%. Judging the worth of the project depends on the choice of discount rate. If the Government's recommended test discount rate (TDR) of 6% is used, then a reef built of quarry rock would be judged to have passed the test since the NPV is positive at £28 152.

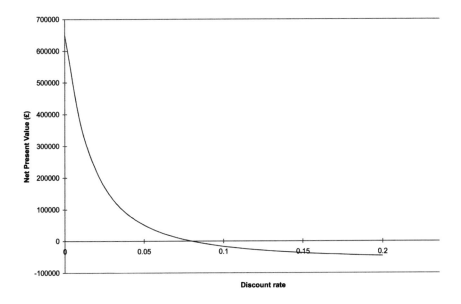

Figure 2. NPV curve for a 5 000 tonne quarry rock reef.

Sensitivity Analysis

The sensitivity of these results to changes in the key assumptions must now be tested. Table 2 presents the results in terms of a sensitivity indicator, which shows the percentage change in NPV that would occur as a result of a 1% change in a given parameter from its base level. It is, in effect, a measure of the elasticity of project worth with respect to a given variable. In the case of lobster prices the figure of 7.02 means that if lobster prices were to rise by 1%, other things being equal, NPV would rise by 7.02%. This measure of sensitivity highlights those aspects of the production process where attention should be focused in order to achieve an improvement in project viability. Apart from lobster prices, which for all practical purposes are outside the control of policy makers, the main areas of importance appear to be levels of sustainable production and juvenile costs.

Risk Analysis

The single-figure estimates of NPV and IRR presented above are potentially mis-leading inasmuch as they may be taken to mean that a particular project option is either uniquely profitable or unprofitable; in other words, either it passed the threshold of acceptability or it did not. As explained previously, however, the correct way to conceptualize the results of a project appraisal is as a probability distri-bution of likely outcomes. For this reason a risk analysis has been undertaken using the range of parameter values given earlier. These are summarized in Table 3.

The analysis has been conducted using a computer-based simulation model in which NPV is repeatedly recalculated across 500 iterations with different combi-nations of the parameter values. The model uses a stratified sampling technique (Latin Hypercube), which is superior to the traditional method of Monte Carlo simulation in that it is more efficient at forcing convergence with a given number of samples. The results of the risk analysis are presented in Table 4, which is based on a 6% TDR. The expected value of NPV approximates to that given previously, but there is a wide distribution of outcomes around the expected NPV. In fact, the results show that for an artificial reef constructed of quarry rock, even though on average it is expected to yield a positive NPV, there is a significant chance (33%)

Table 2. Sensitivity analysis of reef project for lobster production based on a reef constructed of quarry rock.

Parameter	Base case	Sensitivity indicator (%)
Lobster prices (£/lb)	4.14	7.02
Catches: 85–90 mm from year 5 (no.)	1350	3.02
Catches: 90–100 mm from year 6 (no.)	900	2.6
Juvenile costs (£ per lobster)	1.00	2.95
Fishing costs (£ per lobster)	1.08	1.16
Rock supply (ex-quarry) (£/tonne)	5.63	1.0
Sea transport and placing (£/tonne)	5.1	0.91

that it will fall below the zero line. In the light of this it is something of an over-simplification to suggest that, at a 6% discount rate, the reef project will necessarily pass the test of acceptability.

Table 3. Parameter values used in the risk analysis of reef project for lobster production based on a reef constructed of quarry rock.

Parameter	Distribution	Expected level	Variation
Lobster price	Normal	4.14	s.d. = 0.83
Catch: 85–90 mm (years 1–2)	Triangular	120	60, 120, 180
Catch: 90–100 mm (years 1–2)	Triangular	80	40, 80, 120
Catch: 85–90 mm (years 3–4)	Triangular	300	150, 300, 450
Catch: 90–100 mm (years 3–5)	Triangular	200	100, 200, 300
Catch: 85–90 mm (from year 5)	Triangular	1350	675, 1350, 2025
Catch: 90–100 mm (from year 6)	Triangular	900	450, 900, 1350
Juvenile costs	Triangular	1.00	0.90, 1.00, 1.10
Juveniles released	Triangular	5000	4750, 5000, 5250
Fishing costs	Triangular	1.08	0.864, 1.08, 1.296
Rock supply (ex-quarry)	Triangular	5.63	4.25, 5.625, 7.00
Sea transport	Triangular	5.10	2.20, 5.10, 8.00

Table 4. Results of the risk analysis.

a) Summary results
Number of iterations used in the simulation = 500
Expected NPV = 27,600
Chance of a positive NPV = 67%
Chance of a negative NPV = 33%
b) Percentile probabilities and NPV
Probabilities (%)

	NPV (£)
0	– 95 928
5	– 56 754
10	– 41 614
15	– 28 395
20	– 20 914
25	– 12 874
30	– 3 779
35	+ 1 695
40	+ 9 937
45	+ 15 425
50	+ 23 475
55	+ 30 402
60	+ 35 291
65	+ 47 011
70	+ 56 738
75	+ 64 282
80	+ 73 207
85	+ 85 926
90	+100 772
95	+121 120
100	+231 677

Conclusions and Policy Implications

Economics provides an essential analytical framework for evaluating capital investment projects and offers a set of criteria by which projects may be selected. The
basic decision rule is that a project should get the go-ahead if, at a pre-determined
discount rate, the calculated NPV is positive. Where a project is shown to have a
positive NPV it implies that the BCR is greater than 1.0 and the IRR exceeds the
discount rate. Risk and uncertainty can be accounted for using sensitivity analysis
and risk analysis, the latter requiring a computer simulation model in order to
assess the probability distribution of project outcomes.

Policy makers may thus use economic appraisal to help decide on the suitability
of artificial reef deployment. Their judgement will, however, be made in the light
of politico-legal constraints as well as objectives related to wider issues such as
coastal zone planning and management. Economic appraisal may therefore be
seen as providing some but not all of the information that policy makers need.
Despite this limitation, it nevertheless helps to narrow the area for debate about
the relative merits of a project. If policy makers choose to veto the deployment of
artificial reefs despite evidence suggesting that such reefs are likely to be an
efficient use of resources, they presumably do so in the belief that forgoing these
economic benefits is justified on other criteria (e.g. environmental politics).
Conversely, to argue in favour of artificial reef deployment in the face of evidence
showing that an investment of this kind would fail to generate an adequate
economic return means that an explicit decision to go ahead with the investment
is being taken in the full knowledge that an economic loss will be incurred and that
the loss is outweighed by other considerations (i.e. non-quantifiable benefits). In
either case, economic appraisal has served its basic purpose: to provide policy
makers with key information in order to help them arrive at a final decision.

Acknowledgements

In assembling data for the case study we are particularly grateful for the assistance
given by the following organizations: Ministry of Agriculture, Fisheries and Food
(MAFF); Scottish Office Agriculture and Fisheries Department (SOAFD);
University of Southampton, School of Ocean and Earth Science; Sea Fish Industry
Authority (SFIA); North Western and North Wales Sea Fisheries Committee.

References

Addison, J.T. and C.A. Bannister. 1994. Re-stocking and enhancement of clawed lobsters:
a review. *Crustaceana.* 67(2): 131–155.
Bockstael, N., A. Graefe, I. Strand and L. Caldwell. 1986. Economic Analysis of Artificial
Reefs: A Pilot Study of Selected Valuation Methodologies. Technical Report No. 6,
Artificial Reef Development Center, Sport Fishing Institute, Washington D.C.
Burton, C.A. 1992. The United Kingdom lobster stock enhancement experiments. In
Baine, M. (ed.) *Artificial Reefs and Restocking*: proceedings of a conference held on
September 12 (1992) at International Centre for Island Technology, Stromness,

Orkney Islands, Scotland, pp. 22–35.

Collins, K.J., A.C. Jensen, A.P.M. Lockwood and J.J. Mallison. 1992. The Poole Bay artificial reef project. In Baine, M. (ed.) *Artificial Reefs and Restocking*: proceedings of a conference held on September 12 (1992) at the International Centre for Island Technology, Stromness, Orkney Islands, Scotland, pp. 7–14.

Grossman, G.D., G.P. Jones and W. Seaman. 1997. Do artificial reefs increase regional fish production? A review of existing data. *Fisheries.* **22**(4): 17–23.

H.M. Treasury. 1991. Economic Appraisal in Central Government: a Technical Guide for Government Departments. HMSO, London.

Jensen, A.C. and K.J. Collins. 1996. The use of artificial reefs in crustacean fisheries enhancement. In Jensen, A.C. (ed.) *European Artificial Reef Research*. Proceedings of the 1st EARRN Conference, 26–30 March, Ancona, Italy, pp. 115–122.

Milon, J.W. 1988. Travel cost methods for estimating the recreational use benefits of artificial marine habitat. *Southern Journal of Agricultural Economics.* **20**: 87–101.

Milon, J.W. 1989a. Economic evaluation of artificial habitat for fisheries: progress and challenges. *Bulletin of Marine Science.* **44**(2): 831–843.

Milon, J.W. 1989b. Contingent valuation experiments for strategic behaviour. *Journal of Environmental Economics and Management.* **17**: 293–308.

Milon, J.W. 1991. Social and economic evaluation of artificial aquatic habitats. In Seaman, W. and L.M. Sprague. (eds.) *Artificial Habitats for Marine and Freshwater Fisheries*. San Diego: Academic Press Inc., pp. 237–270.

Milon, J.W., S.M. Holland and D. Whitmarsh. (2000: in press). Social and economic evaluation methods. In Seaman, W. (ed.) *Artificial Reef Evaluation with Application to Related Marine Habitats*. CRC Press (forthcoming).

Perkins, F. 1994. *Practical Cost Benefit Analysis*. Macmillan, Melbourne.

Pickering, H. and D. Whitmarsh. 1997. Artificial reefs and fisheries exploitation: a review of the 'attraction versus production' debate, the influence of design and its significance for policy. *Fisheries Research.* **31**(1–2): 39–59.

Radford, A., A. Hatcher and D. Whitmarsh. 1991. An Economic Evaluation of Salmon Fisheries in Great Britain. Report in three volumes submitted to Ministry of Agriculture, Fisheries and Food. University of Portsmouth CEMARE Research Report R.16.

Roberts, K. and M.E. Thompson. 1983. Petroleum Production Structures: Economic Resources for Louisiana Sport Divers. Louisiana Sea Grant College Programme, LSU-TL-83-002, Baton Rouge.

Samples, K.C. 1986. A Socio-economic Appraisal of Fish Aggregation Devices in Hawaii. Marine Economic Report No. 33, University of Hawaii Sea Grant College Program. Honolulu.

Savvides, S. 1994. Risk analysis in investment appraisal. *Project Appraisal.* **9**(1): 3–18.

Sproul, J.T. and O. Tominaga. 1992. An economic review of the Japanese flounder stock enhancement project in Ishikari Bay, Hokkaido. *Bulletin of Marine Science.* **50**(1): 75–88.

Ungson, J.R., Y. Matusda, H. Hirata and H. Shiihara. 1995. An economic benefit-cost estimation of the red sea bream ranching in Kagoshima Bay, Japan. In Liao, D.S. (ed.) *International Cooperation for Fisheries and Aquaculture Development*. Proceedings of the 7th biennial conference of the International Institute of Fisheries Economics and Trade. 18–21 July 1994. Taipei, Taiwan. R.O.C., pp. 189–201.

Whitmarsh, D. 1994. Economic Analysis of Lobster Stock Enhancement. Unpublished Briefing Paper for MAFF Directorate of Fisheries Research.

Whitmarsh, D. 1997. Cost benefit analysis of artificial reefs. In Jensen, A.C. (ed.) *European Artificial Reef Research*. Proceedings of the 1st EARRN Conference, 26–30 March, Ancona, Italy, pp. 175–193.

Whitmarsh, D. and H. Pickering. 1995. Economic Appraisal of Artificial Reefs: a Case Study. Paper presented at the *VIIth Annual Conference of the European Association of Fisheries Economists Portsmouth*, April (1995).

Wickins, J.F. 1994. Consideration of strategies for lobster cultivation. Paper presented at Workshop on Lobster Biology, Management, Cultivation and Stock Enhancement, April 18–22 (1994), Galway, Ireland.

28. Legal Framework Governing Artificial Reefs in the European Union

HELEN PICKERING

Centre for the Economics and Management of Aquatic Resources, University of Portsmouth, Milton Campus, Locksway Road, Southsea, PO4 8JF, UK.

Introduction

There are numerous legal questions associated with artificial reefs, relating to their location, construction, operation and decommissioning. Unlike other disciplines where the state of knowledge is the result of specific research, law exists and applies irrespective of any explicit mention or consideration of artificial reefs or their requirements in its provisions.

Within Europe, the rapid growth of interest in artificial reefs has outpaced the development of law applicable to such structures. Christy (1991) apportions this to the 'lack of firm scientific evaluation of the devices, leaving fisheries administrators and law-makers uncertain what sort of property and use rights, incentives, controls and other legal measures would be appropriate'. With the possible exception of a few countries, such as Spain, there is a general lack of explicit legal provision within Europe. It is probable that in many countries the scale of reef projects and the scope of existing legal provisions have been such that specific legislation to regulate reefs has not been deemed necessary. However, while the explicit mention of artificial reefs is rare, there are a plethora of international and national regulations which, while primarily serving other activities, govern the placing and use of artificial reefs, including some which have profound positive and negative effects (Idyll, 1986).

Ownership and Commercial Exploitation Rights

The ability to deploy and utilize the attributes of an artificial reef is essentially a matter of property rights: the rights governing the ability to occupy, utilize and dispose of land and other resources. The basis of ownership and exploitation rights within the marine environment lies within the provisions of the United Nations Convention on the Law of the Sea 1982 (UNCLOS), which came into force on 16th November 1994. While not all European nations are signatories to the Convention, many nations have pragmatically enacted a considerable number of laws and participated in various regional agreements relevant to UNCLOS, without formal acceptance of the Convention itself. Many of the widely accepted elements of the Convention have now passed into international customary law (Birnie, 1994), including the concepts of internal waters, a 12 nm (nautical mile) territorial sea, a

200 nm exclusive economic zone (EEZ) or exclusive fishing zone (EFZ) and up to a 350 nm continental shelf (depending on seabed topography). These concepts assign to a coastal state an array of sovereign and jurisdictional rights in respect of the construction and operation of artificial reefs in both inshore and offshore locations.

Within internal waters (waters lying landwards of the baseline from which the breadth of the territorial sea is measured), such as ports, river mouths, creeks, bays and semi-enclosed inland seas, and the territorial sea, a coastal state has full sovereignty over the water column, the bed and subsoil for the exploration, exploitation, conservation and management of the natural resources therein, including the deployment, use and management of artificial structures (Honein, 1991).

Within an exclusive economic zone or exclusive fishing zone, sovereign rights extend solely to the economic exploration for and exploitation of natural resources (excluding sedentary species). However, jurisdictional rights are afforded for the establishment and use of artificial structures for economic purposes, marine scientific research and the protection and preservation of the marine environment (Hayashi, 1992). A coastal state also has the exclusive right to authorize and regulate the construction, operation and use of any installations and structures which may interfere with the exercise of its rights in the zone. Its powers of regulation include provisions to enforce, among other aspects, its customs, fiscal, health, safety and immigration laws and regulations.

The designation of a continental shelf extends the coastal state's sovereign and exclusive rights to the exploration for, and exploitation of, sedentary species and jurisdictional rights in respect of artificial structures to the outer edge of the continental margin (Brown, 1992).

To adopt such rights conveyed by international law, states must incorporate the provisions of such conventions into national law. In respect of the UN Convention on the Law of the Sea 1982, state practice in this area has varied, with implications for the extent of coastal state rights to deploy, operate and manage artificial reefs. Table 1 provides a breakdown of national claims within Europe, distinguishing between the north-east Atlantic Ocean and the Mediterranean and Baltic seas.

Many European Community countries with Atlantic or Baltic coasts initially established exclusive fishing zones rather than exclusive economic zones when the decision was taken to extend European coastal state claims to 200 nm in the Hague on November 3, 1976 (the Hague Resolution; Freestone and Ijlstra, 1991; Ijlstra, 1992). In doing so they adopted only a subset of the rights and duties provided by UNCLOS for this area, in the main those pertaining to the exploitation of living resources. At the Third Intergovernmental North Sea Conference held in 1992, the North Sea states (European Community member states plus Norway and Sweden) reviewed this position, committing themselves 'to co-ordinate action, with the aim of increasing coastal state jurisdiction in accordance with international law, including the possibility of establishing exclusive economic zones in areas of the North Sea where they do not exist' (Dirnie, 1994). The effect of this would be the strengthening of jurisdiction to control pollution and the

Table 1.   Maritime claims of Coastal European Nations (source: Ijlstra, 1992)

Country	Atlantic TS	EFZ	EEZ	Mediterranean TS	EFZ	EEZ	Baltic TS	EFZ	EEZ
Belgium	12 nm	1977	n						
Denmark	3 nm	1977	n				3 nm	1978	n
Finland							4* nm	1981	n
France	12 nm	n	1977	12 nm	n	n			
Germany	3/12 nm	1977	n				3	1978	
Greece				6	n	n			
Ireland	12 nm	1976	n						
Holland	12 nm	1977	n						
Portugal	12 nm	n	1977						
Spain	12 nm	n	1978	12 nm	n	n			
Sweden	12 nm	1978	n				12 nm	1978	n
UK	12 nm	1977	n						

TS, territorial sea; EFZ, exclusive fishing zone; EEZ, exclusive economic zone; *for Aaland Islands; n, not claimed; nm, nautical miles

environmental consequences of such marine development projects as artificial reefs. However, enthusiasm towards this end varies between states. For example, the UK is reluctant to establish an exclusive economic zone, considering that it is already able to exercise all the powers it needs without doing so (Birnie, 1994).

The Mediterranean Sea was excluded from the Hague Resolution (Churchill, 1987) and so the pattern of extended maritime claims and rights seen in the Atlantic is not found in the Mediterranean Sea. While France and Spain have established exclusive economic zones off other coasts, they have chosen not to extend their jurisdiction in the Mediterranean. This is also true for many other Mediterranean states. The reasoning for these decisions is attributed to a combination of delimitation problems (Ijlstra, 1992), the importance of distant water fishing fleets and concerns over the potential impact on navigation of declaring exclusive economic zones in the Mediterranean (Scovazzi, 1981).

Along with the adoption, or otherwise, of the concepts of territorial sea, EEZ/EFZ and continental shelf by the coastal state, the other principal factor determining the property rights regime governing artificial reefs is the interpretation of 'sovereign' and 'jurisdictional' rights by the coastal state (Gibson, 1978). The interpretation of sovereign rights, for example, can lie between claiming a proprietary title or claiming purely jurisdictional rights. Such interpretations typically draw their basis from historical precedence (Seabrooke and Pickering, 1994). In most legal systems, the seabed up to the outer limit of the territorial sea belongs to the state, often in the form of property that cannot be entirely disposed of, as in Portugal and the UK (Gibson, 1978; Seabrooke and Pickering, 1994). It can, however, be privately occupied for a substantial period under a concession in European law or a lease in English law (Christy, 1991). In Italy the seashore, beach, bays and ports belong to the state and form part of the public domain. Property that forms part of the public domain is inalienable and cannot be the object of rights in favour of third parties, except in ways and within limits established through legislation. Protection of property in the public domain is entrusted to administrative authority (Fabi, personal communication).

With the seabed inalienable, there remain two basic forms of property rights applicable to assigning rights in respect of reef construction (albeit that the terminology may differ: a licence, permit, lease and concession). A lease involves the demise of land or property for a term of years to a lessee, who is afforded the right of exclusive possession for a certain period. In general, subject to the prescription of statute, the terms attached to a lease and the obligation not to alter, damage, or neglect the land to the prejudice of the lessor, the lessee may use the premises for any lawful purpose. A licence, in contrast, passes no interest in the land but makes lawful what would otherwise be unlawful. It is distinct from a lease in that exclusive possession is not necessarily conferred and there are usually strictly limited purposes attached (Pickering, 1994). It should be noted that the scope, duties and rights attached to these concepts vary between countries.

Within Italy the state retains ownership of the seabed, with a concession relating only to the use of the area and, in the context of artificial reefs, of any submerged and/or floating structures (Fabi, personal communication). As such, the concession is similar in nature to a licence. A concession is only required when mariculture is to be associated with a reef. The placing of a reef to prevent illegal trawling requires only a simple authorization (Fabi, personal communication). In Portugal, the difference between deploying a reef and operating a mariculture facility is also distinguished, a concession and permit being required for the use of public water and land and an authorization and permit being required to set up a mariculture unit (van Houtte *et al.*, 1989). This situation is repeated in the UK, although the distinction here is drawn by the exercise of proprietary rights by the Crown Estate Commissioners in issuing a lease to allow the lessee to occupy a portion of the seabed whilst the rights to fish a reef are conveyed by Statute. In Spain, the state also retains ownership of the seabed for the public benefit, a permit or lease being made for the use of the area depending on whether the applicant is a government ministry or other organization, respectively. In this instance, the lease or permit rules on both the placing of a reef and the right to fish (Revenga *et al.*, 1997). There is no preferential right to fish, but a right to harvest organisms growing on the reef may be granted, subject to the right being exercised purely 'for the exclusive purpose of repaying the cost of the installation' and being limited in quantity and duration (Christy, 1991).

Legal Framework Governing the Allocation of Property Rights for Artificial Reef Construction

The selection of seabed areas to which leases, licences, permits or concessions are likely to be awarded is governed by a number of administrative and legal institutions, in addition to the exercise of property rights. In respect of artificial reefs, prior to the award of a licence etc., administrative decisions are taken as to where reef building for various purposes will be permitted, often with extensive prior consultation. Where reefs are to be constructed for mariculture, Council Directive 79/923/EEC on the quality of shellfish-growing waters requires that Member States designate growing areas based on their freedom from contamination and provide protection for those areas. This has been reflected in such as the Spanish Law No. 23 of 25 June 1984, Marine Cultures, in which 'The Government is empowered to declare areas as being of interest for mariculture which, because of their excellent condition for the production of crustacea, molluscs and shellfish in general, need to receive special protection' (Art. 5). The designation of such areas is likely to make it difficult to obtain licences for mariculture (and associated reef operations) outside these areas (OECD, 1989). This is also likely to be true in French waters. In addition to land-use plans, France has a number of development documents (e.g. Schémas d'Aptitude et d'Utilisation de la Mer (SAUM), the decision of the Inter-Ministerial Committee for National Development, dated 26.10.1972 and the Schémas de la Mise en Valeur de la Mer (SMVM), a Law dated 7.1.1983, and largely replacing SAUM) which designate zones around the coastline for certain activities (such as industrial and harbour development, leisure activities and marine agriculture) and lays down aims for those zones. In Ireland, similar implications exist. The Minister for the Marine designates areas (subject to local inquiry) in which it shall be lawful to engage in aquaculture activities and only in those areas may licences be granted (OECD, 1989).

In Spain and Italy, many artificial reefs are being used in conjunction with marine restocking programmes and marine reserves set up for this purpose. Research on artificial reefs is undertaken as part of the management of these reserves and their placement is therefore subject to the legislation under which the reserves were established. In Italy, artificial reefs are provided for under the Law of 17 February 1982 regulating fishery planning. It provides for the possibility of, among other enterprises, 'establish[ing] areas of biological resting and active repopulation (restocking), realisable ... through the immersion of artificial structures' (Art 1(7)). This possibility is exercised through sequential Triennial Plans on Fishery and Aquaculture and the funding associated with them (Fabi, personal communication). In Spain, a number of marine protected areas have been established (in four categories: hunting refuges, fishery preserved zones, marine reserves and national marine–terrestrial parks) in both regional and national waters. These zones or reserves have been declared under different legislation and managed (along with associated artificial reef structures) by different authorities, largely due to the different jurisdiction over inner and outer waters (regional and national, respectively) and the considerable variation in the objectives of designation. Their development is, however, coordinated by a committee of agencies 'JACUMAR' (Junta Nacional Asesora de Cultivos Marinos). The main legislation governing artificial reefs rests with the 'law of coasts' which addresses development in the coastal zone, and fishery preserved zones (zonas vedadas a la pesca), the regulation of marine restocking activity and the protection of seagrass beds (*Posidonia oceanica*, *Cymodocea nodosa*) (Revenga *et al.* 1997).

Fishery preserved zones were created by Decree in 1980 to control or forbid certain fishing methods, species, areas and periods. These zones include the

Balearic Islands (declared by the regional government, 1982), Islas Medas (by the Catalonian government, 1983) and Punta de Sonabia, Isle of Mouro (Cantabrian government, 1986); all established under regional law. Fishery preserved zones aim to conserve sites in order to sustain harvests of commercially and recreationally important species or to protect important spawning sites (Ramos-Espla and McNeill, 1994). The regulation of marine restocking activity falls under the Order of 11 May 1982 of the Agriculture and Fishing Ministry which declared marine reserves for the purpose of prohibiting the extraction of some or all of the species of marine fauna and flora from those areas, thereby providing a natural source of repopulation for surrounding areas (Ramos-Espla and McNeill, 1994). Reserves in this category include: Tabarca marine reserve (Alicante, 1986), Columbretes Islands (Castellón, 1990), San Antonio Cape (Alicante, 1993) and Ses Negres (Catalonia, 1993) (Ramos-Espla and McNeill, 1994). The protection of seagrass beds was also, initially, under this legislation. However, in 1993, after the European Community declared seagrass beds to be preferred habitats for conservation, designation for this purpose fell under another 1993 Decree (Ramos-Espla and McNeill, 1994). The use of artificial reefs as a management technique within Spain's marine protected areas programme has been supported by EEC Regulations (No. 4028/86 of 18 December 1986) concerning the actions of the community for the improvement and adaptation of the fishing sector and aquaculture (Gómez-Buckley and Haroun, 1994; Guillén et al., 1994). Spain received a total investment of ECU 34.2 million for the construction of 78 artificial reefs to protect marine areas under the European Commission's multi-annual guidance programme 1992–1996 pursuant to these Regulations. The UK and Portugal have also been offered financial support for artificial reef programmes under this multi-annual guidance programme, albeit for different purposes.

   Other factors in the selection of areas of the seabed over which leases, licences, permits or concessions are likely to be awarded for artificial reefs are the environmental implications of reef building. While the United Nations Conference on the Law of the Sea conveys rights in respect of artificial reefs, it does so subject to certain duties imposed on coastal states and, among these are responsibilities towards the environment. Within internal waters, the rights afforded to the coastal state are subject to the duty to protect and preserve the marine environment and the State is required to take 'all measures consistent with [the] Convention that are necessary to prevent, reduce and control pollution of the marine environment from any source' (Art. 194(1)). During the consultation process (prior to the issuing of permission to deploy a reef) such international requirements and the dictates of good governance ensure that marine agencies are consulted to incorporate environmental considerations into the decision-making process. The potential environmental effects, however, are such that the construction and locating of a reef could well justify going further and conducting an environmental impact assessment. Although it does not necessarily apply to artificial reefs, French Act No. 83-630 of 1963 illustrates a public enquiry procedure that could be used. It establishes a general requirement that a public enquiry be conducted before any works and installations are realized which, because of their nature or the characteristics of the zone concerned, are likely to affect the environment. The operations covered and the criteria of assessment are defined by the State Council (Conseil d'Etat) and adapted to the environment under consideration (Christy, 1991). Even where environmental impact legislation itself is not applicable or does not exist, a similar enquiry could well be required under navigation, public domain and other laws governing permission to construct artificial reefs. For example, artificial reefs in Spain, while not requiring an environmental impact assessment as such, do require an ecological study supporting the suitability of the site and species and an engineer's report on the stability of the structure in the light of its construction and local currents (Christy, 1991). Further, with the growth of marine and estuarine conservation designations, such as special areas of conservation, artificial reef projects are likely to be closely scrutinized in terms of their environmental implications and, potentially, banned in certain areas.

As has proven so useful in protecting seagrass beds and spawning grounds, artificial reefs essentially pre-empt the use of the seabed, and in many areas the use of the overlying water column, by other activities. However, while the construction of artificial reefs is an exclusive right, it is not a sovereign right. Unlike sovereign rights, exclusive rights cannot over-ride the rights of other states (Nadelson, 1992). Where an artificial reef would constitute an impediment to international navigation in recognized shipping lanes, then shipping would hold clear priority. However, the issue is one of magnitude and interpretation. The United Nations Conference on the Law of the Sea does not state that installations cannot be built in areas of international traffic within the territorial sea. As long as the installation does not entirely hamper (Mouton, 1952) or unreasonably interfere with innocent passage and there is no other choice or possibility, Honein (1991) is of the opinion that artificial reef construction would not be precluded. It would be a matter of establishing whether any inconvenience is reasonable and manageable and does not represent a hindrance.

Following on from this argument, where navigation is rare or an occasional use of ocean space, the precedence of navigation may again not be absolute (Seymour, 1975). It should be noted, however, that the coastal state is legally responsible for the safety of navigation, even where permission is granted to foreign states or companies to construct artificial reefs (Honein, 1991). The requirement is for a coastal State to regulate shipping in its waters and to implement traffic management measures to ensure that the safety of navigation is maintained (Honein, 1991). In this respect, as well as being entitled to construct, maintain and operate artificial reefs, the coastal State is entitled 'to establish safety zones (up to 500 m) around such installations and devices and within those zones to take measures necessary for their protection' (UNCLOS, Art. 5(6)).

In harbour areas, the placing of any sort of installation, even moorings, falls under the jurisdiction of harbour authorities, whose permission is required prior to the placing of any such structures (see Pickering, 1994 for UK situation). In countries with substantial commercial and private traffic, reflecting in part international commitments but also for the safety of national vessel movements, the placing of artificial structures in wider coastal waters also requires permission from the state authorities responsible for navigation. Conditions attached to that permission will prescribe measures to be taken for the safety of navigation, including the precise location at which the structure is to be placed and the lights and buoyage required to mark it. While liability for damage to vessels is governed by various navigational rules and rules of tort (such as negligence), the general rule is that persons who place or abandon objects so that they become a hazard to navigation are responsible for damage so caused (Christy, 1991). Hence, care is needed in placing a reef on the seabed, and designing the reef for structural integrity and locational stability as well as maintaining and monitoring the structure. Incorrect positioning, movement, break-up and a lack of marking can all pose hazards to navigation (Christy, 1991; Seymour, 1975). In the UK the provision of lights, signals and aids to navigation are covered in conditions attached to any consent given by the Secretary of State for Transport (and any amendments to those conditions) and are the responsibility of the reef operator. The conditions bind not only the person to which the consent is given, but also any person who subsequently owns, occupies or enjoys the use of the reef. In contrast, in Spain, the local authorities provide for the marking of reefs (Christy, 1991).

In the event of multi-purpose structures being proposed, such as through the combining of artificial reefs with harbour works or coastal protection structures, certain other legal provisions also come into play. In terms of the former, while the constitution and powers of harbour authorities differ (Thomas, 1994), the latter generally have the power to authorize and construct works in their harbours in connection with the exercise of their functions. This includes offshore break-waters and jetties, which could potentially have dual roles as artificial reefs. As in the UK (Douglas and Geen, 1993), with all works potentially presenting a hindrance or risk to navigation, the exercise of this power may be subject to the prior consent of the Minister for transport or navigation. It may also be subject to obtaining a lease from the owner of the seabed and foreshore, a licence for the deposit of structures into sea and, where the works extend above the limit of land planning control, permission from the appropriate authority. Apart from the latter requirement for planning permission, the above consent regime would also apply to combined coastal protection and artificial reef projects, albeit that the main authority for such structures lies with coastal protection authorities, such as local authorities in the UK and the municipal, provisional and national Waterstaat in the Netherlands. It should be noted that large-scale harbour works may also fall within planning consent procedures to which the European Community's Directive (EEC 85/337) on Environmental Assessment applies.

Construction and Installation of Artificial Reefs

One of the most debated legal issues with respect to deployment of artificial reefs made from man-made material centres on whether the placement of artificial reefs constitutes habitat enhancement or waste disposal. Present international law treats artificial reefs from the perspective of 'ocean dumping' (Macdonald, 1994).

'Dumping' is defined in the United Nations Convention on the Law of the Sea as any deliberate disposal of waste or other matter from vessels, aircraft, platforms or other man-made structures at sea' (Art. 1(1)). The most significant set of international regulations guiding ocean dumping are those of the London Convention on the Prevention of Marine Pollution by Dumping of Wastes and Other Matter (London Dumping Convention) 1975, the scope of which has been extended by the coming into force of the United Nations Law of the Sea Convention in 1982. The latter obliges those who had not signed the London Dumping Convention to abide by its standards (Macdonald, 1994) and requires states to protect the marine environment from all sources of pollution including dumping (Boyle, 1992; The Greenwich Forum, 1994). For the seas around Europe, the decisions and agreements of the Oslo Convention for the Prevention of Marine Pollution by Dumping from Ships and Aircraft 1972, (the Convention for the Protection of the Marine Environment of the North East Atlantic, its recent replacement), the declarations made at the Second and Third Conferences on the Protection of the North Sea in 1987 and 1990, the Convention on the Protection of the Marine Environment of the Baltic Sea Area (Helsinki Convention) and the Convention for the Protection of the Mediterranean Sea against Pollution 1977 (Barcelona Convention) also apply.

The London Dumping Convention aims to comprehensively and uniformly regulate ocean dumping (Macdonald, 1994) and requires all parties to the Convention 'to take all practicable steps to prevent the pollution of the sea by the dumping of waste and other matter that is liable to create hazards to human health, harm living resources and marine life, to damage amenities or to interfere with other legitimate uses of the sea' (Bates, 1992). It also specifically incorporates within its definition of dumping the deliberate disposal at sea of platforms or other man-made structures (Boyle, 1992). The United Nations Convention on the Law of the Sea 1982 also incorporates such structures within its definition of dumping ('any deliberate disposal of vessels, aircraft, platforms or other man-made structures at sea' (Art. 1(5)). This has implications not only for purpose-built reef units but also for the rigs-to-reef concept. Whereas parts of a platform may be left *in situ* under the 1982 Convention and under international guidance on this matter, the construction of artificial reefs from elements of a platform not left *in situ*, including those toppled, would be regarded as dumping and then subject to the same provisions as other materials (MacDonald, 1994). The London Dumping Convention agreed a three-part classification of substances with regulations governing the disposal of materials in each category. The first category is the Annex I 'black list' of substances which are absolutely prohibited and should be released into the ocean only 'in emergencies posing unacceptable risk relating to human health and admitting no other feasible solution' (Art. v2). Annex II, the 'grey list', details substances that can be dumped with special permits and Annex III materials are allowed to be dumped under a general permit for all other wastes.

Under the London Dumping Convention, dumping 'does not include: ... placement of matter for a purpose other than the mere disposal thereof, provided that such placement is not contrary to the aims of the convention.' (Art. III (1)); a position also adopted by the European Community Waste Framework

Directive (91/156/EEC) and the associated UK Waste Management Licensing Regulations 1994, where an item (or material) discharged but not so that it is no longer part of the normal commercial cycle or chain of utility, is not regarded as waste. The concept of artificial reefs may not, therefore, offend the rules of dumping and may be a permissible form of 'disposal' in certain circumstances. However, this is largely an issue of the materials used. Whether the use of black list substances for artificial reefs is 'against the aims of the [London Dumping] Convention' is open to debate, with the weight of argument yet to be clearly determined (Macdonald, 1994). In 1992, the Scientific Group of the London Dumping Convention recommended that substances contained in the black and grey lists were not to be used in artificial reefs unless the requirements of the Convention and its guidelines were applied. This clearly applies to certain types of fly ash being considered for reef construction globally (Macdonald, 1994). There are a number of countries, including the UK, which oppose the London Dumping Convention guidelines, noticeably those countries with artificial reef programmes (Macdonald, 1994) and numbers of offshore installations potentially requiring disposal over the next few decades. With the long-term effects of certain substances unknown, it needs to be ascertained whether an adequate scientific basis exists for assessing the consequences of such dumping, a consideration the controlling agency is directed to address in issuing the special permit for Annex II substances and the general permits for Annex III substances (Macdonald, 1994).

A somewhat similar approach to the London Dumping Convention, although regional in application, was that adopted by the Oslo Convention. The Convention required substances not prohibited by Annex I to have either a special permit (Annex II substances) or authorization before dumping, with permits or authorization being determined on the basis of the composition of the substance to be dumped and certain environmental and other criteria contained within Annex III to be taken into account (Bates, 1992). Administered under the Oslo Commission, all proposals to dispose of materials at sea were subject to the Prior Justification Procedure to establish that there were no practical alternatives on land and no adverse effects on the marine environment.

Based very much on the concept of the Oslo Convention, the Convention of the Protection of the Marine Environment of the Baltic Sea Area (Helsinki Convention) and the Barcelona Convention for the Mediterranean have very similar characteristics (Boyle, 1992; Tromp and Wieriks, 1994). However, unlike the Oslo Convention, the Helsinki Convention, with limited exceptions, prohibits dumping in the Baltic Sea, including the Kattegat as far as Skagen and internal waters (Boyle, 1992; Ehlers, 1994). It also applies the precautionary principle ('where there are threats of serious or irreversible damage, lack of full scientific certainty should not be used as a reason for postponing measures to prevent environmental degradation' (1990 Bergen Ministerial Declaration)), covering any substances which are liable to create harmful effects, with the potential risk being the decisive factor rather than proof of causality (Ehlers, 1994).

The precautionary principle also has strong manifestations in the new Convention for the Protection of the Marine Environment of the North East Atlantic 1992 which does away with the concept of grey and black lists of substances and adopts a reverse listing approach: a general prohibition to dump, accompanied by a list of exceptions which includes inert materials of natural origin. The convention also has a wider scope, not only covering 'classical' pollution but also the adverse effects of human activities on the maritime area. It sets ambitious targets centred around the precautionary principle and requires contracting parties to take all possible steps to prevent or eliminate pollution through an action plan subject to annual review (Strathclyde, 1994). These action plans along with those of the North Sea Conferences have far-reaching requirements, including actions to reduce or prevent pollution from sea-based sources (including dumping) (Strathclyde, 1994). The Second North Sea Conference in 1987 agreed that the dumping of industrial waste, other than inert natural materials or those proven harmless, was to be phased out. This stance was strengthened at the Third North Sea Conference in 1990 with commitments to end the dumping of industrial waste at sea, which included the UK (by 1992).

While dumping is regarded as acceptable under certain circumstances, with the exception of the Baltic, there is a trend towards its elimination. The decision of the London Dumping Convention's consultative parties in 1990 to work towards the progressive elimination of dumping at sea and to phase it out entirely by 1995 and the similar position adopted by the parties to the Oslo Convention in 1989 demonstrate this move. The prior informed consent requirements in respect of marine dumping significantly limit the freedom of states to undertake this form of disposal. The emergence of the precautionary principle indicates a new priority for measures of environmental protection which are in the process of influencing the law of the sea and also European legislation. There is a growing consensus internationally, that, irrespective of the materials used, current trends in artificial reef programmes are in fact 'dumping'; the 'wilful, direct disposal of material at sea' (Macdonald, 1994). As recent discussions (February 1996 and continuing into 1999) at the OSPAR Commission's working group on sea-based activities (SEBA) have noted, several countries (particularly Germany (SEBA 96/11/2) and Sweden (SEBA 96/11/3)) are concerned that it should not be possible to redefine waste as building material and thereby circumvent the objectives of the international dumping Conventions (PRAM 98/4/9-/E).

While proponents of waste-to-reefs point out that waste is immobilized or stabilized once integrated into concrete blocks and its potentially harmful components fixed, opponents argue that 'immobilisation does not, however, render these contaminants inert – it merely prevents them from leaching while the structure remains intact'. The reefs '...will eventually degrade and disintegrate. When this occurs, the pollutants will be released' (Macdonald, 1994). The balance of argument is likely to increasingly favour the latter with the continued development of the precautionary principle. MacDonald (1994) highlights that it may be a good idea for nations with controversial artificial reef programmes to reconsider those programmes or proceed with caution in the light of the precautionary principle. Guidelines for the Construction of Artificial Reefs are currently in preparation under the auspices of the OSPAR Commission and, while their final form is difficult to predict, it is likely that the stance adopted by the Guidelines will err on the side of precaution. The Helsinki Commission already recommends that no artificial reefs should be installed except in close vicinity to existing urban development and that other existing constructions should be removed.

The position of the individual European States in the reefs-versus-dumping debate varies. In the UK, the Ministry of Agriculture, Fisheries and Food has adopted a conservative stance. As the criteria for awarding a permit require a thorough examination of the effectiveness of the reef for the specified purpose(s) along with satisfaction of a number of other considerations, and the scientific evidence is lacking in certain respects, it is difficult to obtain the necessary approval (Bowles, personal communication). Any indication of artificial reefs being viewed as a waste disposal option would jeopardize any chance of them being awarded a licence. A different emphasis is obviously taken by States with prolific reef-building programmes, such as Spain and Italy, although the ultimate interpretation is affected by the materials used in the construction of the reef components. Irrespective of a State's position in the debate, under Article 210 of the United Nations Conference on the Law of the Sea 1982, States are required to adopt laws and regulations to prevent, reduce and control pollution in the marine environment by dumping. They must also formulate laws, regulations and measures to ensure that dumping is not carried out without the permission of the competent authorities of the coastal State (Brown, 1992; Honein, 1991).

In Spain this requirement for a management framework falls within the elaborate consultations that precede the issue of a lease or permit to deploy a reef. In the UK, it is subject to separate authorization from the Ministry of Agriculture. Fisheries and Food in the waters off England and Wales and the Scottish Office Agriculture and Fisheries Department in the waters off Scotland, under the 1985 Food and Environment Protection Act.

Operating Artificial Reefs

Once a structure is deployed, the regulatory regime changes to that body of law which governs the operation and management of the artificial reef. One of the major elements of this is the regulation of fishing activities.

The ability to fish a reef or control fishing on a reef depends very much on which of four property rights regimes exist: open access, private property, communal/common property or state property rights (Bromley, 1989, 1991). Open access pertains to where there is an absence of well-defined property rights. Private property rights are usually exclusive and transferable. Under communal/common property, the resource is held by an identifiable community of independent users often with equal access to the resource, but with the absence of the exclusivity and transferability attached to private property rights. Under state property rights, the ownership of the resource is vested exclusively in the government which determines the allocation of access and the level and nature of exploitation (Feeny et al., 1990).

Since Holland in 1609 adopted the doctrine of *mare liberum* (freedom of the seas) and began to impose it on the world, under both national and international law, fisheries have been attributed open access status. The reasoning, supported by ancient Roman law, was that the sea is by nature common and not susceptible to possession, including the fisheries within it. Neither individuals nor governments could claim it and its resources were, therefore, open to everyone and only reduced to possession through their being caught (Pearse, 1994). With the industrialization of the fishing industry, the growing awareness of the consequences of this open access concept for fisheries (specifically over-fishing and the depletion of stocks) has lead to the widespread adoption of a number of property rights and regulatory regimes to manage the situation. Open access, however, remains a basic underlying principle within fisheries and one of the key legal issues in respect of artificial reefs. At both the international and the national levels, most fishing regulation is quite specific about what it regulates, is often gear specific and is usually defined in a way that would not include such indirect activities as artificial reefs. As a consequence, unless specific legislation is made to the contrary, anyone can fish around a reef (Christy, 1991).

This leads to a number of management issues. With the potential enhancement and aggregating role of reefs and the size of such structures, it would seem essential that artificial reefs be managed and, in particular, be incorporated within fisheries management plans and systems of licensing. This is essential in order to limit the amount of effort and to control fishing practices (Christy, 1991). For commercial exploitation, the reef owner will also want a measure of exclusivity, to reap a return on investment, while at the same time, where open or state access prevails, other fishermen may claim a right of access based on a traditional freedom of fishing or the probability that they would have had access to the fish aggregated by a reef in the absence of that reef (Christy, 1991). These issues need to be addressed in any reef-building programme. It should also be noted that similar issues will need to be addressed even where the reef is laid for the purpose of preventing fishing or for scientific research. Potential solutions to the access problem include, in increasing order of control:

(1) the control of access by the fisheries authorities (either by area, species or types of fishing) along traditional lines of fisheries management;
(2) the licensing of access (with or without a fee and/or quota) (Christy, 1991);
(3) the creation of a right in the reef-owner, or some other body, to the resource or to restrict or permit access and charge fees.

Where reefs are not privately funded, and where community rights are well established and institutions exist to manage local affairs, control of access can be granted to such an institution. The control of access by area, species or type of fishing historically was one of the first approaches employed in fisheries management. However, while it controls the type of activity conducted it does not control effort. This is also true of the 'licence limitation' strategies employed in capture fisheries up to the 1970s (Pearse, 1994). The problem of controlling effort levels to prevent over-exploitation required the introduction of the 'quotas' in the late

1970s, in conjunction with the determination and allocation of the total allowable catch: the allocation of quantitatively defined rights in the resource. The third approach, the creation of a right in the reef-owner or some other body, employs what is termed in British law a several fishery or Regulatory Order. A several fishery can be created by an Order under the Sea Fisheries (Shellfish) Act 1967, conveying an exclusive right for depositing, propagating, dredging, fishing for and taking of shellfish specified in the order and an exclusive right to make and maintain beds for such shellfish (including structures floating on, or standing or suspended in, water for the propagation or cultivation of shellfish). The 1967 Act also makes provision for the award of a right of regulating a fishery for a specified area (Pickering, 1994).

The adoption of such potential solutions, however, is not straightforward, as can be seen in respect of the use of property rights. Within both European civil and common law systems there are distinctions drawn between sedentary shellfish and finfish and between domestic animals (including fish in pens), which are subject to private property rights and protection, and free-swimming fish (wild animals), which are open access property freely available to anyone undertaking their exploitation under appropriate regulations (Vicuña, 1991). The central problem in assigning private rights lies in distinguishing between wild and ranched/released stocks and apportioning rights accordingly, easier for certain species than others. In the UK, in respect of shellfish farming, this problem has been partly overcome by the law for shellfish having been developed to encompass both farming and the balanced management of naturally occurring shellfish stocks, which are essentially sedentary (Howarth, 1990). For finfish ranching, however, the problem remains.

The ability of a State to employ access management measures is also governed to a large extent by international legislation. Since the member states of the European Community declared EEZs or EFZs, the European Community has acquired significant competence over 'exploitation activities involving aquatic resources, and aquaculture, as well as the processing and marketing of fishery and aquaculture products where practised on the territory of Member States or in Community fishing waters or by Community fishing vessels' (Art.38, the Treaty of Rome). The Community has the competence to establish zones in which fishing activities are prohibited or restricted, limit exploitation rates, or set quantitative limits on catches. These will apply to fishing activities over artificial reefs as much as in other coastal waters, setting the framework within which any management of fishing over an artificial reef would be required to operate. The regulations forthcoming from the common or Community fisheries policy have legal force in all Member States, with European Regulations taking priority over national legislation. Member States retain some responsibility for measures for the conservation and management of resources in waters under their sovereignty or jurisdiction, however, only in respect of strictly local stocks which are entirely of interest to fishermen from the Member State concerned; where they apply solely to the fishermen of the Member State concerned; where they are compatible with and no less stringent than the measures adopted by the common fisheries policy; or where they concern matters for which the Community has not yet adopted fishery conservation and management measures in the Mediterranean (Churchill, 1992).

In addition to such generic issues as assigning rights in the structure and the resource and the management of fishing over artificial reefs, the operation and management of artificial reefs are also governed by a number of legal provisions associated with the purpose to which the reef is to be dedicated (research, conservation, enhancement or mariculture). One such category is an offshoot of the legal framework governing fisheries and their exploitation: the body of law governing mariculture. While mariculture legislation may not specifically mention artificial reefs, as in many other respects, the provisions may be interpreted to include or set a precedent for artificial reefs, in this case their commercial exploitation (Seymour, 1975).

Reflecting the evolutionary paths of the regulatory systems, there is great variation in legal provisions (their scope (particularly in the waters and species covered), the right and duties conveyed and their institutional arrangements) for mariculture among European States (Christy, 1991; van Houtte *et al.*, 1989). The legal frameworks pertaining to mariculture do, however, have a number of common elements in that they (a) facilitate, making available land on sufficiently long (albeit varying) time scales to enable entrepreneurs to make the necessary investments, (b) protect the rights of farmers in the products they raise, and (c) control, in terms of health, disease and environmental issues. The most widely used technique for exercising legal and administrative control over mariculture operations is through an authorization system, whereby a government entity permits a person/company to operate a mariculture operation, either provided for under general fisheries legislation (as in Norway, UK and Sweden) or aquaculture specific regulations (van Houtte, 1994). In Italy, for example, mariculture operations on artificial reefs are subject to authorization by the harbourmaster's office or, if a concession is required for longer than 15 years, by the Central Fishery Direction (Fabi, personal communication). The form of the authorization varies between authorization, licence, permit, lease and concession. In France and Spain the term concession is used, while Norway adopts the terms licence and permit. Many countries specify information requirements and qualifications prior to awarding any licence or other authorization (often to be accompanied by a fee). This information forms the basis of an extensive consultation procedure accompanying and preceding any award (van Houtte *et al.*, 1989). In France, for example, the application for a concession is circulated to member organizations of the Mariculture Committee for the district in which the concession would be operated, including interested government officials and departments (e.g. maritime prefect, inland revenue, health and welfare) and the local representative of the Marine Fisheries Scientific and Technical Institute. A public enquiry will also take place (van Houtte *et al.*, 1989).

Once authorization has been granted, its terms and obligations will determine the nature of the operation, including the term or duration of the licence and the associated security and stability of the operation (van Houtte *et al.*, 1989). In Spain the concession for marine culture lasts 10 years with a possible renewal for up to a maximum of 50 years. The terms and conditions are extended with renewal without change. In most countries the terms and conditions attached to a licence apply to the subsequent use of the authorized area for culture and to the rights of alienation (van Houtte *et al.*, 1989).

It should be noted that as with other elements of artificial reef operations, the regulation of mariculture is not confined to only mariculture specific legislation. Operations are directly affected by land and water laws (including the use of the public domain such as the foreshore or for access to water), environmental laws, natural resource conservation and fish laws, many of which have been discussed previously. They are also affected by regulatory regimes pertaining to animal health, public health, fiscal and import and export arrangements and other issues (van Houtte, 1994).

Decommissioning

The final legal issue surrounding artificial reefs, but by no means one to be underestimated, is that of decommissioning and abandoning marine installations and structures, which applies not only to oil and gas production platforms used in the rigs-to-reefs concept, but also to artificial reefs themselves. Article 60(3) of the United Nations Convention on the Law of the Sea provides that "[a]ny installations or structures which are abandoned or disused shall be removed to ensure safety of navigation, taking into account any generally accepted international standards established in this regard by the competent international organization[, the IMO]. Such removal shall also have due regard to fishing, the protection of the marine environment and the rights and duties of other States. Appropriate publicity shall be given to the depth, position and dimensions of any installations or structures not entirely removed" (Brown, 1992). The IMO adopted its Guidelines and Standards in 1989 (IMO, 1989) and in 1991 the Oslo Commission followed suit, adopting Guidelines for the Disposal of Offshore Installations at Sea. While created with oil and gas platforms in mind, the wording of these provisions and guidelines is not specific to such installations, potentially including other artificial reef structures as well. At the present time these standards are not mandatory, States are not bound to follow the standards, but are required to take them into account. Current guidance is that anything in less than 75 m of water depth should be removed. The essence of the obligation is to ensure the safety of navigation and states are obliged to take whatever measures are necessary to achieve this end (Brown, 1992). While the concept of artificial reefs is based on the long-term development of ecological communities it needs to be recognized that artificial structures have an effective life span. Care needs to be taken in considering the long-term implications and costs of reef construction, including potential requirements for removal.

Conclusions

There are numerous legal questions associated with artificial reefs; their location, construction, operation and decommissioning. Many issues are common throughout

Europe, although their detail varies between countries. The key issues would appear to be: the creation and allocation of property rights in respect of the seabed and the products of any enhancement or mariculture operations; the determination of an appropriate regulatory and administrative regime; the complex balance to be achieved between competing and potentially incompatible uses and the associated legal obligations; the interpretation and development of the dumping versus reefs debate and the development of the precautionary principle. Within each of these main categories, there are numerous other questions and considerations to be taken in to account when contemplating the construction and operation of artificial reefs. Some of these are peculiar to certain reef operations, such as mariculture or multi-purpose structures, others apply across the range of purposes to which artificial reefs can be put. Reflecting the diversity of legal inheritance within Europe, States also have a number of legal issues peculiar to themselves. Further work, however, will be needed to identify these and their implications fully.

As with the issues to be addressed, while solutions are to some degree common throughout Europe, there are great variations between States, reflecting the nature of the legal systems in place and the status afforded to artificial reefs specifically and to such related activities as mariculture. Few countries have specifically incorporated artificial reefs into their legal regime, deeming the scale of reef projects and the scope of existing legal provisions to be such that specific legislation to regulate reefs is not necessary. This ad hoc approach leaves a number of issues unresolved. It may be that this situation will change over time as artificial reefs become more widespread and the economic significance of artificial reefs in mariculture increases. As stated in the introduction, the rapid growth of interest in artificial reefs has outpaced the development of the law applicable to such structures. The legal system may well catch up. However, a number of questions will need to be addressed if it is to do effectively. For example, the identification of what sort of property and use rights, incentives, controls and other legal measures would be appropriate will be a key priority for research over the next few years.

References

Bates, J. 1992. *UK Waste Law*. London: Sweet and Maxwell.

Birnie, P. 1994. Maritime policy and legal issues: impact of the LOS Convention and UNCED on UK Maritime Law and Policy. *Marine Policy*. **18**(6): 483–493.

Boyle, A.E. 1992. Protecting the marine environment: some problems and developments in the law of the sea. *Marine Policy*. **16**(2): 79–85.

Bromley, D.W. 1989. Property relations and economic development: the other land reform. *World Development*. **17**(6): 867–877.

Bromley, D.W. 1991. *Environment and Economic: Property Rights and Public Policy*. Oxford: Basil Blackwell.

Brown, E.D. 1992. The significance of a possible EC EEZ for the law relating to artificial islands, installations, and structures, and to cables and pipelines in the exclusive cconomic zone. *Ocean Development and International Law*. **23**: 115–144.

Christy, L. 1991. Artificial reefs and fish aggregating devices (FADs): Legal issues. In
  Symposium on *Artificial Reefs and Fish Aggregating Devices as Tools for the Management
  and Enhancement of Marine Fishery Resources*, held by the Indo-Pacific Fishery
  Commission (IPFC), Colombo, Sri Lanka, 14–17 May 1990. Bangkok: FAO, pp. 105–115.
Churchill, R.R. 1987. *EEC Fisheries Law*. Dordrecht: Martinus Nijhoff.
Churchill, R.R. 1992. Fisheries and EZ-easy! *Ocean Development and International Law*.
  23: 145–163.
Douglas, R.P.A. and G.K. Green. 1993. *The Law of Harbours and Pilotage*. London:
  Lloyd's of London Press.
Ehlers, P. 1994. Convention on the Protection of the Marine Environment of the Baltic
  Sea Area (Helsinki Convention) 1974 and the Revised Convention of 1992. *Marine
  Pollution Bulletin*. **29**(6–12): 617–621.
Feeny, D., F. Berkes, B.J. McCay and J.M. Acheson. 1990. The tragedy of the commons:
  twenty-two years later. *Human Ecology*. **18**(1): 1–19.
Freestone, D. and T. Ijlstra. 1991. *The North Sea: Basic Legal Documents on Regional
  Environmental Co-operation*. Dordrecht: Martinus Nijhoff.
Gibson, J. 1978. The ownership of the sea bed and British territorial waters. *International
  Relations*. **6**: 474–477.
Gómez-Buckley, M.C. and R.J. Haroun. 1994. Artificial reefs in the Spanish coastal zone.
  Bulletin of Marine Science. **55**(2–3): 1021–1028.
Guillén, J.E., A.A. Ramos, L. Martínez and J.L. Sánchez Lizaso. 1994. Antitrawling reefs
  and the protection of *Posidonia oceanica* (L.) Delile meadows in the western
  Mediterranean sea: demand and aims. *Bulletin of Marine Science*. **55**(2–3): 645–650.
Hayashi, M. 1992. The role of national jurisdictional zones in ocean management. In
  Fabbri, P. (ed.) *Ocean Management in Global Change*. Barking, Essex: Elsevier Science,
  pp. 209–226.
Honein, S.E. 1991. *The International Law Relating to Offshore Installations and Artificial
  Islands*. London: Lloyds of London Press.
Howarth, W. 1990. *The Law of Aquaculture*. Oxford: Fishing News Books.
Idyll, C.P. 1986. Aquaculture legislation and regulations in selected countries. In Bilio,
  M., H. Rosenthal and C.J. Sindermann (eds.) *Realisms in Aquaculture: Achievements,
  Constraints. Perspectives: Review Papers*. World Conference on Aquaculture, Venice,
  Italy 21–25. September 1981. Bredene, Belgium: European Aquaculture Society, pp.
  545–564.
Ijlstra, T. 1992. Development of resource jurisdiction in the EC's regional seas: national
  EEZ policies of EC Member States in the northeast Atlantic, the Mediterranean sea,
  and the Baltic Sea. *Ocean Development and International Law*. **23**: 165–192.
IMO. 1989. Guidelines and Standards for the Removal of Offshore Installations and
  Structures on the Continental Shelf and in the Exclusive Economic Zone.
  International Maritime Organisation 16th Ass. Res. A672. MSC 57/27/Add.2, Annex
  31.
Macdonald, J.M. 1994. Artificial reef debate: habitat enhancement or waste disposal.
  Ocean Development and International Law. **25**: 87–118.
Mouton, M.W. 1952. *The Continental Shelf*. The Hague.
Nadelsen, R. 1992. The exclusive economic zone: state claims and the LOS Convention.
  Marine Policy. **16**: 463–487.
OECD. 1989. Aquaculture: Developing a New Industry. Paris: Organisation for
  Economic Cooperation and Development.
Pearse, P.H. 1994. Fishing rights and fishing policy: the development of property rights as
  instruments of fisheries management. In Voigtlander, C.W. (ed.) *The State of the
  World's Fisheries*: Proceedings of the World Fisheries Congress Plenary Sessions. New
  Delhi: Oxford and IBH Publishing Co., pp. 76–91.
Pickering, H. 1994. Property Rights in the Coastal Zone. Working Papers in Coastal Zone
  Management No. 1. Portsmouth: Centre for Coastal Zone Management, University of
  Portsmouth.

Ramos-Espla, A.A. and S.E. McNeill. 1994. The status of marine conservation in Spain. *Ocean and Coastal Management.* **24**: 125–138.

Revenga, S., F. Fernández, J.L. González and E. Santaello. 1997. Artificial reefs in Spain: the regulatory framework. In Jensen, A.C. (ed.) *European Artificial Reef Research*: Proceedings of the 1st EARRN Conference, Ancona, Italy March 1996. Pub. Southampton Oceanography Centre, pp. 161–174.

Scovazzi, T. 1981. Implications of the New Law of the Sea for the Mediterranean. *Marine Policy.* **5**: 302–312.

Seabrooke, W. and H. Pickering. 1994. The extension of property rights to the coastal zone. *Journal of Environmental Management.* **42**: 161–179.

Seymour, J.L. 1975. Preliminary legal considerations in developing artificial reefs. *Coastal Zone Management.* **2**(2): 149–169.

Strathclyde (Lord) 1994. Coastal water quality in the North Sea. *Marine Policy.* **18**(2): 161–164.

The Greenwich Forum. 1994. Implications for the UK of the Entry into Force of the UN Convention on the Law of the Sea. Conference to Mark the Entry into Force of the Law of the Sea Convention 1994. London: Greenwich Forum.

Thomas, B.J. 1994. The need for organisational change in seaports. *Marine Policy.* **18**(1): 69–78.

Tromp, D. and K. Wieriks. 1994. The OSPAR Convention: 25 years of North Sea protection. *Marine Pollution Bulletin.* 29: 622–626.

van Houtte, A. 1994. The legal regime of aquaculture. *FAO Aquaculture Newsletter.* **7**: 10–15.

van Houtte, A.R., N. Bonucci and W.R. Edeson. 1989. A Preliminary Review of Selected Legislation Governing Aquaculture. Rome: UN Development Programme, FAO.

Vicuña, F.O. 1991. International cooperation in salmon fisheries and a comparative law perspective on the salmon and ocean ranching industry. *Ocean Development and International Law.* **22**: 133–151.

29. Current Issues Relating to Artificial Reefs in European Seas

ANTONY JENSEN, KEN COLLINS and PETER LOCKWOOD

School of Ocean and Earth Science, University of Southampton, Southampton Oceanography Centre, European Way, Southampton SO14 3ZH, UK.

Introduction

European artificial reef research has now been active for about three decades. For much of that time research has been conducted within national programmes, focusing on national or local issues, and has taken place predominantly in the Mediterranean Sea. Over the past 10 years or so interest in artificial reef technology and science has spread into the NE Atlantic and Baltic Seas with an associated variation in aims and ideas. Reef scientists working in European seas have run projects to assess artificial reefs as tools to protect habitat from destruction by trawling (Spain, Italy and France), promote nature conservation (Monaco, Italy and France), aid fisheries (Italy, Spain, Portugal and France), assess novel materials for reef construction (Italy and UK), investigate habitat use for lobsters (UK, Italy and Israel), for aquaculture (Italy), as experimental sites where habitat parameters are known (UK, Holland and Italy) and as biofiltration structures (Finland, Russia, Poland, Ukraine and Romania). This variety of investigation is one of the strengths of artificial reef research in Europe, the community is diverse and there is great scientific value in establishing collaboration and dialogue with colleagues.

The majority of artificial reef investigations have been, and still are, experimental, with Italy dominating the research effort and Spain currently leading the way in the tonnage of reef material deployed, primarily for seagrass habitat protection. Problems associated with old descriptive, qualitative research have led to developments in quantification and comparative studies which have allowed a scientific perspective to be put on artificial reef deployments across Europe. Currently, as part of the EARRN (European Artificial Reef Research Network) initiative, there is an acceptance of the need to standardize some of the ecological methods used. If this is not practicable in some cases then at least the reporting of results will be done in such a way to allow comparison with data gathered elsewhere.

Achievements

Artificial reefs have been built, proving European engineering design and practices, in a variety of habitats. The use of ballast mattresses has allowed substantial reef structures to be placed in areas of relatively soft sediment, providing

A.C. Jensen et al. (eds.), Artificial Reefs in European Seas, 489–499
© 2000 Kluwer Academic Publishers. Printed in Great Britain.

protection against physical disturbance for sensitive habitats such as seagrass beds. The placing of reef units using cranes and barges has generally been found to be more cost effective than techniques using divers, a move away from the low-budget, pilot-experiment-style placement of many initial reefs.

European artificial reefs have been shown to develop as successful ecosystems over prolonged periods of time. Five years seems to be sufficient time for a relatively stable community to develop in water other than the most oligotrophic areas of the Mediterranean. Studies documenting biological colonization of reef surfaces and aggregation of mobile species can be found for both southern and northern European waters (e.g. chapters 1–18, this volume). Colonization characteristics reflect the environment: water quality, larval availability and sedimentation rates strongly influence the 'fouling' communities and these in turn influence, to some extent, the mobile fauna around the reefs.

The variation in community development in response to season of deployment and water quality has led to suggestions that reefs and their communities would make effective environmental quality monitoring sites.

Artificial reef structures have been shown to have a positive impact on fishery yields, especially in the Adriatic Sea where long-term studies have led to reef developments being managed and used by local fishermen's associations (Bombace *et al.*, chapter 3, this volume). Finfish attraction has been the dominant feature studied, but reefs have provided successful habitats for at least two species of lobsters in Europe, and studies of lobster habitat requirements have led to interest in lobster ranching. Fishery reefs may be mixed with mariculture initiatives; the leaders in this field are testing reef designs where wild fisheries, bivalve culture and finfish culture are integrated, a significant move from the traditional philosophy that fishermen hunted and others farmed the sea, and one that may prove to be economically very significant.

Reefs have been used as effective habitat protection devices, so called 'anti-trawling reefs', especially in Spain and Italy, effectively enforcing a legal prohibition in trawling in waters shallower than 50 m in the Mediterranean and 100 m in the Bay of Biscay. This regulation exists to protect seagrass meadow from physical damage. Such a law enforcement role has led to a development in artificial reef 'field' design, which ensures a maximum deterrent for a minimum cost. Additionally the decrease in trawling has allowed static-gear fishermen to re-enter coastal fisheries without the fear of trawls 'carrying away' their equipment. Such structures may allow, in the future, an increase in previously undeveloped activities such as mariculture. Economically, there is a far greater understanding of the potential bioeconomic implications of reef development. However, the social and economic impacts on coastal communities are, as yet, undetermined suggesting a future line of enquiry. There is also a need for effective reef management practices to be developed to ensure that harvesting pressure on reef populations, wild and cultivated, is maintained at the optimum level.

It is now recognized that reef habitat design is of great importance if a reef is to be successful. Reef deployment will only achieve its targets if appropriate habitat is created. Whilst some deployments are general in concept and can be designed with existing knowledge (e.g. the idea of increasing habitat variety to promote biological diversity requires a long-lasting material with a wide range of niches (shelter sizes), this knowledge is generally not yet detailed enough to allow effective design for a single or group of species. One exception to this has been the growth of red coral in artificial caves; here the habitat required was well known and the species valuable and threatened by overexploitation. Where biological knowledge is lacking there is a tendency for human design aesthetics to dominate reef design; this is a cheaper and faster option than developing targeted research programmes but may lead to ineffective reefs. Well-intentioned yet poorly 'designed' reefs, when monitored and appraised against original expectations, may lead the assessors to conclude that 'reefs don't work' when, with the correct habitat requirement information for the target species, the end result would have been successful.

Attitudes

Within Europe the attitudes of scientists, legislators and administrators to the deployment of artificial reefs vary. Broadly, reefs are much more acceptable in the Mediterranean than in northern Europe, possibly reflecting the longer period of activity and greater volume of deployment of reefs in the Mediterranean than in northern seas. This discrepancy is seen in the two international conventions for the protection of the marine environment (which include artificial reef deployment) that apply to European seas. The Barcelona Convention (Mediterranean Sea) allows for the deployment of artificial reefs within its remit without specific material and deployment guidelines. The OSPAR (Oslo and Paris) Convention (NE Atlantic) is currently (1999) debating a series of artificial-reef-specific guidelines covering materials, deployment and assessment. Many of the signatories to the OSPAR Convention appear to feel that artificial reef deployment has such a potential ill effect on the marine ecosystem that international specific controls are required whilst other coastal developments, such as harbours, jetties, breakwaters, dikes and artificial islands do not. This attitude seems somewhat illogical and conflicts with that of southern Europe as well as opinion in Japan and the USA, the two most active countries in reef deployment. Much of the concern relating to reefs appears to be driven by a desire to prevent the use of oil and gas platforms as artificial reefs in the North Sea (Picken *et al.*, chapter 20, this volume). The arguments are based on philosophy and opinion rather than data as research into the topic is limited and results relating to fish presence and behaviour are only just starting to become available.

Scientists and environmentalists concerned about fisheries often raise the issue of 'attraction versus production', artificial reefs attract fish, so facilitating commercial catches, but (it is asked) do they contribute to a net increase in commercial stock biomass? If reefs cannot be proven to benefit commercial fish populations by significantly increasing numbers they are considered by many to have failed. This requirement ignores the almost insurmountable scientific difficulties of proving that reefs increase stock numbers; conventional fisheries-monitoring techniques using catch data and acoustic survey techniques find assessment of stock biomass difficult, if not impossible (the recent cod stock collapse off eastern

Canada is a good example of how difficult it really is). Given the current small scale of European experiments to date, which have relatively small numbers of fish in association with the reefs, the task truly becomes impossible, especially for stocks that visit reefs for a short period of time. However, indicators of biomass increase may be obtained by assessing factors such as growth rates of fish around reefs, fecundity of reef-associated fish and the survivorship of juveniles around artificial reefs. Physiological studies of the possible bioenergetic advantage provide by a fish gaining sheltering from currents may also be worth of study. Much of this work remains to be addressed.

This expectation for artificial reefs to be net producers of commercial fish biomass often ignores the role that hard habitat plays in a fish's life history and that not all commercial fish species (for example many flatfish or pelagic species such as tuna) utilize rocky habitat. Artificial reefs are not a 'cure all' solution for fishermen and fish stock managers.

There are benefits of habitat provision for commercial fishery species other than fish (lobster, possibly edible crabs, bivalve molluscs and cephalopods) and several essential, subtle elements of the function of an artificial reef within a coastal fishery which may produce benefits which are hard to quantify: the provision of habitat for a wide variety of prey species; shelter for the juveniles of exploited species from trawling and possibly some natural predation; protection of nursery habitat (e.g. seagrass) from physical disturbance and provision of spawning structures for those species which do require a hard substratum on which to lay their eggs. In some cases the value of a reef in the production of new biomass can be inferred to some extent but absolute proof evades scientists at present because of the small scale of experimental structures.

The existence of new hard substrata (a reef) that is colonized by a variety of species suggests that additional settlement beyond that which is possible on the existing natural habitat has occurred. This must be balanced with the loss of biota on and in the seabed on which the reef has been placed. In the case of commercial species requiring hard habitat, such as lobsters, any increase in the numbers of individuals will be proportional to the habitat's complexity and the availability of shelter and food. The impact on the fishery will be related to the scale of an artificial reef; to have a realistic effect structures need to be much bigger than at present.

Similar arguments can be put forward for fish, if artificial reefs provide shelter from predators (including trawl fishing), protect or provide juvenile habitat and/or enhance food supply then juvenile/adult survivorship may be improved and/or adult fecundity increased. Increased survival or increase in egg numbers will only have a measurable effect when the scale of this effect is increased so that it becomes demonstrable in fisheries terms, and means that structures will have to increase in size before final proof can be established.

It is noteworthy that all European reefs are well below the size of large reefs (around 50 000 m³) used by the Japanese (Stone *et al.*, 1991), a country which Simard (1995) estimates will have modified an estimated 12% of its fishing grounds by 2000 to increase production of 'seafood'. Here fishery and aquaculture managers are confident of the positive effects of artificial structures, applying a pragmatic judgement of catch levels over time rather than requiring statistical proof of new biomass production.

European artificial reefs are proving to be quite a complex and subtle technology and are capable of doing much more than just aggregating commercial fish for harvest.

Reef Deployment

There are some inconsistencies of approach and attitudes to reef deployment when compared to 'conventional' coastal engineering in Europe. In some areas of northern Europe engineering works to build coastal structures other than reefs appear to be, in general, acceptable to environmental lobby groups and local government. However the construction of artificial reefs which introduce hard habitat onto a previously sedimentary seabed is, in some countries, considered to be undesirable (e.g. Leewis and Hallie, chapter 10, this volume). Factors such as an increase in local biodiversity and the potential for protecting sedimentary seabed from trawling and dredging are being totally discounted in the apparent political desire to maintain the seabed in a supposed 'pristine state' (apparently unaware of the physical impact that trawling and dredging have in coastal waters and the negative impacts of so-called land reclamation).

The legal requirements for permits and permissions vary widely across Europe; no two countries have the same approach to licensing reef deployment (Pickering, chapter 28, this volume). Some European standardization, at least in overall licensing policy and requirement, would be welcomed by reef scientists and developers. Positive policy statements relating to what constitutes an artificial reef deployment so that no doubt was left in the minds of those who see reefs as a disguised disposal option, coupled with a definition of an artificial reef, possibly based on the EARRN model (a submerged structure placed on the substratum (seabed) deliberately, to mimic some characteristics of a natural reef), would be welcomed by the European artificial reef scientific community who do not wish to see the term 'artificial reef' used for something that is truly a waste disposal option.

Reef Materials

Concerns are often voiced is that reef programmes using 'waste' or 'recycled' materials (in the USA 'materials of opportunity') are, by definition, toxic waste disposal in disguise. Assumption of knowledge, often erroneous, is frequently evident in these cases and European reef scientists are working to clarify matters. Tyre utilization as a reef material provides a classic example. Tyres are often used as a material for artificial reef construction outside of Europe; the Philippines and Australia being two countries where tyres form an important component of reefs. In these countries tyres are seen as being a durable material with the economic benefit of being inexpensive and providing a positive environmental benefit as habitat for commercial marine species. In terrestrial situations scrap tyres can have rather negative effects, as seen when tyres clog land-fill sites, hold water for breeding mosquitoes or release toxic fumes when burnt at low temperatures. In Europe tyres (or leachates from tyres) are considered by many to be toxic in the marine environment (no such definitive proof exists in the published literature; for review see Collins et al., 1995) and, at present, it seems unlikely that a licence/ permit to deploy a large-scale tyre reef would be given. This surety of acquired knowledge (based on perception rather than information) ignores that fact that tyres are the most frequently used fender material in ports and, where left sub-merged, recruit fouling communities. Tyres used on roads wear by producing dust particles which enter our rivers and estuaries in run-off and, apparently, have no obvious toxic effects. If tyres are environmentally acceptable, and can be deployed as an effective, targeted artificial reef (both points need to be clarified in a European context) then opportunities exist to re-use a material that is a problem in the terrestrial environment in a positive fashion in the marine environment. Experimental research has recently started in the UK (1998) (K. Collins and A. Jensen, personal communication) to assess the impact of a tyre reef in the marine environment.

European and Israeli workers have developed expertise in the environmental assessment of waste materials such as cement-stabilized pulverized fuel ash (PFA) (UK, Italy and Israel) (Jensen et al., chapter 16, this volume; Relini, chapter 21, this volume) and dredged harbour muds (Italy) for reef construction, and European experimental protocols exist for reef material trials and assessment (Jensen, 1998a). The utilization of environmentally acceptable materials has the potential to lower reef construction costs, reduce pressure on conventional terrestrial disposal methods and lessen environmental impact of quarrying to produce 'natural' materials for reef construction.

European experiments with the re-use of materials such as PFA have shown (as did the coal waste artificial reef program in the USA) that such materials can be stabilized with cement and used in artificial reef structures and support bio-logically indistinguishable communities when compared to control surfaces. The adoption of high-fly-ash-content cement by the Japanese (Suzuki, personal com-munication), who have consistently promoted use of 'prime materials' (cement, steel and rock) against 'materials of opportunity' in reef construction confirms the belief of some European reef scientists that benign waste materials are worthy of evaluation as components of artificial reefs. A 'high ash' cement (35% fly ash) has also been used successfully in coastal defence breakwaters (Díaz Rato and Martiní Unanue, 1998), a high energy environment. Work focusing on the re-use of material in reefs always attracts criticism from those convinced that the project is just an excuse for dumping waste in the marine environment. Within the European reef community the emphasis behind such work is two-fold; the acceptance of economic reality that artificial reefs are expensive to build and that re-use of materials with a low value may make some programmes cost effective; and that in some cases, re-use can provide a positive benefit to the environment as a whole, effectively recycling materials that cause problems in the terrestrial environment into materials that are benign in the marine environment. It is the intention of the European artificial reef community that re-use of acceptable materials should only be proposed where the requirement for an artificial reef is proven and that the material can be used within the design parameters. The creation of so-called artificial reefs as a disposal option where any other outcome is, at best, secondary is not acceptable to the European artificial reef community.

Other issues have impinged on the consideration of re-using materials for reef construction. One such example is the negative publicity surrounding the deep sea disposal of the Brent Spar oil storage facility (made to store oil and constructed from concrete) which has been used to colour any logical consideration of the re-use of the steel jackets from North Sea oil and gas production platforms (made of steel and designed as a lattice to support the 'topsides' of platforms. Lattice structures are frequently used in Japanese fish reefs.) Discussions should be focused on whether there would be value in using these large objects to establish fishing lanes to exploit fish attracted to the steel jackets or indeed as a method of excluding trawlers from 'no-take' areas which would act as reserves for fish and also benthic species whose populations have been affected by the physical distur-bance of trawling. The benefits of reducing fishing pressure in this way require evaluation which can then be entered into the decision-making process. At present it appears that this is unlikely to happen on a significant scale, as the political debate is taking precedent over the provision of scientific data.

In a similar vein, the use of artificial reefs as tools to 'mitigate' environmental impacts of coastal developments are looked upon with suspicion. Many, although not all, are concerned that artificial reefs for mitigation purposes will reduce the pressure on cynical developers to fully assess and minimize any negative impacts of coastal projects when building a 'mitigation reef' would be a cheaper option. Regulation, guidance and a holistic approach to coastal zone management would minimize this risk.

Progress

The scientific community has been making progress in assessment of artificial reefs in several areas of coastal zone management, the dominant success being that of seagrass habitat protection in the Mediterranean sea. The Spanish lead in this area at present, creating efficient deployment patterns to minimize trawler intrusion into the sensitive habitat (workers in Sicily report re-colonization of seagrass in protected areas, which suggests that reefs may have a role to play in habitat remediation as well as protection from further damage) (S. Riggio, personal communication). An interesting result of habitat protection from illegal trawling has been the effective division of the fisheries resource; with the decrease in the threat of damage to equipment, static-gear fishermen are again exploiting the seagrass habitat. These fishermen, generally artisanal in scale and effort, are using techniques that are more discriminating in their catch and less physically damaging than trawling. Income from fishing can now be generated by coastal communities using fishing grounds in their locality rather than by trawlers from a distant port. This has socio-economic consequences beyond the scientific habitat protection issues.

The use of artificial reefs in such a habitat management role has wide application throughout the Mediterranean where fisheries legislation and enforcement are unable to prevent destruction of seagrass habitat by trawling. There does seem to be a role for artificial reefs, or possibly a much simpler structure, in the enforcement of suggested 'no-take' zones in northern Europe. These would provide simple, effective and positive 'on-the-ground' enforcement, ensuring damage to the nets of those who entered such areas illegally, rather than a later legal prosecution which would be the outcome of the use of 'black box' navigation and winch activity recording equipment. Design to maximize enforcement would be unlikely to exclude elements that would provide some biological benefits beyond protection of habitat.

Italian workers (Bombace *et al.*, chapter 3, this volume) are showing how reef structures can be used to promote aquaculture, using the reefs as surfaces for mussel settlement and growth, protection for finfish and lobster cages, and anchor points for conventional suspended bivalve culture. They have also pioneered the use of PFA for settlement and on-growing of piddocks, a high-value, burrowing bivalve. This combination of uses for reef technology is novel and one that shows much promise for the future. As aquaculture develops, the availability of 'traditional', sheltered inshore sites that can be used without significant environmental damage is decreasing, forcing new entrepreneurs to consider moving facilities into less sheltered waters. Artificial reefs may serve as a focus for rope, cage or seabed culture of bivalves and fish; the natural reef biological community may act as a partial 'sink' for excess food flushed from fish cages, possibly also providing habitat for labrids (which may help to control fish lice) and/or lobster. There is considerable scope for developing such structures; collaboration between scientists, mariculturalists and engineers will be essential.

The success of experimental reefs in providing lobster habitat in Israel (Spanier, chapter 1, this volume), the UK (Jensen *et al.*, chapter 16, this volume) and Italy has raised the possibility of developing habitat for lobster ranching (Jensen *et al.*, chapter 23, this volume), either creating entirely new lobster habitat or augmenting existing habitat to provide a full range of shelter sizes. Here the habitat requirements of the target species must be well understood if design is to maximize stocking density. Much of this information is lacking, even in such a well-researched species as *Homarus gammarus,* the European clawed lobster; less is known about spiny lobsters.

The value of purpose-designed habitat construction can also be seen in the artificial caves off Monte Carlo into which red coral 'stubs' have been transplanted and allowed to grow (Allemand *et al.*, chapter 9, this volume). This 'farming' of a valuable, overexploited and threatened Mediterranean species is an example of how provision of habitat by an artificial structure may offer opportunities for both conservation and harvesting

Italian reef scientists (among others) have used reefs to promote local fisheries, and fishermen in the Adriatic are now initiating reef development to promote catches in their area. At a time when dissemination of research results to the end-user are of political importance in the assessment of marine technologies, this serves as a good example of how investigation into fisheries management can result in an appreciable positive result for the fishermen.

The Future

European reef scientists have not, generally, succeeded in communicating their results to a wider audience outside the scientific community. If reefs are to become accepted management tools within Europe the scientists and economists working within the field will have to become more proactive in informing managers, administrators and the public what function artificial reefs can fulfil in our coastal zones. Currently it seems that there is potential for mariculture, fishery management, habitat protection, nature conservation, coastal habitat mitigation, ranching and tourism. The latter is a new feature to the European reef research agenda, brought about by realization that artificial structures may be used by surfers, SCUBA divers and anglers to create conditions suitable for their recreational activity. Whilst surfing reefs are always likely to remain specialist structures to promote wave breaks (possibly integrating with coastal protection schemes) the concept of multipurpose artificial reefs that would serve the need of commercial fishermen, anglers, divers and nature conservation is an attractive one from an economic and social view and a considerable design challenge for artificial-reef scientists and engineers.

Effective reef design is one of the research topics of the future. Understanding the requirements of species with commercial and conservation value will become more important as managers develop a holistic approach to fisheries and nature conservation within the coastal zone. Using reefs to manage habitat could be an integral part of the whole process. Reefs could be designed to encourage specific fishing techniques and provide additional income from tourism. It is this latter aspect that has been so well developed in the USA which remains almost unknown in Europe. The socio-economic benefits of such structures have yet to be assessed (although a start has been made) but diversification of coastal fishing community income sources appears, on a general level, to be a sensible goal.

The problem of scale and functionality of artificial reefs has yet to be addressed. It has become obvious as discussion within EARRN has progressed that we have no idea how large an artificial reef needs to be if it is to function as a self-sustaining ecosystem. We are aware that the European structures have not reached that scale as yet. The Japanese have an arbitrary volume figure (2500 m³) below which they consider a fishing reef to be ineffective and a volume of 150 000 m³ for a regional reef development (Simard, 1995). Research to establish the effective size of artificial reefs to accomplish a specific aim will be needed soon.

In the UK it has been made clear by the regulatory authorities that any reefs deployed for other than experimental purposes will have to be multipurpose. There is significant interest in blending habitat provision with new ideas in 'soft' coastal engineering, 'offshore' reefs which have some portion submerged at all states of the tide. Whilst this may have limitations for some fisheries applications, as these structures move offshore the potential for significant habitat engineering combined with coastal protection will increase.

Currently artificial reef science continues to develop in Europe. Greece deployed their first major artificial reef in summer 1998, Denmark is considering artificial reefs seriously for habitat replacement, and there is considerable interest in the UK and Norway in re-using steel jackets in a positive manner in the North Sea. There is renewed interest in France in developing artificial reefs. In the southern Mediterranean Tunisia has an interest in artificial reefs and in the Black Sea, Romania and Ukraine have developed artificial structures as biofilters to help in solving pollution problems. The established reef research countries are also pushing ahead with new ideas for aquaculture, habitat design and protection,

Table 1.   Summary of future research topics recommended by EARRN.

Aquaculture	A1 Development of reef based aquaculture systems for coastal waters
	A2 Economic and social analysis of developing coastal mariculture
	A3 Development of equipment and methodology
Ranching	R1 An understanding of the habitat requirements
	R2 Reef design
	R3 Economic appraisal
	R4 Legal assessment
Biomass Production	BP1 Survival of juveniles
	BP2 Linked to BP1 would come a consideration of food availability and value
	BP3 Energetic advantage
	BP4 Scale of habitat
Fisheries	F1 Fishery exploitation strategies
	F2 Protection of habitat
	F3 Fishery resource partitioning
	F4 Impact of a reef on existing fisheries
Reef System	RS1 Understand why reefs prove attractive to fish and other mobile species
	RS2 Predicting reef performance
	RS3 Energy flow through a reef system
Monitoring & Appraisal	MA1 Evaluation of socio-economic and technical performance
	MA2 Prove proposed EARRN monitoring programme in the field
	MA3 Appraisal and assessment of physical, biological and chemical parameters around artificial reefs
Recreation & Tourism	RT1 Design. Reef design will have to maximize the needs of the user community
	RT2 Socio-economic benefits
Materials	M1 Use of scrap tyres in artificial reefs
	M2 Use of shipwrecks
	M3 Re-use of steel jackets from oil production platforms
	M4 Development of concrete mixtures
Reef Design	RD1 Design to prevent trawling and/or encourage other fishing methods
	RD2 Design to promote availability of food species (sessile or mobile)
	RD3 Design to provide specific habitat
	RD4 Design to promote tourist benefit
Nature Conservation	NC1 Biodiversity development
	NC2 Scale of reef area how big to have a measurable impact?
	NC3 Environmental assessment

tourism and the use of reefs as test beds for scientific experiments. All of this activity is aimed at producing a greater understanding of how artificial reefs can be used as an integrated management tool within the European coastal zone. In its final report to DG XIV the EARRN (Jensen, 1998b) has outlined research topics (Table 1) important future research proposals.

Many of these aspects interrelate; any single research project would involve a variety of differing topics. Research projects in the future should seek to produce quantified, comparable data that will lead to the construction of planned, targeted, designed and assessed artificial reefs. The development of such structures should involve socio-economists, engineers, scientists, local communities and users as well as those with responsibility for coastal management. For European artificial reefs to progress, researchers must strive to reveal how reef systems work and how they may be manipulated to provide desired biological and socio-economic end-products. Artificial reefs are starting to be used as tools in Italy and Spain, but there is some way to go before reefs are accepted throughout Europe as effective and responsive tools in habitat management. The key to acceptance is the effective dissemination of knowledge gained through good quality research.

References

Collins, K.J., A.C. Jensen and S. Albert. 1995. A review of waste tyre utilisation in the marine environment. *Chemistry and Ecology*. **10**: 205–216.

Díaz Rato, J.L. and F.J. Martiní Unanue. 1998. Recovery of armour layer slopes on 'Principe de Asturias' break water in the Port of Gijón, Spain. In Allsop, N.W.H. (ed.) *Coastlines, Structures and Breakwaters*. Thomas Telford, London, pp. 188–198.

Jensen, A.C. 1998a. Report of the results of EARRN workshop 4: Reef design and materials. European Artificial Reef Research Network AIR3-CT94-2144. Report to DGXIV of the European Commission, SUDO/TEC/98/10.

Jensen, A.C. 1998b. Final report of the EARRN, European Artificial Reef Research Network. AIR3-CT94-2144. Report to DGXIV of the European Commission, SUDO/TEC/98/11.

Simard, F. 1995. Réflexions sur les récifs artificiels au Japon. *Biologia Marina Mediterranea*. **2**: 99–109.

Stone, R.B., J.N. McGurrin, L.M. Sprague and W. Seaman Jr. 1991. Artificial habitats of the world: synopsis and major trends. In Seaman, W. and L.M. Sprague. (eds.) *Artificial Habitats for Marine and Freshwater Fisheries*. Academic Press, London, pp. 31–60.

Index

501